Energy, Environment, and Sustainability

Series Editor

Avinash Kumar Agarwal, Department of Mechanical Engineering, Indian Institute of Technology Kanpur, Kanpur, Uttar Pradesh, India

AIMS AND SCOPE

This books series publishes cutting edge monographs and professional books focused on all aspects of energy and environmental sustainability, especially as it relates to energy concerns. The Series is published in partnership with the International Society for Energy, Environment, and Sustainability. The books in these series are edited or authored by top researchers and professional across the globe. The series aims at publishing state-of-the-art research and development in areas including, but not limited to:

- Renewable Energy
- Alternative Fuels
- Engines and Locomotives
- Combustion and Propulsion
- Fossil Fuels
- Carbon Capture
- Control and Automation for Energy
- Environmental Pollution
- Waste Management
- Transportation Sustainability

Review Process

The proposal for each volume is reviewed by the main editor and/or the advisory board. The chapters in each volume are individually reviewed single blind by expert reviewers (at least four reviews per chapter) and the main editor.

Ethics Statement for this series can be found in the Springer standard guidelines here https://www.springer.com/us/authors-editors/journal-author/journal-author-helpdesk/before-you-start/before-you-start/1330#c14214

Gurunathan Baskar · Veeramuthu Ashokkumar ·
Samuel Lalthazuala Rokhum ·
Vijayanand Suryakant Moholkar
Editors

Circular Bioeconomy Perspectives in Sustainable Bioenergy Production

Springer

Editors
Gurunathan Baskar
Department of Biotechnology
St. Joseph's College of Engineering
Chennai, Tamil Nadu, India

Veeramuthu Ashokkumar
Saveetha Institute of Medical and Technical Sciences (SIMATS)
Saveetha University
Chennai, Tamil Nadu, India

Samuel Lalthazuala Rokhum
Department of Chemistry
National Institute of Technology Silchar
Silchar, Assam, India

Vijayanand Suryakant Moholkar
Department of Chemical Engineering
Indian Institute of Technology Guwahati
Guwahati, Assam, India

ISSN 2522-8366 ISSN 2522-8374 (electronic)
Energy, Environment, and Sustainability
ISBN 978-981-97-2522-9 ISBN 978-981-97-2523-6 (eBook)
https://doi.org/10.1007/978-981-97-2523-6

This Springer imprint is published by the registered company Springer Nature Singapore Pte Ltd.
The registered company address is: 152 Beach Road, #21-01/04 Gateway East, Singapore 189721, Singapore

Preface

The developing economies in the world, throughout their progress and diversification in the last one century, have followed a fundamental feature established in the nineteenth century during industrial revolution: the linear model of using the resources viz. the strategy of "take-make-dispose". This model followed the pattern of obtaining raw materials from nature, converting them into useful products (through physical/chemical/biological processes) to be used by the consumers and finally discarding them as waste back into the nature. This one-way flow of materials or resources produced high adverse long-term environmental and social impacts. With fast growth of industrial sector, coupled with rising population and modernisation of infrastructure, there is an urgent need to pursue and realise the Sustainable Development Goals (derived from 2015 United Nations agreement) that aim to decouple the resource utilisation (or material input) from revenue (or economic growth). A solution to meet this goal is transition from "take-make-dispose" model to "reduce-repair-reuse-recycle" model—which forms the basis of circular economy.

The International Society for Energy, Environment and Sustainability (ISEES) was founded at the Indian Institute of Technology Kanpur (IIT Kanpur), India, in January 2014 to spread knowledge and awareness and to catalyse research activities in the fields of energy, environment, sustainability, and combustion. Society's goal is to contribute to the development of clean, affordable, and secure energy resources and a sustainable environment for society, spread knowledge in the areas mentioned above, and create awareness about the environmental challenges the world is facing today. ISEES aimed to break the conventional silos of specialisations (engineering, science, environment, agriculture, biotechnology, materials, fuels, etc.) to tackle the problems related to energy, environment, and sustainability holistically. This is evident by the participation of experts from all fields to resolve these issues. The ISEES is involved in various activities, such as conducting workshops, seminars, conferences, etc., in the domains of its interests. The society also recognises the outstanding works of young scientists, professionals, and engineers for their contributions by conferring awards in various categories.

The Seventh International Conference on "Sustainable Energy and Environmental Challenges" (VII-SEEC) was organised under the auspices of ISEES from 16–18

December 2022 at the Indian Institute of Technology, Banaras Hindu University, Varanasi (IIT BHU), India. This conference provided a platform for discussions between eminent scientists and engineers from several countries, including India, the USA, Spain, Poland, Austria, the Czech Republic, and Korea. At this conference, eminent international speakers presented their views on energy, combustion, emissions, and alternative energy resources for sustainable development and a cleaner environment. The conference presented two high-voltage plenary talks by Dr. Ajay Kumar, Former Union Defense Secretary, Government of India and Dr. S. S. V. Ramakumar, Director (R&D), Indian Oil.

The conference included 29 technical sessions on energy and environmental sustainability topics, including two plenary talks, 15 keynote talks, 80 invited talks from prominent scientists, and 118 contributed talks by students and researchers. The conference included technical sessions on advanced engine technologies, air pollution monitoring, anaerobic digestion, combustion and flames, air pollution control, biodegradation of toxic chemicals, energy and exergy, desalination and wastewater treatment, environmental bioengineering, pollution and climate change: challenges and priorities, alternative transportation fuels and materials, emerging environmental contaminants, sustainable processing of biomass, human health and environmental sustainability, sprays and atomisation, solid waste: challenges and mitigation, solid waste management, sustainable food and agri-biotechnology, modelling and simulations, renewable energy technologies, bioremediation, biofuels and biorefineries, engine emissions and control, cleaner technologies for pollution mitigation, microbial processes, energy and environment, coal and biomass gasification, and environmental challenges mitigation.

About 250+ participants and speakers attended the conference, where 14 ISEES books published by Springer, Singapore, under a dedicated series, *Energy, Environment and Sustainability*, were released. This conference laid the roadmap for technology development, opportunities, and challenges in the energy, environment, and sustainability domains. These topics are relevant for the country and the world in the present context. We acknowledge the support from various funding agencies and organisations for the successful conduct of the VII-SEEC, where the idea of these books germinated. We would therefore gratefully like to acknowledge IIT BHU (Special thanks to Prof. Akhilendra Pratap Singh); SERB, Government of India; Department of Scientific and Industrial Research (DSIR) (Special thanks to Dr. Vipin Shukla) and our publishing partner Springer (Special thanks to Swati Mehershi).

The editors would like to express their sincere gratitude to a large number of authors from all over the world for submitting their high-quality work on time and revising it appropriately at short notice. We want to express our special gratitude to our prolific reviewers: Dr. Lepakshi Barbora, Dr. Pankaj Kalita, Dr. Debarshi Mallick, Dr. Ritesh Malani, Dr. Rakesh Kumar Sharma, Dr. Anita R. Warrier, Dr. Venkatesa Prabhu Sundramurthy, Dr. Sivaprakash Sundaresan, Dr. Zainul Akmar Zakaria, Dr. P. Radha, Dr. Selvakumar P., Dr. M. Anil Kumar, Dr. M. Jayakumar, Dr. Vinoth Kumar V., Dr. R. Rajesh Kannan, Dr. Vidhya C. V., Dr. S. Raja, Dr. J. Iyyappan, Dr. Shijie Liu, Dr. Anh Tuan Hoang, Dr. Irfan Turhan, Dr. Josiel Martins Costa, Dr. Rodrigo S. Vieira, Dr. Steven Lim, Dr. Shangeetha Ganesan, Dr. Mobolaji

Shemfe, Dr. Chuan Chen, Dr. Ramesh Kumar, Dr. Debora Cynamon Kligerman, Dr. Jun Yang, Dr. Bahram Barati, Dr. Bhupendra Singh Chauhan, and Dr. Zibin Yin, who reviewed chapters of this monograph and provided their valuable suggestions to improve them.

This monograph has addressed the subject of circular bioeconomy for bioenergy production, which application of circular economy principles for sustainable energy production from renewable or bio-resources. Eighteen chapters in this book have touched upon multiple and diverse aspects of production of renewable and sustainable energy. These aspects include: utilisation of lignocellulosic biomass, use of industrial organic waste and byproducts, resources from domestic sector, such as waste cooking oil for biodiesel production, use of hospital-based plastic waste, bioethanol production from biomass feedstocks, use of urban solid waste through biorefinery, thermochemical (pyrolytic) conversion of waste biomass, bio-jet fuel generation from waste resources, and supply chain analysis of waste biomass for bioenergy production. The book also contains case studies on biogas generation from organic wastes. Chapters include recent results and are focussed on current trends in sustainable bioenergy production from waste resources and the related technical issues. We expect that this book will serve as a source of state-of-the-art information and knowledge about the sustainable energy production—in the form of both gaseous and liquid biofuels—from waste resources. The case studies presented in the book are likely to be practical examples and first-hand information for entrepreneurs that demonstrate feasibility and viability of sustainable bioenergy production. We hope that this book will be a handy resource for academicians, researchers, industrial professionals, and postgraduate students working in the area of bioenergy and sustainable development.

Chennai, Tamil Nadu, India — Gurunathan Baskar
Chennai, Tamil Nadu, India — Veeramuthu Ashokkumar
Silchar, Assam, India — Samuel Lalthazuala Rokhum
Guwahati, Assam, India — Vijayanand Suryakant Moholkar

Contents

Editors and Contributors

About the Editors

Dr. Gurunathan Baskar has 24 years of teaching and research experience, published more than 200 research and review articles in reputed National and International Journals, 37 book chapters and edited 6 books in various fields of Biotechnology. He is Fellow of the Institution of Engineers (India) and International Society for Energy Environment and Sustainability; Active Life Member of various national and international professional bodies; listed in top 2% Scientist in the world consecutively in 2020, 2021 and 2022 by Elsevier BV and Stanford University, USA; and received ISTE-Periyar Best Engineering College Teacher Award 2020 from Indian Society for Technical Education-Tamil Nadu Section, Prof. S. B. Chincholkar Memorial Award 2019 from Biotech Research Society, India, for the outstanding work in the area of Biofuels and Food Biotechnology, ISTE-Syed Sajid Ali National Award 2016 for his outstanding research on renewable energy from Indian Society for Technical Education, New Delhi, and Young Scientist Award 2015 from International Bioprocessing Association, France. He has been actively researching for creating sustainable bioenergy solutions while contributing toward a greener and circular bioeconomy environment.

Dr. Veeramuthu Ashokkumar received his Ph.D. from the University of Madras, Chennai, India, in 2014. During his Ph.D. he was awarded an Indo-Australian fellowship and visited Adelaide University, Australia, as a visiting scientist. He received his M.Phil. and M.Sc. in Plant Biotechnology from Indian universities. Currently, he is working as a professor at Saveetha Institute of Medical and Technical Sciences (SIMATS), Saveetha University, Chennai, India. He was the visiting professor at Chulalongkorn University, Thailand. He worked as a postdoctoral scientist in Microalgal Biorefinery at the Department of Mechanical Engineering, University Technology Malaysia (UTM), Malaysia. Also, he worked as a postdoctoral scientist at the Department of Aeronautics and Astronautics, National Cheng Kung University, Taiwan. He was also a visiting scientist and visited European Universities under the Marie Skłodowska-Curie Actions Fellowship scheme

Dr. Samuel Lalthazuala Rokhum is Assistant Professor in the Department of Chemistry, National Institute of Technology Silchar, Silchar. His research interests encompass synthetic chemistry, heterogeneous catalysis, green chemistry, and biofuel synthesis. Dr. Rokhum has completed two SERB-funded research projects and currently running projects funded by CSIR and DST. He has so far published 99 papers in journals of international repute and has authored and edited two books. In addition, he has delivered a lecture at several national and international conferences and has been granted a German patent.

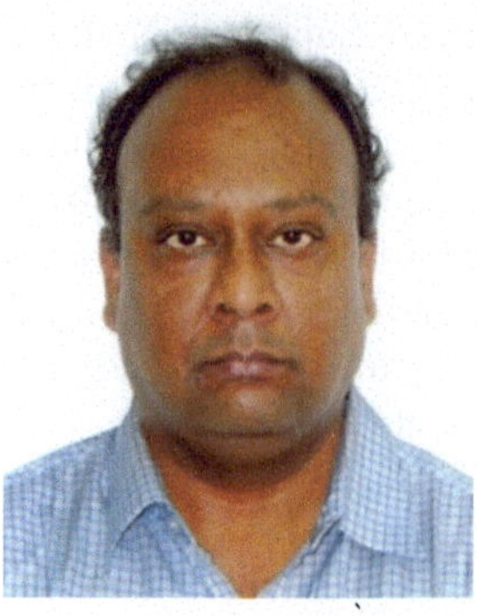

Dr. Vijayanand Suryakant Moholkar is Professor (HAG) of Chemical Engineering and Adjunct Faculty at School of Energy Science and Engineering (erstwhile Centre for Energy) at Indian Institute of Technology Guwahati. He has been Head of the Chemical Engineering Department, IIT Guwahati between 2012 and 2015, and Head of Centre for Energy, IIT Guwahati between 2017 and 2020. His main research interests are sonochemistry, cavitation assisted physical, chemical, and biological processing, and thermo and biochemical routes to biofuels. As of January 2023, he has published more than 200 papers in renowned international journals that have received more than 11000 citations (with *h*-index of 62 and *i*-10 index of 182). He is Co-inventor of 3 US patents (issued to CTI Nanotech, CA, USA) on application of hydrodynamic cavitation reactors for biomass pretreatment and bioalcohol synthesis.

Contributors

M. Afrrin Department of Food Technology, Faculty of Engineering, Karpagam Academy of Higher Education, Coimbatore, India

Muhammad Aliyu Institute of Nanoscience and Nanotechnology, Universiti Putra Malaysia (UPM), Serdang, Selangor, Malaysia;
Department of Material Engineering, Graduate School of Engineering, Kyushu Institute of Technology, Tobata-ku, Kitakyushu, Fukuoka, Japan

M. Anil Kumar Centre for Rural and Entrepreneurship Development, National Institute of Technical Teachers Training and Research, Chennai, Tamil Nadu, India

Belete Tessema Asfaw Department of Chemical Engineering, Haramaya University, Haramaya Institute of Technology, Haramaya, Ethiopia;
College of Biological and Chemical Engineering, Addis Ababa Science and Technology University, Addis Ababa, Ethiopia

Veeramuthu Ashokkumar Saveetha Institute of Medical and Technical Sciences (SIMATS), Saveetha University, Chennai, Tamil Nadu, India

V. S. Avanthi Bioprocess and Bioseparation Laboratory, Department of Biotechnology, School of Bioengineering, College of Engineering and Technology, SRM Institute of Science and Technology, Kattankulathur, Chengalpattu District, Tamil Nadu, India

Gurunathan Baskar Department of Biotechnology, St. Joseph's College of Engineering, Chennai, Tamil Nadu, India;
School of Engineering, Lebanese American University, Byblos, Lebanon

Bunushree Behera Bioprocess Laboratory, Department of Biotechnology, Thapar Institute of Engineering and Technology, Patiala, Punjab, India

Bhalchandra Bhanage Department of Chemistry, Institute of Chemical Technology, Mumbai, India

Pratyush Kumar Das Department of Phytopharmaceuticals, School of Agricultural Engineering and Bioprocessing (SoABE), Centurion University of Technology and Management, Paralakhemundi, Odisha, India

Patitapaban Dash Centre for Biotechnology, Siksha 'O' Anusandhan (Deemed to be University), Bhubaneswar, Odisha, India

Swapna Gade Chemical Engineering and Process Development Division, National Chemical Laboratory, Pashan, Pune, India

Meroda Tesfaye Gari College of Biological and Chemical Engineering, Addis Ababa Science and Technology University, Addis Ababa, Ethiopia

Wan Azlina Wan Ab Karim Ghani Department of Chemical and Environmental Engineering, Faculty of Engineering, Universiti Putra Malaysia, Serdang, Selangor, Malaysia

K. S. Giridharan Centre for Rural and Entrepreneurship Development, National Institute of Technical Teachers Training and Research, Chennai, Tamil Nadu, India

Desta Getachew Gizaw Department of Chemical Engineering, School of Mechanical, Chemical and Materials Engineering, Adama Science and Technology University, Adama, Ethiopia

Louella Concepta Goveas Department of Biotechnology Engineering, NMAM Institute of Technology, Nitte (Deemed to be University), Mangalore, Karnataka, India

Balkis Hazmi Institute of Nanoscience and Nanotechnology, Universiti Putra Malaysia (UPM), Serdang, Selangor, Malaysia

P. Jakulin Divya Mary Centre for Rural and Entrepreneurship Development, National Institute of Technical Teachers Training and Research, Chennai, Tamil Nadu, India

Mani Jayakumar Department of Chemical Engineering, Haramaya University, Haramaya Institute of Technology, Haramaya, Dire Dawa, Ethiopia;
Department of Biotechnology, Faculty of Engineering, Karpagam Academy of Higher Education, Coimbatore, Tamilnadu, India;
Centre for Natural Products and Functional Foods, Karpagam Academy of Higher Education, Coimbatore, Tamilnadu, India

Annette Liisa Kariis Energy and Environmental Systems Group, Department of Engineering and Chemical Sciences, Karlstad University, Karlstad, Sweden

P. Karthik Department of Food Technology, Faculty of Engineering, Karpagam Academy of Higher Education, Coimbatore, India

Snehi Kinger Bioprocess Laboratory, Department of Biotechnology, Thapar Institute of Engineering and Technology, Patiala, Punjab, India

Maria Kristoffersson Energy and Environmental Systems Group, Department of Engineering and Chemical Sciences, Karlstad University, Karlstad, Sweden

P. Loganathan Department of Electrical and Electronics Engineering, Vinayaka Missions Kirupananda Variyar Engineering College, Vinayaka Missions Research Foundation (Deemed to be University), Salem, Tamil Nadu, India

Medha Maitra Bioprocess and Bioseparation Laboratory, Department of Biotechnology, School of Bioengineering, College of Engineering and Technology, SRM Institute of Science and Technology, Kattankulathur, Chengalpattu District, Tamil Nadu, India

K. Manikanda Bharath Centre for Rural and Entrepreneurship Development, National Institute of Technical Teachers Training and Research, Chennai, Tamil Nadu, India

S. Mari Selvam Department of Biotechnology and Medical Engineering, National Institute of Technology Rourkela, Rourkela, Odisha, India

Chirasmita Mohanty Centre for Biotechnology, Siksha 'O' Anusandhan (Deemed to be University), Bhubaneswar, Odisha, India;
Department of Biotechnology, School of Bio Sciences and Technology, Vellore Institute of Technology, Vellore, Tamil Nadu, India

Vijayanand Suryakant Moholkar Department of Chemical Engineering, Indian Institute of Technology Guwahati, Guwahati, Assam, India

Pareshkumar G. Moradeeya Department of Environmental Science and Engineering, Marwadi University, Rajkot, Gujarat, India

S. Nandhini Department of Microbiology, Ethiraj College for Women, Chennai, India

C. Nirmala Department of Biotechnology, Paavai Engineering College, Paavai Institutions, Namakkal, Tamil Nadu, India

Yuvraj Patil School of Health Sciences and Technology, Dr. Vishwanath Karad MIT World Peace University, Pune, India

Selvakumar Periyasamy Department of Chemical Engineering, School of Mechanical, Chemical and Materials Engineering, Adama Science and Technology University, Adama, Ethiopia;
Department of Biomaterials, Saveetha Dental College and Hospitals, SIMATS, Saveetha University, Chennai, India

P. Radha Bioprocess and Bioseparation Laboratory, Department of Biotechnology, School of Bioengineering, College of Engineering and Technology, SRM Institute of Science and Technology, Kattankulathur, Chengalpattu District, Tamil Nadu, India

R. Rahul Department of Food Technology, Faculty of Engineering, Karpagam Academy of Higher Education, Coimbatore, India

M. Rajasimman Department of Chemical Engineering, Annamalai University, Chidambaram, Tamil Nadu, India

Devi Sri Rajendran Integrated Bioprocessing Laboratory, Department of Biotechnology, Faculty of Engineering and Technology, School of Bioengineering,

SRM Institute of Science and Technology (SRMIST), Tamil Nadu, Kattankulathur, Chengalpattu District, India

R. Rajesh Kannan Department of Chemical Engineering, Annamalai University, Chidambaram, Tamil Nadu, India

Pavada Madhusudan Rao Bioprocess and Bioseparation Laboratory, Department of Biotechnology, School of Bioengineering, College of Engineering and Technology, SRM Institute of Science and Technology, Kattankulathur, Chengalpattu District, Tamil Nadu, India

Umer Rashid Institute of Nanoscience and Nanotechnology, Universiti Putra Malaysia (UPM), Serdang, Selangor, Malaysia;
Center of Excellence in Catalysis for Bioenergy and Renewable Chemical (CBRC), Faculty of Science, Chulalongkorn University, Pathumwan, Bangkok, Thailand

Zinnabu Tassew Redda University of Applied Sciences (HTW) Berlin, Berlin, Germany;
School of Chemical and Bio Engineering, Addis Ababa Institute of Technology, Addis Ababa University, Addis Ababa, Ethiopia

M. Reenaa Department of Microbiology, Ethiraj College for Women, Chennai, India

Samuel Lalthazuala Rokhum Department of Chemistry, National Institute of Technology Silchar, Silchar, Assam, India

Prangya Ranjan Rout Department of Biotechnology, National Institute of Technology Jalandhar, Jalandhar, Punjab, India

Joseph V. L. Ruatpuia Department of Chemistry, National Institute of Technology Silchar, Silchar, India

Debasish Sahoo BioInnovale Lifescience (P) Ltd, Bengaluru, Karnataka, India

Ali Salisu Department of Chemistry, Faculty of Natural and Applied Science, Sule Lamido University, Kafin Hausa, Jigawa State, Nigeria

Muhamad Amran bin Mohd Salleh Department of Chemical and Environmental Engineering, Faculty of Engineering, Universiti Putra Malaysia, Serdang, Selangor, Malaysia

Maria Sandberg Energy and Environmental Systems Group, Department of Engineering and Chemical Sciences, Karlstad University, Karlstad, Sweden

Panchamoorthy Saravanan Department of Petrochemical Technology, UCE—BIT Campus, Anna University, Tiruchirappalli, Tamil Nadu, India

V. Saravanan Department of Chemical Engineering, Annamalai University, Chidambaram, Tamil Nadu, India

Raja Selvaraj Department of Chemical Engineering, Manipal Institute of Technology, Manipal Academy of Higher Education, Manipal, Karnataka, India

J. Mary Sheela Department of Microbiology, Ethiraj College for Women, Chennai, India

Ibrahim Garba Shitu Department of Physics, Faculty of Natural and Applied Science, Sule Lamido University, Kafin Hausa, Jigawa State, Nigeria

Joseph M Anto Simon Department of Biotechnology, Sri Krishna Arts and Science College, Bharathiar University, Coimbatore, Tamil Nadu, India

M. Sridevi Department of Biotechnology, Vinayaka Missions Kirupananda Variyar Engineering College, Vinayaka Missions Research Foundation (Deemed to be University), Salem, Tamil Nadu, India

S. Sruthi Bioprocess and Bioseparation Laboratory, Department of Biotechnology, School of Bioengineering, College of Engineering and Technology, SRM Institute of Science and Technology, Kattankulathur, Chengalpattu District, Tamil Nadu, India

Kolli Venkata Supraja Waste Treatment Laboratory, Department of Biochemical Engineering and Biotechnology, Indian Institute of Technology Delhi, Hauz Khas, New Delhi, India

Vinoth Kumar Vaidyanathan Integrated Bioprocessing Laboratory, Department of Biotechnology, Faculty of Engineering and Technology, School of Bioengineering, SRM Institute of Science and Technology (SRMIST), Tamil Nadu, Kattankulathur, Chengalpattu District, India

Amirthavalli Velmurugan Department of Petroleum Engineering, Academy of Maritime Education and Training, Chennai, Tamil Nadu, India

G. Venkatesh Energy and Environmental Systems Group, Department of Engineering and Chemical Sciences, Karlstad University, Karlstad, Sweden

Swethaa Venkatraman Integrated Bioprocessing Laboratory, Department of Biotechnology, Faculty of Engineering and Technology, School of Bioengineering, SRM Institute of Science and Technology (SRMIST), Tamil Nadu, Kattankulathur, Chengalpattu District, India

C. V. Vidhya Department of Microbiology, Ethiraj College for Women, Chennai, India

S. M. Vidya Department of Biotechnology Engineering, NMAM Institute of Technology, Nitte (Deemed to be University), Mangalore, Karnataka, India

Ramesh Vinayagam Department of Chemical Engineering, Manipal Institute of Technology, Manipal Academy of Higher Education, Manipal, Karnataka, India

Anita R. Warrier Nanophotonics Research Laboratory, Department of Physics, Academy of Maritime Education and Training, Chennai, Tamil Nadu, India

Hlawncheu Zohmingliana Department of Chemistry, National Institute of Technology Silchar, Silchar, India

Chapter 1
Introduction to Circular Bioeconomy Perspectives in Sustainable Bioenergy Production

Gurunathan Baskar, Veeramuthu Ashokkumar, Samuel Lalthazuala Rokhum, and Vijayanand Suryakant Moholkar

Abstract Energy security and climate change risk are the two daunting issues faced by developing economies across the globe. The industrial, agriculture and transport sectors in developing nations follow a linear model of economy based on a "take-make-dispose" pattern. The circular economy model that follows the pattern of "reduce-repair-recycle-reuse" offers a potential solution. This book has dealt with the application of circular economy principles to the energy sector. Basically, this book has covered diverse aspects and issues related to the generation of energy from waste and sustainable resources. The topics covered by the book chapters include energy generation (in the form of liquid and gaseous fuels) from agro- and forest residues, municipal solid wastes, industrial organic wastes and food wastes. Technical and economic viabilities of the thermochemical and biochemical technologies for sustainable energy generation as well as issues related to supply chain management of waste resources have been assessed. On a whole, this book has attempted to identify both opportunities and challenges for application of circular bioeconomy models for bioenergy generation, which could be useful for various stakeholders from academia and industry.

G. Baskar
Department of Biotechnology, St. Joseph's College of Engineering, Chennai 600119, Tamil Nadu, India

V. Ashokkumar
Saveetha Institute of Medical and Technical Sciences (SIMATS), Saveetha University, Chennai 602105, Tamil Nadu, India

S. L. Rokhum
Department of Chemistry, National Institute of Technology Silchar, Silchar 788101, Assam, India

V. S. Moholkar (✉)
Department of Chemical Engineering, Indian Institute of Technology Guwahati, Guwahati 781039, Assam, India
e-mail: vmoholkar@iitg.ac.in

G. Baskar et al. (eds.), *Circular Bioeconomy Perspectives in Sustainable Bioenergy Production*, Energy, Environment, and Sustainability,
https://doi.org/10.1007/978-981-97-2523-6_1

Keywords Advanced engine techniques · Alternative fuels · Methanol · Emission control

The twenty-first century has witnessed globalization and tremendous development in industrial, transportation, agriculture and domestic sectors across all developing economies in world. Present annual global energy consumption is estimated at 580 million terajoules, which is expected to grow to about 740 million terajoules (~ 30% rise) by 2040. Fossil fuels—in the form of oil, coal and gas—currently supply about 80% of the total global energy demand. The use of fossil fuels for energy generation results in huge greenhouse gas (GHG) emissions, and in 2022, these emissions were estimated at 34 billion tonnes of CO_2 equivalent. Like all other industries, the power sector has been following the linear economy model of "take-make-dispose" pattern. The circular economy model respects the environment and follows the pattern of "reduce-repair-reuse-recycle" to achieve the sustainable development. Circular economy model aims to maximize resource efficiency, and minimize waste generation by the promoting reuse, recycling, and regeneration of materials and products. Circular bioeconomy essentially emphasizes upon the carbon cycle. Circular bioeconomy aims to close the carbon cycle and creates an additional carbon sink in the realm of technology by utilizing biogenic carbon for making products and other anthropogenic activities. The circular bioeconomy can be deemed as a circular carbon economy that stresses upon capturing atmospheric carbon via photosynthesis to produce renewable bioresources, devising sustainable supply chains and developing new efficient technologies for sustainable transformation of the bioresources to value-added products and fuels for electricity generation, transportation and domestic use such as cooking. The circular bioeconomy model in the context of energy offers simultaneous solutions to the two daunting issues, viz. energy security due to fast depletion of fossil fuel reserves and climate change risk due to emissions from use of fossil fuels. For a developing economy like India, which is highly dependent on oil imports to meet its energy needs, the implementation of a circular bioeconomy is crucial for achieving sustainable economic growth. During the G20 presidency of India, a new and first-of-its-kind initiative was launched entitled "Resource Efficiency and Circular Economy Industry Coalition (RECEIC)" with participation of 39 global companies.

This book has addressed multiple and diverse aspects related to generation of energy from waste, renewable and sustainable sources. Eighteen chapters of this book present comprehensive compilation and analysis of state-of-the-art information on technologies for sustainable energy production from renewable and waste resources and also discuss the opportunities and challenges for commercialization of these technologies. This chapter gives a critical SWOT analysis of the role of bioenergy in circular economy. Chapter 2 has focussed on the opportunities and technical/economic challenges related to production of liquid biofuels from lignocellulosic biomass. Basically, the technical issues related to biomass pretreatment and fermentation for alcoholic biofuels (bioethanol and biobutanol) production are discussed. Chapter 3 presents an overview of bioenergy production from industrial organic wastes and byproducts as sustainable feedstocks. The industrial organic wastes considered in this chapter include sugarcane bagasse, wood chips, potato/orange peels, oil cake, sawdust and molasses. These wastes can be converted to liquid and gaseous fuels through biochemical/thermochemical conversions.

Biodiesel has emerged as a potential renewable liquid fuel to be blended with petroleum-derived diesel. Last two decades have seen intense research on synthesis of biodiesel from different feedstocks using variety of processes. Waste cooking oil produced in domestic kitchens and restaurants is also a potential waste feedstock for biodiesel production. Chapter 4 discusses the potential and challenges of biodiesel synthesis using waste cooking oil. Large-scale biodiesel production requires green and efficient catalysts. Several chapters of this book have addressed this important issue. Chapter 5 presents a review of green and non-toxic catalysts for biodiesel production that are derived from biomass. These carbon- or calcium-based catalysts have large surface area, improved porosity and stability, which contributes towards improving biodiesel yield. Biochar produced during biomass pyrolysis forms a potential green base for catalysts due to high surface area and porosity. Physical and chemical activation of biochar for further increase of surface area and enlargement of the pores can enhance its catalytic activity. In addition, functionalization of the catalyst surface by addition of specific functional groups for a specific chemical reaction can make these catalysts suitable for particular applications. Chapter 7 extensively discusses the functionalization of biochar to synthesize green and sustainable catalysts for biodiesel production. Heterogeneous hydrochar-based catalysts have potential applications in biodiesel production. Chapter 14 has described the synthesis of hydrochar from sustainable feedstocks like agricultural/forestry/food wastes. The chapter also discusses improving the catalytic performance of the hydrochar through addition of metal nanoparticles or acid/base functional groups to the surface of the catalyst. These catalysts have been utilized for improving yield and selectivity during biodiesel production.

Chapter 6 addresses the matter of valorization of hospital-generated plastic waste for bioenergy production, as an alternative to conventional disposal through landfills. Pyrolysis of this plastic waste in fixed or fluidized bed reactors can produce char and hydrocarbons, which can be used for bioenergy production. Bioethanol has found applications as a green and sustainable alcoholic transportation fuel. India has already achieved the target of 10% ethanol blending in petrol. Chapters 8 and 16 discuss production of 2G ethanol from lignocellulosic biomass. Especially, the pretreatment techniques for improving the bioethanol yields have been analysed. The solid waste produced in the urban sector is a potential resource for the production of biofuels and other value-added chemicals. Chapter 9 presents an overview of biorefinery avenues for processing of urban solid waste into value-added products. Natural waste biomass produced from rural sector can also be potential feedstock for bioenergy production. Chapter 10 discusses opportunities for the valorization of this sustainable resource for bioenergy production through pyrolysis. Chapters 11 and 12 present an important case study on biogas generation from organic waste (like food waste, manure and silage) at Biogasbolaget AB in Karlskoga, Sweden. This case study also describes enhancement in biogas production by addition of hydrochar produced from forestry sector wastes and municipal organic wastes. An economic evaluation of the profitability of this process (after integration with hydrothermal carbonization) has also been presented. Chapter 14 discusses the potential of heterogeneous hydrochar-based catalysts biodiesel. Various feedstocks for hydrochar production, production technologies and factors affecting the properties of hydrochar have been discussed.

The aviation industry has flourished remarkably in recent years and so has the demand for aviation fuel. To achieve sustainable growth in the aviation industry, there is a pressing need for manufacture of aviation fuel through renewable resources. Chapter 15 presents a review of the research in the synthesis of bio-jet fuel from sustainable waste resources. Various technologies for conversion of waste resources into jet fuel have been discussed from the perspective of a circular bioeconomy. Economically feasible bioenergy production also requires a continuous and uninterrupted waste biomass supply chain. Different social/technical aspects and issues related to this supply chain have been discussed in Chap. 17. Lastly, Chap. 18 presents a review and technical/economic assessment of a food waste-based biorefinery for bioenergy production. This chapter concludes that economic viability of a food waste-based biorefinery relies on the utilization of the organic waste for the synthesis of a diverse range of green products within the circular economy framework.

We, the editors of this monograph, believe that the contents of this book form a compendium of social, technical and economic facets of bioenergy from the viewpoint of circular bioeconomy. We hope that academic and industrial fraternities working in the areas of sustainable development, policy planning, renewable energy and waste-to-wealth will find this book a valuable source of state-of-the-art information and critical analysis.

Chapter 2
Bioenergy-Based Sustainable Bioeconomy—Perspectives and Challenges

C. V. Vidhya, S. Nandhini, J. Mary Sheela, and M. Reenaa

Abstract Bioenergy is a renewable energy that can be obtained, from biological resources in abundance. The resources include microorganisms, plants, animals and recently, biomass. The applications of bioenergy are wide. It can be used for cooking purpose, power generation, etc. It can also be used as a substitute for petroleum products in both liquid and gaseous forms. Bioenergy is been considered as Nextgen fuel, replacing the other fossil fuels, either partially or completely. Bioeconomy refers to a prudential usage of bioenergy obtained from biological resources. It also deals with conservation of bioenergy, thereby, aiming at sustainable economy for the human welfare. The precise meaning of biological resources not only includes plants, animals, microorganisms or biomass but also the relatable scientific and technological knowledge and innovative ideas based on the same. In the early twenty-first century, the European Union (EU) and Organization of Economic Cooperation and Development (OECD) provided a spotlight on bioeconomy concept. The strategies of bioeconomy, in a holistic approach, aim at sustainable development with circular economy. Sustainable bioeconomy strongly focuses in mitigating global challenges like depletion of fossil fuels, endangered biodiversity, unpredictable inconsistent climatic changes, environmental pollution, etc. It paves a new approach by promoting green-based employment. Thereby, our country's economic status will escalate. Its paramount focus is on the manufacture and consumption of world's finite resources the bioenergy, in an equitable proportion with circularity. This chapter deals with the bioenergy, bioeconomy, circular economy, perspectives, and challenges of the same.

Keywords Bioenergy · Sustainable bioeconomy · Circular economy

---

C. V. Vidhya (✉) · S. Nandhini · J. M. Sheela · M. Reenaa
Department of Microbiology, Ethiraj College for Women, Chennai, India
e-mail: vidhya_cv@ethirajcollege.edu.in

G. Baskar et al. (eds.), *Circular Bioeconomy Perspectives in Sustainable Bioenergy Production*, Energy, Environment, and Sustainability,
https://doi.org/10.1007/978-981-97-2523-6_2

2.1 Bioenergy and Bioeconomy—Introduction

Bioenergy is a renewable energy. It can be procured, in abundance, from biological resources that includes microorganisms, plants, and animals. The Intergovernmental Panel on Climate Change 2019 states that bioenergy with carbon capture and storage, in short BECCS, has the ability to control the emission of greenhouse gases. Bioenergy is one among the important decarbonization source in the energy transition. It can be used as a near zero-emission fuel. Besides being responsible for the release of carbon into the atmosphere, bioenergy is still considered to be less harmful when compared with other fossil fuels on being burnt. This is because it utilizes and releases carbon currently, whereas fossil fuels release carbon that has been stored for millions of years. Biofuels are the most common form of bioenergy. Bioethanol and biodiesel are usually the common types of biofuels—green diesel, biogasoline, biobutanol, biogas, and jet biofuel [1].

Bioenergy is been considered as Nextgen fuel, replacing the other fossil fuels, either partially or completely. There is a great demand in the production of bioenergy as there is a great flexibility in its usages. For example, bioenergy in its solid form can be used in domestic combustion purpose and industrial as well. The liquid form of biofuels can be used in motor cars, airway, and seaway transport. Natural gas can be replaced by biomethane. Bioenergy and carbon capture and storage (BECCS) is a novel upcoming idea as it can be an economic strategy to gain negative emission that may reduce atmospheric carbon [2].

Bioeconomy refers to a prudential usage of bioenergy from biological resources, and their conservation, thereby, aiming at sustainable economy for the human welfare. In the early twenty-first century, it was European Union (EU) and Organization of Economic Cooperation and Development (OECD), who initially adopted the concept of bioeconomy, in their framework, that promoted the wider applications of biotechnology. The strategies of bioeconomy, collectively, aim at sustainable development with circular economy. Sustainable bioeconomy intensely focuses in mitigating global challenges like exhaustion of fossil fuels, endangered biodiversity, erratic climatic changes, environmental pollution, etc. It also supports a country's economy by promoting green-based employment. Its ultimate focus is on the production and consumption of world's finite resources in an equitable proportion with circularity. This chapter deals with the bioenergy, bioeconomy based on bioenergy, bioeconomy with circular economy, perspectives of the same, the challenges to build it, and the challenges dealt by the same.

2.2 Sources of Bioenergy

Bioenergy plays a crucial role in the global pursuit of sustainable development and the transition to a low-carbon economy. Bioenergy is derived from organic materials, which can be continuously replenished through natural processes. Unlike fossil fuels, which are finite and contribute to climate change, bioenergy sources have the potential to be sustainable and help reduce greenhouse gas emissions. It offers an opportunity to diversify energy sources, reduce dependence on imported fossil fuels, and enhance energy security.

Many countries can produce biomass locally, reducing their vulnerability to international energy price fluctuations and geopolitical tensions. Bioenergy has the potential to significantly reduce carbon dioxide emissions when compared to conventional fossil fuels. It can be easily integrated into existing energy infrastructure, such as power plants, heating systems, and vehicles. This adaptability makes it a feasible option for reducing carbon emissions in various sectors.

It is derived from biological material known as biomass. Biomass has the potential to be an increasingly cost-competitive renewable energy source and be used to make a valuable contribution to the overall energy supply system, mainly because of its very low cost and the fact that it is renewable. There is still considerable scope for making better use of the existing biomass energy supplies and for developing new supplies.

Biofuels like bioethanol and biodiesel can be used as alternatives to fossil fuels in the transportation sector, reducing emissions from the transportation industry, which is a significant source of greenhouse gases. Some bioenergy technologies, such as biogas and solid biomass power plants, can provide a stable base load of electricity and contribute to grid stability, particularly when paired with intermittent renewable energy sources like solar and wind. The development and implementation of bioenergy technologies drive research and innovation in the renewable energy sector, leading to advancements in efficiency, cost-effectiveness, and sustainability.

2.2.1 Biomass

Biomass is a versatile and sustainable source of renewable energy that plays a crucial role in meeting our needs. Biomass refers to any organic material derived from plants or animals that can be used as an energy source. This can include wood, crop residues, agricultural and forestry byproducts and organic waste. Biomass feedstocks are diverse and can be tailored to suit regional and environmental conditions. This flexibility allows for the utilization of different types of biomasses depending on local availability and economic considerations.

Ongoing research and development in the field of bioenergy are leading to technological advancements that enhance the efficiency and environmental performance of biomass conversion. The economic viability of bioenergy production depends on factors such as feed-stock costs, energy market prices, and government support. Biomass is a valuable resource for bioenergy production with the potential to reduce carbon emissions, enhance energy security, and support rural economies. However, its sustainable requires careful consideration of feed-stock selection, technological innovations, and supportive policies to ensure its long-time environmental and economic benefits.

2.2.1.1 Types of Biomass

Feed-stock signifies raw substances or inputs used in various processes, particularly in the context of bioenergy production. Feed-stocks are broadly classified into three generations based on the kind of feed-stock which is being used for its production (Fig. 2.1).

First-Generation Feed-Stock

First-generation feed-stock refers to those sources that have been traditionally used for biofuel production. These feed-stocks are typically derived from food crops or crops that are primarily grown for human consumption. The production of biofuels from first-generation feed-stock has been the starting point for the development of bioenergy technologies. The types of first-generation feed-stock are as follows.

Corn (Maize) is a widely utilized feed-stock for ethanol production due to its high starch content. In this process, the starch found in corn kernels is transformed into sugars, which are subsequently fermented to produce ethanol.

Sugarcane, on the other hand, is another prominent source of bioenergy, particularly ethanol. Sugars derived from sugarcane juice or molasses undergo fermentation to yield ethanol. This feed-stock is prized for its high sugar content, efficient conversion rate, and the presence of well-established production systems.

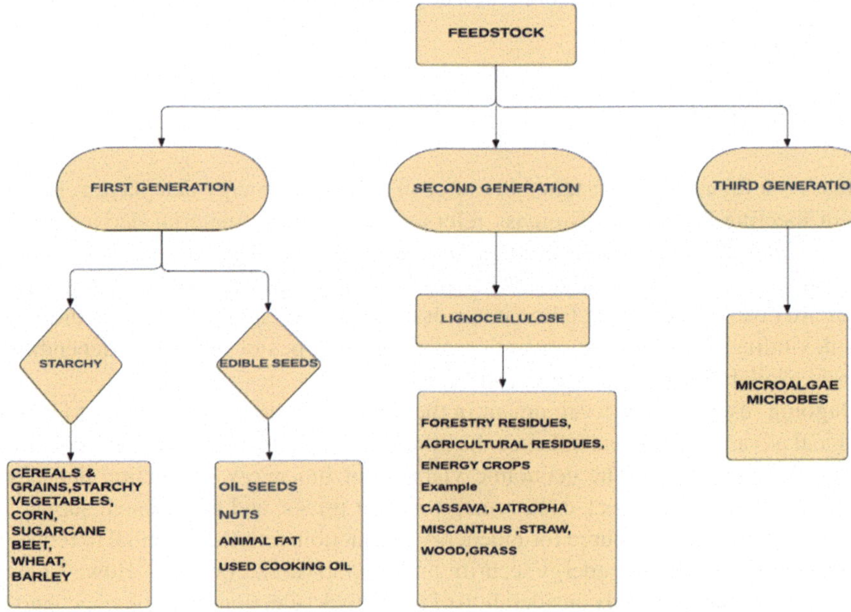

Fig. 2.1 Classification of biomass feed-stocks

Palm oil, derived from the fruit of the oil palm tree, serves a dual role as a versatile and widely consumed vegetable oil and as a prominent feed-stock for biodiesel production. The oil undergoes intricate processing to yield biodiesel, which has found extensive use in transportation and industrial applications.

Wheat, a staple cereal grain known for its versatility and widespread cultivation, serves as a crucial first-generation feed-stock in the production of bioethanol. It is rich in starch, which can be converted into ethanol through a process that mirrors the production of bioethanol from corn.

Canola is a versatile oilseed crop cultivated primarily for its oil-rich seeds. It is known for its high oil content, particularly in the form of oleic acid, linoleic acid, and linolenic acid. Canola oil is not only used in the culinary industry but also serves as a valuable feed-stock for the production of biodiesel.

Barley is a cool-season cereal grain crop that has been cultivated for thousands of years, primarily for use in human and animal nutrition. However, in recent years, barley has gained recognition as a valuable first-generation feed-stock for bioethanol production. This is due to its high starch content, which can be enzymatically converted into ethanol, making it an ideal for sustainable biofuel production.

Soybean plays a vital role in biodiesel production through a transesterification process. Soybean oil is the primary component converted into biodiesel, with glycerol as a valuable byproduct of this transformation. This diversification of feed-stocks underscores the versatility and sustainability of bioenergy production, offering multiple pathways to reduce our reliance on traditional fossil fuels.

Production Process

Ethanol Production The production of ethanol from various feed-stocks involves a series of essential steps. First, there is the pretreatment phase, which encompasses feed-stock preparation and milling to make the starch or sugars readily accessible for subsequent processes. Following pretreatment, enzymatic hydrolysis comes into play, where complex carbohydrates are broken down into simpler sugars, a crucial step in the ethanol production chain. Once the sugars are available, fermentation takes center stage. This phase employs yeast or bacteria to transform these sugars into ethanol, marking a pivotal point in the conversion process. After fermentation, the distillation step comes into action, serving the purpose of separating ethanol from the fermentation mixture. The final step in ethanol production is dehydration, aimed at removing excess water to achieve anhydrous ethanol. These meticulously orchestrated processes together contributed to the successful production of ethanol, a versatile biofuel with the potential to reduce our dependence on conventional fossil fuels.

Biodiesel Production Biodiesel production follows a well-defined sequence of steps, with each phase contributing to the creation of this sustainable biofuel. Biodiesel, a renewable and eco-friendly alternative to conventional diesel fuel, is primarily produced through a chemical process known as transesterification. This process involves converting triglycerides, which are typically found in vegetable oils or animal fats, into biodiesel (fatty acid methyl esters) and glycerin. Transesterification is a crucial step in the biodiesel production process, as it alters the chemical structure of the feed-stock to create a fuel suitable for use in diesel engines.

Transesterification Reaction: The transesterification reaction is catalyzed and typically involves the following steps:

Feed-stock Preparation: Vegetable oils or animal fats are first heated to reduce their viscosity and enhance their reactivity.

Alcohol Addition: Methanol or ethanol is added to the heated feed-stock. These alcohols serve as the reactants and work to replace the glycerol component in the triglycerides.

Catalyst Introduction: A catalyst is introduced to facilitate the reaction. Catalysts play a pivotal role in speeding up the process and enhancing its efficiency.

Stirring and Heating: The reaction mixture is stirred and heated to promote the complete conversion of triglycerides into biodiesel and glycerin.

Product Separation: Product separation in biodiesel production is a crucial stage, as it enables the isolation of the desired biodiesel product from the less valuable byproducts, especially glycerin. It is during this phase that the biodiesel can be further refined to meet quality standards, ensuring it is free from contaminants and impurities that could hinder its performance in diesel engines. This step also emphasizes the sustainable nature of biodiesel production, as both the biodiesel and glycerine can find applications in various industries, reducing waste and maximizing the value of the feed-stock used in the process. Proper separation, followed by thorough purification, is essential to produce high-quality biodiesel that meets regulatory requirements and maintains the integrity of diesel engines while simultaneously contributing to a more environmentally friendly and renewable energy source.

Role of Catalyst in Transesterification

Catalyst Function: Catalysts are fundamental to the transesterification reaction because they significantly accelerate the reaction rate, reduce the reaction time, and improve the overall efficiency of biodiesel production. They achieve this by lowering the activation energy required for the reaction, allowing it to occur more readily.

Types of Catalysts: Historically, sodium hydroxide (NaOH) and potassium hydroxide (KOH) have been widely used as catalysts in transesterification reactions. These alkali catalysts are effective in promoting biodiesel formation, but they have some drawbacks, including the potential for soap formation and the need for water removal. Acid catalysts, such as sulfuric acid, are also used but require careful handling due to their corrosive nature.

Current Innovations in Catalysts for Transesterification

Enzymatic Catalysts: One of the exciting innovations in biodiesel production is the use of enzymatic catalysts. Enzymes, like lipases, can catalyze the transesterification reaction under milder conditions, reducing energy consumption and the need for excessive stirring. They also produce fewer impurities, making the purification process easier.

Solid Catalysts: Heterogeneous solid catalysts, such as metal oxides and zeolites, are gaining attention due to their reusability and the potential for reducing waste in the biodiesel production process. Solid catalysts can facilitate the transesterification reaction effectively while minimizing the formation of soaps and glycerin.

Nano-Catalysts: Nano-catalysts are on the horizon as a promising avenue for further enhancing the efficiency of transesterification reactions. These tiny particles offer a high surface area, which can lead to faster reaction rates and greater control over the biodiesel production process.

Ionic Liquids: Ionic liquids are another innovative approach. They can serve as both solvents and catalysts, eliminating the need for additional chemicals and simplifying the reaction process. Ionic liquids are also highly tunable, allowing for precise control over reaction conditions.

These advancements in catalyst technology are improving the sustainability and cost-effectiveness of biodiesel production, making it a more viable option for reducing carbon emissions and promoting a cleaner, renewable energy source. Researchers and industry experts continue to explore and refine these catalyst innovations to enhance the efficiency and environmental friendliness of biodiesel production.

First-Generation Environmental and Social Impacts

The production and utilization of bioenergy have both positive and negative environmental and social implications. On the positive side, bioenergy offers several advantages. It contributes to the reduction of greenhouse gas emissions when compared to conventional fossil fuels, making it a valuable tool in combatting climate change. Moreover, when managed sustainably, bioenergy sources can be renewed over time, providing a potentially long-term and environmentally friendly energy solution. Additionally, bioenergy can enhance energy security by reducing the dependency on imported fossil fuels, thus increasing a region's energy self-reliance.

However, bioenergy also comes with its share of disadvantages and challenges. One of the main concerns is land use competition, which arises when land is allocated for bioenergy production at the expense of food production or natural habitats. This competition can lead to conflicts and issues related to food security and biodiversity conservation. Additionally, bioenergy production often requires substantial water resources, which can result in increased water consumption and potential pollution if not managed properly. Lastly, social concerns related to land rights, displacement of local communities, and fluctuations in food prices are important considerations when implementing bioenergy projects. Balancing these advantages and disadvantages is crucial in ensuring that bioenergy contributes positively to our sustainable energy future while mitigating potential adverse impacts. First-generation feed-stock has been essential in the early development of bioenergy technologies. However, concerns over sustainability and competition with food production have spurred research into more advanced feed-stock options.

Second-Generation Feed-Stock

Second-generation feed-stock marks a substantial progression in the realm of bioenergy production. In contrast to first-generation feed-stock, which originates from food crops, second-generation feed-stock predominantly centers on non-food biomass sources. Bioenergy production relies on a diverse range of feed-stocks, and two prominent categories are agricultural residues and woody biomass. Agricultural residues, including materials like wheat straw, corn stover, and rice husks, are a valuable resource for bioenergy due to their availability as byproducts of agricultural activities. The utilization of these residues reduces competition with food production, making it a sustainable option.

On the other hand, woody biomass comprises forestry residues, sawdust, and the cultivation of dedicated energy crops like willow and poplar. This category provides another substantial source of biomass for bioenergy production, tapping into the potential of wood-related materials while promoting sustainable forestry practices. Both agricultural residues and woody biomass offer environmentally friendly alternatives to traditional fossil fuels, contributing to the diversification of our energy sources and the reduction of greenhouse gas emissions.

Second-Generation Production Process

The conversion of biomass into valuable biofuels involves a series of intricate processes, each contributing to the transformation of biomass feed-stock. Initially, pretreatment methods are employed, encompassing physical, chemical, or biological processes. The primary objective here is to break down the complex structure of biomass, increasing the accessibility of cellulose and hemicellulose for subsequent processing. Following pretreatment, enzymatic hydrolysis comes into play, where specialized enzymes play a pivotal role in breaking down cellulose and hemicellulose into fermentable sugars. This step requires enzymes capable of targeting the intricate biomass structures effectively.

The journey continues with fermentation, where microorganisms like bacteria or yeast are utilized to convert these sugars into biofuels, such as ethanol. This biological process is a cornerstone in the production of biofuels from biomass. Additionally, thermochemical conversion methods, such as pyrolysis, gasification, or hydrothermal liquefaction, can also be employed to extract biooil, syngas, or biocrude from biomass. These sophisticated processes underscore the versatility and potential of biomass as a renewable and environmentally friendly source of bioenergy.

Second-Generation—Environmental and Social Impacts

The utilization of non-food biomass in bioenergy production offers several advantages worth considering. Firstly, it reduces competition with food crops, alleviating concerns related to food security and resource allocation. Additionally, this approach has the potential for higher energy yields and lower greenhouse gas emissions, contributing to a more environmentally sustainable energy landscape. Furthermore, bioenergy production from non-food biomass provides an opportunity for waste valorization, effectively reducing waste disposal and promoting a circular economy.

However, it is essential to acknowledge and address the challenges associated with this approach. One primary challenge lies in the technological complexity and higher processing costs often associated with non-food biomass. These complexities can impact the economic viability of bioenergy projects. Sustainable sourcing of non-food biomass and responsible land use considerations for energy crops are also key challenges that must be carefully managed to prevent adverse environmental and social impacts. Additionally, there is a need to mitigate the potential effects on biodiversity and ecosystems when converting non-food biomass into bioenergy, ensuring a balanced approach to sustainable energy production. Second-generation feed-stock represents a promising avenue for sustainable bioenergy production. Understanding the potential of these non-food biomass sources and the technological advancements in their conversion processes can contribute to the advancement of renewable energy solutions and address the challenges posed by conventional fossil fuels. The future of bioenergy lies in these innovative and environmentally responsible feed-stock options.

Third-Generation Feed-Stocks

Third-generation feed-stocks represent a remarkable advancement in the realm of bioenergy production, focusing on unconventional and eco-friendly sources that bring unique advantages. Diverging from the paths of first and second-generation feed-stocks, third-generation feed-stocks are charting alternative courses that hold promise for overcoming various challenges inherent to traditional biofuels. A thorough understanding of these cutting-edge feed-stocks and their potential applications is pivotal for the advancement of sustainable bioenergy technologies.

Among the diverse array of third-generation feed-stocks, algae take center stage. These aquatic organisms can be cultivated in various settings, including ponds, tanks, or specialized photo-bioreactors. Algae stand out due to their rapid growth rates and the potential to produce lipids suitable for biodiesel and other valuable compounds.

Within the realm of algae, micro-algae, microscopic aquatic organisms proficient in photosynthesis, come into play. These microorganisms are prized for their high oil content, rendering them suitable candidates for biodiesel production. They can be cultivated in diverse environments, ranging from open ponds to closed photobioreactors and even wastewater treatment systems.

Macro-algae, commonly known as seaweeds, form another branch of third-generation feed-stocks. These large aquatic plants flourish in marine environments, characterized by their rapid growth rates and significant carbon capture capacity. This group exhibits potential for various bioenergy applications, including bioethanol, biogas, and the extraction of other valuable compounds. The exploration of these third-generation feed-stocks represents a dynamic shift in the bioenergy landscape, with the potential to redefine sustainable energy production in a more environmentally conscious and efficient manner.

Third-Generation—Production Process

The production process for third-generation feed-stocks, such as micro-algae and macro-algae, involves distinct cultivation methods and harvesting techniques. Microalgae cultivation necessitates essential resources like light, water, and nutrients, typically nitrogen and phosphorus. Large-scale production is achieved through photobioreactors or open ponds, depending on the scale of the operation. The harvesting of micro-algae employs various methods, including centrifugation, filtration, and flocculation, to separate the biomass efficiently.

In contrast, macro-algae cultivation primarily occurs in coastal areas or specialized offshore farms. The harvesting of macro-algae can be conducted manually or using mechanical methods, which depends on the specific context and scale of the operation. One notable advantage of macro-algae cultivation is its minimal demand for freshwater and land resources, making it a sustainable choice for bioenergy production that aligns with environmental conservation efforts. This distinctive cultivation and harvesting processes highlight the adaptability and potential of third-generation feed-stocks in bioenergy production, offering a glimpse into a more sustainable and efficient future for renewable energy sources.

Third-Generation—Environmental and Social Impacts'

The production of bioenergy from third-generation feed-stocks like micro-algae and macro-algae presents a range of environmental and social implications. On the positive side, there are several advantages to consider. These feed-stocks can thrive in non-arable land, reducing conflicts over land use for food production. Furthermore, depending on the cultivation and conversion methods employed, micro-algae and macro-algae have the potential to be carbon–neutral or even carbon-negative, which can significantly contribute to mitigating greenhouse gas emissions. Additionally, they offer the opportunity for integration with wastewater treatment processes, aiding in reducing pollution and nutrient runoff, thereby benefiting aquatic ecosystems.

However, it is essential to recognize and address the challenges associated with third-generation feed-stock production. High cultivation and processing costs can limit the commercial viability of these bioenergy sources, necessitating further research and development to optimize cost-effectiveness. Technical challenges, such as achieving high yields and maintaining continuous cultivation, require innovative solutions to maximize the potential of these feed-stocks. Additionally, if not properly managed, the cultivation and harvesting of micro-algae and macro-algae can have adverse impacts on aquatic ecosystems, emphasizing the need for sustainable and responsible practices in their production. Balancing these advantages and challenges is vital in harnessing the potential of third-generation feed-stocks for sustainable and environmentally friendly bioenergy production.

2.2.2 Miscellaneous Sources of Bioenergy (Fig. 2.2)

Energy Crops: Dedicated crops grown specifically for bioenergy production, such as switchgrass and miscanthus.

Municipal Solid Waste: Waste materials from households, businesses, and industries can be processed to extract biogas through anaerobic digestion or incineration for energy recovery.

Landfill Gas: Organic waste in landfills generates methane and other gases, which can be collected and used as a source of renewable energy.

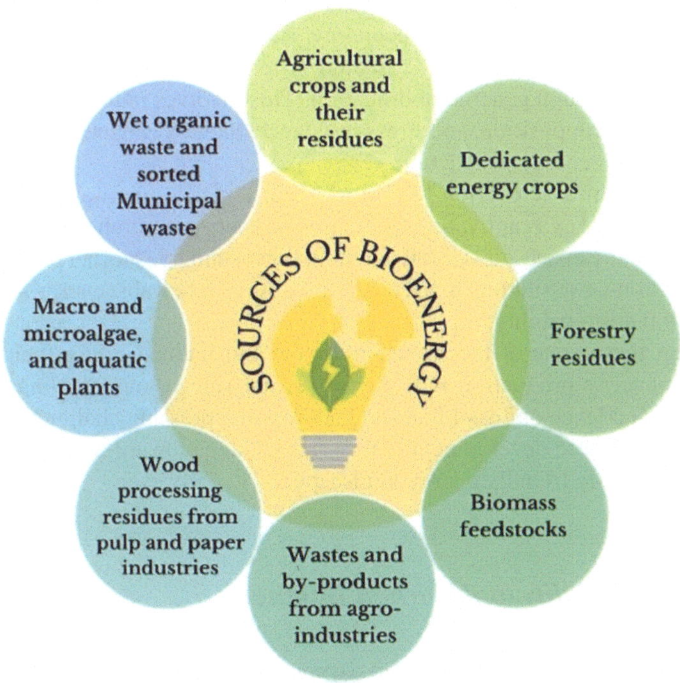

Fig. 2.2 Different sources of bioenergy

Wastewater Treatment: Organic matter in sewage and wastewater can be treated anaerobically to produce biogas, which is a sustainable energy source.

Animal Waste: Livestock and poultry waste can be processed through anaerobic digestion to produce biogas, reducing greenhouse gas emissions and providing energy.

Energy from Crop Residues: Agricultural residues, such as sugarcane bagasse and rice husks, can be utilized for energy generation.

Bioenergy Derived from Food Processing Residues: Energy can be generated from the byproducts and waste produced from food processing industries.

2.2.3 *Technological and Policy Developments*

Technological advancements and evolving policy frameworks are driving significant developments in the field of bioenergy production. One notable innovation is the emergence of biorefineries, which are integrated systems designed to maximize the extraction of value from various feed-stocks. These biorefineries facilitate the production of not only biofuels but also a range of bioproducts and chemicals, promoting a more diversified and sustainable bioeconomy.

In parallel, genetic engineering research is making strides in enhancing feed-stock properties through genetic modification. This involves improving factors such as biomass yield, composition, and stress resistance, thereby enhancing the efficiency and viability of bioenergy production.

Furthermore, third-generation feed-stocks, particularly micro-algae and macro-algae, have emerged as a captivating frontier in bioenergy research and development. These feed-stocks possess substantial potential for sustainable biofuel production and other applications, contributing to the diversification of renewable energy sources and environmental sustainability.

Exploring these state-of-the-art technologies and understanding their implications for bioenergy production will play a pivotal role in our ongoing efforts to reduce our dependence on fossil fuels and address the pressing challenges posed by climate change. These developments mark crucial steps toward a more sustainable and environmentally friendly energy landscape.

2.2.4 *Sustainability and Carbon Balance*

Sustainability and achieving a balanced carbon footprint are paramount considerations in the realm of bioenergy production. Sustainable bioenergy practices prioritize environmentally responsible sourcing and production, aiming to minimize adverse impacts on ecosystems. This entails practices that avoid deforestation and other harmful effects while ensuring that the entire bioenergy life cycle is managed with a commitment to preserving our environment.

In the context of carbon neutrality, bioenergy systems strive to maintain equilibrium in carbon emissions. Carbon-neutral bioenergy systems are designed to offset the carbon dioxide released during biomass combustion by reabsorbing an equivalent amount of carbon dioxide through the growth of new biomass. This approach not only reduces the net emissions of greenhouse gases but also helps mitigate the effects of climate change. By embracing sustainable and carbon-neutral bioenergy practices, we move closer to a future where renewable energy sources play a pivotal role in mitigating environmental challenges and advancing a more sustainable energy landscape.

2.2.5 Classification of Bioenergy

Bioenergy categorization can vary depending on multiple factors, such as the type of biomass employed, the conversion technique, and the ultimate purpose. Figure 2.3 explains the classification of bioenergy.

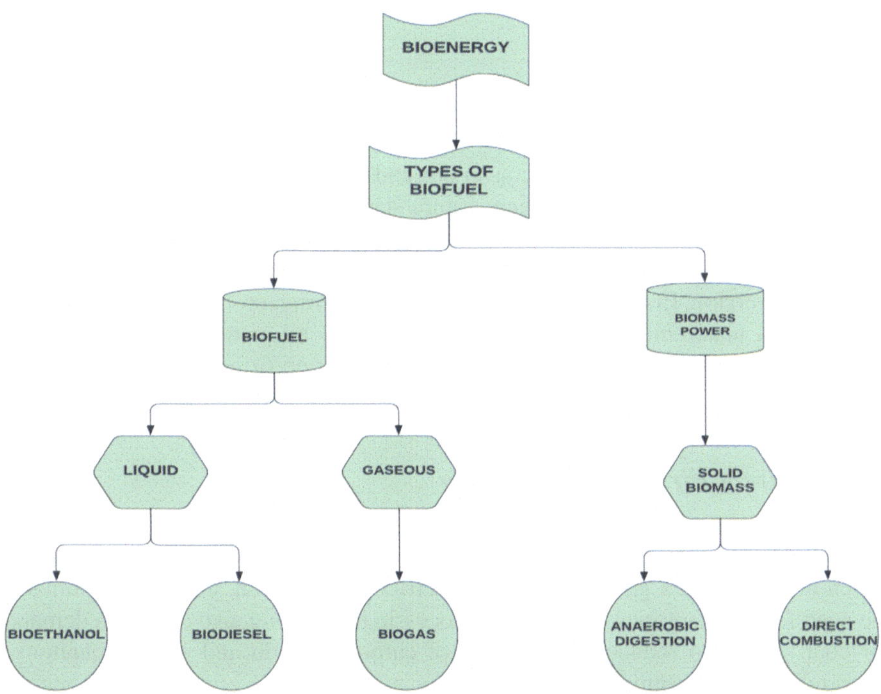

Fig. 2.3 Classification of bioenergy

2.3 Applications of Bioenergy Products

The bioenergy can contribute for more sustainable, safe, secure and economically powerful applications as it can supply domestic clean energy sources, revitalize rural economies, reduce the burden of countries to depend on foreign oil and petroleum products, generate green-based jobs, etc.

The applications are wider starting from simple stove cooking to rocket fuels. The applications are as follows:

- Applications of biomass energy for residential purpose especially, cooking with small stoves.
- It can also be used in small scale to large-scale power plants with centralized utilities for the production of electricity.
- It can be used for the production of day-to-day products
- It is used to produce transportation fuels
- It can also be used for production of aviation fuels
- It increases bioenergy-based job opportunities.

For the abovementioned vast applications, the bioenergy products are the sole resources. The prudential uses of bioenergy products will lead to sustainable clean energy supply. The bioenergy products with their application are as follows:

2.3.1 Thermal Conversion

Biomass can be burned by thermal conversion and used as energy source. This thermal conversion includes a very high temperature starting from 180° (carbonization) to 1000 °C (gasification). Raw materials such as municipal solid waste (MSW) and scraps from paper or lumber mills are the most familiar biomass feed-stocks for thermal conversion. Energy is created through direct firing, co-firing, pyrolysis, gasification, and anaerobic decomposition. Compressed briquettes have high-energy density and are generally easy to burn during direct or co-firing.

2.3.2 Direct Firing and Co-firing

Steam produced during the firing process of briquettes is used as power for turbine which in turning on a generator and in electricity production. Co-firing eliminates the need for new factories for biomass processing. Co-firing also eases the demand for coal. This also reduces the amount of carbon dioxide and other greenhouse gases released by burning fossil fuels.

2.3.3 Syngas

Syngas is a combination of hydrogen and carbon monoxide majorly. It also has fewer percentage of methane and hydrocarbons. This can be purified. It is cleaned of sulfur, particulates, mercury, and other pollutants. The clean syngas is processed into transportation biofuels.

2.3.4 Biochar

Biochar is a type of charcoal and a carbon-rich solid which is specifically useful in agriculture. Biochar enriches soil and prevents it from leaching. Carbon sinks are reservoirs for carbon-containing chemicals, including greenhouse gases, and biochar acts as an excellent carbon sink. Biochar also helps in enriching the soil.

2.3.5 Anaerobic Decomposition

Anaerobic decomposition is a process where microorganisms, usually bacteria, break down the available material in the absence of oxygen. Anaerobic decomposition is an important process in landfills, where biomass is crushed and compressed, creating an anaerobic (or oxygen-poor) environment. Manure and other animal waste can be converted to sustainably meet the energy needs.

2.3.6 Biofuel

Biomass when converted into liquid biofuels such as ethanol and biodiesel acts as a suitable renewable energy source. Biomass that is high in carbohydrates, such as sugarcane, wheat, or corn when fermented gives ethanol. Biodiesel is made by combining ethanol with animal fat, recycled cooking fat, or vegetable oil. Though biofuels help in operating vehicles, they do not operate as efficiently as gasoline. However, they can be blended with gasoline to make use of it efficiently to power vehicles and machinery, and in turn do not release the emissions associated with fossil fuels. Ethanol has become a popular substitute for wood in residential fireplaces. When burned, it gives off heat in the form of flames, and water vapor only and not smoke [3].

2.3.7 *Black Liquor*

Processing of wood into paper produces a high-energy, toxic substance called black liquor. With the invention of the recovery boiler in the 1930s, black liquor could be recycled. In the USA, paper mills use nearly all of their black liquor formed to run their mills, and because of this forest industry is one of the most energy-efficient in this nation. Sweden has experimented very recently in gasifying black liquor to produce syngas, which is further used to generate electricity.

2.4 Bioeconomy

The terms bioeconomy means the economy that is developed by the production and usage of bioenergy, developed through biomass resources, from earth, like microbes, plants, and animals. This bioenergy, thus produced, is utilized for the production in industries and health sector. In simple terms, the economy based on bioenergy is defined as bioeconomy. The perspectives of bioeconomy completely rely on the properly structured protocols and mechanisms, in manipulating the genomic and molecular level, so as to improve the industrial production to produce, improved and novel products.

In the report "The Knowledge Based Bio-Economy (KBBE) in Europe: Achievements and Challenges" by Albrecht et al. [4], the term bioeconomy was defined as follows: The bioeconomy is the sustainable production and conversion of biomass, for a range of food, health, fiber and industrial products and energy, where renewable biomass encompasses any biological material to be used as raw material." [4]. According to a 2013 study, "the bioeconomy can be defined as an economy where the basic building blocks for materials, chemicals and energy are derived from renewable biological resources." [5]

In Berlin (November 2015), the first global Bioeconomy Summit was organized. It defined bioeconomy as "knowledge-based production and utilization of biological resources, biological processes and principles to sustainably provide goods and services across all economic sectors." According to the summit, bioeconomy comprises of three important elements: renewable biomass, enabling and converging technologies, and integration across applications concerning primary production (living natural resources), health (medical and pharma-related devices), and industries [6].

The ideology of bioeconomy is to reduce and finally stop the dependence on exhausting renewable resources like coal, petroleum, and other fossil fuels. More than 60 countries have bioeconomy-based strategies, out of which twenty of them have published dedicated bioeconomy strategies in Asia, America, Africa, Oceania, and Europe [7].

A suitable incident was the implementation of the Energy Independence and Security Act (EISA) by the USA, in 2007. The only goal of EISA was to escalate the availability of renewable energy. And it was planned to be through the production of biofuel [8]. Fifth Fuel Policy was introduced by the Government of Malaysia, in the 8th Malaysia Plan structured for years 2001–2005.

The aim was to motivate the industrial field for the production of bioenergy, worldwide. An overall share of 20% renewable energy has been set as a target in EU's transport consumption by the European Commission in the year 2020 [9].

2.4.1 Bioeconomy: Concept, Policies, and Future Perspectives

The bioeconomy/bio-based economy refers to the actions that are driven by research and innovations related to the production, application, and conservation of biological resources with the ultimate aim of a sustainable environment and economy in the future. It includes all commercial and industrial sectors that rely on consuming, creating, or processing biological resources derived from forestry, agriculture, aquaculture, and organic waste. The bioeconomy seeks to operate according to the circular economy's guiding principles of reuse, repair, and recycling, as opposed to the traditional linear or fossil-based economy (Fig. 2.4).

In the first decade of the twenty-first century, the term "bioeconomy"—a "leapfrogging" strategy became well known as a result of its adoption by the European

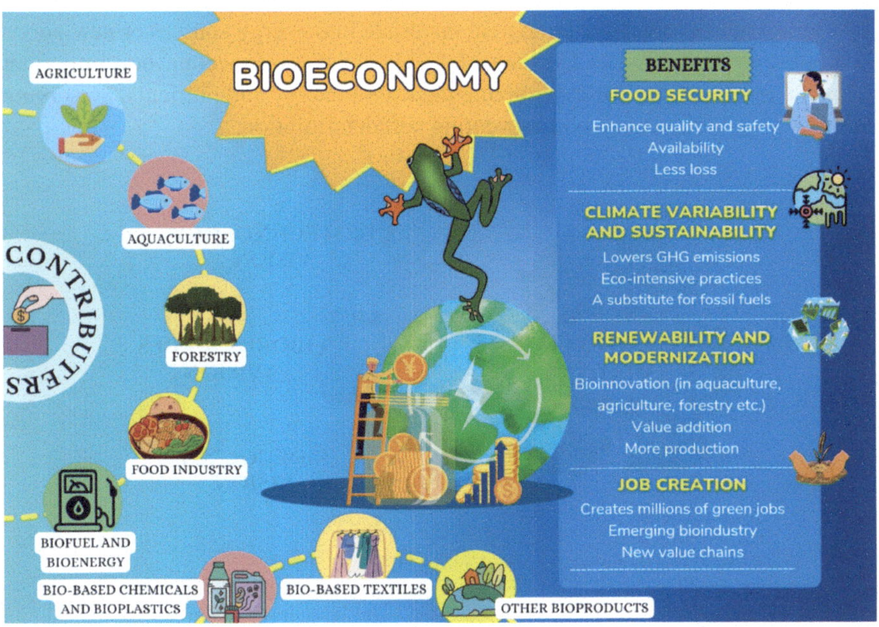

Fig. 2.4 Concept of bioeconomy

Union (EU-2012) and the Organization for Economic Cooperation and Development (OECD-2006) as a framework to promote new goods, procedures, and markets by accelerating biotechnology innovation. The multi-sector perspective of EU bioeconomy policies is to safeguard the environment, prevent overexploitation of natural resources, and increase biodiversity by adopting strategies dependent on biological resources (plants, animals, microorganisms, and their products, or biomass).

Numerous nations support and implement comprehensive policies and initiatives to advance their bioeconomies as a means of addressing environmental concerns like climate change, food security, and energy independence.

The United Nations (UN) adopted the Bioeconomy 2030 Agenda in 2015, which included 17 Sustainable Development Goals (SDGs). According to the FAO studies, the following four connected aspects of the bioeconomy are crucial:

1. The elimination of poverty, hunger, and the reduction of inequality.
2. Sanitation and clean water mission.
3. Encourage environmentally friendly business and infrastructure.
4. Benefit for climate action and the health of land and aquatic ecosystems.

Although not mutually exclusive, the National Academies of Science, Engineering, and Medicine (NASEM) classifies the bioeconomy into three main visions as shown in Fig. 2.5.

According to NASEM, the activities of the bioeconomy revolve around improving scientific knowledge by comprehending mechanisms and processes at the genetic and molecular levels, as well as synthesizing and manufacturing those commercial items, such as vaccines, biopesticides, etc. The vision for bioresources entails the transformation of biomass or biological materials into energy sources or new goods like bioplastics, biofuels, etc. The bioecology vision takes into account how important ecological processes are for maximizing the use of energy and nutrients, preserving and promoting biodiversity, and preventing soil deterioration.

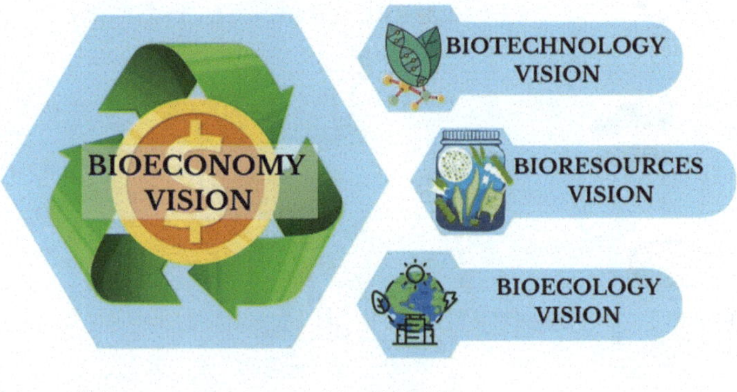

Fig. 2.5 Visions of bioeconomy as proposed by NASEM

2.4.2 Sectors Involved in Bioeconomy

In accordance with the European Union Bioeconomy Strategy, which was updated in 2018, the bioeconomy encompasses all industries that rely on biological resources (including organic waste, animals, plants, microbes, and derived biomass), as well as its principles and functions. This multidisciplinary topic encompasses the subsectors of biohealth, bioenergy, bioagriculture (agri-bio), and bioindustrial (chemicals and materials). All subsectors of the bioeconomy share the objective of commercializing the resulting bioproducts, processes, and intellectual property. A few important sectors are discussed below.

2.4.2.1 Agriculture and Food-Based Bioeconomy

Technology advancements in agriculture have enhanced crop production and farming efficiency while reducing the environmental impact. Examples of these advancements include climate monitoring, nutrition management, vertical farming, pest and disease control, precision and regenerative agriculture, and digital market information.

The notion of "cellular agriculture" is utilized to create agricultural goods from cell cultures utilizing genetic engineering and synthetic biology methods. Animal products such as meat, milk, and eggs are produced using cell culture as a step toward animal welfare, food security, and human health. Spider silk similar plant protein has been used as an alternative to plastics. Crops have been genetically engineered to protect against pests and increase the yield. Genomic-assisted breeding improves bioenergy crops, e.g., in shrub willow, a crop that is gaining popularity as a source of bioenergy, researchers at Cornell University have discovered genes that confer resistance to leaf rust. Fungi-based industries are considered a major part of a growing bioeconomy, e.g., mushroom farming, mycoprotein, etc. Algae cultivation has potential uses in the production of pharmaceuticals, bioplastics, nutraceuticals (algal oil), food colorants and dyes, protein-rich animal feed, and algal fuels.

2.4.2.2 Blue/Ocean Economy

The term "blue/ocean economy" refers to the wise use of ocean resources to support economic viability and the health of the marine environment. According to estimates, it is currently valued more than US$ 1.5 trillion annually and might double in size by 2030. The ocean offers a wide range of possible blue energy sources, including wave, thermal, and biomass (like marine algae). Because they provide enormous financial opportunities and food security, fishing and aquaculture are essential to society. Through interstate collaboration and trade, the ports and maritime transportation can boost economic growth. Diverse sorts of minerals are found in the sea bed, and it is also a source of many innovative substances that could be used to create pharmaceuticals, nutraceuticals, and personal care items that would boost the economy. The production of enzymes, biopolymers, and biomaterials could be accomplished using items of marine ancestry. Economic growth may benefit from sustainable seaside tourism. In order to improve livelihood, lower risk from natural catastrophes, and mitigate climate change, the Indian Ocean Blue Carbon Hub (IORA) aims to develop knowledge to maintain and restore blue carbon ecosystem throughout the Indian Ocean.

2.4.2.3 Forest Bioeconomy

The forest bioeconomy is based on natural resources from forests and encompasses a number of industries and industrial methods that put the idea of the circular economy into practice. The modern forest industry uses every branch of science and technology in its effective industrial value chains, including oil management, drones, satellite imaging, sophisticated harvesting equipment, robust mechanical engineering, innovative process automation, nanotechnology, and biochemical applications. The use of woodchips or wood pellets in automated heating system is based on robust technology. In order to further advance the circular bioeconomy, techniques utilized for industrial conversion of wood into value-added goods can be refined and applied to agricultural leftovers like straw. Recreation and nature tourism are essential components of the bioeconomy, in addition to carbon sequestration and ecosystem services. Many nations, including Finland, are converting their forest-based businesses into biorefineries. Forest biomass is utilized in many processes to make textiles, chemicals, cosmetics, fuel, coatings, food, and animal feed [10].

2.4.2.4 Bioenergy-Based Bioeconomy

The ever-rising global energy crisis, environmental degradation, and climate change issues due to extensive usage of fossil fuels (FFs) are pushing the world to find an alternative green energy source to attain Sustainable Development Goals (SDGs) via decarbonization of the global energy system (GES). Due to the expanding energy needs of contemporary economies and the physical interactions between energy- and non-energy-based processes and products, energy can be seen as the "glue" in the bioeconomy. Bioenergy is described as clean energy generated from renewable sources that can displace fossil fuels. Wide range of natural resources as shown in Fig. 2.3 is used to produce bioenergy. Utilizing processing methods like anaerobic digestion, fermentation, pyrolysis, or torrefaction, the bio-based economy utilizes first-generation biomass (crops), second-generation biomass (cellulosic crop), third-generation biomass (algae based), and fourth-generation (genetically modified algae) to get the most out of biomass.

Regardless of the source, bioenergy is considered as a sustainable source of power providing heat, gas and fuel. The International Energy Agency (IEA) has projected that by 2060, bioenergy will supply roughly 17% of global energy demand and can cut greenhouse gas emissions by 20% overall [11, 12]. The biofuel capacity in India is expected to grow from 5.2 billion liters in 2021 to 10.1 billion liters in 2025 and may generate US$ 20 billion bioeconomy.

2.4.3 Benefits of Bioeconomy

Making the switch to a bioeconomy has several advantages. The bioeconomy can boost crop and livestock productivity while ensuring food security. It prevents overexploitation as it can offer sustainable management of natural resources and encourages the use of waste materials and products to create value-added products by reducing waste and promoting a circular economy. It can result in improved efficiency in the use of agricultural wastes and biomass. For the production of energy, chemicals, and other materials, it is conceivable to substitute renewable biomass or bio-based raw materials for fossil fuels. Adopting bioeconomy strategies can enhance public health while reducing greenhouse gas emissions. A few bioeconomy techniques, such as afforestation and deforestation, can help reduce climate change by removing carbon dioxide from the atmosphere. Rural development can be improved by diversifying income sources and creating jobs in agriculture, forestry, biotechnology, and related services. Innovation in areas like genetic engineering, synthetic biology, and bioprocessing is sparked by the bioeconomy and results in new goods with diverse uses. The creation of novel medications, treatments, and medical innovations like personalized medicine may be facilitated byproducts obtained from the bioeconomy.

2.4.4 Ways to Strengthen the Bioeconomy

The benefits of bioeconomy will not become a reality without the support of local, national and supranational organizations across the world. Guidelines for the effective creation and marketing of bioeconomy-related goods and services include promoting alignment and coherence in policy across a variety of involved sectors; assuring that all people have access to goods and services relevant to the bioeconomy. Improving the rigidity of current production systems and stimulating customer demand as well as acceptance will pave the way to strengthening the bioeconomy.

2.5 Bioeconomy—A Support System of Circular Economy

Bioeconomy and circular economy go hand in hand (Fig. 2.6). Circular economy is a concept of commercial production and proper utilization of the same which involves the 6 Rs concept of sustainability that includes, reduce, rethink, refuse, reuse, repair, and recycle. Circular economy aims to tackle the global challenges like endangered biodiversity, climatic change, pollution, etc., by the implementation of the three basic principles—concept of reuse, control pollution, recycle or regenerate the natural sources [13]. In a circular economy, the resources are either applied or utilized in circular fashion rather than linear. Many business models promoting circularity are those that share platforms, provide products as service, optimize the utilization of resources, etc. The ultimate aim of these models is to promote circular economy [14].

The historical approaches of circular economy convey the ideas that would link biomimicry, cradle-to-cradle design principles and industrial ecology. Biomimicry mimics time-tested designs that formulates human system pattern. Industrial ecology acts as the base of circular economy as it is the study of flow of energy and material in the industrial systems. Cradle-to-cradle design principles are those that device a holistic approach to create a system that deals with procuring raw material to waste disposal. The ideology is to decrease waste and pollution and to increase the efficiency of resources. These intertwined concepts are stepping stones for developing the circular economy and implementing the same.

William McDonough is named as the father of circular economy as he enlightened the society about circular economy through his popular book namely Cradle to Cradle: Remaking the Way We Make Things. He also received Fortune Award

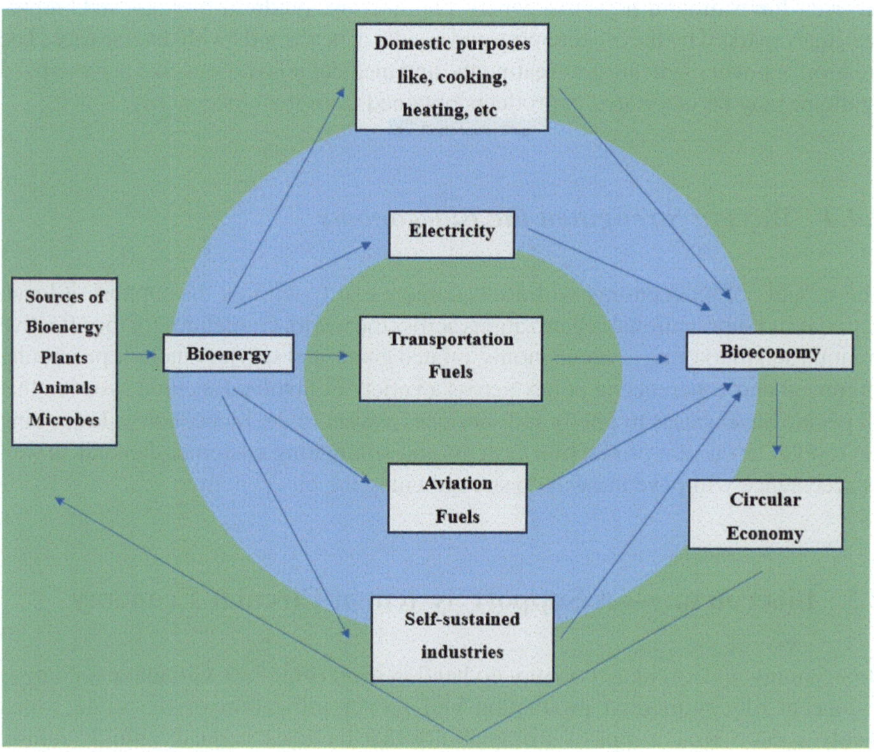

Fig. 2.6 Design of bioenergy-based bioeconomy with circular economy

for Circular Economy Leadership in the 2017, at Davos during the World Economic Forum.

The Centre for Bioenergy Sustainability (CBES) deals with the environmental impacts and the sustainability of bioenergy. This is a leading organization under the Department of Energy Oak Ridge National Laboratory, USA. The ultimate goal of the organization is to analyze the methods for enhancing the sustainability of current and potential bioenergy production and distribution (environmental, economic, and social) so as to serve and support the stakeholders and decision makers with independent source of data and analysis for bioenergy production [15].

The CBES goals aim at considering bioenergy as an important option from a system's perspective; socioeconomic and environmental concerns leading to sustainable bioenergy system; appropriate scales for addressing the same; bartering between land use and land management. It also lays emphasis on quantification of socioeconomic and environmental ideas of selecting appropriate bioenergy source (ex: lignocellulosic feed-stock) and to implement the same at regional to global scale.

As far as the application of bioeconomy in circular economy is concerned, the achievement of "Net Zero" to "True Zero" carbon emission is on progress. "Net Zero" carbon emissions achievement is because of consumption of fossil fuel thereby completely removing the carbon from the atmosphere. Related to this context, Steve Hoy, a global smart grid technologist, says that he believes the concept of creating a circular economy, True Zero concept has to be adapted rather than Net Zero concept wherein fossil fuel consumption is eliminated, entirely, energy will be produced only from renewable sources [16]. *In the* current scenario, to maintain these renewable systems, the renewable energy industry sector is intending to manufacture a significant amount of energy and raw.

Due to the emissions attributed to fossil fuel electricity generation, the overall carbon footprint of renewable energy technologies is significantly lower than for fossil fuel generation over the respective systems lifespan [16]. The other applications include agriculture, textile industry, construction industry, automotive industry, furniture and logistics industry, plastic waste management.

2.6 Indian Bioeconomy Scenario

India has a fantastic chance to rebuild its economy since it uses a variety of feedstock to manufacture various bioproducts. The bioeconomy of India surpassed US $1 billion in 2021, and the country has set an ambitious goal to reach US$ 150 billion by 2025. Bio-agri, which includes genetically modified (GMO) crops, insecticides, marine biotechnology, and animal biotechnology, has the potential to treble from US$ 10.5 billion to US$ 20 billion by 2025. The Indian government has decided to hasten the launch of ethanol-blenched petrol, enhance biofuel output, and adopt modifications to the national biofuel policy. The modifications include enabling more feed-stocks for biofuel production, exporting in certain circumstances, promoting research for indigenous techniques, and creating more jobs. In the National Policy on Biofuels-2018 (NPB), a target of 20% ethanol in gasoline and 5% biodiesel in diesel has been suggested by 2030 [17].

India currently supports specialized bioenergy incubators, infrastructure funding in the fields of synthetic biology, algal biofuels, biohydrogen, bioenzymes, CO_2 capturing and utilization, and biovalorization, to develop value-added products in an efficient manner. India and the USA are sharing leadership of the Mission Innovation Collaborative platform initiative on "Innovation for Sustainable Aviation Fuel" (ISAF) as part of Mission Innovation 2.0 [18]. Through the production of biogas from the processing of solid agricultural waste and livestock manure, the Galvanizing Organic Bioagro Resources (GOBAR)-DHAN [19] project aims to enhance sanitation in Indian village communities. Biomanures from biogas plants have been used as organic manure under fertilizer category for retail sale. Hydrogen-blended HCNG has been used as a clean fuel in the mobility sector to reduce pollution. Along with flagship initiatives like "Make in India," "Swachh Bharat," "Doubling Farmers Income," etc., the concept of the bioeconomy also needs to have a solid political foundation.

Bioeconomy policies should be aligned with international initiatives such as Paris Agreement on climate change, United Nations 2030 Sustainable Agenda or the Aichi Agreement on biodiversity to protect nature and life systems. It is crucial to emphasize that while the bioeconomy offers many advantages, it also brings challenges in terms of preserving biodiversity, effectively exploiting land, competing with food production, and ensuring that bio-based industries are socially just and sustainable. A balance between economic expansion and environmental protection is necessary for the bioeconomy concept to be successfully implemented.

2.7 Challenges of Bioenergy-Based Bioeconomy

Bioenergy, in spite of being Nextgen fuel, still faces challenges in its production and bioenergy-based bioeconomy. The major concern areas with respect to environmental impacts of bioenergy production include the water quality and quantity ranking first (16%) in the year of 2017, emission of greenhouse gases (6%), SOC and biodiversity (5%), and soil erosion (0.8%). This indicates that the bioenergy production has a direct and great impact on water resources and its pollution [20].

2.7.1 Water Quality and Quantity

Bioenergy production has its great impact on water quality and its quantity as well. It is due to the reason that high water consumption is required for such crop plantation. For an instance, USA with the support of EISA (2007) started a large corn ethanol production unit to produce potential water stress both local and regional levels [21–23]. This is due to the reason that when compared with soyabean and wheat, the corn plantation requires higher quantities of water and its uses a water level which equals the amount used by a community of 5000 people. (Service, 2009). Other disadvantages in bioenergy crop cultivation include nutrient pollution due to groundwater pollution, surface runoff, and usage of more fertilizers.

The bioenergy crop plantation in erodible and marginal areas reduces the streamflow and nutrients loss. This was identified by scenario analysis with Soil and Water Assessment Tool (SWAT) [24]. But these problems can be solved by following certain concepts like the selection of proper crop species and proper management (irrigation, appropriate fertilization, harvest rate, etc.) which will build possible balance between bioenergy production and water resource protection [25, 26].

2.7.2 Greenhouse Gas Emission

The decrease in greenhouse gas (GHG) emissions is also an equally important concern in the bioenergy production. Carbon dioxide and nitrous oxide are the two primary components as they are in high quantity. Quantity of net CO_2 emissions is comparatively less when biofuels are used than in the usage of fossil fuels [25, 27]. In a pilot study, Liu et al. quantified that the maximum production of switchgrass on marginal land would decrease emissions by 29 million tons $CO_{2\text{-eq}}$/year. This was done by replacing fossil fuel. The end results suggested that for transportation use in the USA, 40–85% of GHG emissions can be decreased when is replaced by ethanol on a per megajoule (MJ) energy basis. The efficiency of GHGs reduction always varies with usage of different feed-stocks [28].

Harris et al. from their study concluded that the conversion of arable land to the bioenergy crop plantation (second generation) will definitely lead to decreasing CO_2 emissions. Whereas, when native grasslands were converted to plantations of first-generation bioenergy crops and short rotation coppice (SRC), as well, showed a remarkable hike in CO_2 emissions. Hence, it is indispensable to select an appropriate bioenergy crop types as well as its management practices, to mitigate carbon dioxide emissions [29].

Emission of N_2O is also equally important as it is the second important greenhouse gas and second prime reason for global warming (298 times that of CO_2). One major source of N_2O is agriculture [30]. For an instance, corn cultivation requires high nitrogen-rich fertilizer, as it is the substrate for denitrification process. This directly increases the emission of N_2O. Hence, wise selection of the type of bioenergy plant and planting directly influences in controlling the N_2O emission.

2.7.3 Biodiversity

Bioenergy production indirectly abrades the biodiversity—Landscape choices, landscape configuration, land use conversion, and bioenergy production system types.

For example, conversion of normal plantation may interfere with local ecosystem but conversion of grasslands to bioenergy production plantation will increase local productiviwty and help maintain the ecosystem functions [31, 32].

Additionally, many research studies laid emphasis on growing *Miscanthus*, (comparing with other annual crops) which had reduced negative impact on biodiversity. This is due to the reason that cultivation of perennials provides relatively stable habitats for wildlife [33, 34].

2.7.4 Soil Organic Carbon

Soil organic carbon (SOC) is the index of soil quality. High soil organic carbon content improves retention of soil water and increases crop productivity and thereby soil biodiversity. The production of bioenergy influences SOC directly. This is through three major pathways. They are removal of residues, tillage, and conversion of land usage.

The decrease in SOC could be controlled, through a proper residue management like removal of little amount of residues and increased manure applications [26, 35, 36]. The biochar application improves the carbon sink of agricultural sector. It increases the SOC but it also absorbs the inorganic carbon in the atmosphere thereby improving the quality of air [37, 38].

Secondly, the loss of SOC is also due to soil disturbances and management practices. For example, in his study, Drewniak et al. [39] observed that tillage would always cause SOC loss. Wise choice of most suitable land area, plant types, and the management practices can prevent SOC [39].

2.7.5 High Cost

The expense that is involved in building up and maintenance of bioenergy plant is relatively high as other renewable technologies do not need to accounted for. Certain of bioenergy types of plantations or its management practices are sometimes on par or still cheaper, than fossil fuels, as the former does not require expensive drilling and mining processes which involve large capital.

2.7.6 Land Requirement

Socioeconomical and environmental pressure may be created in biomass production unit as there is increased demand for biomass. The reason is due to low surface power density of biomass as larger land areas are needed to produce the same amount of energy, compared to fossil fuels [40].

Bioenergy plantation requires land just like any other plantation. But if the same is built exclusively for an industry or company, then it has to be built in close proximity to cut off storage and transportation costs. If this condition is not possible, high capital and investment are required for the same: as it is easily possible only for those industries that are newly being built. Additional land requirement is essential, and if plantation is for the sole purpose of electricity purpose, then this adds on larger land footprint/unit of production power supply.

2.7.7 *Transportation*

Distant transportation of biomass is considered as prodigal and unsustainable. It is advantageous if a plantation and resource-consuming industry are located nearby as there will be considerably less expenditures on transport.

2.7.8 *Bioenergy-Based Bioeconomy—Nextgen Challenge: Carbon Management*

The bioenergy and bioeconomy has a fundamental challenge. Is it possible for bioenergy and bioenergy-based bioeconomy to feed the demand of the entire carbon-based value chains? Recently, it was found that the annual consumption of fossil carbon for oil, polymer industries, and textiles was more than 1 billion tons. To replace aviation fuel alone may require a many million tons for bioenergy fuel. A part of bioenergy has to be modified by removing oxygen and moisture so that it can be widely used in chemical and textile industries. All these will lead not only the hike in demand for natural resources but also reinstate the sources of bioenergy as primary asset for international trading.

Like fossil fuel, biomass is also not equally distributed on the earth, and there may be national discord. If the bioenergy-based bioeconomy turns out to be inefficient in supplying adequate amount of biomass or only with inappropriate bartering for other sustainability goals like biodiversity and food security, then it requires add-ons like artificial photosynthesis, direct air capture, etc. But still science and technology and industrial sector are working with full efficiency in bringing about the successful replacement of fossil fuels with bioenergy and hold the nation's economy status high with bioenergy-based bioeconomy.

2.8 Practical Approach

The bioenergy is not a new topic neither it is only a topic of theoretical view but much applicable for practical applications. In Estonia, Finland, and Denmark, application of bioenergy is well represented by the production of electricity (> 15%). The production was mainly by the combination of heat and power—CHP. Next to these three countries stand UK, Germany, Brazil and Sweden, Germany. Whereas in a few other countries, it is only 2–5% [41]. The practical approach is evident from the following companies that either produce bioenergy or use it for their production.

In India, the government of India supports bioenergy and this is evident from the following statement. Direct supply of biodiesel (B-100) blended along with diesel was to high-end consumers like Railways and State Road Transport Corporations and was permitted by the Government of India on August 10, 2015 [42].

One of the leading manufacturers of biodiesel (B-100) in India is Emami Agrotech Limited (EAL). They manufacture and supply biodiesel (B-100) to many industry sectors and applications-oriented sectors, in our country. Emami Group has its state-of-the-art biodiesel plant in Haldia, West Bengal. Its production capacity is 350 tons/day. It has a state-of-the-art NABL-accredited laboratory at Haldia. This biodiesel plant has ISO 9001-2008, ISO 14001, OHSAS 18001 and International Sustainability and Carbon Certification (ISCC) Certifications. Biodiesel produced by EAL conforms to Indian and European Quality Standards IS 15607-2016 and EN 14214 [43].

In the USA, one of the leading biodiesel producers is renewable energy group (REG). The annual production capacity is 432 million gallons as per January 2023 data. There are eight biodiesel plants under REG. The largest plant is in Washington.

These are the ten leading bioenergy producers in the world by 2021. The companies were enlisted in **Fortune Business Insights,** 2021 [44]. ADM is an MNC. It was established in Chicago in 1902. **Amersco Inc.** is one of the leading bioenergy assets developers. It was established in Massachusetts, USA, in 2000. **EnviTec Company** is leading biogas producer in Germany. It is a multiservice company that provides a range of services. **Enviva** in the USA is the largest producer of coal's alternate—viable wood pellets. **Drax group** in the USA is a pellet manufacturing mill that is trying out to have sustainable zero carbon level. **STARBAG** is a pioneer in innovation and field strength with respect to designing and constructing biogas plant, through its European technology. **Pinnacle Renewable Energy Inc.** is one of the leading wood pellet manufacturers. The production capacity is 2.83 million metric tons/year.

A Canadian Company **Enerkem** ranks first in the commercial production of renewable ethanol and methanol from non-recyclable municipal solid waste. **Green Plains Inc.** (established in 2007) is a US-based company pioneering in biorefining. The company has successfully brought down the carbon emission by over 34.3 MMT, since its establishment. POET, a Scotland-based company, is also one among the largest biorefinery. The company apart from being a biorefinery is also aiming at producing alternatives for petrochemicals.

These are few biomass energy producing and utilizing Companies Premium. Harp Renewables Limited, California; Granutech-Saturn Systems, USA;, EnviTec Biogas AG, Germany; Talbotts Biomass Energy Limited, UK; Taylor Biomass Energy, New York, Green Vinci Biomass Energy Co., Ltd. China; Vattenfall AB, Sweden.

Suzuki Motor Corporation, Toyota Motor Corporation ENEOS Corporation, Subaru Corporation, Toyota Tsusho Corporation and Daihatsu Motor Co. Ltd., collaboratively established the Research Association of Biomass Innovation for Next Generation Automobile Fuels (Research Association). This association was created to explore and analyze various possible ways to successfully optimize the production of best quality biofuel [45].

2.9 Conclusions

In this chapter, the production and applications of bioenergy, the application of bioenergy in raising bioeconomy, the association of bioeconomy and circular economy, the interrelationship among bioenergy, its resources, bioeconomy and circular economy and the collaborative effect of the same, on solving the mitigating environmental changes are explained. In general, bioenergy expansion fell by 50% in 2020. Only Europe and China reported significant expansion in 2020, adding 2 GW and 1.2 GW of bioenergy capacity, respectively. Under Net Zero Scenario, total global bioenergy usage, in 2030, is only going to be about 20% higher than in 2021. In 2021, about 35% of the bioenergy was from biomass, which was utilized for traditional practices that were inefficient, polluting and unsustainable. In the Net Zero Scenario, the usage of this traditional biomass may plummet to zero by 2030 in order to achieve the UN Sustainable Development Goal 7 on Affordable and Clean Energy. The usage of modern bioenergy, which does not include traditional uses of biomass, nearly doubles up from about 42 EJ in 2021 to 80 EJ in 2030. There will be escalation in the production of bioenergy so as to cater the estimation by Net Zero Scenario in 2030. Following scenarios are expected by the year 2030: In generation of electricity, the use of bioenergy will nearly double up. 750 TWh of electricity (about 2.5% of total demand) in 2021 will nearly reach about 1350 TWh (about 3.5% of total demand) in 2030. The usage of biojet kerosene as aviation fuel was nearly zero in 2021 and will increase in 2030 accounting for over 7% of all aviation fuel demand. In road transport, the consumption of liquid biofuel will quadruple from 2.1 mboe/d in 2021 to over 8 mboe/d in 2030.

The usage of bioenergy in industrial sectors (mainly paper, cement, and light) was moderately above 11 EJ of energy in 2021 and may increase considerably to more than 17 EJ in 2030. Bioenergy with carbon capture and storage (BECCS) plays a critical role in bioenergy management. It will capture and store bioenergy emissions that are already carbon–neutral and create negative emissions. In 2021, BECCS captured and stored 2 Mt of CO_2. There will be a hike to around 250 Mt of CO_2 in 2030, counterbalancing the emissions from the sectors where abatement will not be an easy condition.

Hence to head toward Sustainable Development Goals (SDGs), bioenergy production and usage has to profoundly increase to meet up with a wide variety of its implementations, by 2030, so as to get on track with the Net Zero Scenario, according to UN Sustainable Development Goal 7.

References

1. Vidhya, C. V. (2022). Microalgae—The ideal source of biofuel. In *Biofuels and bioenergy opportunities and challenges*. Elsevier Publications. 978-0-323-85269-2.
2. Babin, A., Vaneeckhaute, C., & Iliuta, M. C. (2021). Potential and challenges of bioenergy with carbon capture and storage as a carbon-negative energy source: A review. *Biomass and Bioenergy, 146*, 105968.
3. Kumar, R., & Kumar, P. (2017). Future microbial applications for bioenergy production; a perspective. *Frontiers in Microbiology*. https://doi.org/10.3389/fmicb.2017.00450
4. Albrecht, J., Carrez, D., Cunningham, P., Daroda, L., Mancia, R., Máthé, L., Raschka, A., Carus, M., & Piotrowski, S. (2010). *The knowledge based bio-economy (KBBE) in Europe: Achievements and challenges.*
5. McCormick, K., & Kautto, N. (2013). The bioeconomy in Europe: An overview. *Sustainability., 5*(6), 2589–2608.
6. Food and Agriculture Organization of the United Nations Rome. (2016). *How sustainability is addressed in official bioeconomy strategies at international, national and regional levels? An overview.*
7. Food and Agriculture Organization of the United Nations. (2022). *Document card*. www.fao.org. Retrieved 16 September 2022.
8. United states Environmental Protection Agency. (2007). *Summary of the Energy Independence and Security Act.*
9. Van Dam, J., & Junginger, M. (2011). Striving to further harmonization of sustainability criteria for bioenergy in Europe: Recommendations from a stakeholder questionnaire. Author links open overlay panel *Energy Policy, 39*(7), 4051–4066.
10. Arasto, A., Koljonen, T., & Similä, L. (2018). Growth by integrating bioeconomy and low-carbon economy: scenarios for Finland until 2050. [Espoo]. ISBN 978-951-38-8699-8. OCLC 1035157127.
11. Organization for Economic Cooperation and Development. (2009). *The Bioeconomy to 2030: Designing a policy agenda* (p. 3). OECD Publishing.
12. Scarlet, N., Dallemand, J. F., Monforti-Ferrario, F., & Nita, V. (2015). The role of biomass and bioenergy in a future bioeconomy: Policies and facts. *Environment and Behaviour, 15*, 3–34.
13. European Parliament News. (2021).
14. Urbinati, A., Chiaroni, D., & Chiesa, V. (2017). Towards a new taxonomy of circular economy business models. *Journal of Cleaner Production., 168*, 487–498.
15. Chung, J. N. (2013). Grand challenges in bioenergy and biofuel research: engineering and technology development, environmental impact, and sustainability. *Front Energy Research: Bioenergy and Biofuels, 1.*
16. Gallagher, J., Basu, B., Browne, M., Kenna, A., McCormack, S., Pilla, F., & Styles, D. (2019). Adapting stand-alone renewable energy technologies for the circular economy through eco-design and recycling. *Journal of Industrial Ecology., 23*(1), 133–140.
17. Das, S. (2020). *The National policy of biofuels of India—A perspective. Energy Policy, 143.*
18. Fuelling India's future with bioenergy. (2023). https://www.pwc.in/research-and-insights-hub/fuelling-indias-future-with-bioenergy.html
19. Galvanizing Organic Bio-Agro Resources Dhan (GOBARdhan). (2023). https://www.myscheme.gov.in/schemes/gobardhan

20. Wu, Y., Zhao, F., Liu, S., Wang, L., Qiu, L., Alexandrov, G., & Jothiprakash, V. (2015). Bioenergy production and environmental impacts. *Official Journal of the Asia Oceania Geosciences Society, 5* (Article number: 14).
21. Gasparatos, A., Stromberg, P., & Takeuchi, K. (2011). Biofuels, ecosystem services and human wellbeing: Putting biofuels in the ecosystem services narrative. *Agriculture, Ecosystems & Environment, 142*(3–4), 111–128.
22. Hoekman, S. K., Broch, A., & Liu, X. (2018). Environmental implications of higher ethanol production and use in the US: A literature review. Part I—Impacts on water, soil, and air quality. *Renewable and Sustainable Energy Reviews, 81*, 3140–3158.
23. Zhou, X., Clark, C. D., Nair, S. S., Hawkins, S. A., & Lambert, D. M. (2015). Environmental and economic analysis of using SWAT to simulate the effects of switchgrass production on water quality in an impaired watershed. *Agricultural Water Management, 160*, 1–13.
24. Guo, T., Cibin, R., Chaubey, I., Gitau, M., Arnold, J. G., Srinivasan, R., Kiniry, J. R., & Engel, B. A. (2018). Evaluation of bioenergy crop growth and the impacts of bioenergy crops on streamflow, tile drain flow and nutrient losses in an extensively tile-drained watershed using SWAT. *Science of the Total Environment, 613–614*, 724–735.
25. Qin, Z., Zhuang, Q., Cai, X., He, Y., Huang, Y., Jiang, D., Lin, E., Liu, Y., Tang, Y., & Wang, M. Q. (2018). Biomass and biofuels in China: Toward bioenergy resource potentials and their impacts on the environment. *Renewable and Sustainable Energy Reviews, 82*, 2387–2400.
26. Wu, Y., Liu, S., Young, C. J., Dahal, D., Sohl, T. L., & Davis, B. (2015). Projection of corn production and Stover-harvesting impacts on soil organic carbon dynamics in the US temperate prairies. *Science and Reports, 5*, 10830.
27. Dunn, J. B., Mueller, S., Kwon, H.-y, & Wang, M. Q. (2013). Land-use change and greenhouse gas emissions from corn and cellulosic ethanol. *Biotechnology for Biofuels, 6*, 51.
28. Liu, T., Huffman, T., Kulshreshtha, S., McConkey, B., Du, Y., Green, M., Liu, J., Shang, J., & Geng, X. (2017). Bioenergy production on marginal land in Canada: Potential, economic feasibility, and greenhouse gas emissions impacts. *Applied Energy, 205*, 477–485.
29. Harris, Z. M., Spake, R., & Taylor, G. (2015). Land use change to bioenergy: A meta-analysis of soil carbon and GHG emissions. *Biomass and Bioenergy, 82*, 27–39.
30. Williams, A. G., Audsley, E., & Sandars, D. L. (2010). Environmental burdens of producing bread wheat, oilseed rape and potatoes in England and Wales using simulation and system modelling. *International Journal of Life Cycle Assessment, 15*(8), 855–868.
31. Correa, D. F., Beyer, H. L., Possingham, H. P., Thomas-Hall, S. R., & Schenk, P. M. (2017). Biodiversity impacts of bioenergy production: Microalgae vs. first generation biofuels. *Renewable and Sustainable Energy Reviews, 74*, 1131–1146.
32. Sang, T., & Zhu, W. (2011). China's bioenergy potential. *GCB Bioenergy., 3*(2), 79–90.
33. Rowe, R. L., Street, N. R., & Taylor, G. (2009). Identifying potential environmental impacts of large-scale deployment of dedicated bioenergy crops in the UK. *Renewable and Sustainable Energy Reviews, 13*(1), 271–290.
34. Werling, B. P., Dickson, T. L., Rufus, I., Hannah, G., Claudio, G., Gross, K. L., Heidi, L., Malmstrom, C. M., Meehan, T. D., Ruan, L., Roberston, B. A., Roberston, G. P., Schmidt, T. M., Schrotenboer, A. C., Teal, T. K., Wilson, J. K., & Landis, D. A. (2013). Perennial grasslands enhance biodiversity and multiple ecosystem services in bioenergy landscapes. *PNAS, 111*(4), 1652–1657.
35. Sheehan, J. J., Adler, P. R., Del Grosso, S. J., Easter, M., Parton, W., Paustian, K., & Williams, S. (2014). CO2 emissions from crop residue-derived biofuels. *Nature Climate Change, 4*(11), 932–933.
36. Robertson, G. P., Grace, P. R., Izaurralde, R. C., Parton, W. P., & Zhang, X. (2014). CO_2 emissions from crop residue-derived biofuels. *Nature Climate Change, 4*(11), 933–934.
37. Li, W., Dang, Q., Brown, R. C., Laird, D., & Wright, M. M. (2017). The impacts of biomass properties on pyrolysis yields, economic and environmental performance of the pyrolysis-bioenergy-biochar platform to carbon negative energy. *Bioresource Technology, 241*, 959–968.
38. Pourhashem, G., Rasool, Q. Z., Zhang, R., Medlock, K. B., Cohan, D. S., & Masiello, C. A. (2017). Valuing the air quality effects of biochar reductions on soil NO emissions. *Environmental Science and Technology, 51*(17), 9856–9863.

39. Drewniak, B. A., Mishra, U., Song, J., Prell, J., & Kotamarthi, V. R. (2015). Modelling the impact of agricultural land use and management on US carbon budgets. *Biogeosciences, 12*(7), 2119–2129.
40. Daley, J. (2018). The EPA declared that burning wood is carbon neutral. It's actually a lot more complicated. *Smithsonian Magazine*. Archived from the original on 30 June 2021. Retrieved 14 September 2021.
41. International Energy Agency. (2021). *World energy outlook*.
42. Ministry of Petroleum and Natural Gas, Government of India. (2015). *Annual report*.
43. Emami Agrotech—Emami Group. https://www.emamigroup.com/group-of-companies/agrotech/biodiesel
44. Fortune Business Insights. (2021). https://www.fortunebusinessinsights.com/blog/10-leading-bioenergy-producers-in-the-world-10598
45. Green Car Congress—Energy Technologies Issues and Policies for Sustainable Mobility. (2022). *Six Japan-based companies established research association of biomass innovation for next generation automobile fuels*.

Chapter 3
Lignocellulosic Biomass for Sustainable Production of Renewable Fuels: Embracing Natural Resources

Medha Maitra, S. Sruthi, Pavada Madhusudan Rao, V. S. Avanthi, and P. Radha

Abstract The green lignocellulosic biomass industry is now gaining recognition as an appealing substitute for fossil resources, seeking to balance the world's escalating dependency on petroleum while being economical. The resurgence of the most abundant lignocellulosic resources into effective goods is an intriguing option for diminishing greenhouse gas emissions. Aromatic lignocellulose is a complex carbohydrate polymer composed of polysaccharides derived from xylose, glucose, and lignin. The interior framework of lignocellulosic matter is responsible for plant cell walls' hydrolytic resilience and rigidity, rendering them immune to microbial destruction. Recycling of residues leads to less dumping of waste. The chapter mainly concentrates on commercializing lignocellulosic waste biomass in the contemporary period to develop alternative biofuels, which encompasses diverse processes such as prior treatment, saccharification, and oleaginous microbe-mediated fermentation. Different means of prior treatment are implemented to conquer the inflexibility of LCB and advance its breakdown into hemicellulose, lignin, and cellulose. Components such as bioethanol, biodiesel, and biobutanol from the lignocellulosic feedstock act as inhibitors of detoxification and play major roles in strain enhancement, process integration, and optimization, which are the main fundamentals for boosting biofuel synthesis. The re-engineering of lignocellulosic material to extract fuel alternatives has tremendous potential because of its enticing features, comprising low density, affordability, green credentials, and biodegradability.

Keywords Lignocellulosic biomass · Biofuels · Pretreatment · Bioethanol · Biobutanol · Biodiesel

---

M. Maitra · S. Sruthi · P. M. Rao · V. S. Avanthi · P. Radha (✉)
Bioprocess and Bioseparation Laboratory, Department of Biotechnology, School of Bioengineering, College of Engineering and Technology, SRM Institute of Science and Technology, Kattankulathur, Chengalpattu District, Tamil Nadu 603203, India
e-mail: radhap@srmist.edu.in

G. Baskar et al. (eds.), *Circular Bioeconomy Perspectives in Sustainable Bioenergy Production*, Energy, Environment, and Sustainability,
https://doi.org/10.1007/978-981-97-2523-6_3

Abbreviations

LCB	Lignocellulosic biomass
LC	Lignocellulose
LCS	Lignocellulosic substrate
PHB	Poly hydroxybutyrate
CBP	Consolidated bioprocessing
DESs	Deep eutectic solvents
IL	Ionic liquid
LHW	Liquid hot water
ABE	Acetone-butanol-ethanol synthesis
BTL	Biomass to liquid
OMs	Oleaginous microorganisms
DMC	Direct microbial conversion
PCR	Polymerase chain reaction
SHF	Separate hydrolysis and fermentation
SSF	Separate saccharification and fermentation
CRISPR-Cas	Clustered regularly interspaced short palindromic repeats
HBA	Hydrogen bond acceptor
HBD	Hydrogen bond donor
DNA	Deoxyribose nucleic acid
HMF	Hydroxymethylfurfural
EG	Endoglucanase
AFEX	Ammonia fiber expansion
PSA	Pressure swing adsorption
CHP	Combined heat and pressure
MESP	Minimum selling price
LCEE	Life cycle energy efficiency

3.1 Introduction

The lignocellulosic substance is vital for feedstock utilized by several technological advances and a source of eco-friendly green energy in the upcoming days. Although the exploitation of fossil fuels has been extremely beneficial to the advancement of humankind, it has also ended up with several complications, including adverse ecological effects and a fuel shortage. This decade has seen an unsurpassed boom in the global economy and industrial revolution, an accomplishment that is primarily defined by the implementation of various energy assets. In addition, due to extraordinary human population expansion, the reliance on fossil fuel resources has risen [1]. According to the United Nations, the worldwide inhabitants will continue to evolve rapidly, exceeding 8.5 billion in 2030, especially 9.7 billion by the year 2050. Because of these reasons, the severe exploitation of petroleum-based non-renewable resources has resulted in acute scarcity, unanticipated price increases, and global greenhouse gas emissions. As a result, international governments and academics have set the objective of discovering and producing sustainable and renewable fuels [1]. As a result, administrations across the nation are quickening initiatives toward improving the task and commercialization of lignocellulose as a renewable resource, intending to address concerns such as energy shortages, environmental impact, and equitable growth. On average, the sheer quantity of LCB manufactured through photosynthesis on Earth surpasses 100 billion metric tons. However, only a tiny percentage has a job for human endeavors, leading to a substantial wastage of assets. The quantity of LCB capable of being harvested is anticipated in the USA each year, with the utilized amount being over 1.3 billion metric tons. However, China has roughly 800 million metric tons.

The consumption of lignin as a material for processing in the manufacturing of bioethanol as a replacement bioenergy for fossil fuels is an indispensable aspect of averting the hazards that accompany the combustion of fossil fuels [2]. Lignocellulosic resources are feedstock that could potentially be exploited to get regenerative biofuels. They are composed of vegetative biomass and are ubiquitous in nature. They consist mostly of the following substances: cellulose, lignin, tanning agents, saponin, and hemicellulose from agricultural, forestry, or wood debris. Cellulose comprises the entirety of the polysaccharide framework of the outermost layer of plant cells and accounts for between 30 and 50% of the total weight in the dry form of lignocellulose and is composed of an ordered sequence of 14 corresponding D-glucose units. A digestive enzyme, cellulose, may digest to yield glucose. Hemicellulose, unlike cellulose, incorporates an extensive list of heteropolymers, notably glucomannan, glucuronoxylan, xyloglucan, xylan, and arabinoxylan [2].

The key elements of the feedstock for lignocellulosic plants are hemicelluloses and cellulose (about most of a plant's dry volume), which break down into glucose via thermochemical and biological processes. There are three kinds of lignocellulosic feedstocks: herbaceous and woody energy crops, waste from agriculture, and forest leftovers. Cellulose in the cell barrier should dissipate and release glucose, resulting in can later be devoured into ethanol or another substance like biodiesel or butanol. It should be emphasized, however, that the plant's cell exteriors hinder cellulosic ethanol generation due to their structural characteristics. The conversion of feedstock is a vital stage. It predominantly comprises two treatment approaches for synthesizing biofuel via lignocellulosic components: the biochemical and thermochemical paths [3]. The enzymatic breakdown of lignocellulosic materials is often executed via the following phases: lignin elimination, which provides availability of both hemicellulose and cellulose; degradation of polymers composed of carbohydrates to collect unconstrained sugar; and brewing of the liberated sugars to ethanol from biomass. Every year, an enormous quantity of lignocellulosic residue created by harvesting and food service businesses can be employed as a source for the next generation [2].

Bioenergy cultivars ought to be cultivated on the outskirts, adopting advances in genomics along with reproductive technology breakthroughs to offer broadened nutrients, especially starches (which are subsequently converted into energy). These cultivars developed on vacant land will ultimately result in the manufacturing of environmentally conscious fuels. An extensive selection of plant-based feed fluctuates in structural and substance composition, leading to the industrial production of a vast range of biofuels [3]. After considering where they originate, biofuels can be distinguished as first-generation and fourth-generation biofuels. Second-generation biofuel is an increasingly proficient biofuel created from diversified non-food biomass of plant or animal provenance. Lignocellulosic biological fuel is made by using unpalatable vegetation or plant components. Non-edible lignin-rich biomass (such as vegetable weeds, forestry wastes, crop residue) appears widespread among living things and might be exploited to execute just like an unprocessed component during the synthesis of bioenergy. Lignocellulosic ethanol, butanol, amalgamated alcohols, and additional biofuels of the second kind are the latest possibilities to overcome the scarcity [4]. Basis substances for second-generation lignocellulose-based energy tend to comprise biomass unsuitable for human dietary use, so it has nothing to do with agronomy. The biomass industry from leftovers from farming, not alimentary agricultural products, jatropha, timber scraps, and even microorganisms are all viable components for biofuels of a subsequent generation. One of the most prevalent methodologies for offering sustainable aviation energy derived from natural sources is the conversion of LCB to ethanol. Due to the built-in susceptibility of lignin derivatives to decay by bacteria alongside their substantial thermal needs, distilling the combined production of ethanol and electricity corresponds to a modest general efficiency of 50–55% [5]. The efficient utilization of lignocellulose as a bioenergy source could encounter various setbacks, encompassing the difficulty of evolving an inexpensive method enabling cellulosic breakdown and sugar recovery. Depending on the biological component of the lignocellulose substrate, this needs certain pretreatment workflows. This type of pretreatment can boost production costs. Aside from that, the employment of different chemicals in lignocellulose pretreatment produces waste products containing chemicals or substances with inhibitory properties, which could diminish the efficacy of fermentation and hydrolysis methodologies [2].

On the contrary, advancing lignocellulosic biomass (LCB) transformation to refined chemicals and polymers remains an urgent hurdle. Lignocellulose has evolved to be less susceptible to depletion. Because of this intrinsic trait, lignocellulosic materials are resilient to enzymatic and chemical synthesis. The preliminary treatment of LCB, which is an exorbitant approach when considering both expense and vitality, is required to change the chemical and physical attributes of the lignocellulosic framework [4]. The fundamental variations in plant cell walls and deviations in ultrastructure extensively affect the biomass's preprocessing and hydrolysis (dissolution) rate. To circumvent the problems associated with chemical pretreatment, enzymatic pretreatment of lignocellulose is thought to be more efficient and produce less chemical waste. Conventional hydrolysis approaches are costly, labor-intensive, and not so eco-friendly. More treatments (pretreatment, neutralization, etc.) are recommended, and liberated carbs break down under difficult hydrolysis conditions [6]. Although enzymatic pretreatment is selective, accurate, and environmentally benign, it is costly to sustain, posing difficulty in impoverished developing nations that lack the financial means to support or maintain the technology. Because the enzymes are temperature-sensitive, the process parameters must be carefully monitored. Various microorganism strains have been examined for their ability and uniqueness to operate as viable sources of lignocellulolytic enzymes [2]. These biomass are diverse and convoluted. Its solid structures render it difficult to transform them into an assortment of valuable goods. Since biomass has a lower energy density than typical hydrocarbon fuels, it needs more biomass feedstocks to produce the same quantity of energy. Ample moisture might make methods of conversion unsuccessful. Because of their high humidity, the fuels are challenging to combust. In accordance with the biomass species, these biopolymers' topologies and/or degree of polymerization can change [6].

So, LCB sources have been regarded as exciting fuel reserves not solely because they are sustainable but also because they generate a net reduction in greenhouse gases released into the surroundings. Lignocellulose, among the most prevalent green energy organic substances, constitutes almost half of the plant matter generated by photosynthesis. This type of power resource can be summarized as using discarded feed as an environmentally friendly power source. Bioenergy, frequently called biomass energy, is any viable power supply deriving from non-fossil substances. However, bioenergy can alleviate the worldwide financial system's concerns regarding energy safety due to its insistence on minimizing fossil fuel availability [7]. Different biomass feedstock sources differ in content, homogeneous shape, size, etc., which enormously affects the conversion processes. As it turns out, the standardization of biomass feedstocks has become necessary for biorefineries. Biomass feedstocks must be standardized before being brought to the facility for a biorefinery to efficiently convert a given product. The technical aspects of the chosen distinct lignin biomass conversion approaches, which comprise biochemical, chemical, and biological pathways and numerous pretreatments to create liquid fuels, have been highlighted in the present investigation. Densified organic matter and LCB are the most beneficial choices for addressing handling, storage, and moisture issues. The biological transformation treatments of different biofuels, which comprise bioethanol, biobutanol, PHB esters, biomethanol fermentation, and accessible biomass pretreatment, have been analyzed, as have the techno-economic assessments and different hurdles in utilizing LCB. In broad terms, the overarching objective of this endeavor is to provide divergent techniques to generate liquid fuels using lignocellulosic waste and to discuss the potential constraints and alternatives to fossil fuels.

3.2 Exploring Woody and Non-woody Biomass as a Prominent Lignocellulosic Resource

Lignocellulosic materials, such as horticulture waste, microalgae, forestry, and metropolitan solid wastes, are significant, inexpensive, and widely accessible. These are some of the nutrients from biomass used to generate fuels and chemical-based goods. In the case of agro-residue, three general categories of lignocellulosic biomass are virgin biomass, discarded biomass, and renewable energy crops in India. The plants are included in the virgin feedstock. Waste biomass is created as a poor-quality byproduct of several distinct sectors of industry, especially forestry (sawmill and paper mill discards) and agricultural (corn stover, sugarcane bagasse, straw, etc.). Biomass comprises forest residues like pine bagasse, agricultural wastes like cereal straw, and short rotation coppices with specialized cultivars like switchgrass, miscanthus, and poplar. Significant quantities of LCB can be obtained by energy crops, such as switchgrass (*Panicum virgatum*) and elephant grass, which are used as a foundational component for developing second-generation biofuel. These sources of sustainable energy help reduce reliance on fuels such as coal, petroleum, greenhouse gas emissions, and, therefore, environmental pollution. Pretreatment, hydrolysis, fermentation, and product recovery are numerous steps required to convert LCB into biofuels and bioproducts.

Several obstacles must be addressed to successfully convert into valuable commercial goods while minimizing the generation of byproducts [8]. Lucerne, Miscanthus, Napier grass, giant reed, and canary grass are some herbaceous plants considered important sources of LCB, of which common grasses are regarded as the major source. They are thought to be economically feasible and environmentally friendly over food crops for ethanol production in addition to avoiding the "food versus fuel" dilemma. The primary source of cellulosic biomass is the forest, where plants grow quickly and rotate quickly. Despite their uneven distribution, forests are essential for preventing landslides, limiting CO_2 entering our atmosphere, and maintaining the balance between people and wildlife. In the meantime, the USA produces over 370 million tons of LCB from its forests each year; other nations with abundant forests include Canada, Russia, China, and Brazil. As mentioned earlier, half of the world's total LCB has been gathered in the nations, including sources such as timber, branches, wood shavings, and wastes from dead trees like sawdust, chips, etc. Non-woody biomass refers to biomasses with lower lignin content, like rice husks, grass straw (wastes from fields), and sugarcane oil wastes (processing wastes). They are large and have a low energy content [9].

3.3 Structural Composition of Lignocellulose

A typical LCB structure, predominantly composed of cellulose, hemicellulose, and lignin, is represented in Fig. 3.1. The nature and kind of plant feed, like agricultural, processed, or forestry, determine efficient feedstock used to produce nanocellulose by selecting appropriate compositions [10]. The polymers cellulose, hemicellulose, and lignin combine to form microfibrils, giving stability to the cell wall, which is further compacted to form microfibrils [11]. Cellulose is the main element, which consists of 1,4-linked glucose. The hydrogen bonds prevent degradation between the polysaccharide chains. Hemicellulose is the second main element, comprising five and six-carbon polymers like mannose, glucose, arabinose, xylose, etc. Lignin is made up of chemicals composed of sinapyl alcohol (S), *p*-coumaric alcohol (H), and coniferyl alcohol (G) [12]. These polymers are polymerized in different compositions and ratios based on plant and tissue type, cell wall, wood quality, and layers. Cellulose has a polymerization degree of about 10,000, consisting of glucose molecules joined by water-insoluble 1,4-glycosidic linkage. Each glucose unit contains adequate hydroxyl groups to generate both intra- and inter-chain hydrogen bonds, which stabilize the molecular structure and combine, forming microfibrils. The structure of cellulose is made up of several hydrogen bonds, which are inter- and intramolecular, forming a complex three-dimensional arrangement and linking crystalline and non-crystalline areas. Crystallinity denotes the percentage of crystalline regions in the polymer cellulose, whose structure is broken down into an amorphous form to boost its enzymatic hydrolysis conversion.

Hemicellulose is mostly made up of two or more glycans, like xylose, arabinose, mannose, and galactose, which are non-homogeneous. Hemicellulose is bonded by H bonds, whereas Van der Waals forces are used to link cellulose and lignin. Hemicellulose can have a variety of structures depending on the plant: typically, hexosan and pentosan predominate in coniferous and broad-leaved wood, respectively. Hemicellulose varies greatly in its concentration and composition, even within the same plant, especially in the bark, stems, roots, and branches. Hemicellulose is more soluble and vulnerable to chemical assault than cellulose because it has a lower molecular weight and a larger amorphous area. Lignin is the only natural, renewable, and aromatic polymer found in plants' secondary cell walls, which are made up

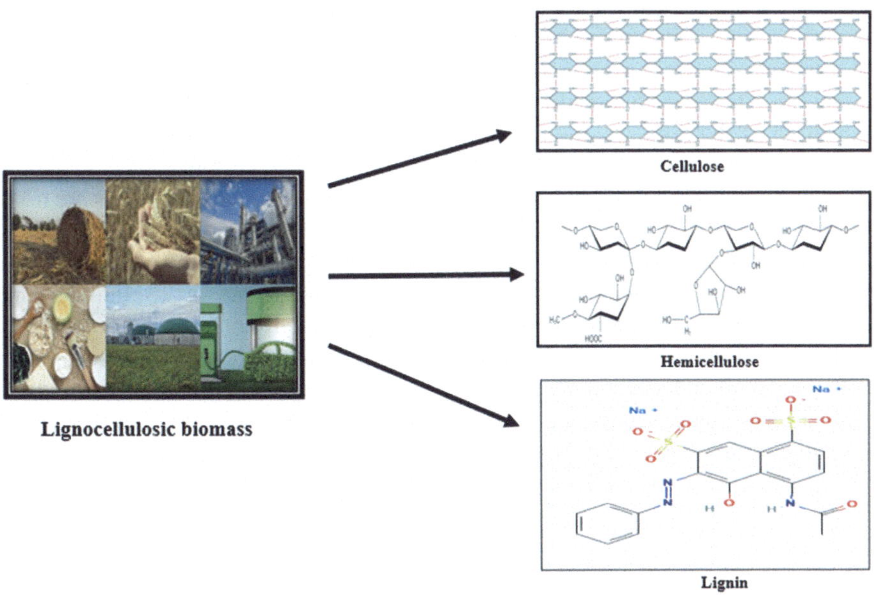

Fig. 3.1 Structure of lignocellulosic feedstock

of hydroxycinnamic alcohol, cetearyl alcohol, and mustard alcohol. Syringyl structure (S), *p*-hydroxyphenyl structure (H), and mustard alcohol are the three lignin units. These three structural units constitute the majority of lignin polymers and are joined together by ether and carbon–carbon bonds, such as β-O-4 β-β, β-5, as well as α-O-4. The quantity of lignin and its structural makeup differs substantially between different types of lignocelluloses; coniferous wood only has *G* structural units and contains 25–31% lignin. Broadleaf wood, containing *G* and *S* units, on the other hand, has 16–24% lignin. The lignin content of herbaceous species ranges from 16 to 21%; it contains all three structural units. Thus, lignin is divided into three types: *G*-type (coniferous lignin), GSH-type (graminaceous lignin), and *GS*-type (broadleaf lignin), based on structural unit contents [13].

3.4 Physicochemical Properties of Lignocellulose

LCB has some physical properties, including its ability to absorb moisture, compactness, flow behavior, heat properties, and particle size. It contains the fixed carbon of biomass samples, ash, moisture, and volatile matter [14]. The amount of moisture present in biomass is referred to as moisture content. It has a significant impact on the harvest and preparation, in addition to the transportation, storage, processing, and final goods. The biomass specimen, when entirely burned, results in the formation of solid residue. The amount of solid residue left over is measured by the amount of ash. The essential constituents of biomass residue contain calcium, iron, potassium, magnesium, silica, titanium, and aluminum. The condensable vapor and stable gases that biomass releases during heating are considered volatile. It is influenced by the heating limitations, such as residence time, temperature, and heating rate. Following the heating of the biomass, the labile stuff is liberated, and the fixed carbon is the solid combustible residue left behind. Optimum analysis determines the concentration of nitrogen, carbon, oxygen, hydrogen, and sulfur. Elemental analyzers are used to burn a weighted specimen of biomass within a managed atmosphere and analyze the gaseous byproducts. Energy content is often measured as the heat of combustion, which is the total energy released as heat when it goes through an absolute burning in the presence of oxygen within controlled conditions. It contains the quantity of energy present inside the particular specimen of biomass. The lignocellulosic feedstocks can be taken based on their different physical–chemical properties: structural composition, moisture content, electrical conductivity, pH, nutrient content, specific size, and various shapes of the raw materials.

3.5 Different Pretreatment Approaches

The crucial first stage in the breakdown of LCB's three components and their consequent high-efficiency conversion to bioenergy is pretreatment. Factors influencing pretreatment include energy use and the country's economy. The pretreatment step is another important factor for determining the processes' industrial feasibility. Furthermore, the pretreatment process must not alter the components of LCB's natural structure. For example, the natural structure of lignin should be altered during delignification, rendering it susceptible to valorization. Additionally, when the crystallinity of cellulose reduces during pretreatment, it is less accessible to the enzymes. It is crucial to maintain the various components of a successful LCB deconstruction to ensure the efficacy of a pretreatment procedure. The efficiency of pretreatment is determined by its capacity to (1) delignify the LCB without significantly changing the native structure of lignin; (2) less energy utilization; (3) operate at a low cost; (4) lower the cellulose's crystallinity index; (5) reduce the LCB's particle size in order to increase its surface area to improve enzymatic hydrolysis; (6) avoid the production of enzyme inhibitors (7) pretreat several LCB feedstocks; and these pretreatment methods can be of three categories: physical, chemical, and physicochemical (Fig. 3.2) [15].

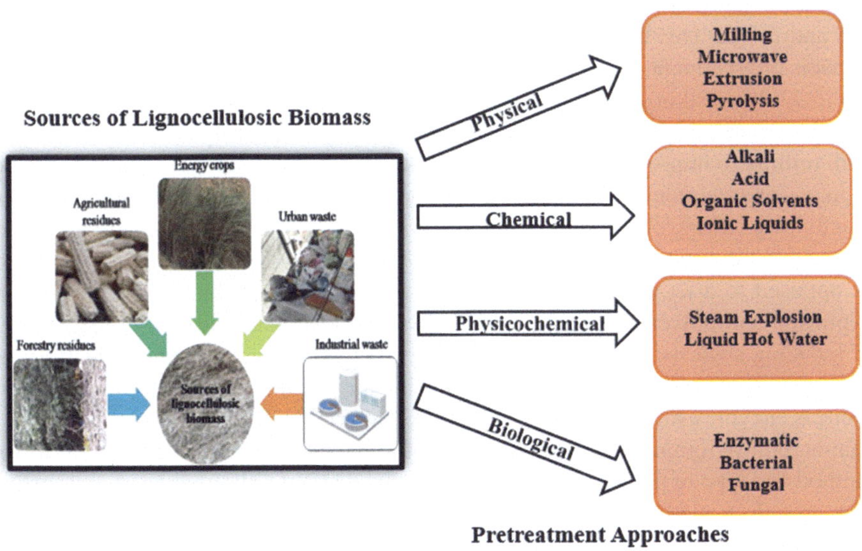

Fig. 3.2 Methods of pretreatment of lignocellulose

3.5.1 Physical Pretreatment Methods

Prior to using additional pretreatment techniques, LCB must first undergo physical pretreatment. It is primarily done to reduce the size of particles by increasing surface area and decreasing polymerization and crystallinity levels. As a result, future processes are improved and made simpler. These processes are environmentally beneficial and rarely result in the production of harmful materials. The significant energy consumption of physical pretreatment is one of its main drawbacks. The type of LCB being used typically affects energy consumption [16].

3.5.1.1 Milling

A common pretreatment approach for diminishing biomass particle size is milling. Milling can enhance enzymatic degradation since it reduces the amount of crystallinity and the size of the material. By milling and grinding, particles can be reduced in size by up to 0.2 mm. To an extent, biomass particle size can also be reduced; beyond that point, the pretreatment process is unaffected by the reduction in particle size. A stove for corn, with a little maize stover with small particles ranging from 53 to 75 m, is more effective than a maize stover with broad particles of size between 475 and 710 m. This variation in particle size demonstrates how productivity greatly impacts pretreatment. Ball milling significantly reduces the crystallinity index from 4.9 to 74.2%, making it suitable for saccharifying straws in mild hydrolytic conditions to produce highly fermentable sugars. Enzymatic hydrolysis can be combined with milling to improve hydrolysis results, kinetic energy, and mass transfer. Unlike biomass pretreatment, which is done without milling, the ball beads number in the reactor plays a key role in cellulose hydrolysis due to its less enzyme loading requirement and ability to obtain a complete 100% hydrolysis rate. When a fluidized bed, as opposed to a jet mill, is loaded with rice straw for grinding finely, after steam explosion and pulverization, the maximum hydrolysis rate with the highest output of reducing sugar is produced. Ball milling is an expensive alternative for biomass preparation, consuming high energy, which is a major limitation. Furthermore, lignin reduces enzyme availability to the substrate, and grinding is less desirable because it cannot remove lignin. Crystallinity decline, biomass type, the process's duration, the method and type of pretreatment used, etc., can all affect polymerization and surface area growth. A method used to bring down and reduce crystallinity and digestibility is vibratory ball milling. Wet disk milling generates fibers that promote cellulose hydrolysis and uses less energy than hammer milling, which results in finer bundles of material. Because of this, wet disk milling is favored when milling is the option [17]. The size of 0.4 mm and lesser biomass particles have no effect or impact on the hydrolysis rate. Based on the kind of powered machinery, vibratory milling, two-roll milling, hammer milling, ball milling, colloid milling, etc., are some of the numerous milling techniques used. The type of milling technique utilized, the amount of time processed, and biomass type all affect the decrease in the size of particles and crystallinity [15]. This technique has several significant benefits: (1) The crystallinity of cellulose decreases; (2) cellulose polymerization decreases to an extent; (3) an increase in the surface that is available for enzymatic hydrolysis; and (4) advancement in mass transfer, as a result of particle size reduction [14].

3.5.1.2 Microwave

Microwaves are electromagnetic radiation of wavelengths between 1 mm and 1 m, which consist of nonionizing radiation with a frequency range of 300–300,000 MHz that selectively transmits energy to various substances. Energy-efficient microwave heating increases hydrolysis, oxidation, alkylation, and esterification processes, and using microwaves has gained new interest. Over the past 30 years, studies have reported effective lignocellulosic pretreatment utilizing microwave radiation and have steadily scaled up from lab to pilot scale. There are now two primary categories for LCB in microwave-assisted pretreatment technologies: (1) The pyrolysis of lignin is carried out at high temperatures, greater than 400 °C, in the absence of oxygen. Microwaves are used to depolymerize biomass and convert it to bio-oil or gases. (2) Microwave-assisted solvolysis produces high-value compounds on depolymerization of biomass.

When compared to conventional heating, microwave radiation offers many benefits, including (a) quick heat transfer, (b) fewer reaction times, (c) simple process, and (d) little degradation or creation of harmful compounds and extraneous goods. Additionally, microwave hydrothermal pretreatment eliminates most acetyl groups that are elevated by the hot spot effect of microwave irradiation from hemicellulose. During traditional heating, energy is transported through conduction from the material's exterior surface to its inside. So, while the interior remains cold, overheating on the exterior surface is still possible. By directly converting the electromagnetic energy in microwaves into heat, on the other hand, they produce heat at the molecular level. As a result, energy is dispersed evenly throughout the material [18]. Microwave irradiation causes explosions within the particles of a substance, facilitating the disintegration of resistant structures. Great interest has been sparked in biorefinery and its uses by introducing microwaves as a thermal technique for processing LCB. The existence of crystalline, ionic, and nonionic amorphous areas in a typical LCB makes it function like a biological conductor. Hot spots are produced in the biomass due to polar groups present in the cellulose [15].

3.5.1.3 Extrusion

One of the most popular physical preparation techniques used for LCB is extrusion. This method's foundation is a tight barrel with temperature control and one or two screws that spin into it. The refractory structure of lignocellulose is disrupted by the screw blades of the barrel, which are rotated under high temperature and pressure [16]. Materials that fit through a die with a specific cross section come out with a rigid, definite profile. This extrusion method is well-known for recovering sugar from biomass. Advantages of this technique include flexibility to modifications, zero requirements of degradation products, a controlled environment, and high output.

Extrusion machines are divided into twin-screw extruders and single-screw extruders, composed of small cylindrical parts known as screw elements [16]. The screw design has a significant impact on the decomposition of LCB. To boost the generation of biogas from LCB, the impact of various screw configurations on wheat straw and deep litter during the extrusion pretreatment process was studied [19]. They examined five screw configurations for pretreatment: mild, long, reverse, and kneading with reverse. This resulted in efficient sugar yield for each arrangement, causing a sizable methane output. However, the key issue in this study is the high energy consumption, which calls for additional research to upgrade the pretreatment process and make it highly feasible. Several variables, including screw design, speed, and barrel temperature, influence the extrusion pretreatment. The pretreatment was explored in 2017, employing a twin-screw extruder of sugarcane bagasse and straw, and optimized the factors, including the type of additives, the biomass, additive ratio, the extrusion passes no., the barrel temp, the screw speed, shape, size, and its configurations. Different additives, such as water, ethylene glycol, glycerol, and Tween R80, were used for pretreatment in varied amounts. When pretreatment is carried out using glycerol, bagasse, and straw, with varying glycerol ratios of 1:0.75 and 1:0.5, experiments revealed that glycerol was considered the optimal addition factor [16].

3.5.1.4 Pyrolysis

Pyrolysis is regarded as another efficient method for pretreating biomass. Char and certain gases are produced on breaking down cellulose, which occurs slowly at low temperatures. This can be improved by supplying oxygen and mild acid hydrolysis [20]. LCB is thermally degraded during pyrolysis at extremely high temperatures without an oxidizing agent. Pyrolysis was carried out at temperatures ranging from 500 to 800 °C. The rapid breakdown of cellulose led to charcoal and pyrolysis oil production. Several variables, including biomass type, pyrolysis type, and reaction parameters, influence end products. Nowadays, thermal industries are adjusting to this method because it produces products of high value and energy, which makes retrofitting transport management simple, allows for combustion and storage, and allows for flexibility in utilization and marketing. This mechanism is more effective when oxygen is present, and the temperature is lower.

Gasification, pyrolysis, and direct combustion are the three thermochemical processes that can convert biomass to biofuels. The various pyrolysis modes are the cause of the various product yields. Water and polar organics are the major components of bio-oil. Where bio-oil generation is necessary, pyrolysis is used. The generation of liquid products (fuels) is caused by rapid pyrolysis in a controlled atmosphere. Emerging methods are now termed mild pyrolysis or torrefaction. Other thermochemical processes that occur at 200–300 °C temperatures differ from pyrolysis. In this process, biomass partially decomposes, resulting in terrified biomass. Pyrolysis produces vapor, aerosols, and char, breaking down plant biomass. Torrefaction has been divided into two types: wet and dry torrefaction. An inert atmosphere, thoroughly dried biomass, and normal atmospheric pressure are requirements for dry torrefaction, whose main byproduct is biochar. In wet torrefaction, the pretreatment is performed in a pressurized water vessel. Wet torrefaction uses biomass that contains moisture, which is later dried. Other names comprise hydrothermal carbonization/torrefaction. Wet torrefaction requires a pressure between 1 and 250 MPa to be carried out. Hydro-char is the principal end product of the wet torrefaction processing of biomass [18].

3.5.1.5 Ultrasonication

Pretreatment using ultrasonication is based on the cavitation theory and uses ultrasonic energy. The compact structure of LCB is cleaved by the shear forces produced by cavitation, which also facilitates the extraction of desirable constituents like cellulose, lignin hemicellulose, etc. [17]. Chemical processing uses ultrasound at frequencies between 20 kHz and 1 MHz, whereas medical and diagnostic processes use higher frequencies. Ultrasound processing of biomass alters the exterior structure and generates oxidizing radicals, damaging the lignocellulose matrix's chemical structure. Additionally, lignin's connections can be broken with ultrasound by producing small bubbles, enabling the splitting of polysaccharides and lignin. After reaching an assertive critical size, the generated bubbles become unstable, violently disintegrating at pressures of 1800 atm and temperatures between 2000 and 5000 °K. As a result, ultrasonic disruption is a useful environmentally friendly approach for pretreating LCB [19]. The ether bonds between lignin and hemicelluloses are broken down by ultrasonic waves, affecting cell wall stability, which makes the hemicelluloses more accessible and extractable. This method enhances the lignin degradation rate and saccharification of enzymes by severing hydrogen bonds between lignocellulosic molecules and decreasing the crystallinity, but it cannot produce enough energy to alter the conformation of the surface of biomass. The effectiveness of sonication is determined by various variables, including ultrasonic frequency, residence duration, biomass type, solvent utilized, reaction kinetics, and reactor configuration/geometry [19].

3.5.1.6 Pulsed Electric Field

In this process, the cell membrane is punched with holes using a high voltage of 5–20 kV/cm for a few nano-milli seconds to permit invading substances to enter the cell to activate disintegration. The sample is placed between 2 parallel plate electrodes, where the electric field strength, E, is expressed. Mass permeability and tissue rapture dramatically increase when the electric field is shown. Then, electric pulses are applied in the mode of exponential decay or square waves. This method comprises components such as a control system, material handling equipment, control system [17]. As a result of the pulsed electric field's disruption of the biological membrane and local structural changes that cause a loss of semi-permeability, intracellular chemicals can pass through the biological membrane and solve in the area. As a result, hydrolytic enzymes can enter the treated plant cell membrane more easily through its holes. To improve the hydrolysis of cellulose, lignocellulosic materials (wood chips and switchgrass) are pretreated with 2000 pulses with a field strength of about 10 kV/cm [18].

The physical approaches to pretreatment are frequently utilized to decrease the biomass's particle dimension. Mechanical instruments and methods including millers, grinders, screws, and UV or microwave radiations are typically used during physical pretreatments. Furthermore, the entire procedure is aided by using secondary or tertiary sources such as chemicals, solvents, or enzymes. An evaluation of several pretreatment methods indicates that the optimal procedure should overcome the LCB's stubborn character, boost cellulosic crystallization, and guarantee the highest possible yield of sugars and other utilitarian value bioproducts.

3.5.2 Chemical Pretreatment Approaches

Highly complex substrates can be broken down efficiently by chemically treating them with acids, organic solvents, and alkalis. This process also increases the bioavailability of carbohydrates by eliminating some of the inhibitors—lignin and reducing cellulose crystallinity or polymerization levels [21].

3.5.2.1 Alkali Pretreatment

One of the earliest attempts at biomass preparation involved the use of alkaline. While most of the cellulose remains intact after the alkaline pretreatment, a significant quantity of lignin and hemicellulose are solubilized. Enzymes with a high hydrolysis rate and potential yield could further hydrolyze the cellulose stored in the residue solid. The prohibitive factor for alkaline application in industry is its high cost [20]. The hydroxides of sodium, potassium, calcium, and ammonium are alkaline reagents generally utilized for alkali pretreatment. The most efficient of these was discovered to be sodium hydroxide [17]. During the alkali pretreatment procedure, a saponification reaction breaks the intermolecular ester bonds between hemicelluloses and lignin. This causes the lignin and hemicellulose fragments to solubilize in the alkali solution and introduces the cellulose into the enzyme interaction, causing it to swell. Hence, the degree of polymerization and crystallinity is brought down due to the modification in the structure of lignocellulose, which also broadens the internal surface area. In this method, uronic acid and acetyl groups are eliminated in hemicellulose, improving the accessibility of carbohydrates during enzyme hydrolysis [16].

A few bases are also utilized, in addition to acids, while biomass is pretreated. The amount of lignin significantly impacted the alkaline treatment's result.

Alkali treatment requires less pressure and temperature than other pretreatment techniques. However, it also requires several hours to days of work. Alkali treatment results in less sugar degradation than acid treatment, and it is also practical and simple to remove and recover caustic salt. High efficiency is demonstrated by alkaline pretreatment, particularly in the delignification procedure. Alkaline pretreatment causes the cell wall to swell, increasing the interior surface area while lowering the cellulose's polymerization and crystallinity. Crucial process parameters, including alkaline loading, reaction temperature, and the amount, must be optimized to obtain alkaline effects entirely. Alkaline reagent at high concentrations causes polysaccharides to degrade and decompose. Hence, lower concentrations of atmospheric pressure and temperatures are advised. Additionally, it does not produce harmful substances like furfurals and hydroxymethylfurfural (HMF), leading to improved efficiency in the biomethanation process [21].

3.5.2.2 Acid Pretreatment

Acids are employed in this pretreatment to pretreat LCB. Acid pretreatment loses appeal as a pretreatment method due to the production of inhibitory compounds. After acid pretreatment, the inhibitory chemicals produced in large quantities include furfurals, aldehydes, 5-hydroxymethylfurfural, and phenolic acids. Concentrated acids are also used to treat LCB. Acids like sulfuric acid and hydrochloric acid are most commonly preferred. Fermented sugars are liberated from the LCB by this method, as it enhances hydrolysis. Sulfuric acid pretreatment for spruce, switchgrass, poplar, and corn stover is typical. After Bermuda grass and rye were subjected to acid pretreatment, 19.71% and 22.93% reducing sugars were generated, respectively. According to studies, diluted sulfuric acid is most frequently used to treat LCBs before usage. The impact of diluted acid and alkali pretreatment on wild rice grass (*Zizania latifolia*) for enzymatic hydrolysis was compared [17].

3.5.2.3 Organosolv Procedure

In order to isolate each component of the LCB and produce relatively pure lignin that may be sold as a byproduct or transformed into higher-value products in a biorefinery concept, organosolv pretreatment rises as the only available method. With or without adding a catalytic reagent, organic solvents such as methanol, ethanol, acetone, acetic acid, peracetic acid are utilized. Oxalic, acetylsalicylic, salicylic, and mineral acids—hydrochloric, sulfuric, and phosphoric acid—are catalysts. Although using a catalyst can increase pretreatment efficiency, it can also adversely affect the environment. For example, using a chemical catalyst as acid can cause lignin condensation reactions, with the formation of reactive aldehydes and furfural and 5-hydroxymethyl furfural due to the acid-catalyzed degradation of monosaccharides [21]. The features of pretreatment biomass, such as crystallinity, fiber length, and degree of polymerization, are influenced by many variable parameters, including catalyst type, temperature, solvent concentration, and reaction time. Long reactions, high temperatures, and acid concentrations all lead to the production of inhibitors. The effectiveness of various catalysts was examined for the production of ethanol. Sulfuric acid, sodium hydroxide, and magnesium sulfate were found to be the most effective catalysts, but sodium hydroxide is potent for improving digestibility. Although sulfuric acid is a useful catalyst, it is less frequently used due to its toxicity and inhibitory properties.

This method is not cost-effective since the catalysts are of high cost. The costs, however, can be reduced by recovering and recycling solvents. Because the presence of solvents affects fermentation, microbial growth, and enzymatic hydrolysis, their removal is crucial. Handling such abrasive chemical solvents comes with an additional risk. Acid aids in the depolymerization and hydrolysis of lignin. When lignin cools, it dissolves in phenol, and sugars are present in the aqueous phase. Lignin is soluble in Formasolv, an organosolv that uses formic acid, water, and hydrochloric acid, allowing low-temperature processing [17]. Due to its inherent benefits, such as the simplicity with which the solvents can be recovered by distillation, the ability to recycle the solvents back into the pretreatment process, and the utilization of high-quality lignin extracted from this process act as value-added byproducts for industrial applications. Hence, this pretreatment method is flourishing in the present times.

3.5.2.4 Deep Eutectic Solvents (DESs)

The adoption of eco-friendly solvents is a critical factor for determining the overall renewability of a process, according to the fundamentals of green chemistry. The traditional solvents, now utilized in the industry, are hazardous to the environment and constitute a major risk. As a result, the search for new eco-friendly solvents has intensified. By linking a hydrogen bond donor (HBD) and hydrogen bond acceptor (HBA) in the right proportions, DESs are formulated. DESs are less expensive, hazardous, and more biodegradable than traditional solvents and ionic liquids (ILs). Moreover, the synthesis of DESs is straightforward and does not require elaborate purifying procedures. One of their key benefits is the extensive tuning ability of the physicochemical characteristics of eutectic solvents, which are similar to ionic liquids (ILs). This method has some unique features, such as high polarity, large miscibility with water, and the ability to mix with other co-solvents. Hence, it is used widely for the effective delignification of various biomass feedstocks. A new and eco-friendly approach for the pretreatment of LCB under mild reaction conditions has been created by the DESs' intriguing ability to selectively eliminate lignin from the LCB complex. DESs with acidic hydrogen donors are superior to ILs and can effectively remove lignin and hemicellulose. Zhang provides more evidence for the findings, and his collaborators looked into how corn cobs may be pretreated using three DESs: monocarboxylic acid (ChCl), dicarboxylic acid (ChCl), and polyalcohol (ChCl). It was discovered that the strength and quantity of acid and the type of hydrogen bond acceptors significantly impacted delignification and cellulose accessibility. The pretreatment of polyalcohol: ChCl mixture can be enhanced when the free hydroxyl groups of polyalcohol and the free and etherified hydroxyl groups of lignin interact [16].

3.5.2.5 Ionic Liquids (ILs)

ILs have long piqued the curiosity of scientists who want to pretreat lignocellulose. A novel family of solvents, known as ILs, containing cations or anions, has low melting points, negligible vapor pressure, and good thermal stability and polarity. Ionic liquids typically amount to big organic cations and small inorganic anions. The degree of anion charge delocalization and cation structure greatly impact ionic liquids' physical, chemical, and biological properties. Pretreatment time, temperature, cations, and anions influence the ionic liquids and biomass interaction. Compared to conventional solvents, ionic liquids offer special characteristics, such as low vapor pressure, that prevent atmospheric emissions by preventing solvent losses during evaporation. Also, they have reasonably high thermal stability, low flammability, and recycling potential, resulting in less waste segregation [15]. Certain factors must be considered in this pretreatment method, such as temperature, ionic liquid and its physicochemical properties, biomass type, reaction duration, and amount of water in the sample. Because extremely viscous solutions are created during pretreatment and have a considerable negative impact on the commercial application of ionic liquids, physiological properties like viscosity are the most detrimental feature of the process. Despite large changes in biomass structure, the content of the biomass underwent only minor alterations after ionic liquid pretreatments. Because of its great thermal and chemical stability, less hazardous processing conditions, low solvent vapor pressure, and ability to persist in the liquid state even under the widely varied range of temperatures, this technique is not likely to be favored over alternative techniques. These ionic liquids rarely degrade and can be recycled with ease. The disadvantage of utilizing ionic liquid pretreatment is that cellulase will unfurl and become inactive as a result of the incompatibility between the two substances [21].

Chemical pretreatment has drawn more attention since it is typically less expensive, has faster breakdown rates for complex organic compounds, and has superior efficiency overall. A succinct synopsis of the multiple biochemical approaches to pretreatment and how they relate to the LCB components is provided by Pandey et al. 2021; nonetheless, it is still incomplete. Dedicated studies employing acidic, alkaline, and organosolv techniques to comprehend the aforementioned procedures can expand the potential of DESs for use in upcoming bio-based refineries.

3.5.3 Physicochemical Pretreatments

Certain physicochemical approaches can be used to solubilize the LCB components based on temperature and moisture content and to make the lignocellulose readily accessible during the hydrolysis stage, preventing inhibitors' production. Despite being more difficult to apply, pretreatment of lignocellulose feedstock using these techniques promises good yield in the following bioprocesses. Physical pretreatment is typically more expensive and requires more energy, making it high-priced on an industrial scale. The amount of time and energy consumed in this technique can be lowered by process optimization [21].

3.5.3.1 Steam Explosion

Biomass is exposed to extremely hot steam in a reactor during the steam explosion pretreatment procedure. The biomass is penetrated by the high-pressure steam, which starts an autohydrolysis reaction. Hemicellulose is hydrolyzed to soluble sugars by utilizing organic acids as catalysts, produced initially from acetyl groups in the biomass. Highly fermented sugars are produced in this technique, compared to biomass hydrolysis with cellulases. Reaction time, temperature, particle size, and moisture content are crucial parameters to consider in this process. Good performance may result from either high temperature and a quick reaction time (270 °C, 1 min) or low temperature and a lengthy reaction time [20].

Softwoods cannot be effectively pretreated with the steam explosion, but adding an acid catalyst throughout the process is necessary to make the substrate available to hydrolytic enzymes. Steam is utilized to maintain the biomass at a specific temperature, eliminating the requirement for a specific dilution. After the process, the temperature is lowered by a sudden pressure release. Rapid heat expansion causes the biomass's particle structure to break down, which is engaged to stop the reaction. Certain parameters, such as moisture content, residence duration, chip size, and temperature, impact the steam explosion. Some advantages of this pretreatment technique include less energy requirement since 70% less energy is required compared to the mechanical techniques to obtain the same particle size. This method, which involves a catalyst, has been studied and reported that it could be scaled up to large commercial levels because of its cost-effectivity [17].

3.5.3.2 Liquid Hot Water (LHW)

Analogous to a steam explosion, LHW pretreatment employs water at high pressure (up to 5 MPa) and temperature conditions (170–230 °C), replacing steam. In contrast to the steam explosion, quick pressure release is not necessary for LHW, and pressure is merely applied to stop water from evaporating. By freeing hemicellulose's acetyl groups and removing lignin, LHW hydrolysis hemicellulose increases the explosibility of the cellulose fibers. The processed slurry's liquid portion still contains detached hemicellulose, and very little monomeric sugar is formed after the treatment. However, the LHW pretreatment is carried out at a regulated pH between 4 and 7 to impede the production of inhibitors and sugar degradation [16]. The disadvantage of LHW is that it requires a large amount of energy for the subsequent procedure because there is a large quantity of water involved. The advantage of this method is that no inhibitor is produced, and no chemicals or catalysts are needed [22]. Since no chemicals or catalysts are needed for LHW pretreatment, harmful material production is nearly nonexistent, and solvent costs are minimal for large-scale applications. Furthermore, while the particles are broken down during the treatment, biomass size has no effect because the process is more striking when done on a large scale. However, the process requires considerable energy because a large amount of water is used [16].

3.5.4 Biological Pretreatment

Prior to enzymatic saccharification, LCB can be treated using a low-cost, environmentally acceptable method known as biological pretreatment. This method is promising since it uses less energy, produces fewer inhibitors, and is environmentally benign. By using entire cells or enzymes from lignin-degrading bacteria or fungi, LCB is pretreated. The enzymes lignin peroxidase, laccases, manganese peroxidase, and versatile peroxidase are employed to break down lignin. The greatest candidates for these uses are fungi since they can break down lignin, hemicelluloses, and cellulose. Biological pretreatment is utilized to eliminate certain components, such as antibacterial compounds, in addition to lignin. With relatively little impact on cellulose, white-rot, soft-rot, and brown fungi are renowned for removing lignin and hemicellulose since these organisms contain lignin-degrading enzymes like peroxidases and laccases, which help in the breakdown of lignin. The control of these enzymes that degrade substances involves carbon and nitrogen supplies. While white and soft-rot specifically target plant biomass's lignin and cellulose components, brown rot more frequently attacks cellulose [16].

The inclusion of microbes like mold and bacteria in a naturally occurring pretreatment is becoming increasingly prevalent throughout lignocellulose pretreatments due to its positive effects on both the natural world and the economy. Arguably, the most important step is carefully choosing the appropriate microbial consortia. Due to their high enzyme activity, co-cultivating bacteria and/or fungus during consolidated bioprocessing (CBP) is advantageous in breaking down complicated biological polymers [23].

3.5.5 Combined Pretreatment Approaches

Gasification, pyrolysis, and direct combustion are the three thermochemical processes that can convert biomass into biofuels. The various pyrolysis modes are the cause of the various product yields. Pyrolysis is mostly used when bio-oil is required, whose major components are water and polar organics. The generation of liquid products (fuels) is caused by rapid pyrolysis in a controlled atmosphere. Emerging methods are termed mild pyrolysis or torrefaction. The thermochemical process that occurs at temperatures between 200 and 300 °C differs from pyrolysis. In this process, biomass partially decomposes, resulting in terrified biomass. An inert atmosphere, thoroughly dried biomass, and normal atmospheric pressure are requirements for dry torrefaction, whose byproduct is biochar. Wet torrefaction is done in a pressurized vessel containing moisture and is dried after the torrefaction process. Wet torrefaction requires a pressure between 1 and 250 MPa to be carried out. Hydro-char is the principal end product of the wet torrefaction processing of biomass [18]. Due to the shortcomings of individual pretreatment strategies, experts have been trying to fix these issues and boost effectiveness by combining various techniques. Plenty of studies have been conducted using an assortment of pretreatment techniques from a few decades back. For instance, the primary drawback of fungal pretreatment is its extended duration of action. This limitation can be addressed by integrating the fungal treatments with other chemical as well as physical methods. For example, compared to the acid pretreatment alone, the fungal and diluted acid pretreatment of the olive tree elevated the amount of sugar recuperation to 51% and expanded the hydrolysis by enzyme output by 34%. Microbiological combination diluted acid and microbiological combined diluted alkaline pretreatments were recently employed by Zhang et al. [24].

In the case of physical pretreatment measures, colloid mills and extruders are advised for pretreating LCBs with moisture contents of more than 15–20%, whereas hammer and knife mills are advised for LCBs with moisture contents of 10–15%. As the target particle size lowers and the moisture content rises, more energy is needed in normal circumstances. Grinding hemp to improve methane output by 15% needed more energy than steam pretreatment procedures [25]. In the field of pretreatment, acid pretreatment, which is not completely eco-friendly, becomes more significant. In acid-catalyzed pretreatment, temperature and acid concentration are two of the foremost processing aspects. Although the C5 sugars are similarly degraded by high temperatures and an excess of acid in the organosolv treatment, it is an effective pretreatment approach. Alkaline prior treatments are the most feasible treatment approach under all these facts. Ammonia-assisted alkaline treatment is an original method; however, it could not tackle the underlying cause of hemicellulose decomposition. This procedure also requires a high temperature. Limestone and sodium hydroxide are fairly affordable. The primary benefit of this procedure is that it does not turn the C5 sugars into inhibitors produced from furans. This alkaline approach is an efficient and successful way of attaining delignification. When all the benefits and drawbacks are considered, lime-assisted or sodium hydroxide treatment can be regarded as the most environmentally secure and feasible business approach [26].

3.6 Various Applications of Biomass for Bioenergy Production

3.6.1 Production of Second-Generation Bioethanol Using LCB

Bioethanol can be obtained from carbohydrates as raw materials such as fruit juices, leftovers from breweries, soya, peanuts, wheat-rice, and particularly waste, corn waste, trees, rice wastes which contain lignocellulose that can be used as feedstocks in traditional methods (Fig. 3.3).

The pretreatment of the lignocellulose is an important stage, followed by hydrolysis and fermentation [27]. Physical, thermochemical, physicochemical, and biological techniques are the most frequently employed pretreatment procedures for LCB conversion. These approaches have been thoroughly studied but have several drawbacks, like limited yield, high processing costs, and detrimental environmental effects. As a result, more effective green technologies are always being investigated to address these issues [28]. In the case of the physical pretreatment method, IL pretreatment and DESs pretreatment have been used. According to numerous researches, LCB, which has been pretreated using ionic liquid treatment, showed surface area expansion with decreased cellulose crystallinity and lignin content.

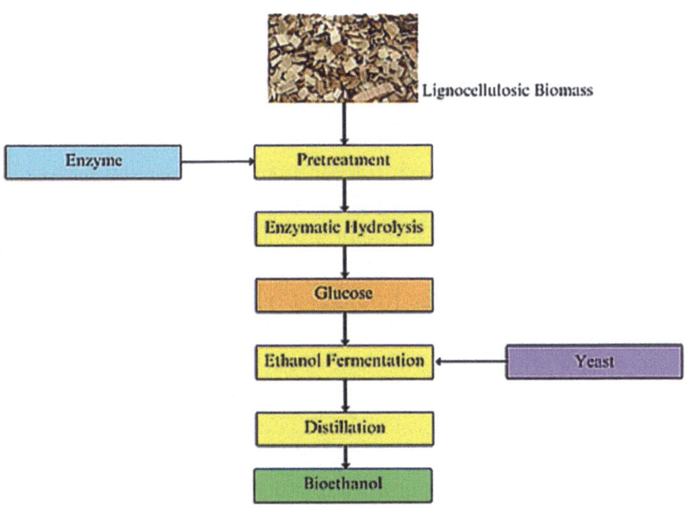

Fig. 3.3 Bioethanol production

Although ILs pretreatment methods are cost-effective and environmentally beneficial, their implementation in large-scale biorefineries may be constrained. DESs, a new class of environmentally friendly and designer solvents, were brought to the scientific community's attention by [29] as an alternative to ILs. For the production of bioethanol, lignocellulose is subjected to pretreatment procedures before being hydrolyzed and fermented because of the following:

- Improves the cellulose amorphous regions, which can be dissolved more easily than the crystalline regions.
- Enhances the porosity of fibers by perforating chemicals and enzymes into the structure.
- Free the cellulose from its lignin and hemicellulose contents [27].

Another appealing method that promises affordable equipment and operation costs is using microorganisms to pretreat lignocellulose, which is the biological approach. Combining two or three pretreatment methods is common in producing bioethanol derived from lignin cellulose. The biomass must be mechanically milled, sliced, and crushed to reduce the size. Subsequently, it must undergo a steam explosion, and finally, it must be immersed in an acidic or alkaline solution. Then, the pretreated material must be washed and neutralized [27].

3.6.1.1 Bioethanol from Pretreated Lignocellulose

In addition to fermentation, direct microbial conversion (DMC), hydrolysis, and the biomass that has been pretreated can all be utilized for producing bioethanol. The DMC procedure is time-consuming, the conversion yields are only moderately high, and there is a considerable risk of contamination. Enzymatic hydrolysis, combined with microbial fermentation, is a method with higher preference and has been demonstrated to work significantly better [27].

Lignocellulosic saccharification: After being processed with lignocellulose, the polysaccharide material is degraded by enzymes into monosaccharides. A combination of many enzymes isolated from microorganisms makes up the cellulose, which is commercially accessible and used to hydrolyze cellulose and hemicellulose. To break down cellulose enzymatically, the following three enzymes are used: Exo-1,4-β-D-glucanases, including 1,4-β-D-glucan glucohydrolase, to free D-glucose from 1,4-β-D-glucan and slowly hydrolyze D-cellobiose, and enzyme 1,4-β-D-glucan cellobiohydrolase, to free cellobiose from 1,4-β-glucan. Also, Endo-1,4-glucanases (EG) or 1,4-D-glucan 4-glucanohydrolases which cleave 1,4-glucan linkages and β-D-Glucosidase or β-D-glucoside glucohydrolase forms D-glucose from cellobiose [27].

LCB saccharification using enzymes: LCB is converted to sugar monomers, but their industrial yields are one of the primary obstacles to produce second-generation biofuels, as it is challenging to obtain high process efficiencies. Hydrolyzing the polysaccharide is done using acid to obtain monomers because it completely transforms the crystalline form of the polysaccharide into an amorphous state by dissolving the hydrogen bonds between the cellulose chains. Cellulase, hemicellulases, and additional microbial lignocellulolytic enzymes have been discovered in various biological settings, including woodlands, composting soils, compost piles, sewage sludge, and rumens [30]. Aerobic bacteria and filament-containing fungi, along with some enzyme complexes, form the cellulosomes, which are abundant in fungal strains, and anaerobic bacteria belong to common lignocellulolytic systems.

3.6.1.2 Strain Development for Improved Bioethanol Fermentation

Lignocellulolytic enzymes are very costly; this is a major bottleneck in biofuel-producing industries to decrease the enzymes' price and enhance the enzymes' yields with the required productivity. Recently, several techniques have been employed for enhancing the expression of enzymes using microbial strains, including genome and metabolic engineering, CRISPR-Cas9, protoplast fusion, protein engineering, cell surface engineering, genetic engineering, and recombinant DNA technology. Random mutagenesis is widely used to alter microorganisms' natural makeup and functionality. Numerous papers have described cyclic mutagenesis that combined physical (UV radiations) and chemical (ethanomethane, acriflavine, and *N*-methyl-*N* Nitro-*N*-nitrosoguanidine) factors. Site-directed mutagenesis is another widely used method that causes precise changes in the recognized DNA sequences. Site-directed mutagenesis and error-prone PCR (EP-PCR) have been demonstrated to increase EGIII's alkaline tolerance in *Trichoderma reesei* and increase enzyme activity in *Bacillus amyloliquefaciens* DL-3 (by up to 7.93 times) [31]. To produce alcoholic fuel, the microbial genes are altered through adaptation, selection, mutation, and protoplast fusion to improve fermentation and saccharification efficiency. Choosing genetically modified ethanol-resistant bacteria involves transposon-mediated mutation collection and deletion mutant library screening [32, 33].

3.6.1.3 Fermentation Procedures for Producing Bioethanol

The simple sugars produced during enzymatic hydrolysis can be converted to bioethanol by microorganisms. There are two processes, including separate as well as simultaneous fermentation and hydrolysis.

Ninety-five percent of soluble carbohydrates undergo fermentation in order to yield CO_2 and ethanol. Pentose and hexose sugars are mixed in the hydrolysate substrate of LCB, in contrast to the processing of starch hydrolysate, which solely includes glucose. Numerous microbes that occur naturally can easily ferment hexose carbohydrates, including galactose, glucose, and sugar mannose. Typically, *Zymomonas mobilis* and *Saccharomyces cerevisiae* are the foremost ones. Nevertheless, pentose sugars like arabinose as well as xylose cannot be processed by natural microbes. Conversely, several yeast strains, such as *Kluyveromyces marxianus, Pichia stiipitis, Pachysolen tannophilus,* and *Candida shehatae* ferment xylose [34].

Separate hydrolysis and fermentation (SHF): After hydrolyzing the sugars, bacteria are added to the mixture to ferment them. Contamination, the introduction of inhibitors, and the need for more time and equipment are a few of its inherent weaknesses [27].

Simultaneous saccharification and fermentation (SSF): Hydrolysis using enzymes and fermentation using microbes occur simultaneously in individual machinery. This mixture contains microorganisms and enzymes. This method has been demonstrated to be substantially more effective than the SHF mentioned above, with less equipment, less time, and a lesser risk of contamination [27]. This step is very important to produce chemicals and fuels that are industrially relevant and involves the conversion of sugar monomers produced by hydrolyzing the raw materials into products by the activity of microorganisms. *Calonectria brassicae, Zymomonas mobilis, Saccharomyces cerevisiae, Pichia stipitis, Pachysolen tannophilus, Mucorindicus, Escherichia coli,* and *Candida shehatae* are the wild-type microorganisms investigated in ethanol fermentation [35]. Some variables that may impact the fermentation process for producing bioethanol include salt concentration, pH, carbohydrate content, aeration rate, ethanol concentration, and temperature. The three basic fermentation processes used to make ethanol are batch, fed-batch, and continuous fermentation. These methods have their own merits and demerits [35].

Batch fermentation: The high-concentration substrate is converted into a desirable product in this fermentation. It is a very old and popular fermentation operation. Repeated batch fermentation is a variation of this operation, which increases the system's effectiveness by using immobilized microbial cells rather than free ones.

Fed-batch fermentation: Unremoved products are intermittently added while using the fed-batch method, which combines batch and continuous modes of fermentation. This fermentation has some merits over batch fermentation, which involves high ethanol productivity, lower toxicity of media components, short fermentation period, and higher dissolved oxygen in the media.

Continuous fermentation: The most prevalent kind of fermentation, known as continuous fermentation, has been utilized to produce industrial quantities of bioethanol since it is simple to regulate the process, does not require as much time to clean, and is less expensive to invest [36, 37].

Ethanol generation using multiple LCBs is mostly governed by the extent to which glucan hydrolyzes into soluble sucrose. The order of glucan digestibility among the residues from agriculture was stated to be as follows: bagasse (47%) > crop maize stover (23%) > hemp for industrial use hurd (14%) > wheat residue (11%) > rice stalks (10%). Conversely, the ability to be digested by bioenergy crops was found to be lesser than that derived from agricultural wastes, with common reed ranking highest (19%), followed by switchgrass (17%), Miscanthus (8%), and bamboo (3%). Poplar, one type of untreated biomass from trees, has an estimated glucan digestibility of about 7%. A comparison analysis was conducted employing sugarcane, maize, and sugar beetroot as the feedstocks implemented in the industrialization of ethanol in Brazil, the USA, and Europe [38]. This study shows that sugarcane is the most affordable fodder when viewed alongside the other two since it reduces carbon dioxide emissions by 30 and 40% when compared to maize and sugar beetroot. According to estimates, sugarcane has more possibilities for ethanol output than sugar beet (5500 L/ha) and maize (4180 L/ha) [34].

3.6.2 Production of Second-Generation Biobutanol Using LCB

Due to the depletion of petroleum reserves and the fluctuating price of gasoline, replacing fossil fuels with alternative energy sources is a critical concern. One of the problematic liquid fuels is bioethanol, but its use is restricted by its higher volatility and corrosiveness compared to petrol and its lower energy content. Biobutanol is a biofuel with a significant potential that may be utilized directly in car engines without modification due to its similarities to petrol characteristics [39]. By using lignocellulosic feedstock, we can obtain biobutanol (Fig. 3.4).

3.6.2.1 Renewable and Carbohydrate-Rich Lignocellulosic Feed for Producing Biobutanol

Organic municipal wastes, which include fibrous and wood waste, agricultural wastes, and waste generated from industries, are all included in the feedstock, which contains lignocellulose; also, plants are comprised of cellulose, lignin, and hemicellulose [40]. Typically, LCB has 15–25% lignin, 38–50% cellulose, and 23–32% hemicellulose. The proportions of lignin, cellulose, and hemicellulose in different

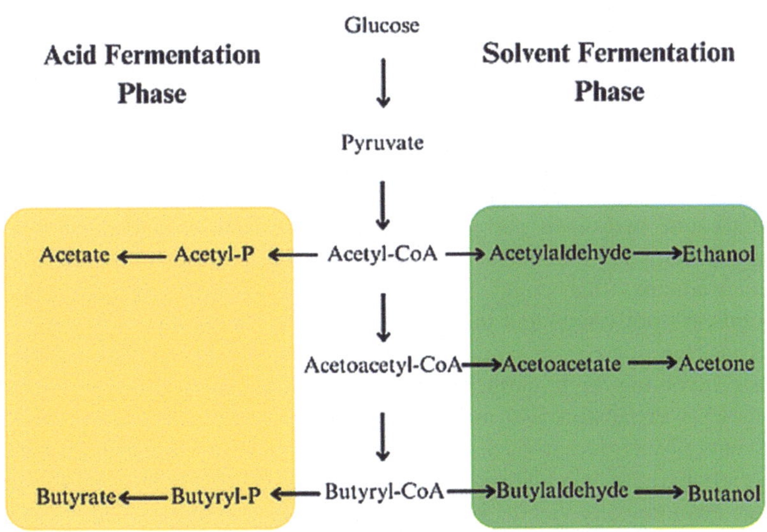

Fig. 3.4 Acetone-butanol-ethanol synthesis by *Clostridium acetobutylicum*

lignocellulosic materials vary. In comparison to maize cob, which has 14–15% lignin, 42–45% cellulose, and 35–39% hemicellulose, rice straw has 12–14% lignin, 28–36% cellulose and 23–28% hemicellulose [41–43]. Because of its crystallinity and the resultant intense inter- and intramolecular hydrogen bonding, it greatly resists enzymatic hydrolysis. Xylans, mannans, and glucomannans are only a few polysaccharides found in hemicellulose, another polymer in plant cell walls. By forming hydrogen bonds with lignin or cellulose, it contributes to the sturdiness of the cell wall. P-hydro-xycinnamoyl, coniferyl, and sinapyl alcohols make up the three alcohol groups that make up lignin, along with a variety of interunit connections such as biphenyl and other bonds [43, 44]. It contributes to the stability of the structure while protecting it from oxidative stress and microbial attack [45]. The most potential feedstocks are LCB and other abundant and sustainable raw resources. However, most solventogenic bacteria, including *Clostridia*, could not use lignocellulose because they lacked the expression of enzymes that break down lignocellulose [46].

3.6.2.2 The Traditional Method of Producing Biobutanol from Lignocellulose

The most popular methods for releasing sugars are thermochemical and enzymatic hydrolysis procedures [47]. Although acid hydrolysis is a less expensive and quicker technique, it produces inhibitors that are hazardous to cell growth and are not ecologically friendly. However, the complex structure of LCB necessitates using many enzymes that work in concert to degrade lignocellulose. While hemicellulose is a complex polymer of carbohydrates, which contains significant amounts of hydrolyzable subunits, it also requires a large number of enzymes for degradation, such as glucuronidase, exo-xylanase, acetyxylan esterase, endo-xylanase, xylosidase [48, 49]. The biobutanol manufacturing process is less efficient due to the costly enzymes, which contribute around 50% of the production expenses. In general, enzymatic hydrolysis and chemical methods were used to pretreat lignocellulosic feedstock and modify the LCB's structure and chemical makeup to accelerate the conversion of biomass feedstock to obtain biobutanol [50]. However, the inhibitors formed during the lignocellulose hydrolysis and enzyme cost restrict the utilization.

Acid/alkaline hydrolysis and ozonolysis are currently used pretreatment techniques involving large amounts of chemicals, including powerful oxidizing agents, acids, and alkalis, that could hinder hydrolysis. Additionally, because these compounds are poisonous and dangerous, corroding pretreatment is more expensive because anti-corrosion equipment is required. Similarly, physical–chemical processes involving high temperature and pressure, such as steam explosion, ammonia fiber expansion, and moist air oxidation, render the process too expensive and energy-intensive to be practical. Thus, the main pretreatment problem is to prevent carbohydrate degradation or loss while reducing treatment costs to make the procedure economically viable [39]. In the case of acetone-butanol-ethanol fermentation, *Clostridium Saccharoperbutylacetonicum, Clostridium saccahrobutylicum, Clostridium beijerinckii, Clostridium Acetobutylicum* have been used. The two stages of acetone-butanol-ethanol synthesis involve acidogenesis and solventogenesis. Oxidizing agents, diluted acid, alkaline processing, and liquid hot water are used during the chemical pretreatment of lignocellulose to manufacture biobutanol. The commonly solventogenic *C. beijerinckii* is used to produce butanol, which is basically used in chemical pretreatment, specifically in alkali pretreatment techniques [51]. To reduce lignocellulose's high resistance to biological transformation, the cross-linked matrices are dissolved before enzymatic hydrolysis, which is required for generating biobutanol [52]. Before fermentation, the enzymes hydrolyze the lignocellulose using substances like crude enzyme extract or commercial cellulase.

3.6.2.3 Fermentation Procedures for Manufacturing Biobutanol

Enzymatic hydrolysis is necessary after pretreatment to obtain butanol by transforming cellulose into glucose and later into butanol. The process of turning LCB into butanol occurs via four major processes that are based on the production of cellulase, cellulose hydrolysis, and butanol fermentation: consolidated biotreatment (CBP), simultaneous saccharification, and co-fermentation (SSCF) and separate hydrolysis and fermentation (SHF) [39].

CBP: This is one of the most significant barriers to SHF and SSF. Therefore, creating affordable, efficient, and effective fermentation and saccharification methods is essential for biobutanol production. It, also known as direct microbial conversion, is a substitute method to minimize this significant cost increase. Cellulase synthesis, fermentation, and cellulose hydrolysis are accomplished during CBP in one phase. This process is cost-effective for low-cost fermentation and biomass hydrolysis. Because enzymatic hydrolysis and cellulase synthesis come at operational and capital costs that are higher than those of SSF and SHF processes, it is anticipated that CBP has the potential to lower expenses by more than 50% [39].

SSCF: During this procedure, cellulose fermentation and hydrolysis occur simultaneously. Also, a process named simultaneous saccharification and co-fermentation operates to yield butanol by hydrolyzing both C5 and C6 sugars. It operates at lower temperatures, usually 35–37 °C, to promote microbial development and butanol production [39].

SHF: Cellulases hydrolyze cellulose to glucose under ideal circumstances at a temperature of 50 °C. The glucose obtained is then transported into a reactor where fermentation occurs with the help of microorganisms that produce butanol at a temperature of 35 °C. It is currently used most frequently in the butanol manufacturing process of LCB. Cellulose hydrolysis and fermentation can occur in this process under their respective ideal circumstances. Cellulases' activity during enzymatic hydrolysis may be inhibited, nevertheless, by the buildup of glucose during the process [39].

3.6.2.4 Techniques for Increasing Butanol Production from Lignocellulose

Detoxification of inhibitors: During hydrolysis and pretreatment, sugar monomers are formed along with many unwanted degradation products that can be produced, including aliphatic acids, furan derivatives, and phenolic compounds. These inhibitors were found to affect biomass conversion efficiency significantly. Therefore, the key to effective biofuel production is the removal of these inhibiting substances prior to fermentation. Adsorption, extraction, precipitation, and combinations are the primary techniques used in traditional inhibitor removal processes. An alternate approach to removing inhibitory substances from biomass hydrolysates is biological control [39].

Strain improvement: The poor butanol yield caused by the production of acetone and ethanol byproducts and the microorganism's low butanol tolerance are the key obstacles to the industrialization of butanol fermentation. To address this weakness, butanol fermentation necessitates the creation of strong strains with increased butanol yield and tolerance. Mutagenesis has been used to modify the strains; for example, both *C. beijerinckii* BA101 and C. *beijerinckii* NCIMB 8052 produced 19 g/L of butanol when treated with Methylnitronitrosoguanidine [39].

Clostridium sp. favors glucose as its major energy source. In batch fermentation, various amounts of glucose (20, 40, 60, and 80 g/L) were employed as a carbon substrate for *C. acetobutylicum* to synthesize biobutanol, and an elevated biobutanol titer of 11.10 g/L was observed. In conformity with other recently reported studies, the ideal level of glucose for batch fermentation employing suspended *C. acetobutylicum* cells was 60 g/L. After glucose, xylose is the second-largest sugar found in lignocellulosic biomass hydrolysates. ABE fermentation has been carried out in a batch approach, with *C. acetobutylicum* implementing xylose as the sole carbon source. At 60 g/L xylose, the maximum biobutanol titer was 8.52 g/L. Since *Clostridium* groups are also amylolytic, they may formulate biobutanol as an alternative to carbon. As a result, 9.10 g/L of reduced concentration xylose is produced, and the biobutanol titer yields 0.42 mol biobutanol/mol glucose. Alkaline pretreated rice straw outperformed sugarcane bagasse as the substrate of choice. In the absence of yeast extract, the use of immobilized cells during the phase of fermentation leads to the high biobutanol generation, efficiency, and concentration of 13.80 g/L, 0.90 g/L/h, and 0.58 mol biobutanol/mol glucose, respectively [47].

3.6.3 Production of Second-Generation Biodiesel Using LCB

Biodiesel is sustainable, eco-friendly, harmless, sulfate-free, biodegradable, combustible, and free of aromatic chemicals. Compared to normal diesel, it has a greater flash point and cetane number. The intrinsic lubricate nature improves engine performance [53], negating the need for additional engine lubrication. To cut the cost of production, it is a good idea to use cooking oil, oils from non-edible crops, and animal fats as raw materials. Additionally, this approach is unable to meet the current energy demands for renewable fuels. Therefore, in order to generate biodiesel, experts are seeking new oil sources. Due to its similarity in fatty acid composition to vegetable oils, single-cell oil, a lipid generated by microorganisms like yeast, bacteria, molds, and algae, is one of many resources that are thought to be a suitable feedstock for the generation of biodiesel. These microorganisms can synthesize oils in their cellular compartments using organic carbon (Fig. 3.5) [33, 54].

3.6.3.1 Traditional Feedstocks Used to Produce Biodiesel

Lignocellulose-containing feedstocks are the greatest sustainable resource for biodiesel production. It should be emphasized that not all available wastes are often used for producing biodiesel; some are used for producing fodder, manure, and paper or are directly burned as fuels [47]. Edible plants are a source of vegetable oils that serve as a starting material for biodiesel production. The objective of employing vegetable and non-edible oils as feedstocks for biodiesel manufacturing has been defeated by their exorbitant price, acreage need, and low oil output [48]. Oleaginous microorganisms (OMs) can use low-cost feedstocks, such as agricultural wastes, lignocellulosic substrates (LCSs), and waste substrates to accumulate more lipids [55]. LCB is converted to enriched bioproducts that help generate electricity through alternative methods in place of direct burning. Most commonly, waste cooking oils, animal fats, and vegetable oils, either edible or inedible, are used in a trans-esterification procedure to create biodiesel [56]. Due to global food security, oil sourced from food sources is insufficient for large-scale biodiesel manufacturing. Thus, looking for sustainable feedstocks for manufacturing biodiesel, including undigestible lignocellulose-containing raw materials, is essential [57].

Agricultural residues: Numerous agricultural waste products are a source of lignocellulose-containing raw materials that help in lipid production by generating small amounts of sugars. Crop remnants and agro-industrial residues are two general groups into which these residues fall [22]. Crop remnants are parts of edible products like grains crops that are obtained after harvesting, whereas agri-industrial leftovers are byproducts after post-harvesting processes.

Non-edible crop residues: Food crops are grown on the world's most fertile lands. However, due to the population's rapid growth, we will soon be confronted with the issues of a food shortage and an energy catastrophe. Forest-derived woody biomass is still a significant source of raw materials for biofuel production [58–61]. Most non-edible LCB comes from agricultural residues, followed by these byproducts left over after superfluous elements are eliminated to create trees of proper size and shape. Other notable sources of LCBs include sawmill rejects and sawdust, pallets, cardboard boxes, wooden packing crates, and recovered wood obtained from the demolition of buildings [57, 62]. Many LCBs are made from non-edible energy crops, [63] agricultural waste, herbaceous and municipal solid waste, and forest woody feedstocks. The most promising plant leftovers, which help generate energy, are present in the USA, Europe, and Asia [22, 64, 65], including corn stover, rice straws and wheat, sugarcane husk, and bagasse. Lignin, hemicellulose, and cellulose are the three primary polymers that comprise LCBs. These polymers are tightly entangled and linked together by covalent and non-covalent connections. The lignocellulose that exists in plants contains cellulose, which is the primary structural component of cell walls. Hemicellulose and lignin surround the microfibrils that make up celluloses. The repeating unit of cellulose, known as the "clubhouse," contains *D*-glucose subunits that are connected by α-(1,4)-glycosidic linkages. Proper pretreatment is needed to get cellulose units from LCBs, and then the material must be further processed with sulfuric acid or an enzyme cocktail to produce sugars. The hydrolysates include pentoses and hexoses released from the treated cellulose.

Industrial residues: If industrial waste or byproducts are not properly disposed of, they can significantly negatively influence the environment in many ways. In many situations, the expense of treating these pollutants exceeds the cost of production, and they are often handled in ways that are harmful to the environment [30]. Anaerobic digestion of organic wastes demands expertise to assure safety due to the buildup of dangerous gases. Oleaginous bacteria can process the waste materials as an alternative to anaerobic digestion, metabolizing them as lipid droplets inside their cells before further processing them into biodiesel [59]. The biological treatment using microorganisms is more effective at lowering the amount of hazardous organic compounds in treated effluents.

Oleaginous microbes: Under stressful circumstances, such as those involving low nitrogen and high carbon sources, these bacteria can store fat up to 70% or more of their dry weight. These bacteria utilize a variety of renewable resources to create microbial oil, which is then used in the transesterification process to create biodiesel [55]. Production of biodiesel with OM has a number of benefits, including a short incubation period in contrast to animal and plant resources, and the production of lipids is not affected by seasonal, meteorological, or geographic fluctuations. Microbial oil has a low viscosity and is similar in caloric value and composition to oils derived from plant and animal resources. OM are microalgae, yeasts, filamentous fungi or molds, bacteria, and other microbial groups. Because it can supply enough oil as a feedstock for worldwide consumption, it yields far more biodiesel per acre than plant-based feedstock. Biodiesel obtained from microalgae is currently the subject of research interest.

3.6.3.2 The Traditional Method of Producing Biodiesel from Lignocellulose

Some pretreatment methods are undertaken to enhance the digestibility of lignocellulose materials, including oxidative, acid, thermal, alkaline, and mechanical pretreatment. Additionally, CO_2 explosion, acid-triggered steam explosion, liquid hot water, and ammonia fiber explosion are currently used as pretreatment methods. The approach selected is determined by the initial makeup of the materials used [66]. The inhibitors that are generated during the process and present in the medium are what the oleaginous microorganisms use to ferment materials. Mechanical pretreatment is the best method for converting LCB to produce biodiesel. Milling is a process that reduces biomass particle size and eliminates crystallinity [67]. When biomass is sheared, the surface area increases, decreasing polymerization levels and accelerating total hydrolysis [68]. The LCBs are heated over 150 °C during thermal pretreatment to saturate the biomass's hemicellulose portion, followed by saturating lignin [69].

The two primary components of hemicellulose are glucomannan and xylan, being the least thermally stable. When LCB is heated to a high temperature (4160 °C), inhibitors prevent bacteria and yeasts' growth [70, 71]. Alkaline pretreatment involves converting dissolved polysaccharides into xylan while maintaining a low temperature to prevent end groups from peeling off and losing carbohydrates. The biomass itself consumes some of the alkalies used for alkaline pretreatment, with the remaining alkali being utilized in subsequent reactions. Industries frequently use various procedures [70]. The solubilization of the components is the main distinction between these two procedures; while lignin fraction solubilization is seen during alkali pretreatment, hemicellulose solubilization occurs more slowly during acid pretreatment. Adopting an alkali pretreatment method removes the lignin fraction without causing the other biomass constituents to deteriorate [66]. For specific biodiesel production, ozonolysis, liquid hot water, steam explosion, carbon dioxide (CO_2) pretreatment, microwave treatment as well as some biological approaches have been applied. The main disadvantage of utilizing this pretreatment technique, which alters the composition of the hydrolysates, is salt production.

3.6.3.3 Bioconversion of LCB to Biodiesel Production

Complex unit processes, such as depolymerization of lignin, saccharification of biomass, cultivation of OMs for producing lipids, and transesterification, are used in biodiesel production using LCB [55].

Lignin depolymerization: Lignin is removed during pretreatment to enhance the hemicellulose and cellulose hydrolysis components and create fermentable sugars. Through this method, cellulose crystallinity is decreased, porosity is increased, and hydrolytic enzymes and chemicals for efficient hydrolysis of biomass better access hemicellulose. To remove lignin, several pretreatment processes such as physicochemical, physical, biological, and chemical, pretreatments are used, such as pulsed electric field irradiation, extrusion, and pyrolysis to minimize the particle size, crystallization, and degree of polymerization. Pretreatment via physical method is economically infeasible on a larger industrial scale since it requires considerable energy. There are many different chemical pretreatment methods, such as basic (potassium hydroxide, calcium oxide, and magnesium oxide), acidic (hydrochloric acid and sulfuric acid), ionic liquid, and organosol pretreatment. The term "physicochemical pretreatment" refers to the mix of chemical and physical processes, and it encompasses some techniques such as CO_2 explosion, ammonia fiber, explosion, steam explosion, and hydrothermal methods. Additionally, microorganisms or enzymes are used in the biological processing of biomass [55].

LCB saccharification: Prior to saccharification, lignocellulosic materials are pretreated to transform carbohydrate polymers into fermentable sugars effectively. Enzymatic and chemical techniques are used to hydrolyze the biomass cellulose portion. At high temperatures and the presence of acid, the biomass feedstock polysaccharides can hydrolyze chemically. The creation of furfurals, the poor sugar yield, and the method's lower economic feasibility are drawbacks of this approach. Enzymatic hydrolysis ensues under low pressure, weak pH, and low temperatures (45–50 °C) [55].

Transesterification and lipid extraction: Extraction of lipids from the microbe is a primary requirement for cell wall degradation. A variety of processes such as bead mill, subcritical water hydrolysis, pulsed electric field, ultrasound, high-pressure homogenization, osmotic shock, microwave, chemical hydrolysis, enzymatic hydrolysis, autolysis are commonly utilized to disrupt cells. In addition to pressing and integrated solvent extraction, elevated temperature, supercritical extraction, and pressure solvent extraction, ultrasonic extraction is employed to extract lipids from the oleaginous microbes. Dried biomass residues of algae are pyrolyzed into oils that are used for transporting fuel. Biomass containing moisture is treated by anaerobic digestion and hydrothermal means to produce diesel fuel, gasoline, and methane. Once the extraction has been done, lipids are transformed into biodiesel fuel through transesterification; this helps reduce lipids' viscous nature. Glycerol and esters are formed when lipid reacts with alcohol during transesterification (Eq. 3.1). The reactions are chemically catalyzed and divided into homogeneous and heterogeneous reactions. A solid acid or alkaline catalyst is required for the feasibility of heterogeneous reactions.

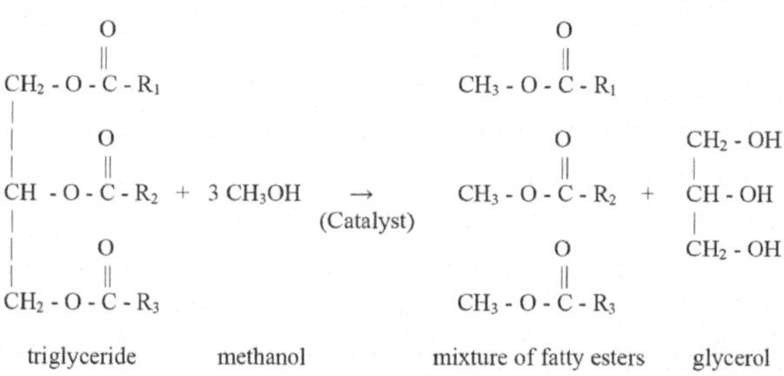

(3.1)

Moreover, a homogeneous reaction takes place using an acid or alkaline catalyst. Because of soap formation, due to higher free fatty acids, alkali-based transesterification is not feasible. The drawbacks of acid-based transesterification include a longer incubation period, corrosive character, higher energy usage, higher glycerin purification expenses, etc. [55].

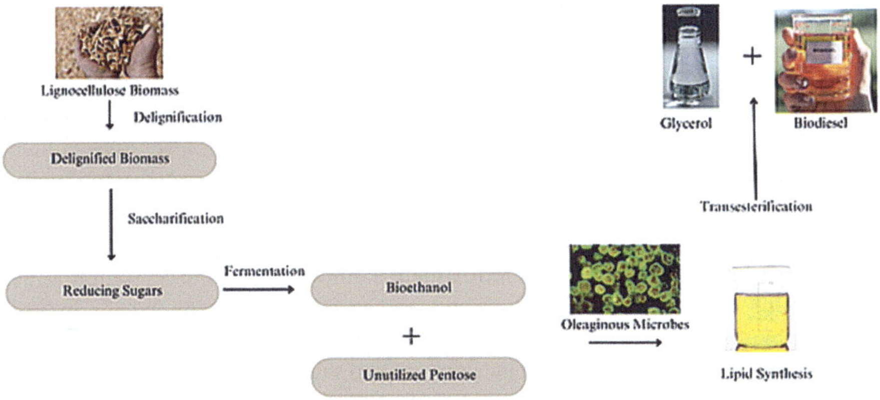

Fig. 3.5 Biodiesel production from LCB

3.6.4 LCB for Second-Generation Polyhydroxybutyrate (PHB) Ester Production Production

The biodegradability of PHB esters makes them environmentally friendly materials. They can be degraded by microorganisms present in soil, water, and composting environments, ultimately breaking down into natural components such as water, carbon dioxide, and biomass. This biodegradability makes PHB esters suitable for various applications where sustainability and environmental compatibility are essential, including packaging materials, agricultural films, disposable products, and biomedical applications. PHB esters are a class of biodegradable polymers derived from polyhydroxybutyrate, which is a naturally occurring polyester. PHB esters exhibit improved thermal and mechanical properties compared to unmodified PHB. They have increased flexibility, toughness, and elongation at break. Additionally, the introduction of ester groups enhances the processability of PHB, allowing for easier melt processing and shaping into various forms such as films, fibers, and molded products [38]. Overall, PHB esters combine the biodegradability of PHB with improved properties, making them versatile and eco-friendly materials for a wide array of uses.

3.6.4.1 Bioconversion of Lignocellulosic Feed to PHB Esters

PHB esters can be obtained from LCB using a two-stage procedure. Initially, lignocellulose is transformed into sugars, and subsequently, these sugars are fermented to produce PHB esters. To manufacture PHB esters, the usual pretreatment approach is significant.

3.6.4.2 Prior Treatments to Enzymatic Hydrolysis

For producing second-generation biofuel, this method aims to disrupt the lignin structure by reducing the crystallinity of cellulose. Lignin serves as a protective blockade, impeding the accessibility of hemicellulose and cellulose. Pretreatment techniques are employed to degrade or alter lignin, enhancing the exposure of cellulose and hemicellulose. This makes them more amenable to enzymatic or chemical hydrolysis. Cellulose, in its native form, has a highly ordered crystalline structure, which limits enzyme accessibility. Pretreatment methods disrupt the crystalline structure of cellulose, increasing its surface area and susceptibility to enzymatic hydrolysis. Removing the hemicellulose is another crucial point to improve its enzymatic digestibility. Eventually, increasing the porosity and surface area enhances the contact between the biomass and the catalysts or enzymes used in subsequent conversion processes [38]. Several chemical, biological, and physical pretreatment methods, including milling, grinding, and size reduction techniques, are available to mechanically break down the biomass structure. In the case of physicochemical methods, they integrate physical as well as chemical workflows, for instance, ammonia fiber expansion (AFEX), organosolv processes, and steam explosion. The choice of the pretreatment method is contingent upon many factors, including biomass feedstock, target products, process economics, and environmental considerations. The choice of pretreatment method can significantly impact the efficiency and cost-effectiveness of the overall biomass conversion process.

Enzymatic hydrolysis: Enzymatic treatment offers several advantages, including specificity, mild operating conditions, and environmental friendliness. Here are some key aspects of enzymatic treatment for LCB [72].

Enzymes used: Enzymes used in the enzymatic treatment of LCB include cellulases, hemicellulases, and lignin-modifying enzymes. Cellulases are responsible for the hydrolysis of cellulose into glucose, while hemicellulases break down hemicellulose into various sugars. Lignin-modifying enzymes, such as lignin peroxidase and laccase, target lignin degradation.

Enzyme production: Enzymes for biomass conversion can be obtained from various sources. Fungi, such as *Trichoderma* species, are commonly utilized for enzyme production as they have high cellulase and hemicellulase activities. *Aspergillus* species can also be used for the same reason.

Enzyme cocktail: A mixture of different enzymes is often used as an enzyme cocktail due to the intricate nature of LCB. The enzyme cocktail contains multiple enzymes that act synergistically to break down different components of the biomass. This approach improves the overall hydrolysis efficiency and reduces the enzyme dosage required.

Process conditions: Enzymatic treatment generally necessitates gentle process conditions, typically involving moderate temperatures, ranging from 40 to 60 °C and a pH level close to neutral. The specific process conditions may vary based on the enzymes employed and the characteristics of the biomass feedstock.

Inhibitors and enzyme stability: LCB can contain inhibitory compounds, such as lignin-derived phenolics and sugar degradation products, which can inhibit enzyme activity. Strategies to mitigate enzyme inhibition include detoxification of the biomass or modification of enzymes for improved stability and tolerance to inhibitors.

Enzyme recycling: Enzymatic treatment processes can be designed to enable enzyme recycling, where the enzymes are recovered and reused in subsequent hydrolysis steps. This reduces the overall expenses associated with enzyme production, rendering the process more economically feasible.

Enzymatic treatment of LCB significantly contributes to the generation of biofuels, biochemicals, and bioproducts. Advances in enzyme technology, enzyme engineering, and process optimization continue to upgrade the efficiency and cost-effectiveness of this pretreatment method, making it a promising approach for sustainable biomass conversion.

3.6.4.3 Fermentation Procedure for Production of PHB Ester

The obtained sugars from enzymatic hydrolysis are then used as a feedstock for fermentation. Here is an overview of the fermentation process (Fig. 3.6):

Microorganism selection: The choice of microorganisms depends on the desired product and process conditions. Bacterial strains capable of producing PHB are selected for fermentation. Examples include *Cupriavidus necator* (formerly *Ralstonia eutropha*) and other PHB-producing bacteria. These bacteria can naturally convert glucose or other fermentable sugars into PHB as a storage material.

Fermentation medium: The sugars derived from LCB are typically present in a hydrolysate solution, which contains various compounds, including sugars, inhibitors, and nutrients. The hydrolysate may need to be treated to remove or detoxify inhibitors and adjust the nutrient composition to support microbial growth and product formation. The fermentation medium for PHB ester production typically contains the necessary nutrients and carbon sources to support the growth of the PHB-producing bacteria and the accumulation of PHB within the cells. The specific composition of the fermentation medium may vary based on the bacterial strain used and the availability of LCB-derived sugars.

Fermentation process: The fermentation process involves the inoculation of the selected microorganism into the fermentation medium containing the sugars from LCB. Fermentation is typically carried out in bioreactors equipped with monitoring and control systems.

3.6.4.4 Procedures for PHB Recovery

Once the fermentation is complete, the PHB must be separated from the broth [63]. The recovery of PHB from the fermentation broth involves several steps to separate and purify the PHB from the bacterial cells and other impurities. Here is a general outline of the PHB recovery process:

Cell harvesting: The first step is to separate the bacterial cells containing PHB from the fermentation broth. This can be achieved through techniques such as centrifugation, filtration, or both. The harvested cells are typically referred to as wet biomass.

Cell disruption: Once the cells are harvested, they must be disrupted to release the intracellular PHB granules. Cell disruption methods encompass mechanical methods like high-pressure homogenization and bead milling, chemical methods like treatment with enzymes or detergents, or a combination of different techniques.

PHB extraction: After cell disruption, the PHB granules are extracted from the disrupted cells. Various extraction methods can be employed, including solvent extraction, where a suitable organic solvent (e.g., chloroform, chloroform–methanol mixture) dissolves the PHB while leaving other cellular components behind. The extraction can be performed using soxhlet extraction, shaking or stirring with the solvent, or continuous extraction methods.

PHB purification: Once the PHB is extracted, it needs to be purified to remove impurities, such as residual cell debris, proteins, and other contaminants. Purification methods can include precipitation, filtration, centrifugation, or chromatography techniques. For example, the extracted PHB can be dissolved in a non-solvent (e.g., ethanol, methanol) to precipitate the PHB, followed by washing and drying steps to obtain purified PHB [72].

Drying: The purified PHB is typically in the form of a wet cake or solid. It needs to be dried to remove residual moisture and obtain dry PHB. Techniques like freeze-drying, vacuum drying, or hot air drying can achieve drying.

PHB characterization: The recovered and dried PHB can be characterized using various techniques to determine its molecular weight, thermal properties, crystallinity, and other physical and chemical properties. This characterization helps ensure the quality and suitability of the PHB for specific applications.

3.6.4.5 Transesterification for PHB Esters Production

A transesterification process can be employed to form a PHB ester from PHB. Transesterification entails the substitution of one ester group with a different alcohol, leading to the creation of a novel ester compound [73].

Preparation of reactants: The PHB polymer is depolymerized into monomeric units, typically 3-hydroxybutyrate (3HB) monomers. The alcohol used for esterification is usually a low molecular weight, like methanol or ethanol.

Reaction setup: The reaction is performed in a suitable reaction vessel, which may be equipped with a condenser to prevent the loss of volatile components. The vessel is often heated or maintained at a specific temperature to facilitate the reaction.

Catalyst addition: An acid or base catalyst is added to the reaction mixture depending on the transesterification method. Common acid catalysts include sulfuric acid (H_2SO_4) or hydrochloric acid (HCl), while common base catalysts include sodium hydroxide (NaOH) or potassium hydroxide (KOH). The catalyst helps to accelerate the reaction and improve the yield of PHB esters.

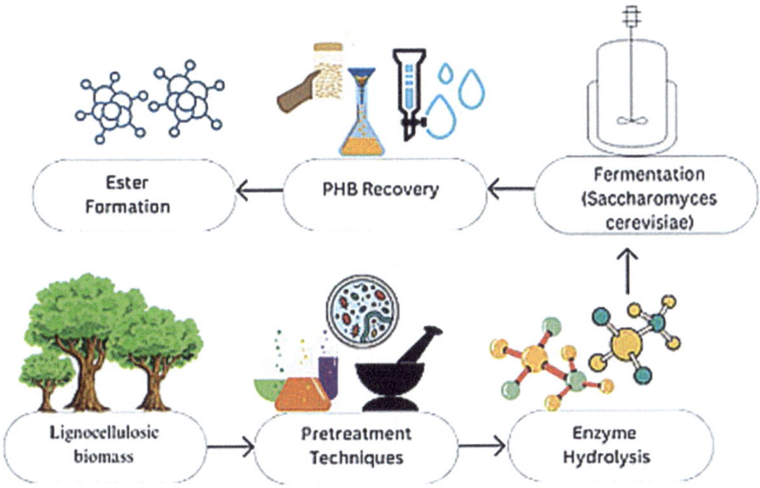

Fig. 3.6 Flow scheme for the production of PHB ester from LCB

Reaction initiation: The reaction mixture is heated or maintained at a specific temperature, typically around 60–80 °C, to promote the transesterification reaction. The reaction time can vary depending on the specific conditions and desired conversion.

Monitoring the reaction: The progress of the transesterification reaction can be monitored by sampling the reaction mixture at regular intervals and analyzing it using strategies such as gas chromatography (GC) or nuclear magnetic resonance (NMR) spectroscopy. This helps determine the conversion and optimize the reaction conditions.

Purification: At the end of the reaction, the reaction mixture is cooled, and the PHB ester product is separated from the remaining reactants, catalysts, and impurities.

In the presence of a 2% concentration of glucose and an available carbon source, Alcaligenes eutrophus MTCC 1285 produces bioplastics. The synthesis of PHB has been demonstrated employing molasses (pH 6.9–8) and sago effluent as feedstock. In the manufacture of media based on molasses, higher output was achieved at pH 8.0. The concentrations of PHB generated in sago, molasses, glucose, and thippi substrate-based medium at pH 6.9 were 0.5, 0.73, 0.8, and 0.4 g/ml, respectively. The expenses of producing biopolymers might be considerably minimized by harvesting leftovers such as starch, damaged food grains, date seeds, dairy waste, and grain crops. The efficient utilization of food and agricultural waste sources for the value-added manufacturing of biopolymers has been the focus of current research. Processing industry byproducts are frequently disposed of as waste because their economic worth is lower than their recovery cost and potential for reuse. These waste fluxes can be efficiently valorized by producing and processing biopolymers [74].

3.6.5 Production of Second-Generation Biomethanol Using LCB

Biomethane production from LCB typically involves a two-step process: pretreatment and anaerobic digestion [75]. Figure 3.7 provides the overview of the steps involved in biomethanol production:

Pretreatment approaches: LCB is often complex and resistant to enzymatic breakdown. Therefore, pretreatment is necessary to make biomass more accessible to microbial digestion. Pretreatment methods can include physical, chemical, or biological processes. The aim is to break down the complex lignin and hemicellulose structures, making cellulose accessible for succeeding enzymatic hydrolysis.

Enzymatic hydrolysis procedure: After pretreatment, the LCB is subjected to enzymatic hydrolysis. Enzymes, such as cellulases and hemicellulases, are added to the biomass to break down the cellulose and hemicellulose into simpler sugars, such as glucose and xylose.

Anaerobic digestion: The hydrolyzed LCB is then made to undergo anaerobic digestion. This microbial process occurs in anaerobic processes. The biomass is placed in an anaerobic digester, a sealed vessel containing a consortium of microorganisms called methanogens. These microorganisms break down the sugars into biogas, primarily methane (CH_4) and carbon dioxide (CO_2).

Biogas upgrading: The biogas produced during anaerobic digestion contains impurities, including CO_2, hydrogen sulfide (H_2S), and trace contaminants. In order to transform the biogas into biomethane, which can be injected into the natural gas grid or utilized as fuel for vehicles, biogas upgrading is performed. Common upgrading techniques encompass water scrubbing, amine scrubbing, pressure swing adsorption (PSA), and membrane separation. These techniques remove impurities and increase the methane content in the biogas.

Gas compression and purification: The biomethane is then compressed to the desired pressure for transportation or injection into a natural gas network. Further purification may be carried out to meet the specific quality necessities of the gas grid.

The overall procedure followed to produce biomethane from LCB is known as anaerobic digestion or biomethanation. It is a renewable and sustainable process that produces biomethane and yields a nutrient-rich digestate that can be used as an organic fertilizer. Biomethane can be utilized in combined heat and power (CHP) systems or natural gas-fired power plants to create electricity and heat. It can also be utilized as a fuel source in various industrial processes, including boilers, furnaces, kilns, and cogeneration systems. Industries such as food and beverage, pulp and paper, chemical, and pharmaceutical sectors can benefit from the use of biomethane, reducing their carbon footprint and dependence on fossil fuels.

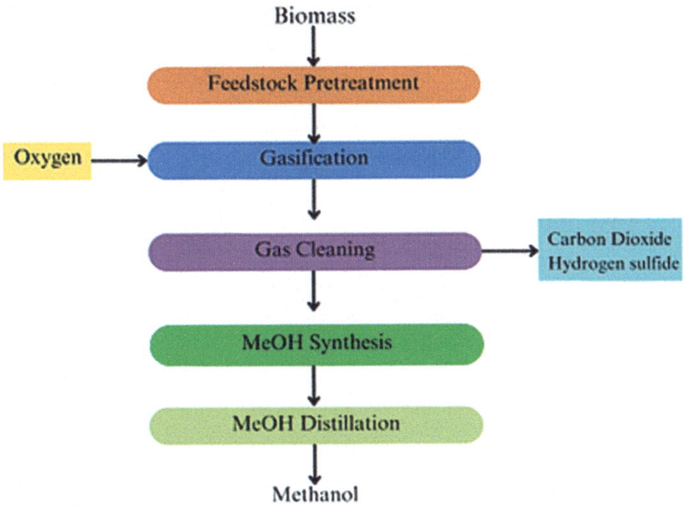

Fig. 3.7 Biomethanol production

3.6.6 Lignocellulose-Derived High-Value Bioproducts

LCB acts as a valuable resource for the production of various high-value bioproducts [4]. Here are some examples of high-value bioproducts that can be derived from LCB (Fig. 3.8):

Bio-based chemicals: LCB can be modified into a range of bio-based chemicals through biochemical and thermochemical processes. For example, furan derivatives like furfural and levulinic acid can be produced from hemicellulose, which has applications in the production of resins, plastics, and solvents. Other chemicals that can be derived include organic acids, platform chemicals, and specialty chemicals.

Bio-based polymers: LCB can be utilized as raw materials for manufacturing bio-based polymers. Polymers like cellulose acetate, cellulose esters, and lignin-based polymers can be obtained from cellulose and lignin components of biomass. These bio-based polymers have applications in packaging, textiles, coatings, and biomedical materials, offering more sustainable alternatives to fossil fuel-based polymers.

Bio-based fuels and energy: Besides biofuels like bioethanol and biobutanol, LCB can be used to produce advanced biofuels such as bio-jet fuel and biohydrogen. These biofuels offer cleaner energy options, reducing greenhouse gas emissions and dependence on fossil fuels. Additionally, LCB can be used for biomass gasification or pyrolysis to produce syngas, bio-oil, or biochar, which can be utilized for the creation of heat and electricity.

Specialty chemicals and natural products: LCB acts as a resource for the production of various specialty chemicals and natural products. For example, lignin, a complex polymer in biomass, can be chemically modified to produce high-value chemicals like vanillin (a flavoring agent), phenolic compounds (used in adhesives

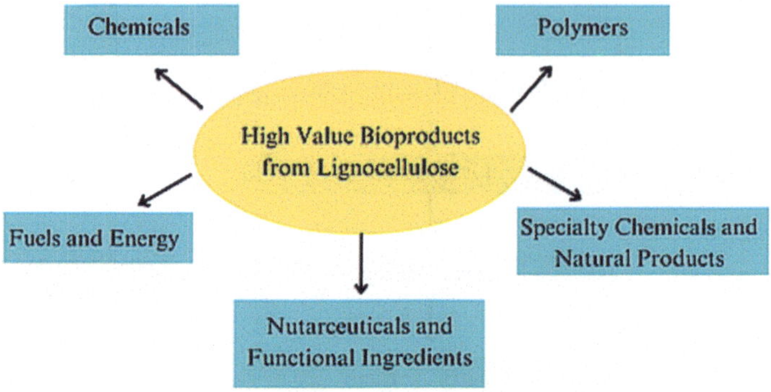

Fig. 3.8 Various bioproducts obtained from lignocellulose

and antioxidants), and aromatic compounds for fragrances. Extractives in biomass, such as essential oils and plant-based extracts, can also be used in cosmetics, personal care products, and flavors.

Nutraceuticals and functional ingredients: LCB can provide a source of bioactive compounds and functional ingredients in the nutraceutical and pharmaceutical industries. For instance, LCB can be processed to obtain dietary fibers, antioxidants, prebiotics, and other bioactive compounds that offer health benefits. These ingredients can be used in functional foods, dietary supplements, and pharmaceutical formulations.

3.7 LCB Conversion: Impact on Techno-Economic Viability

The technological-economic viability of manufacturing bioenergy from substrates made from lignocellulosic materials is the foremost important element determining how viable they are for business. The cost of biomass pretreatment is anticipated to be between 19 and 22% of the total cost of a biofuel manufacturing chain. Several authors have created techno-economic analyses of various pretreatments under the notion of a biorefinery to clarify the profitable circumstances. The dwindling supply of petroleum and gas and the adverse environmental implications of swift urbanization and technological progression are two substantial challenges. Because renewable energy sources can be purchased easily, it is one of the biggest energy sources, encompassing over 70% of the energy used worldwide. Nevertheless, rising electricity production and utilization coincide with elevated GHG emissions and particle pollution levels. Employing thermochemical synthesis using technological advances, the LCB feed can potentially be effectively altered into substantial-quality outcomes such as bioenergy and biofuel [8]. Damartzis and Zabaniotou analyzed the extraction and production of biofuels of the subsequent generation concerning biomass leveraging thermal energy and indicated constraints as well as prospects in process consolidation and execution [9].

The most important life cycle assessment is a widely enacted approach to evaluate the prospective ecological effects of specific goods by integrating quantitative, qualitative, confirmable, and measurable statistics. The life cycle assessment (LCA) inquiry has been split into four sections. The initial stage identifies objectives and dimensions, shortcomings, context, and repercussions for the environment. The following step is to undertake a life cycle supply, encompassing input materials and energy, outcomes including products and byproducts, and release pollutants such as wastewater, gas, and any harmful particle emissions. The next stage, also depicted as life cycle inventory estimation, is an administrative approach that quantifies the possible consequences of different aspects of an outcome. The culminating phase, elucidation, is assessing the product-associated data, leading to conclusions and suggestions for reducing environmental impact. The variety of lignocellulosic substances being used, conversion technique effectivity, and maximum efficiency must be explored to maximize the product yield and profit.

Several factors contribute to uncertainty in the case of using lignocellulosic feedstock, including various viable designs, a lack of understanding about how the feedstock market will behave in the long run in the situation when there is a highly increased demand, ignorance of ethanol market prices, particularly the answer to the big query, regarding how ethanol price and gasoline price interact, when ethanol blend ratio in transportation fuel is high; and most importantly investors' readiness to commit to the manufacturing of subsequent generation bioethanol. Due to a lack of access to trustworthy data and economic evaluations based on numerous process assumptions, techno-economic analyses of the manufacture of cellulosic ethanol have encountered challenges. Cellulosic bioenergy generation facilities could be able to help with these problems. Nevertheless, before the state-of-the-art is broadly disseminated to the community, it can take some time as industrial data and experience are proprietary and confidential. In an effort to reduce the cost of biofuel production and hence investigate the minimum selling price (MESP), numerous studies about the different bioenergy production have been conducted. Likewise, numerous biorefinery facility approaches seek supplementary technological breakthroughs, financing, promotional activities, and lignocellulosic material valorization, which turns off their respective commercial and economic viability positions. Consequently, a comprehensive setup for the reputable and statistical assessment of ecological sustainability must be constructed [76] (Table 3.1).

Table 3.1 Comparison between biobutanol, bioethanol, biodiesel, biomethane

Comparison factor	Biobutanol	Bioethanol	Biodiesel	Biomethane	Citations
Chemical formula	$C_4H_{10}O$	C_2H_5OH	Various mono-alkyl esters	CH_4	[77]
Energy content	Higher energy content compared to ethanol	Lower energy content compared to gasoline	Comparable energy content to diesel fuel	High energy content, similar to natural gas	[77]
Octane rating	Higher octane rating, similar to gasoline	A lower octane rating may require blending with gasoline for higher performance	A lower cetane rating may require blending with petroleum diesel for better combustion	No octane rating, not relevant for gaseous fuel	[78]
Vapor pressure	Lower vapor pressure, less volatile	Higher vapor pressure, more volatile	Varies depending on feedstock and blend	Varies depending on composition and blend	[79]
Density	Higher density	Lower density	Higher density	Lower density	[77]
Combustion characteristics	Combusts like gasoline	Combusts like gasoline	Combusts in a manner similar to diesel fuel	Combusts like natural gas	[78]
Water content	Low water content	Higher water content requires dehydration before use	Low water content	Low water content	[80]
Raw materials required	LCB (crop residues, wood chips, etc.)	Sugar crops (corn, sugarcane, sugar beets)	Vegetable oils (soybean, palm, rapeseed, etc.)	Organic waste materials (agricultural waste, food waste, etc.)	[72–74]

(continued)

Table 3.1 (continued)

Comparison factor	Biobutanol	Bioethanol	Biodiesel	Biomethane	Citations
Production process	1. ***Pretreatment***: Breakdown of LCB into fermentable sugars 2. ***Enzymatic Hydrolysis***: Conversion of complex sugars into simple sugars 3. ***Fermentation***: Conversion of sugars into butanol through microbial fermentation	1. ***Pretreatment***: Extraction of sugars from sugar crops 2. ***Fermentation***: Conversion of sugars into ethanol through microbial fermentation	1. ***Transesterification***: Chemical reaction between vegetable oils and alcohol (usually methanol) to produce biodiesel and glycerol 2. ***Separation and Purification***: Removal of impurities from biodiesel 3. ***Glycerol Purification***: Refining and purification of glycerol byproduct	1. ***Anaerobic Digestion***: Microbial decomposition of organic waste materials in an oxygen-free environment 2. ***Purification***: Removal of impurities such as moisture and hydrogen sulfide 3. ***Upgrading***: Remove carbon dioxide and other trace elements to increase methane concentration	[81–83]

(continued)

Table 3.1 (continued)

Comparison factor	Biobutanol	Bioethanol	Biodiesel	Biomethane	Citations
Typical yield	8.4–12 g/L (Under the influence of different microorganisms. Differ concerning various substrates and sugar content)	The yearly yield of LCB waste worldwide was around 220 billion tons. The following shows the yearly output of bioethanol using non-wood LCB (million tons): The following straws are used to produce bioethanol: sugarcane bagasse (1044.8), rice straw (657.5), wheat straw (472.2), corn straw (376.4), sorghum straw (12.0), and oat straw (10.1). Of these, sugarcane bagasse appears to be an appealing entity	The production volume of biodiesel fuel in India was 185 million liters in 2022	India has 125 billion cubic meters of yearly trash that might be converted into energy by producing biomethane. This can produce 748.59 TWh of electric power and 4.49 EJ of heat energy	[84, 85]
Compatibility	It can be blended with gasoline in various proportions	It can be blended with gasoline in various proportions	It can be used in diesel engines with a few to no modifications	It can be used in natural gas vehicles (NGVs) or injected into the natural gas grid	[86]
Fuel storage	Less volatile and less prone to evaporation compared to bioethanol	More volatile and requires careful handling to minimize evaporation	Similar storage characteristics to petroleum diesel	The gaseous form allows for easier storage and distribution	[87]
Fuel efficiency	Higher energy density can lead to improved fuel efficiency compared to bioethanol	Lower energy density may result in decreased fuel efficiency compared to gasoline	Similar energy density to petroleum diesel, maintaining fuel efficiency	High energy content enables efficient combustion	[87]

(continued)

Table 3.1 (continued)

Comparison factor	Biobutanol	Bioethanol	Biodiesel	Biomethane	Citations
Co-product utilization	Byproduct glycerol can be used in various industrial applications	Byproduct distillers' grains can be used as animal feed or in other applications	Glycerol byproduct can be used in various applications	Digestate byproduct can be used as a fertilizer or soil amendment	[88]
Greenhouse gas emissions	Can contribute to less GHG emission in comparison to fossil fuels	Can contribute to less GHG emission in comparison to fossil fuels	Can contribute to less GHG emission in comparison to fossil fuels	Can contribute to less GHG emission in comparison to fossil fuels	[86, 89]
Applications	Fuel additive, gasoline blend	Fuel additive, gasoline blend	Diesel fuel substitute, heating oil substitute	Electricity generation, heating, transportation fuel	[90, 91]
Advantages	Higher energy density, lower vapor pressure, compatibility with existing infrastructure	Renewable, reduced GHG emissions	Renewable, reduced GHG emissions	Utilizes organic waste, reduces GHG emissions	[86, 89]
Challenges	The complex production process, cost competitiveness, scalability	Competition with food production, limited feedstock availability	Competitiveness with petroleum diesel, feedstock availability	Feedstock availability, purification process, infrastructure compatibility	[92, 93]

3.8 Challenges and Future Directions in Utilization of Lignocellulose-Derived Feedstock

LCB means plant-derived biomass. Due to its abundant availability and potential as a renewable resource, it is a promising feedstock for generating biofuels, biochemicals, and bioproducts. Nevertheless, several challenges are linked with the usage of LCB, and ongoing research is focused on addressing these challenges and exploring future directions. Here are some key challenges and future directions in LCB [94].

3.8.1 Valorization of Lignin

Lignin, a complex aromatic polymer, is a major constituent of LCB. Currently, lignin is underutilized and often treated as a low-value byproduct. Future research aims to develop efficient lignin extraction and valorization processes, such as its conversion into high-value chemicals, materials, and energy products. Finding innovative applications for lignin can significantly enhance the overall economic viability of lignocellulosic biorefineries. Lignin-infused polyurethanes develop cost-effective composites and foams favorable to the environment, advancing efforts in drainage, building, and soil stabilization.

3.8.2 Sustainability and Environmental Impact

As LCB is obtained from agricultural residues, energy crops, and forestry residues, it is crucial to ensure sustainable sourcing practices and minimize the environmental impact of biomass production and conversion. Future directions include optimizing land use, minimizing water and energy requirements, reducing greenhouse gas emissions, and implementing integrated biorefinery concepts to maximize resource efficiency.

Techno-economic feasibility: For LCB to become commercially viable, it is essential to achieve cost competitiveness with fossil-based alternatives. Future efforts focus on reducing the overall production costs through process optimization, scale-up, and developing advanced biorefinery technologies. Improved efficiency and cost reduction in each step of the biomass conversion process are keys to achieving economic feasibility.

3.8.3 Integration of Biorefineries

Integrating biorefineries involves manufacturing multiple valuable products from LCB by efficiently utilizing various components and byproducts. Future directions emphasize the development of integrated biorefinery platforms that can simultaneously produce biofuels, biochemicals, materials, and energy, maximizing the overall value derived from LCB. Also, genetic engineering and breeding techniques can be employed to enhance the properties of lignocellulosic feedstocks, making them more amenable to conversion processes. Future research aims to improve biomass feedstocks' yield, quality, and sustainability [95].

3.9 Conclusions

The findings outline several phases for processing lignocellulosic substrates into feasible biofuels, including biobutanol, biomethane, PHB esters, etc. All the hydrolysis, thermal, and catalytic routes will be crucial in potential biorefineries, where it will ultimately be necessary to combine these methods to refine the entire biomass to have the best possible yield. Resembling a petroleum refinery, a profitable biorefinery may be formed by unifying a number of extraction processes. The economically feasible biorefinery will be able to derive a variety of different substances as well as biofuels in this manner. The overarching objective of LC biomass, the most abundant feedstock, intends to yield a substantial quantity of liquefied fuels to meet the world's energy requirements and produce high-value byproducts, utilizing an assortment of technological innovations. With the amalgamation of plenty of steps into an all-encompassing affordable, along with convenient interpretation, consolidated bioprocessing illustrated tremendous possibilities in lignocellulosic feedstock biorefinery. Scientists must concentrate on enhancing and streamlining catalytic transformation procedures to ensure a firm basis for producing high-priced bi-products such as HMF in tandem with ethanol. Regarding enforceable massive-scale industries, extensive study has been focused on breakthroughs and convergence, the implementation of possible sources, viable environment-friendly approaches, and the formulation of techno-economic and financial frameworks to produce different byproducts and lignocellulose-derived biofuels. Compared to the transesterification process for biodiesel and fermentation for bioalcohol, more sophisticated manufacturing methods such as biomass to liquid (BTL) and Fischer–Tropsch synthesis technology for advanced-generation fuels are emerging at an alarming rate. Furthermore, some properties have been significantly enhanced, which are oxidation durability and thermal efficiency. It is possible to improve the production rates of second or third-generation biofuels to a minimum of two to three times that of first-generation biofuel. Specifically, advanced-generation biofuels offer a far greater potential to replace fossil fuels in transportation than first-generation biofuels. An adequate combination of several feedstock sources is advised to formulate a competitive biofuel and strike a reasonable balance among high fuel quality and cheap manufacturing prices. Improved fuel economy and a higher chance of survival in the global fuel market have been demonstrated by biofuels with higher life cycle energy efficiency (LCEE) values. In order to foster the commercial potential of biofuel production process byproducts like glycerol, more sophisticated conversion and purification methods are also evolving quickly. From the extensive studies, it is inferred that supercritical alcohol technology and microwave radiation are fairly capable of speeding up the rates at which feedstocks from biomass undergo conversion into fuel. Every phase of producing different biofuels has certain drawbacks, either due to inhibitory synthesis or whether the procedure is financially or insufficiently competitive. As a result, constraints need to be resolved through optimized performance, and various proceedings ought to be worked together in order to develop an overall profitable biofuel manufacturing framework. For either the emergence of inhibitors or the fact that it is not viable in the marketplace, almost every phase of the extraction process has some negative aspects. Accordingly, optimization ought to be employed to fix these issues, and these processes need to be incorporated to establish a system that is typically efficient in yielding biofuels.

References

1. Yadav, A., Sharma, V., & Tsai, M. L. (2023). Development of lignocellulosic biorefineries for the sustainable production of biofuels: Towards circular bioeconomy. *Bioresource Technology, 381*, 129145. ISSN 0960–8524.
2. Adewuyi, A. (2022). Underutilized lignocellulosic waste as sources of feedstock for biofuel production in developing countries. *Frontiers in Energy Research, 10*, 741570.
3. Kumar, A., Makishah, N. H. A., Wen, Z., Gupta, G., Pandit, S., & Prasad, R. (2022). Recent developments in lignocellulosic biofuels, a renewable source of bioenergy. *Fermentation, 8*(4), 161.
4. Isikgor, F. H., & Becer, C. R. (2015). Lignocellulosic biomass: A sustainable platform for the production of bio-based chemicals and polymers. *Polymer Chemistry, 6*(25), 4497–4559.
5. Gassner, M., & Maréchal, F. (2013). Increasing efficiency of fuel ethanol production from lignocellulosic biomass by process integration. Energy & Fuels
6. Irmak, S. (2019). Challenges of biomass utilization for biofuels', biomass for bioenergy—recent trends and future challenges. *IntechOpen.*
7. Nanda, S., Mohammad, J., Reddy, S. N., Kozinski, J. A., & Dalai, A. K. (2013). Pathways of lignocellulosic biomass conversion to renewable fuels. *Biomass Conversion and Biorefinery, 4*(2), 157–191 (2013).
8. Inyang, V., Laseinde, O. T., & Kanakana, G. M. (2022). Techniques and applications of lignocellulose biomass sources as transport fuels and other bioproducts. *International Journal of Low-carbon Technologies, 17*, 900–909.
9. Yousuf, A., Pirozzi, D., & Sannino, F. (2020). Fundamentals of lignocellulosic biomass. In *Lignocellulosic biomass to liquid biofuels* (pp. 1–15). Academic Press.
10. Ghaemi, F., Abdullah, L. C., & Ariffin, H. (2019). Lignocellulose structure and the effect on nanocellulose production. In *Lignocellulose for future bioeconomy* (pp. 17–30). Elsevier.
11. Thota, S. P., Bag, P. P., Vadlani, P. V., & Belliraj, S. K. (2022). Plant biomass derived multidimensional nanostructured materials: a green alternative for energy storage. *Engineered Science*

12. Adewuyi, A. (2022). Underutilized lignocellulosic waste as sources of feedstock for biofuel production in developing countries. *Frontiers in Energy Research, 10*
13. Wu, W., Li, P., Huang, L., Wei, Y., Li, J., Zhang, L., & Jin, Y. (2023). The role of lignin structure on cellulase adsorption and enzymatic hydrolysis. *Biomass, 3*(1), 96–107.
14. Cai, J., He, Y., Yu, X., Banks, S. A., Yang, Y., Zhang, X., Yu, Y., Liu, R., & Bridgwater, A. V. (2017). Review of physicochemical properties and analytical characterization of lignocellulosic biomass. *Aston University, 76*, 309–322.
15. Mankar, A. R., Pandey, A., Modak, A., & Pant, K. K. (2021). Pretreatment of lignocellulosic biomass: A review on recent advances. *Bioresource Technology, 334*, 125235.
16. Baruah, J., Nath, B. C., Sharma, R., Kumar, S., Deka, R. C., Baruah, D. C., & Kalita, E. (2018). Recent trends in the pretreatment of lignocellulosic biomass for Value-Added products. *Frontiers in Energy Research, 6.*
17. Qin, Z., Iqbal, I., Riaz, F., Karadağ, A., & Tabatabaei, M. (2019). Different pretreatment methods of lignocellulosic biomass for use in biofuel production. In *IntechOpen eBooks.*
18. Hassan, S. S., Williams, G. A., & Jaiswal, A. K. (2018). Emerging technologies for the pretreatment of lignocellulosic biomass. *Bioresource Technology, 262*, 310–318.
19. Wahid, R., Hjorth, M., Kristensen, S., & Møller, H. (2015). Extrusion as pretreatment for boosting methane production: Effect of screw configurations. *Energy & Fuels, 29*(7), 4030–4037.
20. Wei, H., YingTing, Y., JingJing, G., Wenshi, Y., & Jun-Hong, T. (2017). Lignocellulosic biomass valorization: Production of ethanol. In *Elsevier eBooks* (pp. 601–604)
21. Hernández-Beltrán, J. U., Lira, I. O. H., Cruz-Santos, M. M., Saucedo-Luevanos, A., Hernández-Terán, F., & Balagurusamy, N. (2019). Insight into pretreatment methods of lignocellulosic biomass to increase biogas yield: Current state, challenges, and opportunities. *Applied Sciences, 9*(18), 3721.
22. Jin, M., Slininger, P. J., Dien, B. S., Waghmode, S. B., Moser, B. R., Orjuela, A., Da Costa Sousa, L., & Balan, V. (2015). Microbial lipid-based lignocellulosic biorefinery: Feasibility and challenges. *Trends in Biotechnology, 33*(1), 43–54.
23. Sharma, H. K., Xu, C., & Qin, W. (2019). Biological Pretreatment of lignocellulosic biomass for biofuels and bioproducts: An overview. *Waste Biomass Valor, 10*, 235–251.
24. Rezania, S., Oryani, B., Cho, J., Talaiekhozani, A., Sabbagh, F., Hashemi, B., Rupani, P. F., & Mohammadi, A. A. (2020). Different pretreatment technologies of lignocellulosic biomass for bioethanol production: An overview. *Energy, 199*, 117457
25. Tan, J., Wu, H., Li, H., & Yang, S. (2022). Chapter 7—Pretreatment methods for converting straws into fermentable sugars. In H. Li, S. Saravanamurugan, A. Pandey, S. Elumalai (Eds.), Biomass, biofuels, biochemical (pp. 117–162). Elsevier. ISBN 9780128244197
26. Baksi, S., Saha, D., Saha, S., et al. (2023). Pre-treatment of lignocellulosic biomass: Review of various physico-chemical and biological methods influencing the extent of biomass depolymerization. *International Journal of Environmental Science and Technology, 20*, 13895–13922.
27. Tran, T. H., Le, T. P. Q., Phong, M. T., & Nguyen, D. H. (2020). Bioethanol production from lignocellulosic biomass. In *IntechOpen eBooks.*
28. Kassaye, S., Pant, K. K., & Jain, S. (2017). Hydrolysis of cellulosic bamboo biomass into reducing sugars via a combined alkaline solution and ionic liquid pretreament steps. *Renewable Energy, 104*, 177–184.
29. Abbott, A. P., Capper, G., Davies, D. B., Rasheed, R. K., & Tambyrajah, V. (2002). Novel solvent properties of choline chloride/urea mixtures. *Chemical Communications, 1*, 70–71. (Electronic supplementary information (ESI) available: Spectroscopic data). See http://www.rsc.org/suppdata/cc/b2/b210714g/
30. Sun, Y., & Cheng, J. (2002). Hydrolysis of lignocellulosic materials for ethanol production: A review. *Bioresource Technology, 83*(1), 1–11.
31. Wang, T., Liu, X., Yu, Q., Zhang, X., Qu, Y., Gao, P., & Wang, T. (2005). Directed evolution for engineering pH profile of endoglucanase III from Trichoderma reesei. *Biomolecular Engineering, 22*(1–3), 89–94.

32. Stanley, D., Fraser, S., Chambers, P., Rogers, P., & Stanley, G. (2009). Generation and characterisation of stable ethanol-tolerant mutants of Saccharomyces cerevisiae. *Journal of Industrial Microbiology & Biotechnology, 37*(2), 139–149.
33. Wati, L. (1996). Characterisation of genetic control of thermotolerance in mutants of Saccharomyces cerevisiae. Ulster University. https://pure.ulster.ac.uk/en/publications/characterisation-of-genetic-control-of-thermotolerance-in-mutants-5
34. Akashdeep Dey and R. Camilla Thomson. (2022). India's biomethane generation potential from wastes and the corresponding greenhouse gas emissions abatement possibilities under three end use scenarios: Electricity generation, cooking, and road transport applications. *Royal Society of Chemistry.*
35. Sánchez, Ó. J., & Cardona, C. A. (2008). Trends in biotechnological production of fuel ethanol from different feedstocks. *Bioresource Technology, 99*(13), 5270–5295.
36. Kumar, S., Dheeran, P., Singh, S., Mishra, I. M., & Adhikari, D. K. (2015). Continuous ethanol production from sugarcane bagasse hydrolysate at high temperature with cell recycle and in-situ recovery of ethanol. *Chemical Engineering Science, 138*, 524–530.
37. Olsson, L., & Hahn-Hägerdal, B. (1996). Fermentation of lignocellulosic hydrolysates for ethanol production. *Enzyme and Microbial Technology, 18*(5), 312–331.
38. Suriyachai, N. (2017). Organosolv-based fractionation and conversion of bagasse to sugars and phenolics.
39. Wang, Y., Guo, W., Cheng, C. L., Ho, S., Chang, J. S., & Ren, N. (2016). Enhancing bio-butanol production from biomass of Chlorella vulgaris JSC-6 with sequential alkali pretreatment and acid hydrolysis. *Bioresource Technology, 200*, 557–564.
40. Wen, Z., Wu, M., Lin, Y., Yang, L., Lin, J., & Cen, P. (2014). Artificial symbiosis for acetone-butanol-ethanol (ABE) fermentation from alkali extracted deshelled corn cobs by co-culture of Clostridium beijerinckii and Clostridium cellulovorans. *Microbial Cell Factories, 13*(1).
41. Gottumukkala, L. D., Binod, P., Valappil, S. K., Mathiyazhakan, K., Pandey, A., & Sukumaran, R. K. (2013). Biobutanol production from rice straw by a non acetone producing Clostridium sporogenes BE01. *Bioresource Technology, 145*, 182–187.
42. Murphy, J. D., Drielak, E., Varma, V., Muthusamy, S., & Aslam, M. (2017). Updates on the pretreatment of lignocellulosic feedstocks for bioenergy production—A review. *Biomass Conversion and Biorefinery, 8*(2), 471–483.
43. Kim, J. N., Lee, H., Lee, S., Jae, J., & Park, Y. (2019). Overview of the recent advances in lignocellulose liquefaction for producing biofuels, bio-based materials and chemicals. *Bioresource Technology, 279*, 373–384.
44. Pérez, J. M., Muñoz-Dorado, J., De La Rubia, T., & Martínez, J. A. (2002). Biodegradation and biological treatments of cellulose, hemicellulose and lignin: An overview. *International Microbiology, 5*(2), 53–63.
45. Xin, F., Wu, Y., & He, J. (2014). Simultaneous fermentation of glucose and xylose to butanol by clostridium sp. strain BOH3. *Applied and Environmental Microbiology, 80*(15), 4771–4778.
46. Taherzadeh, M. J., & Taherzadeh, M. J. (2016). A critical review of analytical methods in pretreatment of lignocelluloses: Composition, imaging, and crystallinity. *Bioresource Technology, 200*, 1008–1018.
47. Amiri, H., & Taherzadeh, M. J. (2018). Pretreatment and hydrolysis of lignocellulosic wastes for butanol production: Challenges and perspectives. *Bioresource Technology, 270*, 702–721.
48. Chandel, A. K., Garlapati, V. K., Singh, A. K., Antunes, F. A. F., & Da Silva, S. S. (2018). The path forward for lignocellulose biorefineries: Bottlenecks, solutions, and perspective on commercialization. *Bioresource Technology, 264*, 370–381.
49. Liu, W., Chen, W., Hou, Q., Wang, S., & Liu, F. (2018). Effects of combined pretreatment of dilute acid pre-extraction and chemical-assisted mechanical refining on enzymatic hydrolysis of lignocellulosic biomass. *RSC Advances, 8*(19), 10207–10214.
50. Kumar, P., Barrett, D. M., Delwiche, M. J., & Stroeve, P. (2009). Methods for pretreatment of lignocellulosic biomass for efficient hydrolysis and biofuel production. *Industrial & Engineering Chemistry Research, 48*(8), 3713–3729.

51. De Albuquerque Wanderley, M. C., Martin, C., De Moraes Rocha, G. J., & Gouveia, E. R. (2013). Increase in ethanol production from sugarcane bagasse based on combined pretreatments and fed-batch enzymatic hydrolysis. *Bioresource Technology, 128*, 448–453.
52. Tsai, T. Y., Lo, Y. C., Dong, C. D., Nagarajan, D., Chang, J. S., & Lee, D. J. (2020). Biobutanol production from lignocellulosic biomass using immobilized Clostridium acetobutylicum. *Applied Energy, 277*, 115531. ISSN 0306-2619
53. Ageitos, J. M., Vallejo, J. A., Veiga-Crespo, P., & Villa, T. G. (2011). Oily yeasts as oleaginous cell factories. *Applied Microbiology and Biotechnology, 90*(4), 1219–1227.
54. Pandey, A., Soccol, C. R., & Mitchell, D. L. (2000). New developments in solid state fermentation: I-bioprocesses and products. *Process Biochemistry, 35*(10), 1153–1169.
55. Leung, D. Y., Wu, X., & Leung, M. (2010). A review on biodiesel production using catalyzed transesterification. *Applied Energy, 87*(4), 1083–1095.
56. Sawangkeaw, R., & Ngamprasertsith, S. (2013). A review of lipid-based biomasses as feedstocks for biofuels production. *Renewable & Sustainable Energy Reviews, 25*, 97–108.
57. Kim, S., & Dale, B. E. (2004). Global potential bioethanol production from wasted crops and crop residues. *Biomass & Bioenergy, 26*(4), 361–375.
58. Mosier, N. S., Wyman, C. E., Dale, B. E., Elander, R. T., Lee, Y., Holtzapple, M. T., & Ladisch, M. R. (2005). Features of promising technologies for pretreatment of lignocellulosic biomass. *Bioresource Technology, 96*(6), 673–686.
59. Hendriks, A. T. W. M., & Zeeman, G. (2009). Pretreatments to enhance the digestibility of lignocellulosic biomass. *Bioresource Technology, 100*(1), 10–18.
60. Ericsson, K., Rosenqvist, H., & Nilsson, L. (2009). Energy crop production costs in the EU. *Biomass & Bioenergy, 33*(11), 1577–1586.
61. Hossain, A., & Badr, O. (2007). Prospects of renewable energy utilisation for electricity generation in Bangladesh. *Renewable & Sustainable Energy Reviews, 11*(8), 1617–1649.
62. Limayem, A., & Ricke, S. C. (2012). Lignocellulosic biomass for bioethanol production: Current perspectives, potential issues and future prospects. *Progress in Energy and Combustion Science, 38*(4), 449–467.
63. Kadam, K. L., & McMillan, J. R. (2003). Availability of corn stover as a sustainable feedstock for bioethanol production. *Bioresource Technology, 88*(1), 17–25.
64. Kim, K. H., & Hong, J. (2001). Supercritical CO_2 pretreatment of lignocellulose enhances enzymatic cellulose hydrolysis. *Bioresource Technology, 77*(2), 139–144.
65. Influence of the size reduction of organic waste on their anaerobic digestion (2000). PubMed.
66. Balat, M., Balat, H., & Öz, C. (2008). Progress in bioethanol processing. *Progress in Energy and Combustion Science, 34*(5), 551–573.
67. Ariunbaatar, J., Panico, A., Esposito, G., Pirozzi, F., & Lens, P. N. (2014). Pretreatment methods to enhance anaerobic digestion of organic solid waste. *Applied Energy, 123*, 143–156.
68. Garrote, G., Domínguez, H., & Parajó, J. C. (1999). Hydrothermal processing of lignocellulosic materials. *European Journal of Wood and Wood Products, 57*(3), 191–202.
69. Gossett, J. M., Stuckey, D. C., Owen, W. F., & McCarty, P. L. (1982). Heat treatment and anaerobic digestion of refuse. *Journal of the Environmental Engineering Division, 108*(3), 437–454.
70. Behera, S., Arora, R., Nandhagopal, N., & Kumar, S. (2014). Importance of chemical pretreatment for bioconversion of lignocellulosic biomass. *Sciencedirect, 36*, 91–106.
71. Samir, A., Ashour, F. H., Hakim, A. A., & Bassyouni, M. (2022). Recent advances in biodegradable polymers for sustainable applications. *Npj Mater Degrad 6*, 68
72. Zhang, Y., Wang, H., Sun, X., Wang, Y., & Liu, Z. (2021). Separation and characterization of biomass components (cellulose, hemicellulose, and lignin) from corn stalk. *BioResources, 16*(4), 7205–7219.
73. Rocha-Meneses, L., Ivanova, A., Atouguia, G., Ávila, I., Raud, M., Orupõld, K., & Kikas, T. (2019). The effect of flue gas explosive decompression pretreatment on methane recovery from bioethanol production waste. *Industrial Crops and Products, 127*, 66–72.
74. Dao, C. N., Mupondwa, E., Tabil, L., Li, X., & Castellanos, E. C. (2018). A review on techno-economic analysis and life-cycle assessment of second generation bioethanol production via biochemical processes. *The Canadian Society for Bioengineering*, 22–25.

75. Zhang, Q., Dong, J., Liu, Y., Wang, Y., & Cao, Y. (2016). Towards a green bulk-scale biobutanol from bioethanol upgrading. *Journal of Energy Chemistry, 25*(6), 907–910.
76. Tagami-Kanada, N., Yoshikuni, K., Mizuno, S., Sawai, T., Fuchihata, M., & Ida, T. (2022). Combustion characteristics of densified solid biofuel with different aspect ratios. *Renewable Energy, 197*, 1174–1182.
77. Elkatory, M. R., Hassaan, M. A., & Nemr, A. E. (2022). Algal biomass for bioethanol and biobutanol production. In *Elsevier eBooks* (pp. 251–279).
78. Kulbeik, T., Scherzinger, M., Höfer, I., & Kaltschmitt, M. (2021). Autoclave pre-treatment of foliage—Effects of temperature, residence time and water content on solid biofuel properties. *Renewable Energy, 171*, 275–286.
79. Mujtaba, M., Fraceto, L. F., Fazeli, M., Mukherjee, S., Savassa, S. M., De Medeiros, G. A., Pereira, A. D. E. S., Mancini, S. D., Lipponen, J., & Vilaplana, F. (2023). Lignocellulosic biomass from agricultural waste to the circular economy: A review with focus on biofuels, biocomposites and bioplastics. *Journal of Cleaner Production, 402*, 136815.
80. Cao, G., & Sheng, Y. (2016). Biobutanol production from lignocellulosic biomass: Prospective and challenges. *Journal of Bioremediation and Biodegradation, 7*(4).
81. Awogbemi, O., & Von Kallon, D. V. (2023). Application of biochar derived from crops residues for biofuel production. *Fuel Communications, 15*, 100088.
82. Eloka-Eboka, A. C., Maroa, S., & Taiwo, A. E. (2023). Biobutanol from agricultural and municipal solid wastes, techno-economic, and lifecycle analysis. In *Elsevier eBooks* (pp. 171–198).
83. Basak, B., Kumar, R., Bharadwaj, A. S., Kim, T. H., Kim, J. R., Jang, M., Oh, S., Roh, H., & Jeon, B. (2023). Advances in physicochemical pretreatment strategies for lignocellulose biomass and their effectiveness in bioconversion for biofuel production. *Bioresource Technology, 369*, 128413.
84. Ma, S., Li, Y., Li, J., Yu, X., Cui, Z., Yuan, X., Zhu, W., & Wang, H. (2022). Features of single and combined technologies for lignocellulose pretreatment to enhance biomethane production. *Renewable & Sustainable Energy Reviews, 165*, 112606.
85. Dempfle, D., Kröcher, O., & Luterbacher, J. S. (2021). Techno-economic assessment of bioethanol production from lignocellulose by consortium-based consolidated bioprocessing at industrial scale. *New Biotechnology, 65*, 53–60.
86. Gurunath, R. B., Gobinath, R., & Giridhar, P. V. (2022). Bioethanol production from lignocellulosic biomass: Past, present and future trends. *Research Journal of Biotechnology 17*(10), 124–132.
87. Kumara, B., Raghavendra, C., Naik, S., Kumar, M. D., Shadambi, C. B., & Cholachagudda, N. A. (2023). Performance study of Neem/Soyabean biofuels on 4-stroke diesel engine. *Materials Today: Proceedings, 92*, 406–412.
88. Krishnan, M. G., Rajkumar, S., Thangaraja, J., & Devarajan, Y. (2023). Exploring the synergistic potential of higher alcohols and biodiesel in blended and dual fuel combustion modes in diesel engines: A comprehensive review. *Sustainable Chemistry and Pharmacy, 35*, 101180.
89. Gnansounou, E., & Raman, J. K. (2018). Environmental performances of coproducts. Application of Claiming-Based Allocation models to straw and vetiver biorefineries in an Indian context. *Bioresource Technology, 262*, 203–211.
90. Becerra-Ruiz, J. S., González-Huerta, R., Gracida, J., Amaro-Reyes, A., & Macias-Bobadilla, G. (2019). Using green-hydrogen and bioethanol fuels in internal combustion engines to reduce emissions. *International Journal of Hydrogen Energy, 44*(24), 12324–12332.
91. Mehmood, T., Nadeem, F., Meer, B., Ashraf, H., Meer, K., & Saeed, S. (2023). Biomass valorization to biobutanol. In *Elsevier eBooks* (pp. 151–178).
92. Siegrist, A., Bowman, G., & Burg, V. (2022). Energy generation potentials from agricultural residues: The influence of techno-spatial restrictions on biomethane, electricity, and heat production. *Applied Energy, 327*, 120075.
93. Lin, C. Y., & Lu, C. (2021). Development perspectives of promising lignocellulose feedstocks for production of advanced generation biofuels: A review. *Renewable and Sustainable Energy Reviews, 136*, 110445. ISSN 1364-0321

94. Zhang, R., Gao, H., Wang, Y., He, B., Lu, J., Zhu, W., Peng, L., & Wang, Y. (2023). Challenges and perspectives of green-like lignocellulose pretreatments selectable for low-cost biofuels and high-value bioproduction. *Bioresource Technology, 369*, 128315.
95. Yoo, C. G., Meng, X., Pu, Y., & Ragauskas, A. J. (2020). The critical role of lignin in lignocellulosic biomass conversion and recent pretreatment strategies: A comprehensive review. *Bioresource Technology, 301*, 122784.

Chapter 4
Industrial Organic Waste and Byproducts as Sustainable Feedstock for Bioenergy Production

Desta Getachew Gizaw, Selvakumar Periyasamy, Zinnabu Tassew Redda, and Gurunathan Baskar

Abstract Bioenergy production has developed as a sustainable solution to address the increasing energy demand while decreasing environmental pollution. The industrial sectors are increasing along with the fast population growth and urban developments. Due to various activities, a significant amount of organic waste and byproducts are generated in the industrial sectors. If improperly processed, this organic waste with high nutritional content might cause environmental contamination and speed up climate change. However, with suitable technologies and procedures, waste materials could be used to produce valuable bioenergy sources. Converting industrial organic waste and byproducts into bioenergy is a sustainable and environmentally friendly approach. Besides, it offers several eco-economic benefits, such as reducing reliance on fossil fuels, mitigating environmental pollution and reducing landfilling. This chapter explores the potential of industrial organic waste and byproducts as feedstock for bioenergy generation and its implications for sustainable energy production.

Keywords Organic waste · Byproducts · Industries · Feedstocks · Biofuels

---

D. G. Gizaw · S. Periyasamy (✉)
Department of Chemical Engineering, School of Mechanical, Chemical and Materials Engineering, Adama Science and Technology University, 1888 Adama, Ethiopia
e-mail: selvaa26kumar@gmail.com

Z. T. Redda
University of Applied Sciences (HTW) Berlin, Wilhelminenhofstraße 75A, 12459 Berlin, Germany

School of Chemical and Bio Engineering, Addis Ababa Institute of Technology, Addis Ababa University, King George VI St., P.O. Box 385, Addis Ababa, Ethiopia

G. Baskar
Department of Biotechnology, St. Joseph's College of Engineering, Chennai 600119, India

S. Periyasamy
Department of Biomaterials, Saveetha Dental College and Hospitals, SIMATS, Saveetha University, Chennai 600077, India

G. Baskar et al. (eds.), *Circular Bioeconomy Perspectives in Sustainable Bioenergy Production*, Energy, Environment, and Sustainability,
https://doi.org/10.1007/978-981-97-2523-6_4

4.1 Introduction

Due to rising population growth and changes in living standards, several industries have increased their production capacities to keep up with demand. The number of various organic wastes produced worldwide has dramatically increased as a result of this [37]. Nearly 50% of the average composition of global waste, 3 M tons per day, is organic material [23]. Significant levels of carbohydrates, lipids, proteins, inorganic compounds, and a lot of water are present in many industrial and municipal organic waste and byproduct streams, mostly from industries connected to the food industry [20, 104]. Along with the dearth of virgin resources, environmental issues related to treating and disposing organic-rich waste have long been a cause of worry [56]. For example, the juice industry produced a huge amount of organic waste as peels, the coffee industry produced coffee pulp as waste, and the cereal industry produced husks. Therefore, these organic-rich waste must be treated efficiently to convert them into sustainable bioenergy and value-added products and mitigate their environmental impact.

When industrial organic wastes are disposed of in landfills, they release methane, which is 34 times more potent than carbon dioxide in causing climate change and global warming [14]. As a result, it is essential to create effective waste-to-energy systems to improve energy recovery from the industrial organic wastes that are now disposed of in landfills. Even though more waste is being kept out of landfills as a whole, there is not much energy being extracted from it. Based on the need to reduce our carbon footprint [77, 86] and the security of our fuel supply [79], high-energy fuels such as renewable diesel [76, 107], bioethanol [25], and biogas are being developed. Energy and value-added products recovery from organic waste plays an important role in the urban waste hierarchy, making better use of waste, and meeting government targets. Researchers around the world are developing different approaches to converting organic waste into value-added products. In this regard, industrial organic waste is an interesting and valuable renewable source of biomass due to its biodegradability, water, and organic content, absence of competition with food, the fact that it is unlimited if used sustainably, and, in particular, because of its relatively high-energy density.

There are numerous technical options for generating renewable energy from industrial organic waste. Direct combustion, anaerobic digestion, transesterification, hydrothermal liquefaction, fermentation, pyrolysis, and thermal gasification are some of the technical processes that enable the production of renewable energy [9, 83]. Direct burning is the most popular method for converting trash into heat or power. In this procedure, waste-derived fuel is burned to provide energy in the presence of additional oxygen (oxygen that gathers from the air). At high temperatures and under controlled oxygen conditions, waste gasification transforms organic materials into syngas composed of CO, H_2, CO_2, N_2, CH_4, etc. [35]. According to [1, 64], pyrolysis is the thermochemical breakdown process of organic or inorganic waste at a high temperature (430 °C) without oxygen. The process by which organic waste is broken down to produce biogas and biofertilizers is known as anaerobic digestion. To degrade organic material, microbes must go through many activities without oxygen. This can be applied to managing industrial, agricultural, or domestic waste and creating clean fuels. This chapter provides a comprehensive idea of industrial organic waste and their conversion methods to produce sustainable bioenergy.

4.2 Industrial Organic Wastes as a Feedstock

Finding a sustainable solution to the world's rising energy consumption is now vital because of population growth. The utilization of industrial organic waste for energy production appears to be a solution to provide energy and lower carbon dioxide emissions, as fossil fuels are scarce and non-renewable resources. Energy crops, waste products, and other biological resources that can be utilized to produce renewable energy and other valuable products are all included in the term biomass [38, 108]. Agricultural products like corn, sugarcane, soybean oil, and sunflower are used to make first-generation biofuels [8, 106]. Because these biomasses are also used as food, the phrase "food versus fuel" has been coined to describe the dilemma. Additionally, emissions of greenhouse gases (GHG) are believed to be lesser for second-generation biofuels than first-generation fuels. For these reasons, industrial organic wastes have gained attention due to their disponibility, including residues from the crop, food, and oil industries.

Since they do not directly compete with food crops, industrial organic waste streams represent a sustainable alternative to fossil-based resources. The term "waste" refers to any organic substance that is not the primary material for which the plants were cultivated. It also includes any byproduct derived from biomass for which supply is significantly more than demand [23]. The food processing industries, such as the juice, chips, meat, confectionary, and fruit industries, produce enormous amounts of organic waste and associated effluents each year [12, 36, 100]. Different energy sources can be made from these organic wastes. Industrial organic wastes come from many industries and are as diverse as their origins as a result. For instance, peels, seeds, and other organic waste are frequently produced as waste at fruit juice factories [98]. Industry liquid byproducts like brine from pickling cucumbers and whey from producing cheese are both frequent and act as substrates for different conversions. Animal and post-harvest waste from agriculture is used to produce bioenergy [74, 126]. However, proteins, lipids, carbohydrates, and other components make organic waste a viable alternative for producing lower-cost and more valuable chemicals [90, 138]. Figure 4.1 shows different types of organic waste.

Sawdust, wood chips, and abandoned logs are examples of organic wastes from the wood processing industry that can be used as biofuel feedstocks [47, 110]. For instance, the wood waste and sawdust produced by the paper and saw industries can be used as boiler fuels and feedstock for ethanol manufacturing. The term "straw" refers to the byproducts or residues left over after harvesting food crops, such as rice, wheat, corn, beans, cotton, and sugar cane. The capability to convert corn stover, including

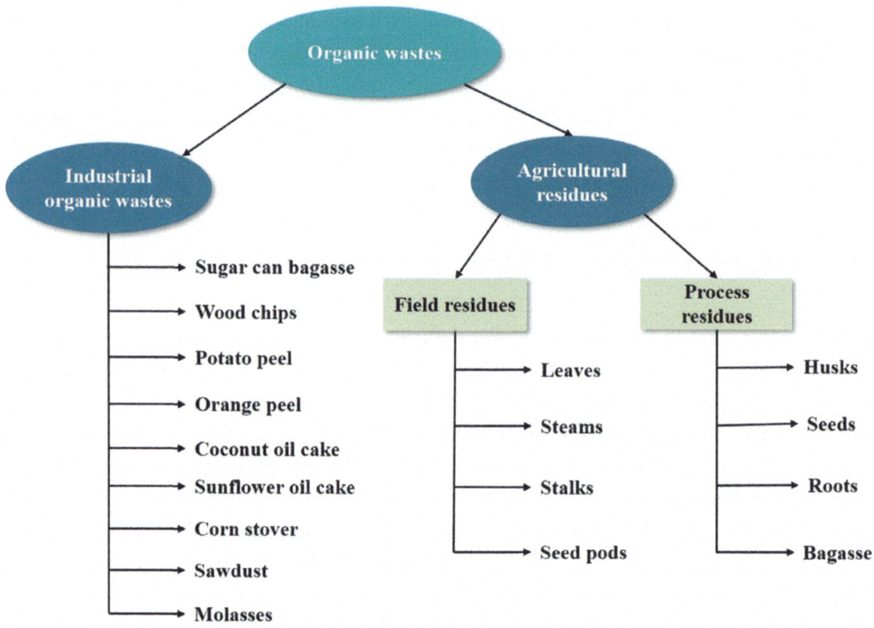

Fig. 4.1 Different types of industrial organic wastes

the stalks, cobs, and leaves, into fermentable sugars for biobutanol generation has also been noted [131]. For the economically viable usage of residual substrates, the sugarcane residues, in particular sugarcane bagasse, can be a good choice for the manufacture of bioethanol and other biofuels like biochar [15, 65].

Besides milling, malting, pearling, fermentation, heat processing, extrusion, and puffing, there are numerous other ways to process grains used in the food industry. Bran, germ, hull, husk, fiber, and other undesirable crop elements are produced in considerable quantities when grains are processed using various processing techniques [88]. Since all grains are preferred for consumption after milling, milling industries account for the majority of the wastage of cereal crops [116]. The main byproduct of grain milling is bran, which separates from the grain endosperm. It has a rough outer shell made of coarse seed with little flour and a high protein and fiber content. Another significant byproduct created during the wet milling of corn and sorghum is a germ whose main component is gluten, which is high in protein [32]. On top of that, there is a significant potential for producing sustainable bioenergy from these organic industrial wastes.

After the oil is extracted from seeds, a significant amount of processed organic residues—known as oil cakes—are created, particularly in the oil industry [125]. Due to the high concentration of fat, oil, grease, suspended solids, and dissolved solids in these residues, these organic wastes pollute the air, water, and solid waste. Canola oil cake, sunflower oil cake, coconut oil cake, sesame oil cake, mustard oil cake, palm kernel cake, soybean cake, groundnut cake, cotton seed cake, olive oil cake, and rapeseed cake are only a few examples of the various types of oil cakes [102, 115]. These agro-industrial organic wastes are studied because they are very inexpensive, contain a large number of elements, and have an endless potential to be used as substitute substrates for fermentation.

On the other hand, enormous quantities of organic waste from the meat, poultry, and fish processing industries are accumulating and contain perishable proteins that may irritate people through an offensive odor [113]. According to [99], non-food-based agro-industries produce effluents that are primarily biodegradable. Except for the textile and tannery sectors, the effluents are non-toxic. The bulk of these wastes are not used or handled, endangering the environment as well as the health of people and animals. To produce biofuel, it is necessary to use these industrial organic wastes.

4.3 Bioenergy Conversation Techniques

Due to the current state of the economy and environment, there is an increasing demand for recycling and energy conservation. Diverse technologies have been developed and tapped into to use industrial organic waste for bioenergy production. The method used to convert these organic wastes to energy entails turning waste material into a variety of fuels that can be used to generate energy. Recent years have seen an increase in the use of environmentally friendly methods to extract and transform organic waste into chemical fuels [6, 45]. The conversion of organic waste to energy is accomplished using one of two primary methods: thermochemical or biochemical conversion. Various organic waste conversion methods, products, process short descriptions, and environmental effects are presented in Table 4.1. Biochemical conversion uses microbes or enzymes to transform organic waste into usable energy, as opposed to thermochemical conversion, which uses heat to decompose the waste's organic components [73]. The thermochemical conversion methods include combustion, gasification, liquefaction, and pyrolysis. The three processes of anaerobic digestion, alcoholic fermentation, and photobiological reaction fall under the category of biochemical conversion, on the other hand [96].

4.3.1 Thermochemical Conversion

Through the use of thermochemical technology, industrial organic waste is converted into biochar (a solid), synthesis gas, and highly oxygenated bio-oil (a liquid). This process needs bond-breaking and high-temperature chemical reformation. Gasification, pyrolysis, and liquefaction are the three primary process options available within thermochemical conversion [82]. The type and quantity of organic waste, the selected energy source, end-use conditions, environmental principles, and financial considerations can all impact the chosen conversion type [5]. According to several research studies, thermal conversion technologies have attracted more attention as a

Table 4.1 Organic waste conversion methods, products, process short descriptions, and environmental effects

Conversion methods	Products	Descriptions	Environmental effects	References
Pyrolysis	Biochar, bio-oil and syngas	Thermal decomposition of biomass/organic waste at higher temperatures in the absence of oxygen	Higher impact on climate change	[31, 87]
Gasification	Biohydrogen and producer gas	Produces gas molecules directly from organic waste at higher temperature	Lower GHG emissions Lower impact on climate change	[112, 121]
Liquefaction	Bio-oil	Absence of oxygen during the thermochemical treatment of organic waste	Higher impact on climate change Higher energy consumption	[4, 66]
Anaerobic digestion (AD)	Biogas and biomethane	Technology for AD of organic waste is used in a one- or two-stage method	Lower impact on climate change Lower carbon emission and GHG	[69, 117, 122]
Transesterification	Biodiesel	Extracted oil from organic waste converted into biodiesel at optimal conditions	Lower impact on air emission compared to incineration Higher energy demand	[71, 117]
Dark fermentation (DF)	Biohydrogen	Microorganisms are used in a single-stage process to convert organic waste into biohydrogen in the absence of light and oxygen	Lower GHG emissions	[121]

result of the availability of an industrial infrastructure to provide highly developed thermochemical transformation equipment, a short processing time, and minimal water use [95]. Additionally, for production, thermochemical conversion is unaffected by ambient conditions. Therefore, it is essential to understand the various thermochemical process choices to evaluate their potential in the future.

4.3.1.1 Pyrolysis

For the production of biofuel from industrial organic waste, pyrolysis is frequently used. It is a thermal degradation process that occurs between 400 and 900 °C without the presence of oxygen [46]. When heated to a high temperature during the pyrolysis process, the polymeric components of macromolecular structures are broken down into smaller molecules and hydrocarbons, the final products. These final products include solid coke (char), pyrolysis gas (syngas), and pyrolysis oil (bio-oil or tar). This technique converts organic wastes into liquid fuel [41]. About 80% of the organic waste is made up of volatile substances, with the remaining 15% being carbon and 5% being ash. Pyrolysis is a relatively new method for turning organic waste into valuable products. Due to the inert atmosphere, pyrolysis plants emit less sulfur and nitrogen oxides than other types of energy production [134]. Additionally, it generates excellent solid residues. Figure 4.2 demonstrates different types of the pyrolysis process.

Slow pyrolysis, fast pyrolysis, flash pyrolysis, and catalytic pyrolysis are the four major types of pyrolysis. Table 4.2 shows different types of pyrolysis methods. Slow pyrolysis is the decomposition of industrial organic wastes at a slow heating rate, which results in greater char generation and less pyrolysis oil and gas production. More pyrolysis oil is made possible by quick pyrolysis [135]. The heating rate is

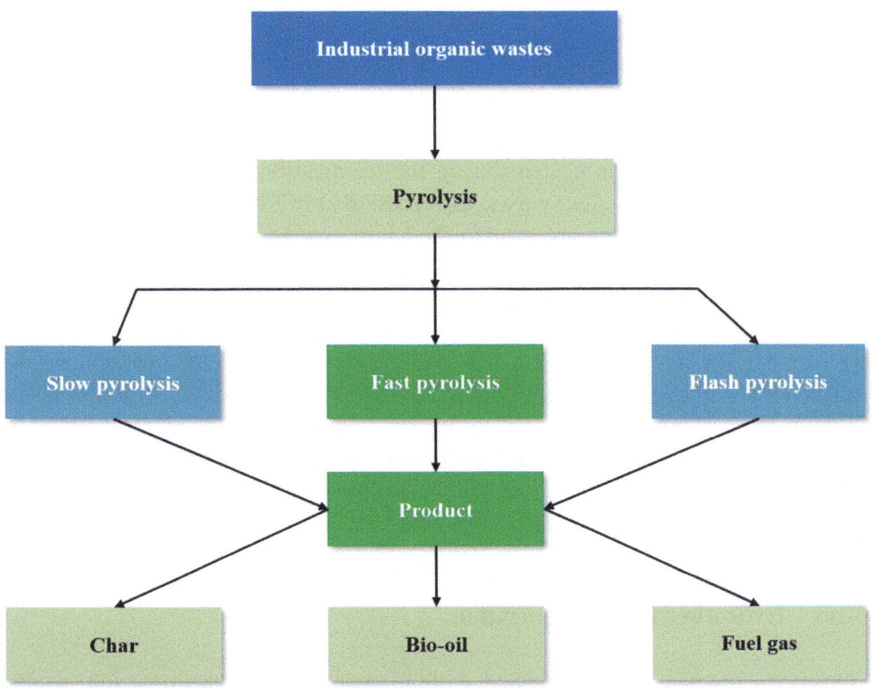

Fig. 4.2 Illustration of different types of pyrolysis process

Table 4.2 Different types of pyrolysis methods and their major features [28, 33]

Pyrolysis methods	Major product	Temperature (°C)	Retention time	Heating rate (°C/min)
Slow pyrolysis	Biochar, bio-oil, syngas	550–600	5–30 min	5–7
Fast pyrolysis	Bio-oil	550–650	1–5 s	300
Flash pyrolysis	Bio-oil	450–650	< 1 s	< 600
Catalytic pyrolysis	Bio-oil	400–600	Hours–days	–
Plasma pyrolysis	Biochar	> 1500	< 1 s	> 1000

extremely high during flash pyrolysis. Only a few seconds or even less pass between actions [21]. Flash pyrolysis primarily aims to increase the output of pyrolysis gas and oil. According to [40], pyrolysis oil produced by slow, quick, and flash pyrolysis procedures cannot be used as fuel for transportation in its pure form. These oils are less miscible than fossil fuels because they contain a lot of water and oxygen. These oils need to be upgraded before being used as fuel for transportation. On the other hand, catalytic pyrolysis improves the pyrolysis oil's quality [141]. In the pyrolysis process, raising the temperature and adding a catalyst enhance the gas yield while lowering the char yield. Since pyrolysis oil has a high moisture content, it cannot be utilized directly as fuel. The quality of the pyrolysis oil can be improved via quick pyrolysis.

4.3.1.2 Gasification

Industrial organic waste is heated during the gasification process along with the gasifying chemicals to create syngas, as represented in Fig. 4.3. In the process of gasification, organic wastes are transformed into syngas, a mixture of CO, CO_2, CH_4, NH_3, HCN, and H_2, at a high temperature (500–1800 °C), while being exposed to a controlled amount of oxygen [48]. Air, oxygen, carbon dioxide, or steam are used as gasification agents in a gasification process. The gasifying agent and the ratio of gasifying agent to organic waste are carefully determined to achieve the requisite chemical composition of syngas and maximize gasification efficiency. The fuel from these syngas can be used to generate heat and/or power. It can also be used as feedstock to create liquid fuel using the Fischer–Tropsch process, electricity using a gas turbine or fuel cell, or chemical compounds like ammonia, hydrogen, and methanol using the fuel cell [97, 111, 137].

4.3.1.3 Hydrothermal Carbonization

In pyrolysis, organic waste is heated without oxygen to break down into usable compounds. The same procedure is known as hydrothermal carbonization or hydrous

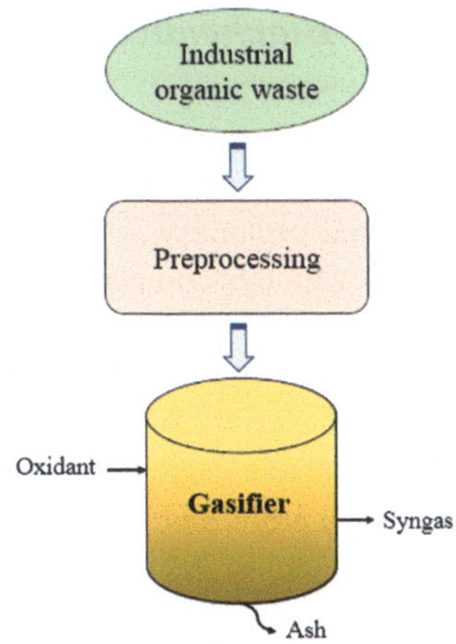

Fig. 4.3 Conversion of organic waste into syngas through gasification

pyrolysis if it is carried out in the presence of subcritical water. According to [62, 109], the main benefit of hydrothermal carbonization is that it may be utilized to efficiently transform wet organic materials into carbonaceous solids, i.e., hydrochar. It does not need to dry the materials before or during the procedure because it can convert most input materials. The conversion of feedstocks, including industrial processing organic wastes, human waste, wet animal manures, sewage sludges, municipal solid waste, aquaculture residues, and algal residues, are examples of the types of feedstocks that this approach may be used for. In high-pressure reactors used for the hydrothermal carbonization process, water preserved in the liquid phase surrounds the solid organic matter [27, 54]. The reaction temperature often governs the kind and quantity of generated products. When the process temperature is kept below 220 °C and matching pressures up to 20 bar are maintained, the majority of the organic matter is transformed into solids (hydrochar) with a very small amount of gas (1–5%). The process of producing more liquid hydrocarbons and gas fractions at greater temperatures up to 400 °C is known as hydrothermal liquefaction [29].

When used as a solid fuel, hydrochar from industrially processed food wastes produces more energy than the gases made when waste is converted through anaerobic digestion and incineration. At a relatively low temperature (180–350 °C) during hydrothermal carbonization, carbohydrates are transformed into a carbonaceous residue (hydrochar) [70]. Hydrochar is useful as a soil amendment, feedstock for carbon fuel cells, and a good absorber for hazardous contaminants. This method is particularly well suited for handling up to 90% wet organic waste. So, industrial organic wastes can be converted effectively using this technology.

4.3.2 *Biochemical Conversion*

4.3.2.1 Anaerobic Digestion

Industrial organic wastes are broken down naturally in landfills by microbes into biogas, which is composed primarily of methane, carbon dioxide, and trace amounts of other gases. This process is known as anaerobic digestion. However, this biogas escapes and contaminates the atmosphere because it is created in an unregulated environment (with oxygen present). The same procedure may be used to produce biogas and other useful products if imitated in a controlled setting (without oxygen) [24, 139]. The four steps of the anaerobic degradation process are hydrolysis, acidogenesis, acetogenesis, and methanogenesis, as indicated in Fig. 4.4. By extracellular enzymes and fermentative bacteria, high molecular weight organic matter (carbohydrates, protein, and lipids) is first broken down into tiny molecular solubility organic matter (long chain fatty acids, glucose, and amino acids) through the hydrolysis stage [48]. Small molecular organic matter is broken down during the acidogenesis stage into volatile fatty acids (acetate, propionate, and butyrate) and additional byproducts such as NH_3, CO_2, and H_2S. The organic materials created in the acidogenesis stage are subsequently digested into acetogenesis, where they produce acetate, H_2, CO_2, and other gases that are then utilized by methanogens to produce biogas [103]. The final products of anaerobic digestion are nitrogen-rich organic waste that can be utilized as fertilizer and biogas that contains 60–70% methane. This technique has been used in Europe and Asia to treat industrial organic wastes, agricultural wastes, and wastewater sludge.

Classification of the AD Process of Organic Waste

The biological, technological, dependability, and overall performance of the AD process are often used to categorize it [72]. Depending on the total amount of solids, the AD process is classified as either dry or wet, according to the feeding method, it is classified as either batch or continuous; according to the operating temperature, it is classified as either mesophilic or thermophilic; according to the number of stages, it is classified as either single-stage or multi-stage; and according to the digester, it is classified as either a fixed dome, floating dome, balloon, or garage type AD process. In contrast to a wet AD system, which has a solid content of > 15%, a dry AD process has a solid composition of between 20 and 40% [136].

While in a continuous system, the organic waste is fed constantly in the digester, in a batch-type AD process, the industrial organic waste is fed once in the digester, and inoculum is introduced when it is closed for a certain amount of time. The system is referred to as mesophilic if the operating temperature is between 20 and 40 °C, and thermophilic if it is between 50 and 65 °C. AD can be carried out in single-stage, two-stage, or multi-stage systems. In a single-stage system, digestion takes place in

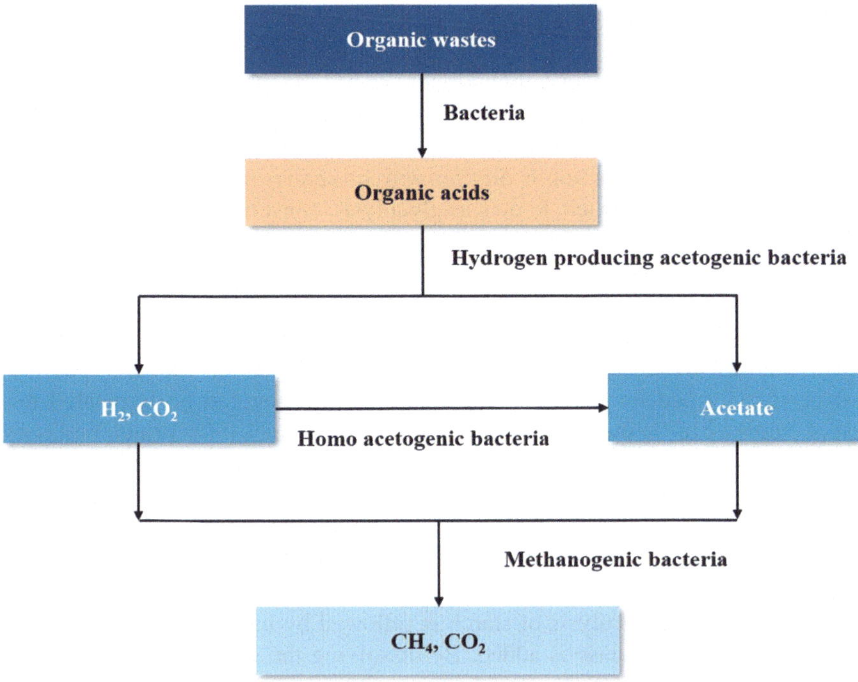

Fig. 4.4 Schematic representation of organic waste conversion through anaerobic digestion process

a single reactor, whereas a multi-stage system uses two or more reactors [127]. A wide variety of trash can be treated effectively with two stages of AD.

Factors Affecting the AD Process of Organic Waste

Many chemical and physical parameters, such as seeding, stirring, temperature, pH, C:N ratio, organic loading rate, hydraulic retention time, and volatile fatty acids, have an impact on the AD mechanism [18, 123]. Any adjustments to these settings run the risk of causing a breakdown in the digester process because they alter the movement and environment of the bacteria inside. Controlling these characteristics is crucial for maximizing biogas output. The biogas plant should operate properly and efficiently by varying these parameters within a certain range. The organic feedstock utilized for the digestion process that produces biogas must provide an environment that allows the microorganisms engaged in this process to grow and function metabolically at their best [72].

4.3.2.2 Fermentation

Industrial organic waste rich in carbohydrates can be fermented to produce hydrogen. Since organic waste is high in carbohydrates, producing hydrogen has two advantages: it recovers energy and reduces waste [118, 128]. The most often employed microbe for ethanol production is *Saccharomyces cerevisiae*. The primary process by which ethanol is fermented, known as glycolysis, converts one glucose molecule into two pyruvate molecules [60]. Under anaerobic circumstances, pyruvate is further converted to ethanol and CO_2.

Any substance that includes sugar can undergo the fermentation process to become ethanol [26, 59]. The three main types of feedstocks utilized to produce ethanol by fermentation are (a) sugar, (b) starch, and (c) cellulose. While starches and celluloses must be transformed into fermentable sugars before they can be fermented into ethanol, sugars can be immediately fermented into ethanol [90]. Fruits, molasses, sugar beets, and sugar cane are all sources of sugar. Molasses, a thick syrup created during the sugar refining process, is the most popular sugar for ethanol fermentation [52]. After being sterilized, molasses are transported to the fermentation facility. Molasses are diluted with water because a high sugar content causes inefficient sugar conversion and prolongs fermentation. Therefore, it needs to be converted into fermentable sugars. Hydrolysis of starch is followed by high-temperature cooking of the starch while -amylase is added. By dissolving the starch polymer, amylase prevents gelatinization [68]. Further fermentation of the resultant dextrose yields ethanol.

Organic waste is a complex mixture of fatty, starchy, and cellulosic components that is challenging for ethanol-producing bacteria to ferment [92]. This is similar to industrial waste that has been produced through food processing. As a result, a pretreatment is frequently required to convert these organic wastes into fermentable sugars, which can then be fermented to make ethanol. Heat treatment, acid hydrolysis, and enzymatic hydrolysis are all possible pretreatment methods that can be used for this purpose. An efficient pretreatment can either directly or indirectly hydrolyze carbohydrates into fermentable sugars, prevent sugar loss, restrict inhibitory activity, and lower energy requirements. Simultaneous saccharification and fermentation (SSF) and separate hydrolysis and fermentation (SHF) are the procedures that are typically used in fermentation. SSF is preferable to the SHF method, which has historically been used to produce bioethanol, in terms of ethanol yields, cost, and the need for an additional reactor [17, 114]. In recent years, genetic engineering has created genetically modified organisms to improve ethanol production by reducing the intake of microbes' oxygen consumption. At the same time, the technology is successful in the microbial utilization of xylose to produce ethanol by altering the metabolic activity of specific yeast species [120].

4.4 Biofuels from Organic Wastes

4.4.1 Liquid Biofuels

4.4.1.1 Biodiesel

In two processes, transesterification and lipid extraction from industrial organic waste can be used to synthesize biodiesel. Organic wastes contain carbon, such as phosphates, lipids, vitamins, amino acids, and carbohydrates [12, 57]. Lipids can be converted into biodiesel. For instance, the lipids that are typically included in 5–30% of organic food processing waste should be extracted first. Therefore, removing lipids from organic waste is the first step in turning it into biodiesel. Lipids can be extracted using one of four methods: evaporation, enzymatic hydrolysis, Soxhlet extraction, or supercritical extraction. Clean lipids recovered from organic waste can be used immediately to make biodiesel through transesterification [57]. Transesterification is a well-known process in which the oils/lipids extracted from various sources can be converted into biodiesel with suitable short-chain alcohol and catalyst, any substance that fastens the rate of the reaction without itself being consumed [105].

Although extraction is challenging at low values, the lipid portion of organic waste is well adapted for conversion to fatty acid methyl ester (FAME). Oils from industrial food wastes have recently been successfully extracted using dimethyl ether (DME) [101]. Due to DME's low boiling point (28 °C) and high pressure, e.g., 0.51 MPa at 20 °C, where extraction occurs, this extraction method is appealing [98]. As a result, the DME may be efficiently recovered. DME could extract coffee ground oils, which comprise around 17% of the total weight [101]. Soybean and rapeseed cakes each contain about 1% oil. Therefore, DME can be used to extract oils further when pressing has achieved its maximum efficiency. Various industrial food waste streams could be processed using the DME extraction method. For instance, the lipids are left more concentrated, at 1.4% of solids post-harvest, after the fermentation of the carbohydrates in potato peel waste [67]. This procedure might also be employed for pomegranate seeds containing up to 20% oil by dry weight [7].

4.4.1.2 Bioethanol

A sustainable liquid fuel called bioethanol can be produced from a variety of industrial organic wastes. Additionally, the cellulose component of industrial food waste exhibits the ability to be converted into ethanol. Citrus peels are among the organic wastes that have the highest cellulose content (37% for oranges) and lowest lignin content (7.5% for oranges) [98]. Pectin (23% for oranges) and D-limonene are among the fermentation inhibitors present, but removing these substances before ethanol fermentation allows for the synergistic creation of high-value byproducts. Through acetone, butanol, and ethanol (ABE) fermentation, readily degradable sugars from industrial organic waste also show potential for conversion to butanol and coproducts.

Butanol has an advantage over ethanol because it can be utilized in standard gasoline vehicles without requiring engine modifications and has an energy density of 83% that of gasoline as opposed to 65% for ethanol [2]. Recent research has focused on the whey from cheese production and waste from baking operations, two organic wastes that are abundant in quickly degradable sugars. Similar to making butanol from other products, manufacturing butanol from organic waste presents the same main challenge: low fermentation concentration (30 g/L) brought on byproduct inhibition [94]. Because of its high boiling point (117 °C) and low vapor pressure (2.3 kPa), butanol might be difficult to separate [2, 124]. These qualities increase the safety of pure butanol while reducing its recovery efficiency by distillation. Simultaneous separation and fermentation have been suggested as a way to circumvent product inhibition, with pervaporation and adsorption being the least energy-intensive separation techniques. In general, methods for turning industrial organic waste into biofuels show the ability to link waste sources to a dependable market. Additionally, combining the extraction of lipids and carbohydrates often has positive synergistic effects and provides markets for both biodiesel and bio-alcohol.

4.4.1.3 Pyrolysis Oil (Bio-oil)

A sustainable liquid fuel called bio-oil can be synthesized from a variety of industrial organic wastes. Bio-oil is a complicated mixture of water, char particles, and oxygenated hydrocarbons. Bio-oil has the potential to replace traditional petroleum fuels as a liquid fuel because of its dark brown color [133]. Bio-oil can be utilized as a sustainable liquid fuel because it is easily transported and stored. Additionally, it can be utilized to create compounds. Due to its weak thermal stability, corrosive behavior, and subpar fuel qualities, bio-oil is not yet commercially feasible. The bio-oil, which mostly contains oxygen-containing structures, can be supplied to the petroleum refinery as a feedstock or utilized directly as fuel in boilers for combustion [42, 44]. Biochar, syngas, and bio-oil are the three main byproducts of pyrolysis. Fast pyrolysis is used to create bio-oil, while slow pyrolysis is typically utilized to create charcoal.

Bio-oil can be produced as either a stable emulsion or a separable liquid that contains both aqueous and oil phases, depending on the feedstock's content and the pyrolysis process parameters. However, bio-oil has a low heating value (often around 50% of the heating value of diesel fuel) because it contains a substantial amount of dissolved water. Although the use of bio-oil has been tested in diesel engines with success, there are certain drawbacks to the bio-oil, including poor cold flowability, corrosion, and the development of gummy deposits [48, 75]. Because bio-oil and diesel are not miscible, it is necessary to create emulsions that require pricey surfactants to stabilize them. Bio-oils cannot be used in contemporary diesel engines due to the corrosion and deposits they cause. As a result, bio-oil needs to be improved before it can be utilized in transportation. Catalytic cracking or hydro-treating can be used to upgrade bio-oil [10, 13].

4.4.2 *Gaseous Fuels*

4.4.2.1 Biogas

Anaerobic digestion is used to produce one of the most significant sustainable biofuels from industrial organic waste: biogas. It mostly consists of 65% methane and 35% carbon dioxide, with water vapor and hydrogen sulfide traces. According to [53], it is lighter than air by roughly 20%. Unlike liquefied petroleum gas (LPG), it cannot be transformed into a liquid at room temperature. Bio-methane (enriched biogas) can be compressed into cylinders for simple transportation after the CO_2 has been removed. Additionally, enriched biogas, which has the potential to be used in all applications where CNG is now used, can be produced using CNG technology [61]. To boost the energy content of biogas and enable high-pressure storage, CO_2 is removed when biogas is utilized as fuel [85]. A fascinating fact is that methane predominates in biogas and natural gas production under anaerobic conditions. However, turning dead biomass into natural gas takes a million years, whereas turning organic waste into biogas takes just a few weeks. The components of the acidogenesis and acetogenesis stages of anaerobic digestion are transformed into methane and carbon dioxide by methanogens (methane-forming bacteria). In other words, acetogenesis further converts alcohols and short-chain volatile acids to acetic acid and hydrogen from VFAs, alcohol, ketones, CO_2, H_2, NH_3, and other compounds.

A few problems exist with biogas. Typically, it has a poor heating value. It generates more pollutants than natural gas when it is burnt in engines. It includes siloxanes and, when burned, forms silica deposits. These deposits of silica harm the components and engines. In addition, methane emissions are produced, which are thought to be stronger greenhouse gases than carbon dioxide [34]. When in touch with human skin, methanol is less safe than gasoline. Limit the exposure hazards, which in turn leads to somewhat higher investment costs, this necessitates safety systems such as closed filling systems.

4.4.2.2 Syngas

Syngas synthesis from industrial organic waste has recently attracted a lot of attention. Synthetic gas, often known as syngas, is produced by extracting hydrogen, carbon monoxide, and carbon dioxide from natural gas or coal. It can be used to make compounds like methanol and ammonia. In addition to creating various byproducts such as methanol, ethanol, hydrogen, and Fischer–Tropsch products, syngas can be utilized directly for power generation in boilers, IC engines, gas turbines, and fuel cells [51, 55, 63]. Syngas can be transformed into synthetic petroleum using Fischer–Tropsch synthesis, which has uses as a fuel or lubricant [30, 39]. Syngas production from waste and biomass has recently attracted a lot of attention. Additionally, syngas can be created directly through gasification or in two stages by pyrolyzing organic waste to create bio-oil, which can subsequently be gasified to create syngas.

The reaction temperature significantly influenced the composition of the produced gas, and the syngas production rate increased as the reaction temperature rose [132]. Water can increase reactivity through the steam reforming and decrease carbon monoxide generation through the water–gas shift reaction. The amount of oxygen enhanced methane production while simultaneously increasing hydrogen and carbon monoxide production. In the study by [81], different agro-food wastes and fruit peels were tested to produce hydrogen-rich syngas using supercritical water gasification at high temperatures (400–600 °C), high pressures (23–25 MPa), and biomass-to-water ratios (1:5 and 1:10) for a reaction time (15–45 min). With a total syngas yield of 15 mmol/g and a higher hydrogen yield of 4.8 mmol/g from coconut shells, a lower heating value of 1595 kJ/Nm^3 could be attained. Under conditions of high pressure and temperature and the presence of a catalyst (hydrogen), syngas can be transformed into methanol [119]. Hydrogen can be produced from syngas via steam reforming. Organic waste with less than 35% moisture level can be gasified using subcritical water. Supercritical water gasification can be employed since organic waste contains more than 60% moisture. After steam reforming, the water–gas shift reaction can improve hydrogen production even more [16, 22]. All carbon content is utilized in the synthesis of methanol when sufficiently enough hydrogen is present under high pressures, doubling the yield of methanol production from the same biomass.

4.4.2.3 Biohydrogen

To decrease our dependence on fossil fuels, researchers are paying more attention to biohydrogen-generating systems that can produce hydrogen from renewable resources such as industrial organic wastes. Due to its numerous benefits, biohydrogen has recently emerged as one of the most potential fuels to replace fossil fuels [50, 93]. Biohydrogen has 2.75 times more energy per unit of weight than hydrocarbon fuels, it has a high-energy content of roughly 122 kJ/g [49]. It creates water during combustion, making it an eco-friendly and clean fuel. For immediate power generation, it can be utilized in fuel cells. Conventional methods for producing hydrogen include partial oxidation of fossil fuels, steam reforming of methane and other hydrocarbons, autothermal reforming, desulfurization, pyrolysis, plasma reforming, aqueous phase reforming, and ammonia reforming. There are other non-reforming techniques for hydrogen production, including gasification and biological processes. Four different methods can be used to create biohydrogen: direct photolysis of water using algae or cyanobacteria, photofermentation using photosynthetic bacteria, dark fermentation using anaerobic bacteria, and hybrid systems [3, 80]. Dark fermentation is regarded as the most practical procedure because it does not require any outside energy or light.

4.4.3 *Biochar*

A carbon-rich byproduct of the pyrolysis process is biochar. Biochar, synthesis gas, and pyrolysis oil are all products of pyrolysis [58, 89]. Due to its high acidity and complicated composition, bio-oil cannot be utilized directly and must be upgraded. Syngas's use is constrained by its low yield and difficult separation and purification process. Contrarily, biochar offers many benefits over bio-oil and syngas. Both fertilizer and soil conditioner can be used with it. By boosting stable soil carbon stocks and soil carbon sequestration while lowering atmospheric CO_2 concentrations, it has the potential to be a technique for mitigating climate change. It raises agricultural productivity, particularly on poor and damaged soils. Additionally, it increases the soil's ability to hold water. In terms of locking up carbon, the generation of biochar is more effective than composting [84]. Additionally, following 10–20 years, microbial activity in compost will liberate the carbon in it. However, the biochar-sequestered carbon would stay stable in the soil. Additionally, it reduces emissions of greenhouse gases such as nitrous oxide and methane [11, 140]. High-production bio-oil is produced when the operating temperature is low, the heating rate is high, and the gas residence time is short [19]. High-production syngas are produced when the operating temperature is high, the heating rate is low, and the gas residence time is long. On the other hand, a low heating rate and low operating temperature result in a high output of biochar [91, 129]. As a result, the high cellulose and hemicellulose feedstocks are quickly pyrolyzed to produce bio-oil and gas as the primary products with a low biochar yield [130, 142]. High-lignin feedstocks can yield the most biochar when pyrolyzed slowly. Therefore, whether or not slow pyrolysis is employed depends on the choice of feedstock and the necessary balance of products, i.e., bio-oil, syngas, and biochar [43].

4.5 Sustainability of Industrial Organic Wastes

The issues posed by industrial organic wastes and recent urbanization in industrialized countries have increased the demand for natural resources and created opportunities for the use of organic waste in a variety of industries [78]. There are now several industrial organic wastes being exploited as renewable bioenergy sources. Various industries produce a lot of organic waste each year, and improper disposal of this waste may pollute the environment and be harmful to human health [12]. Organic waste is typically not processed adequately, resulting in incorrect disposal methods such as open-air burning, landfilling, and dumping in prohibited locations. Untreated waste produces a variety of greenhouse gases, which change the climate. Using non-renewable energy sources also contributes to the ozone layer's depletion. Consequently, advancing other cleaner and renewable energy is now a universal concern.

4.6 Challenges and Future Perspectives

Global population growth, agricultural operations, and food processing have led to increased industrial organic waste generation, hygienic problems, and dangerous gas emissions from decomposing materials. This chapter has provided sustainable and environmentally friendly strategies for effective industrial organic waste conversion and utilization as sustainable bioenergy. It has also been demonstrated that industrial organic wastes may be transformed into bioethanol, biomethane, biohydrogen, biobutanol, bio-oil, biochar, and biodiesel for use in fueling vehicles and generating electricity. The creation of biofuels from organic wastes has advantages in terms of the environment, public health, economy, industry, and technology, and when used in transportation engines, it lowers greenhouse gas emissions, improves engine performance, encourages smoother engine operation, and extends engine life. The difficulties posed by the enormous volume of organic waste generated by industrial activity can be avoided by transforming such organic waste into useful forms. Creating biofuels from this waste is still an economically and environmentally sound solution. To improve conversion rates, use less energy, and produce more goods of higher quality, these organic wastes must first undergo pretreatment. To improve the digestibility and biodegradability of the biomass, the pretreatment procedure breaks down the cellulose, hemicellulose, and lignin components. It is reasonable to predict the direction of future research given the significance of employing industrial organic waste as a feedstock for the production of biofuels.

4.7 Conclusions

Industrial organic waste is one of the potential sources to produce various bioenergy products. The waste conversion methods determine the quality and quantity of the biofuel. However, novel and improved pretreatment procedures are required to ensure better degradation and conversion of organic wastes to valuable products. To better understand the intricate mechanisms and molecular kinetics of organic waste degradation and conversion, interdisciplinary research on cost analysis, sensitivity analysis, energy and exergy analysis, techno-economic assessment, and life cycle assessments is essential. The reactor design and optimization, design parameters, cost, time, and manpower requirements to improve conversion efficiency and product quality, the deployment of appropriate innovative technologies is crucial. These include computational fluid dynamics, machine learning, artificial intelligence, statistical modeling, and optimization tools. The successful conversion of industrial organic waste into bioenergy could be a potential option for a sustainable circular bioeconomy.

References

1. Abbas-Abadi, M. S., Kusenberg, M., Zayoud, A., Roosen, M., Vermeire, F., Madanikashani, S., Kuzmanović, M., Parvizi, B., Kresovic, U., De Meester, S., & Van Geem, K. M. (2023). Thermal pyrolysis of waste versus virgin polyolefin feedstocks: The role of pressure, temperature and waste composition. *Waste Management, 165*, 108–118.
2. Abdehagh, N., Tezel, F. H., & Thibault, J. (2014). Separation techniques in butanol production: Challenges and developments. *Biomass and Bioenergy, 60*, 222–246.
3. Abdelqader, A. A., Abdelsalam, E. M., Attia, Y. A., Moselhy, M., Ali, A. S., Arisha, A. H., & Samer, M. (2022). Application of helium-neon red laser for increasing biohydrogen production from anaerobic digestion of biowastes. *Egyptian Journal of Chemistry, 65*(1), 11–17.
4. Aierzhati, A., Watson, J., Si, B., Stablein, M., Wang, T., & Zhang, Y. (2021). Development of a mobile, pilot scale hydrothermal liquefaction reactor: Food waste conversion product analysis and techno-economic assessment. *Energy Conversion and Management: X, 10*, 100076.
5. Al, A. R., Bikić, S., & Radojčin, M. (2023). Bioenergy conversion technologies: A review and case study. *Journal on Processing and Energy in Agriculture, 27*(1), 30–38.
6. Alaedini, A. H., Tourani, H. K., & Saidi, M. (2023). A review of waste-to-hydrogen conversion technologies for solid oxide fuel cell (SOFC) applications: Aspect of gasification process and catalyst development. *Journal of Environmental Management, 329*, 117077.
7. Al-Maiman, S. A., & Ahmad, D. (2002). Changes in physical and chemical properties during pomegranate (*Punica granatum* L.) fruit maturation. *Food Chemistry, 76*(4), 437–441.
8. Anand, S., Kaur, J., Sarao, L. K., & Singh, A. (2023). Agricultural residues and manures into bioenergy. In *Agroindustrial waste for green fuel application* (pp. 67–87). Springer Nature Singapore.
9. Anca-Couce, A., Hochenauer, C., & Scharler, R. (2021). Bioenergy technologies, uses, market and future trends with Austria as a case study. *Renewable and Sustainable Energy Reviews, 135*, 110237.
10. Arefin, M. A., Rashid, F., & Islam, A. (2021). A review of biofuel production from floating aquatic plants: An emerging source of bio-renewable energy. *Biofuels, Bioproducts and Biorefining, 15*(2), 574–591.
11. Arif, M., Jan, T., Riaz, M., Fahad, S., Adnan, M., Amanullah, Ali, K., Mian, I.A., Khan, B., & Rasul, F. (2020). Biochar; a remedy for climate change. In *Environment, climate, plant and vegetation growth* (pp. 151–171).
12. Ashokkumar, V., Flora, G., Venkatkarthick, R., SenthilKannan, K., Kuppam, C., Stephy, G. M., Kamyab, H., Chen, W. H., Thomas, J., & Ngamcharussrivichai, C. (2022). Advanced technologies on the sustainable approaches for conversion of organic waste to valuable bioproducts: Emerging circular bioeconomy perspective. *Fuel, 324*, 124313.
13. Atanda, L., Fraga, G. L. L., Ahmed, M. H., Alothman, Z. A., Na, J., Batalha, N., Aslam, W., & Konarova, M. (2021). Conversion of agricultural waste into stable biocrude using spinel oxide catalysts. *Journal of Hazardous Materials, 402*, 123539.
14. Atelge, M. R., Krisa, D., Kumar, G., Eskicioglu, C., Nguyen, D. D., Chang, S. W., Atabani, A. E., Al-Muhtaseb, A. H., & Unalan, S. (2020). Biogas production from organic waste: Recent progress and perspectives. *Waste and Biomass Valorization, 11*, 1019–1040.
15. Awasthi, M. K., Sindhu, R., Sirohi, R., Kumar, V., Ahluwalia, V., Binod, P., Juneja, A., Kumar, D., Yan, B., Sarsaiya, S., & Zhang, Z. (2022). Agricultural waste biorefinery development towards circular bioeconomy. *Renewable and Sustainable Energy Reviews, 158*, 112122.
16. Babatabar, M. A., & Saidi, M. (2021). Hydrogen production via integrated configuration of steam gasification process of biomass and water-gas shift reaction: Process simulation and optimization. *International Journal of Energy Research, 45*(13), 19378–19394.
17. Bakari, H., Djomdi, Falama Ruben, Z., Roger, D. D., Cedric, D., Guillaume, P., Pascal, D., Philippe, M., & Gwendoline, C. (2023). Optimization of bioethanol production after enzymatic treatment of sweet sorghum stalks. *Waste and Biomass Valorization*, 1–15.

18. Bella, K., & Rao, P. V. (2021). Anaerobic digestion of dairy wastewater: Effect of different parameters and co-digestion options—A review. *Biomass Conversion and Biorefinery*, 1–26.
19. Bhoi, P. R., Ouedraogo, A. S., Soloiu, V., & Quirino, R. (2020). Recent advances on catalysts for improving hydrocarbon compounds in bio-oil of biomass catalytic pyrolysis. *Renewable and Sustainable Energy Reviews, 121*, 109676.
20. Bilal, M., Mehmood, T., Nadeem, F., Barbosa, A. M., de Souza, R. L., Pompeu, G. B., Meer, B., Ferreira, L. F. R., & Iqbal, H. M. (2022). Enzyme-assisted transformation of lignin-based food bio-residues into high-value products with a zero-waste theme: A review. *Waste and Biomass Valorization*, 1–18.
21. Chen, L., Yang, K., Huang, J., Liu, P., Yang, J., Pan, Y., Qi, F., & Jia, L. (2022). Experimental and kinetic study on flash pyrolysis of biomass via on-line photoionization mass spectrometry. *Applications in Energy and Combustion Science, 9*, 100057.
22. Chen, W. H., Lu, C. Y., Chou, W. S., Sharma, A. K., Saravanakumar, A., & Tran, K. Q. (2023). Design and optimization of a crossflow tube reactor system for hydrogen production by combining ethanol steam reforming and water gas shift reaction. *Fuel, 334*, 126628.
23. Coma, M., Martinez-Hernandez, E., Abeln, F., Raikova, S., Donnelly, J., Arnot, T. C., Allen, M. J., Hong, D. D., & Chuck, C. J. (2017). Organic waste as a sustainable feedstock for platform chemicals. *Faraday discussions, 202*, 175–195.
24. Dahl, R. (2015). A second life for scraps: Making biogas from food waste.
25. Dhandayuthapani, K., Kumar, P. S., Chia, W. Y., Chew, K. W., Karthik, V., Selvarangaraj, H., Selvakumar, P., Sivashanmugam, P., & Show, P. L. (2022). Bioethanol from hydrolysate of ultrasonic processed robust microalgal biomass cultivated in dairy wastewater under optimal strategy. *Energy, 244*, 122604. https://doi.org/10.1016/j.energy.2021.122604
26. Dhandayuthapani, K., Sarumathi, V., Selvakumar, P., Temesgen, T., Asaithambi, P., & Sivashanmugam, P. (2021). Study on the ethanol production from hydrolysate derived by ultrasonic pretreated defatted biomass of chlorella sorokiniana NITTS3. *Chemical Data Collections, 31*, 100641. https://doi.org/10.1016/j.cdc.2020.100641
27. Di Giacomo, G., Gallifuoco, A., & Taglieri, L. (2016). Hydrothermal carbonization of mixed biomass: Experimental investigation for an optimal valorisation of agrofood wastes. In *Proceedings of the 24th European biomass conference and exibition (24th EUBCE)* (pp. 6–9).
28. Dickerson, T., & Soria, J. (2013). Catalytic fast pyrolysis: A review. *Energies, 6*(1), 514–538.
29. Dimitriadis, A., & Bezergianni, S. (2017). Hydrothermal liquefaction of various biomass and waste feedstocks for biocrude production: A state of the art review. *Renewable and Sustainable Energy Reviews, 68*, 113–125.
30. dos Santos, R. G., & Alencar, A. C. (2020). Biomass-derived syngas production via gasification process and its catalytic conversion into fuels by Fischer Tropsch synthesis: A review. *International Journal of Hydrogen Energy, 45*(36), 18114–18132.
31. Elkhalifa, S., Al-Ansari, T., Mackey, H. R., & McKay, G. (2019). Food waste to biochars through pyrolysis: A review. *Resources, Conservation and Recycling, 144*, 310–320.
32. ElMekawy, A., Diels, L., De Wever, H., & Pant, D. (2013). Valorization of cereal based biorefinery byproducts: Reality and expectations. *Environmental Science & Technology, 47*(16), 9014–9027.
33. Fabry, F., Rehmet, C., Rohani, V., & Fulcheri, L. (2013). Waste gasification by thermal plasma: A review. *Waste and Biomass Valorization, 4*, 421–439.
34. Farghali, M., Osman, A. I., Umetsu, K., & Rooney, D. W. (2022). Integration of biogas systems into a carbon zero and hydrogen economy: A review. *Environmental Chemistry Letters, 20*(5), 2853–2927.
35. Frolov, S. M., Smetanyuk, V. A., Shamshin, I. O., Sadykov, I. A., & Frolov, F. S. (2021). Production of highly superheated steam by cyclic detonations of propane-and methane-steam mixtures with oxygen for waste gasification. *Applied Thermal Engineering, 183*, 116195.
36. Gaur, V. K., Sharma, P., Sirohi, R., Awasthi, M. K., Dussap, C. G., & Pandey, A. (2020). Assessing the impact of industrial waste on environment and mitigation strategies: A comprehensive review. *Journal of Hazardous Materials, 398*, 123019.

37. Gizaw, D. G., Periyasamy, P., Redda, Z. T., John, B. I., Mengstie, H. B., & Asaithambi, P. (2023). A comprehensive review on sewage sludge as sustainable feedstock for bioenergy production. *Environmental Quality Management, 1–16*. https://doi.org/10.1002/tqem.22116
38. Guldhe, A., Singh, B., Renuka, N., Singh, P., Misra, R., & Bux, F. (2017). Bioenergy: A sustainable approach for cleaner environment. In *Phytoremediation potential of bioenergy plants* (pp. 47–62).
39. Gupta, P. K., Kumar, V., & Maity, S. (2021). Renewable fuels from different carbonaceous feedstocks: A sustainable route through Fischer-Tropsch synthesis. *Journal of Chemical Technology & Biotechnology, 96*(4), 853–868.
40. Haghighat, M., Majidian, N., & Hallajisani, A. (2020). Production of bio-oil from sewage sludge: A review on the thermal and catalytic conversion by pyrolysis. *Sustainable Energy Technologies and Assessments, 42*, 100870.
41. Hasan, M. M., Rasul, M. G., Khan, M. M. K., Ashwath, N., & Jahirul, M. I. (2021). Energy recovery from municipal solid waste using pyrolysis technology: A review on current status and developments. *Renewable and Sustainable Energy Reviews, 145*, 111073.
42. Hu, X., & Gholizadeh, M. (2020). Progress of the applications of bio-oil. *Renewable and Sustainable Energy Reviews, 134*, 110124.
43. Huang, C., Mohamed, B. A., & Li, L. Y. (2022). Comparative life-cycle assessment of pyrolysis processes for producing bio-oil, biochar, and activated carbon from sewage sludge. *Resources, Conservation and Recycling, 181*, 106273.
44. Ilyin, S. O., & Makarova, V. V. (2022). Bio-oil: Production, modification, and application. *Chemistry and Technology of Fuels and Oils, 58*(1), 29–44.
45. Jalali, M., & Saremi, P. (2020). A study on the impact of energy conversion technologies on food waste from macromolecules on the environment. *Indian Journal of Science Research, 11*(1), 45–56.
46. Jamro, I. A., Khoso, S., Shah, S. A. R., Ali, S., Ali, F., & Hassane, H. U. (2022). Investigation of thermal decomposition and gases release from pre-drying municipal solid waste (PMSW) via pyrolysis technology. *Architecture, Civil Engineering, Environment, 15*(4), 119–131.
47. Japhet, J. A., Luka, B. S., Maren, I. B., & Datau, S. G. (2020). The potential of wood and agricultural waste for pellet fuel development in nigeria—A technical review. *International Journal of Engineering Application Science Technology, 4*, 598–607.
48. Jeevahan, J., Anderson, A., Sriram, V., Durairaj, R. B., Britto Joseph, G., & Mageshwaran, G. (2021). Waste into energy conversion technologies and conversion of food wastes into the potential products: A review. *International Journal of Ambient Energy, 42*(9), 1083–1101.
49. Jeong, Y. S., Park, K. B., & Kim, J. S. (2020). Hydrogen production from steam gasification of polyethylene using a two-stage gasifier and active carbon. *Applied Energy, 262*, 114495.
50. Johari, A., Singh, S., & Vidya, S. (2022). Engine performance analysis for diesel engine using hydrogen as an alternative fuel. *Materials Today: Proceedings, 56*, 342–346.
51. Jones, M. P., Krexner, T., & Bismarck, A. (2022). Repurposing Fischer-Tropsch and natural gas as bridging technologies for the energy revolution. *Energy Conversion and Management, 267*, 115882.
52. Kabeyi, M. J. B., & Olanrewaju, O. A. (2022). Sugarcane molasses to energy conversion for sustainable production and energy transition. In *12th annual Istanbul international conference on industrial engineering and operations management*.
53. Kalsum, L., & Hasan, A. (2022). The effect of the packing flow area and biogas flow rate on biogas purification in packed bed scrubber. *Journal of Ecological Engineering, 23*(11).
54. Kambo, H. S., Minaret, J., & Dutta, A. (2018). Process water from the hydrothermal carbonization of biomass: A waste or a valuable product? *Waste and Biomass Valorization, 9*, 1181–1189.
55. Kapoor, R., Ghosh, P., Tyagi, B., Vijay, V. K., Vijay, V., Thakur, I. S., Kamyab, H., Nguyen, D. D., & Kumar, A. (2020). Advances in biogas valorization and utilization systems: A comprehensive review. *Journal of Cleaner Production, 273*, 123052.
56. Karimi, S., Mahboobi Soofiani, N., Mahboubi, A., & Taherzadeh, M. J. (2018). Use of organic wastes and industrial by-products to produce filamentous fungi with potential as aqua-feed ingredients. *Sustainability, 10*(9), 3296.

57. Karmee, S. K. (2016). Liquid biofuels from food waste: Current trends, prospect and limitation. *Renewable and Sustainable Energy Reviews, 53*, 945–953.
58. Karthik, V., Karuna, B., Jeyanthi, J., & Periyasamy, S. (2023). Biochar production from Manilkara zapota seeds, activation and characterization for effective removal of Cu2+ ions in polluted drinking water. *Biomass Conversion and Biorefinery, 13*(11), 9381–9395. https://doi.org/10.1007/s13399-022-03627-2
59. Kavitha, S., Gajendran, T., Saranya, K., Selvakumar, P., & Manivasagan, V. (2021). Study on consolidated bioprocessing of pre-treated nannochloropsis gaditana biomass into ethanol under optimal strategy. *Renewable Energy, 172*, 440–452. https://doi.org/10.1016/j.renene.2021.03.015
60. Kavitha, S., Gajendran, T., Saranya, K., Selvakumar, P., Manivasagan, V., & Jeevitha, S. (2022). An insight-A statistical investigation of consolidated bioprocessing of *Allium ascalonicum* leaves to ethanol using Hangateiclostridium thermocellum KSMK1203 and synthetic consortium. *Renewable Energy, 187*, 403–416. https://doi.org/10.1016/j.renene.2022.01.047
61. Khan, I. U., Othman, M. H. D., Hashim, H., Matsuura, T., Ismail, A. F., Rezaei-DashtArzhandi, M., & Azelee, I. W. (2017). Biogas as a renewable energy fuel—A review of biogas upgrading, utilisation and storage. *Energy Conversion and Management, 150*, 277–294.
62. Khan, T. A., Saud, A. S., Jamari, S. S., Ab Rahim, M. H., Park, J. W., & Kim, H. J. (2019). Hydrothermal carbonization of lignocellulosic biomass for carbon rich material preparation: A review. *Biomass and bioenergy, 130*, 105384.
63. Khosravani, H., Meshksar, M., Rahimpour, H. R., & Rahimpour, M. R. (2023). Introduction to syngas products and applications. In *Advances in synthesis gas: Methods, technologies and applications* (pp. 3–25). Elsevier.
64. Kim, S., Lee, Y., Lin, K. Y. A., Hong, E., Kwon, E. E., & Lee, J. (2020). The valorization of food waste via pyrolysis. *Journal of Cleaner Production, 259*, 120816.
65. Koul, B., Yakoob, M., & Shah, M. P. (2022). Agricultural waste management strategies for environmental sustainability. *Environmental Research, 206*, 112285.
66. Lee, U., Benavides, P. T., & Wang, M. (2020). Life cycle analysis of waste-to-energy pathways. In *Waste-to-energy* (pp. 213–233). Academic Press.
67. Liang, S., & McDonald, A. G. (2014). Chemical and thermal characterization of potato peel waste and its fermentation residue as potential resources for biofuel and bioproducts production. *Journal of Agricultural and Food Chemistry, 62*(33), 8421–8429.
68. Liu, L., An, X., Zhang, H., Lu, Z., Nie, S., Cao, H., Xu, Q., & Liu, H. (2020). Ball milling pretreatment facilitating α-amylase hydrolysis for production of starch-based bio-latex with high performance. *Carbohydrate Polymers, 242*, 116384.
69. Liu, M., Ogunmoroti, A., Liu, W., Li, M., Bi, M., Liu, W., & Cui, Z. (2022). Assessment and projection of environmental impacts of food waste treatment in China from life cycle perspectives. *Science of The Total Environment, 807*, 150751.
70. Lu, X., Jordan, B., & Berge, N. D. (2012). Thermal conversion of municipal solid waste via hydrothermal carbonization: Comparison of carbonization products to products from current waste management techniques. *Waste Management, 32*(7), 1353–1365.
71. Mahmood, R., Parshetti, G. K., & Balasubramanian, R. (2016). Energy, exergy and techno-economic analyses of hydrothermal oxidation of food waste to produce hydro-char and bio-oil. *Energy, 102*, 187–198.
72. Mahmudul, H. M., Rasul, M. G., Akbar, D., Narayanan, R., & Mofijur, M. (2021). A comprehensive review of the recent development and challenges of a solar-assisted biodigester system. *Science of The Total Environment, 753*, 141920.
73. Manikandan, S., Vickram, S., Sirohi, R., Subbaiya, R., Krishnan, R. Y., Karmegam, N., Sumathijones, C., Rajagopal, R., Chang, S. W., Ravindran, B., & Awasthi, M. K. (2023). Critical review of biochemical pathways to transformation of waste and biomass into bioenergy. *Bioresource Technology*, 128679.
74. Maraveas, C. (2020). Production of sustainable and biodegradable polymers from agricultural waste. *Polymers, 12*(5), 1127.

75. Masudi, A., Muraza, O., Jusoh, N. W. C., & Ubaidillah, U. (2023). Improvements in the stability of biodiesel fuels: Recent progress and challenges. *Environmental Science and Pollution Research, 30*(6), 14104–14125.
76. Mofijur, M., Siddiki, S. Y. A., Shuvho, M. B. A., Djavanroodi, F., Fattah, I. R., Ong, H. C., Chowdhury, M. A., & Mahlia, T. M. I. (2021). Effect of nanocatalysts on the transesterification reaction of first, second and third generation biodiesel sources—A mini-review. *Chemosphere, 270*, 128642.
77. Mofijur, M. G. R. M., Rasul, A. M., Hyde, J., Azad, A. K., Mamat, R., & Bhuiya, M. M. K. (2016). Role of biofuel and their binary (diesel–biodiesel) and ternary (ethanol–biodiesel–diesel) blends on internal combustion engines emission reduction. *Renewable and Sustainable Energy Reviews, 53*, 265–278.
78. Mohamed, B. A., Bilal, M., Salama, E. S., Periyasamy, S., Fattah, I. R., Ruan, R., Awasthi, M. K., & Leng, L. (2022). Phenolic-rich bio-oil production by microwave catalytic pyrolysis of switchgrass: Experimental study, life cycle assessment, and economic analysis. *Journal of Cleaner Production, 366*, 132668. https://doi.org/10.1016/j.jclepro.2022.132668
79. Muhammad, G., Alam, M. A., Mofijur, M., Jahirul, M. I., Lv, Y., Xiong, W., Ong, H. C., & Xu, J. (2021). Modern developmental aspects in the field of economical harvesting and biodiesel production from microalgae biomass. *Renewable and Sustainable Energy Reviews, 135*, 110209.
80. Mukherjee, T., & Mohan, S. V. (2022). Bio-waste to hydrogen production technologies. In *Advanced biofuel technologies* (pp. 389–407). Elsevier.
81. Nanda, S., Isen, J., Dalai, A. K., & Kozinski, J. A. (2016). Gasification of fruit wastes and agro-food residues in supercritical water. *Energy Conversion and Management, 110*, 296–306.
82. Nkosi, N., Muzenda, E., Gorimbo, J., & Belaid, M. (2021). Developments in waste tyre thermochemical conversion processes: Gasification, pyrolysis and liquefaction. *RSC Advances, 11*(20), 11844–11871.
83. Okolie, J. A., Epelle, E. I., Tabat, M. E., Orivri, U., Amenaghawon, A. N., Okoye, P. U., & Gunes, B. (2022). Waste biomass valorization for the production of biofuels and value-added products: A comprehensive review of thermochemical, biological and integrated processes. *Process Safety and Environmental Protection, 159*, 323–344.
84. Oldfield, T. L., Sikirica, N., Mondini, C., López, G., Kuikman, P. J., & Holden, N. M. (2018). Biochar, compost and biochar-compost blend as options to recover nutrients and sequester carbon. *Journal of Environmental Management, 218*, 465–476.
85. Olugasa, T. T., Odesola, I. F., & Oyewola, M. O. (2014). Energy production from biogas: A conceptual review for use in Nigeria. *Renewable and Sustainable Energy Reviews, 32*, 770–776.
86. Ong, H. C., Tiong, Y. W., Goh, B. H. H., Gan, Y. Y., Mofijur, M., Fattah, I. R., Chong, C. T., Alam, M. A., Lee, H. V., Silitonga, A. S., & Mahlia, T. M. I. (2021). Recent advances in biodiesel production from agricultural products and microalgae using ionic liquids: Opportunities and challenges. *Energy Conversion and Management, 228*, 113647.
87. Opatokun, S. A., Lopez-Sabiron, A. M., Ferreira, G., & Strezov, V. (2017). Life cycle analysis of energy production from food waste through anaerobic digestion, pyrolysis and integrated energy system. *Sustainability, 9*(10), 1804.
88. Patel, A., Temgire, S., & Borah, A. (2021). Agro-industrial waste as source of bioactive compounds and their utilization: A review. *Pharma Innov, 10*(5), 192–196.
89. Pawar, A., Panwar, N. L., & Salvi, B. L. (2020). Comprehensive review on pyrolytic oil production, upgrading and its utilization. *Journal of Material Cycles and Waste Management, 22*, 1712–1722.
90. Periyasamy, S., Isabel, J. B., Kavitha, S., Karthik, V., Mohamed, B. A., Gizaw, D. G., Sivashanmugam, P., & Aminabhavi, T. M. (2023). Recent advances in consolidated bioprocessing for conversion of lignocellulosic biomass into bioethanol–A review. *Chemical Engineering Journal, 453*, 139783. https://doi.org/10.1016/j.cej.2022.139783
91. Periyasamy, S., Karthik, V., Senthil Kumar, P., Isabel, J. B., Temesgen, T., Hunegnaw, B. M., Melese, B. B., Mohamed, B. A., & Vo, D. V. N. (2022). Chemical, physical and biological

methods to convert lignocellulosic waste into value-added products. A review. *Environmental Chemistry Letters, 20*(2), 1129–1152. https://doi.org/10.1007/s10311-021-01374-w
92. Prasoulas, G., Gentikis, A., Konti, A., Kalantzi, S., Kekos, D., & Mamma, D. (2020). Bioethanol production from food waste applying the multienzyme system produced on-site by *Fusarium oxysporum* F3 and mixed microbial cultures. *Fermentation, 6*(2), 39.
93. Qazi, U. Y. (2022). Future of hydrogen as an alternative fuel for next-generation industrial applications; challenges and expected opportunities. *Energies, 15*(13), 4741.
94. Quested, T., Ingle, R., & Parry, A. (2013). Household food and drink waste in the United Kingdom 2012.
95. Ramesh, M., Adithya, K., Kumar, C. J., Mohan, C. G., Nalawade, J., & Prakash, R. (2021). Thermochemical conversion methods of bio-derived lignocellulosic waste molecules into renewable fuels. In *Advanced technology for the conversion of waste into fuels and chemicals* (pp. 197–215). Woodhead Publishing.
96. Rasheed, T., Anwar, M. T., Ahmad, N., Sher, F., Khan, S. U. D., Ahmad, A., Khan, R., & Wazeer, I. (2021). Valorisation and emerging perspective of biomass based waste-to-energy technologies and their socio-environmental impact: A review. *Journal of Environmental Management, 287*, 112257.
97. Rauch, R., Hrbek, J., & Hofbauer, H. (2016). Biomass gasification for synthesis gas production and applications of the syngas. *Advances in Bioenergy: The Sustainability Challenge*, 73–91.
98. RedCorn, R., Fatemi, S., & Engelberth, A. S. (2018). Comparing end-use potential for industrial food-waste sources. *Engineering, 4*(3), 371–380.
99. Rosero-Delgado, E. A., Zambrano-Arcentales, M. A., Gómez-Salcedo, Y., Baquerizo-Crespo, R. J., & Dustet-Mendoza, J. C. (2021). Biotechnology applied to treatments of agro-industrial wastes. In *Advances in the domain of environmental biotechnology: microbiological developments in industries, wastewater treatment and agriculture* (pp. 277–311).
100. Sadh, P. K., Duhan, S., & Duhan, J. S. (2018). Agro-industrial wastes and their utilization using solid state fermentation: A review. *Bioresources and Bioprocessing, 5*(1), 1–15.
101. Sakuragi, K., Li, P., Otaka, M., & Makino, H. (2016). Recovery of bio-oil from industrial food waste by liquefied dimethyl ether for biodiesel production. *Energies, 9*(2), 106.
102. Sarkar, N., Chakraborty, D., Dutta, R., Agrahari, P., Sundaram, D., Singh, A., & Jacob, S. (2021). A comprehensive review on oilseed cakes and their potential as a feedstock for integrated biorefinery. *Journal of Advanced Biotechnology and Experimental Therapeutics, 4*(3), 376.
103. Sawyerr, N., Trois, C., Workneh, T., & Okudoh, V. (2019). An overview of biogas production: Fundamentals, applications and future research. *International Journal of Energy Economics and Policy, 9*(2), 105.
104. Selvakumar, P., & Sivashanmugam, P. (2018). Multi-hydrolytic biocatalyst from organic solid waste and its application in municipal waste activated sludge pre-treatment towards energy recovery. *Process Safety and Environmental Protection, 117*, 1–10. https://doi.org/10.1016/j.psep.2018.03.036
105. Selvakumar, P., & Sivashanmugam, P. (2017). Optimization of lipase production from organic solid waste by anaerobic digestion and its application in biodiesel production. *Fuel Processing Technology, 165*, 1–8.
106. Selvakumar, P., Adane, A. A., Zelalem, T., Hunegnaw, B. M., Karthik, V., Kavitha, S., Jayakumar, M., Karmegam, N., Govarthanan, M., & Kim, W. (2022). Optimization of binary acids pretreatment of corncob biomass for enhanced recovery of cellulose to produce bioethanol. *Fuel, 321*, 124060. https://doi.org/10.1016/j.fuel.2022.124060
107. Selvakumar, P., Arunagiri, A., & Sivashanmugam, P. (2019). Thermo-sonic assisted enzymatic pre-treatment of sludge biomass as potential feedstock for oleaginous yeast cultivation to produce biodiesel. *Renewable Energy, 139*, 1400–1411. https://doi.org/10.1016/j.renene.2019.03.040
108. Selvakumar, P., Karthik, V., Kumar, P. S., Asaithambi, P., Kavitha, S., & Sivashanmugam, P. (2021). Enhancement of ultrasound assisted aqueous extraction of polyphenols from waste fruit peel using dimethyl sulfoxide as surfactant: Assessment of kinetic models. *Chemosphere, 263*, 128071. https://doi.org/10.1016/j.chemosphere.2020.128071

109. Sharma, H. B., Sarmah, A. K., & Dubey, B. (2020). Hydrothermal carbonization of renewable waste biomass for solid biofuel production: A discussion on process mechanism, the influence of process parameters, environmental performance and fuel properties of hydrochar. *Renewable and sustainable energy reviews, 123*, 109761.
110. Shibu, C., Chandel, S., & Vats, P. (2023). Scaling up of wood waste utilization for sustainable green future. In *Handbook of research on sustainable consumption and production for greener economies* (pp. 358–383). IGI Global.
111. Shirazi, A., Rahbari, A., Asselineau, C. A., & Pye, J. (2019). A solar fuel plant via supercritical water gasification integrated with Fischer-Tropsch synthesis: System-level dynamic simulation and optimisation. *Energy Conversion and Management, 192*, 71–87.
112. Siddiqui, O., & Dincer, I. (2019). A well to pump life cycle environmental impact assessment of some hydrogen production routes. *International journal of hydrogen energy, 44*(12), 5773–5786.
113. Singh, N. K., Vats, A., Singh, A., Tyagi, A., Mishra, S. K., Kumar, N., & Kumar, S. (2022). Production of bioethanol from lignocellulosic waste parali. In *Role of microbes in industrial products and processes* (pp. 195–222).
114. Singh, R., Das, R., Sangwan, S., Rohatgi, B., Khanam, R., Peera, S. P. G., Das, S., Lyngdoh, Y. A., Langyan, S., Shukla, A., & Shrivastava, M. (2021). Utilisation of agro-industrial waste for sustainable green production: A review. *Environmental Sustainability, 4*(4), 619–636.
115. Siva Sankari, M., Vivekanandhan, S., Misra, M., & Mohanty, A. K. (2022). Oil cakes as sustainable agro-industrial feedstock for biocarbon materials. *ChemBioEng Reviews, 9*(1), 21–41.
116. Skendi, A., Zinoviadou, K. G., Papageorgiou, M., & Rocha, J. M. (2020). Advances on the valorisation and functionalization of by-products and wastes from cereal-based processing industry. *Foods, 9*(9), 1243.
117. Sridhar, A., Kapoor, A., Kumar, P. S., Ponnuchamy, M., Balasubramanian, S., & Prabhakar, S. (2021). Conversion of food waste to energy: A focus on sustainability and life cycle assessment. *Fuel, 302*, 121069.
118. Srisowmeya, G., Chakravarthy, M., & Devi, G. N. (2020). Critical considerations in two-stage anaerobic digestion of food waste—A review. *Renewable and Sustainable Energy Reviews, 119*, 109587.
119. Srivastava, R. K., Sarangi, P. K., Bhatia, L., Singh, A. K., & Shadangi, K. P. (2022). Conversion of methane to methanol: Technologies and future challenges. *Biomass Conversion and Biorefinery, 12*(5), 1851–1875.
120. Sundarsingh, T. J. A., Ameen, F., Ranjitha, J., Raghavan, S., & Shankar, V. (2024). Engineering microbes for sustainable biofuel production and extraction of lipids—Current research and future perspectives. *Fuel, 355*, 129532.
121. Tian, H., Li, J., Yan, M., Tong, Y. W., Wang, C. H., & Wang, X. (2019). Organic waste to biohydrogen: A critical review from technological development and environmental impact analysis perspective. *Applied Energy, 256*, 113961.
122. Tian, H., Wang, X., Lim, E. Y., Lee, J. T., Ee, A. W., Zhang, J., & Tong, Y. W. (2021). Life cycle assessment of food waste to energy and resources: Centralized and decentralized anaerobic digestion with different downstream biogas utilization. *Renewable and Sustainable Energy Reviews, 150*, 111489.
123. Uddin, M. N., Siddiki, S. Y. A., Mofijur, M., Djavanroodi, F., Hazrat, M. A., Show, P. L., Ahmed, S. F., & Chu, Y. M. (2021). Prospects of bioenergy production from organic waste using anaerobic digestion technology: A mini review. *Frontiers in Energy Research, 9*, 627093.
124. Ujor, V., Bharathidasan, A. K., Cornish, K., & Ezeji, T. C. (2014). Feasibility of producing butanol from industrial starchy food wastes. *Applied Energy, 136*, 590–598.
125. Usman, I., Saif, H., Imran, A., Afzaal, M., Saeed, F., Azam, I., Afzal, A., Ateeq, H., Islam, F., Shah, Y. A., & Shah, M. A. (2023). Innovative applications and therapeutic potential of oilseeds and their by-products: An eco-friendly and sustainable approach. *Food Science & Nutrition, 11*(6), 2599–2609.

126. Van Nguyen, T. T., Phan, A. N., Nguyen, T. A., Nguyen, T. K., Nguyen, S. T., Pugazhendhi, A., & Phuong, H. H. K. (2022). Valorization of agriculture waste biomass as biochar: As first-rate biosorbent for remediation of contaminated soil. *Chemosphere*, 135834.
127. Van, D. P., Fujiwara, T., Tho, B. L., Toan, P. P. S., & Minh, G. H. (2020). A review of anaerobic digestion systems for biodegradable waste: Configurations, operating parameters, and current trends. *Environmental Engineering Research, 25*(1), 1–17.
128. Varanasi, J. L., & Das, D. (2020). Maximizing biohydrogen production from water hyacinth by coupling dark fermentation and electrohydrogenesis. *International Journal of Hydrogen Energy, 45*(8), 5227–5238.
129. Velusamy, K., Periyasamy, S., Kumar, P. S., Jayaraj, T., Krishnasamy, R., Sindhu, J., Sneka, D., Subhashini, B., & Vo, D. V. N. (2021). Analysis on the removal of emerging contaminant from aqueous solution using biochar derived from soap nut seeds. *Environmental Pollution, 287*, 117632. https://doi.org/10.1016/j.envpol.2021.117632
130. Velusamy, S., Subbaiyan, A., Murugesan, S. R., Shanmugamoorthy, M., Sivakumar, V., Velusamy, P., Veerasamy, S., Mani, K., Sundararaj, P., & Periyasamy, S. (2022). Comparative analysis of agro waste material solid biomass briquette for environmental sustainability. *Advances in Materials Science and Engineering, 2022*. https://doi.org/10.1155/2022/3906256
131. Veza, I., Said, M. F. M., & Latiff, Z. A. (2021). Recent advances in butanol production by acetone-butanol-ethanol (ABE) fermentation. *Biomass and Bioenergy, 144*, 105919.
132. Wang, B., Gupta, R., Bei, L., Wan, Q., & Sun, L. (2023). A review on gasification of municipal solid waste (MSW): Syngas production, tar formation, mineral transformation and industrial challenges. *International Journal of Hydrogen Energy*.
133. Wang, S., Zhao, S., Uzoejinwa, B. B., Zheng, A., Wang, Q., Huang, J., & Abomohra, A. E. F. (2020). A state-of-the-art review on dual purpose seaweeds utilization for wastewater treatment and crude bio-oil production. *Energy Conversion and Management, 222*, 113253.
134. Yankovsky, S. A., Kuznetsov, G. V., Tolokolnikov, A. A., Cherednik, I. V., & Ivanov, A. A. (2021). Experimental study of the processes of reducing the formation of sulfur oxides during the co-combustion of particles of metalignitous coal and wood processing waste. *Fuel, 291*, 120233.
135. Zadeh, Z. E., Abdulkhani, A., & Saha, B. (2021). A comparative production and characterisation of fast pyrolysis bio-oil from populus and spruce woods. *Energy, 214*, 118930.
136. Zamri, M. F. M. A., Hasmady, S., Akhiar, A., Ideris, F., Shamsuddin, A. H., Mofijur, M., Fattah, I. R., & Mahlia, T. M. I. (2021). A comprehensive review on anaerobic digestion of organic fraction of municipal solid waste. *Renewable and Sustainable Energy Reviews, 137*, 110637.
137. Zang, G., Sun, P., Elgowainy, A. A., Bafana, A., & Wang, M. (2021). Performance and cost analysis of liquid fuel production from H_2 and CO_2 based on the Fischer-Tropsch process. *Journal of CO_2 Utilization, 46*, 101459.
138. Zena, Y., Periyasamy, S., Tesfaye, M., Tumsa, Z., Jayakumar, M., Mohamed, B. A., Asaithambi, P., & Aminabhavi, T. M. (2023). Essential characteristics improvement of metallic nanoparticles loaded carbohydrate polymeric films—A review. *International Journal of Biological Macromolecules*, 124803. https://doi.org/10.1016/j.ijbiomac.2023.124803
139. Zhang, C., Su, H., Baeyens, J., & Tan, T. (2014). Reviewing the anaerobic digestion of food waste for biogas production. *Renewable and Sustainable Energy Reviews, 38*, 383–392.
140. Zhang, Q., Xiao, J., Xue, J., & Zhang, L. (2020). Quantifying the effects of biochar application on greenhouse gas emissions from agricultural soils: A global meta-analysis. *Sustainability, 12*(8), 3436.
141. Zhang, S., Zhang, H., Liu, X., Zhu, S., Hu, L., & Zhang, Q. (2018). Upgrading of bio-oil from catalytic pyrolysis of pretreated rice husk over Fe-modified ZSM-5 zeolite catalyst. *Fuel Processing Technology, 175*, 17–25.
142. Zhang, Y., Fan, S., Liu, T., Omar, M. M., & Li, B. (2022). Perspectives into intensification for aviation oil production from microwave pyrolysis of organic wastes. *Chemical Engineering and Processing-Process Intensification, 176*, 108939.

Chapter 5
Transesterification of Waste Cooking Oil Through Microwave Technology: Recent Advances and Challenges

Bunushree Behera, Kolli Venkata Supraja, S. Mari Selvam, Snehi Kinger, and Prangya Ranjan Rout

Abstract Undeniably higher feedstock costs are hindering the commercialization of biodiesel production. Amidst the mentioned problem, there is a current exploration of alternatives like waste cooking oil (WCO). WCO holds the promise of substantially reducing production costs by 60–90%, attributed to its cost-effectiveness compared to vegetable cooking oils and energy crops. Microwave-assisted transesterification of WCO is a time-efficient process that yields high quality and quantity of fatty acid methyl esters (FAMEs) with a lower molar ratio of methanol to oil. Besides the growing interest in using microwaves for WCO transesterification, a comprehensive review of the technology for analysing factors affecting biodiesel production, and the challenges and limitations for process scale-up and commercialization is essential which is currently limited. The chapter illustrates the significance of microwave heating of WCO for biodiesel production. Further, the influence of parameters on the quality and quantity of biodiesel has also been discussed. The process energy requirement in comparison with the conventional method has been analysed. Challenges and limitations associated with technology commercialization have been annotated. Overall, this chapter enables the researchers to understand the

---

B. Behera (✉) · S. Kinger
Bioprocess Laboratory, Department of Biotechnology, Thapar Institute of Engineering and Technology, Patiala, Punjab 147004, India
e-mail: bunushree.behera@thapar.edu

K. V. Supraja
Waste Treatment Laboratory, Department of Biochemical Engineering and Biotechnology, Indian Institute of Technology Delhi, Hauz Khas, New Delhi 110016, India

S. Mari Selvam
Department of Biotechnology and Medical Engineering, National Institute of Technology Rourkela, Rourkela, Odisha 769008, India

P. R. Rout
Department of Biotechnology, National Institute of Technology Jalandhar, Jalandhar, Punjab 144027, India

G. Baskar et al. (eds.), *Circular Bioeconomy Perspectives in Sustainable Bioenergy Production*, Energy, Environment, and Sustainability,
https://doi.org/10.1007/978-981-97-2523-6_5

technology-driven biodiesel production through microwave irradiation of WCO and identify the potential loopholes to make the process sustainable.

Keywords Biodiesel · Microwave · Sustainability · Transesterification · Waste cooking oil

5.1 Introduction

Rapidly increasing energy demands of the ever-rising global population have led to the over-exploitation of fossil-based resources, which are currently on the verge of depletion. This has also led to an overwhelming increase in carbon dioxide levels in the atmosphere, resulting in global warming and disturbance of ecological diversity. There is a pressing demand to find an alternative form of energy from renewable sources to sort out the above-mentioned issues [1, 2]. Among the available sources, first-generation alternatives like solar, wind, and hydro energy require very high capital investment. The third-generation biofuel from microalgae is currently under investigation due to questionable economics linked with the algal productivity and lipid yield during scale-up [3]. In case of the second-generation feedstocks often claimed to be overshadowed by the feed versus fuel dilemma, used cooking oil, also popularly termed waste cooking oil (WCO), has gained much attention. Cooking oil obtained basically from plant-based sources, when utilized repeatedly due to economic reasons, especially in hostels, food outlets, and restaurants, often results in the mixing of oxygen with unsaturated acylglycerols, thereby forming dimeric/polymeric acids, which causes an increase in viscosity. The change in physicochemical properties, increase in acidity, and gradual browning with an unpleasant smell make it harmful during human consumption. Thus, a considerable amount of WCO (500,000 tonnes) is discarded into the environment annually, causing pollution of water bodies [4]. Alternatively, the high-free fatty acids (FFAs) containing WCO can be trans-esterified and converted into biodiesel.

Utilizing the transesterification process for converting the FFAs in WCO would produce a composite mixture of fatty acid alkyl esters (FAAEs), popularly termed as the "biodiesel", with properties similar to diesel as per the international standards. Acid/alkali-catalysed transesterification reactions under homogenous or heterogeneous conditions have been studied by numerous researchers [5–8] for biodiesel production. Influence of factors like the methanol: oil ratio, reaction time, temperature, catalyst amount, and governing the yield and properties of biodiesel have been reported by Aghel et al. [9], Gaur et al. [10], Mahesh et al. [11] and Martinez-Guerra and Gude [12]. Talebian-Kiakalaieh et al. [13] has enlisted different acid and alkaline homogenous catalysts for production of biodiesel from WCO. Though alkaline catalysts are commonly utilized for transesterification of WCO due to good reaction efficiency, and to obtain better quality biodiesel, high FFA content often might hinder the reaction due to unwanted saponification. This might result in catalyst deactivation and cause tri-acyl glycerol hydrolysis. Thus, it is imperative to use a two-step acid–alkali-based approach [14], where the first process (esterification would reduce the FFA content to 1–2%, while the latter would aid in the transesterification into FAAEs [15]. Several improvements to the existing conventional approaches like the use of ultrasonic [16, 17]; and microwave treatments [18, 19] applied either individually or in combination [12] have been researched to achieve an optimum biodiesel yield with a lower reaction temperature and in lesser time. Ultrasonic processes have shown improved extraction and transesterification efficiency but often have a longer reaction time, consume more solvent, and are energy-intensive than microwave irradiation [20]. Microwave-assisted extraction and transesterification have gained massive attention over the past few years due to uniform volumetric heating that results in better yield and product quality within minimal processing time [21]. Other advanced techniques of super critical solvent extraction are extremely energy intensive [22]. Microwave heating shows promising improved biodiesel production through esterification and transesterification, warranting further research for industrial applications. Studies by Cheng et al. [23] and Zhang et al. [24] reported that microwave-assisted transesterification resulted in uniform heating at a lower temperature of 100 °C within a shorter reaction time of 10–30 min. Using hetero-poly acid as catalysts, the authors in the latter study reported fatty acid methyl esters (FAMEs) yield of 96.2%, and the former reported an 86.7% yield. Lokman et al. [25], using microwave irradiation and heterogeneous carbon-based sulphuric acid catalysts, reported a FAME conversion of 96% in 15 min from palm acid distillate. Microwave-assisted transesterification for different feedstocks has also been investigated by Azcan and Yilmaz et al. [26], Gina et al. [27], and Yadav et al. [28]. Microwave-assisted transesterification consumes 0.075 kWh energy, much lower than conventional heating that requires 0.227 kWh, resulting in high yield of biodiesel with better FAME quality and purity [29]. Fatimah et al. [30] reported that zirconium oxide (ZrO_2) heterogeneous catalyst produced approximately 96% biodiesel yield in case of both conventional and microwave-assisted transesterification, however later consumes only 1/3rd of the time needed in conventional method. To date, most studies carried out for microwave-assisted transesterification have been chiefly done at the laboratory scale under batch operational mode [31]. Gude et al. [32] reported that the microwave process is challenging under batch operational conditions at the industrial scale. Tangy et al. [33] utilized SrO/SiO_2 catalyst in continuous microwave reactor and projected substantial catalytic potential at 24.6 min leading to 99.2% FAMEs from WCO. Yang et al. [34] have utilized magnetic SrO–ZnO/MOF catalyst for one-step biodiesel synthesis. Owing to the presence of active components of $Sr_3Fe_2O_6$ and ZnO, the catalyst showed basic and acidic strength. The details on the existing state-of-art of microwave-assisted transesterification have been reviewed by Mohamad Aziz et al. [21], Motasemi and Ani [35], Nayak et al. [36], Nomanbhay and Ong [31] and Sajjadi et al. [37]. It could be interpreted that microwaves offer high-efficiency biodiesel production, with shorter processing times and improved product quality, paving the way for industrial adoption through better understanding and optimization.

Most of the above-mentioned reviews have summarized the factors influencing the microwave-assisted process and its effect on the quantity of FAMEs and the physio-chemical properties of biodiesel. Except for a few, a comprehensive summary of the economics and environmental impacts is seldom discussed. To expedite the industrial design and safer usage of microwave-assisted transesterification, an in-depth understanding of the influencing parameters along with the underlying mechanism and its associated economic and environmental impacts is necessary. Thus, the present chapter holistically and systematically reviewed the mode of action of microwaves during the transesterification process and enlisted the influencing factors, and FAMEs yield which could be obtained from WCO. The enviro-economic impacts linked with the conversion aspects have been discussed in detail. Challenges encompassing the technical and regulatory requirements are delineated. Overall, the review will act as a baseline platform to guide potential researchers and stakeholders in implementing WCO-based conversion into biodiesel.

5.2 Microwave-Mediated Transesterification: Modes and Mechanism

The conversion of WCO to biodiesel through transesterification process can be assisted through various techniques such as reactive distillation, non-catalytic super-critical method, membrane technology, microwave, ultrasonication, biox process, and reactive distillation [38]. Super-critical transesterification though regarded as a neat approach, the thermal stability of fatty acid methyl esters (FAMEs) is compromised due to high temperature and pressure. Also, co-solvent, co-catalyst, and sub-critical method are undoubtedly efficient in conversion but are often limited by the high cost of chemicals/solvents involved [22]. Among these methods, microwave-assisted process is energy-intensive and shows higher conversion efficiency, product quality in shorter duration. The limited solvent requirements, rapid reaction rates, contactless uniform volumetric heating, and minimal undesired reactions are the advantages of microwave heating over other techniques [39]. Thus, understanding the basic mechanism of microwave technology is essential for optimal design and operation of reactors to achieve process conversion without harmful impacts of microwave leakage.

Microwave irradiation encompasses high-frequency waves (ranging from 0.3 to 300 GHz), including infrared and radio waves, characterized by wavelengths spanning 0.01–1 m. Domestic microwave ovens have a frequency of 2.45 GHz, while industrial-scale microwave ovens have a range of operation of 915–2450 GHz, where rapid molecular vibrations of 2450 million times a second occur due to the change in electric field [21]. Over the years, microwave irradiation has been increasingly employed in biochemical transformations resulting in biodiesel production, hydrothermal synthesis, carbonization, and depolymerization reactions [40]. Microwave irradiation resulting in high-intensity molecular vibrations often increases thermal energy resulting in more uniform heating, facilitating more straightforward and faster biochemical transformations. The energy content of microwaves is usually less than most biomolecules and their intermolecular interactions [32]. This radiation induces molecular and interfacial interactions while preserving the internal structure unchanged [41]. Microwave-based heating has several advantages over the conventional process resulting in uniform volumetric heat transmission, avoiding heat loss without direct contact with the source [40]. This feature results in faster heating of catalysts, and the enhanced interaction between the reactants for transesterification, thereby providing better efficacy and control. The mechanism illustrating the difference between microwave and conventional thermal heat transfer is depicted in the review by Gude et al. [32] and Mohamad Aziz et al. [21]. The charged molecules or dipoles in the medium often collide and re-orient themselves with the alternating electric field applied, resulting in rapid frictional interaction generating thermal energy. The interaction of the thermal energy with the reactant components causes rapid and localized superheating, which is not appropriately measured through bulk temperature but can be quantified by conductive polarization. Microwave technology could be utilized in the extractive process and the transesterification reaction. Further, microwave reactors can also combine oil extraction and its subsequent conversion into a single step, thereby reducing the costs and energy requirements in the process, making it more sustainable. The stoichiometric equation for the transesterification reaction is given by Eq. 5.1 [42]

$$\mathrm{WCO} + 3\mathrm{AC} = G + 3\mathrm{FAE} \tag{5.1}$$

where AC represents alcohol and base catalyst solution, G represents glycerol, and FAE represents a free fatty acid ester. Glycerol produced during microwave-assisted process is much less due to higher FAME conversion efficiency. Glycerine produced is dependent on the type of alcohol used during reaction. Glycerol has been reported to be produced more with methanol, followed by iso-propyl alcohol, iso-butyl alcohol, and iso-pentyl alcohol [43]. Saifuddin et al. [44] reported that even the by-product glycerol obtained can be purified using a combined process of microwave acidification followed by adsorption using yeast cells immobilized on chitosan.

The mechanism of base-catalysed transesterification commences with the generation of alkoxide ions (RO−) in preliminary phase. These alkoxide ions proceed to initiate an attack on carbonyl carbon present in triglyceride resulting in tetrahedral intermediate [42]. This intermediate undergoes a rearrangement, resulting in fatty acid esters production through subsequent reaction with alcohol, thereby augmenting the alkoxide ion population and contributing to the accumulation of fatty acid esters. The pulse and continuous modes of microwave irradiation are effective for transesterification reactions owing to the higher reaction rates [37]. Molecular orbital method was employed to elucidate these phenomena. Under microwave irradiation, the triglyceride's structure transforms into a flatter conformation leading to a reduction in its dipole moment [31]. Furthermore, the molecular structure with the lower dipole moment persists for a brief period following the cessation of irradiation. In the

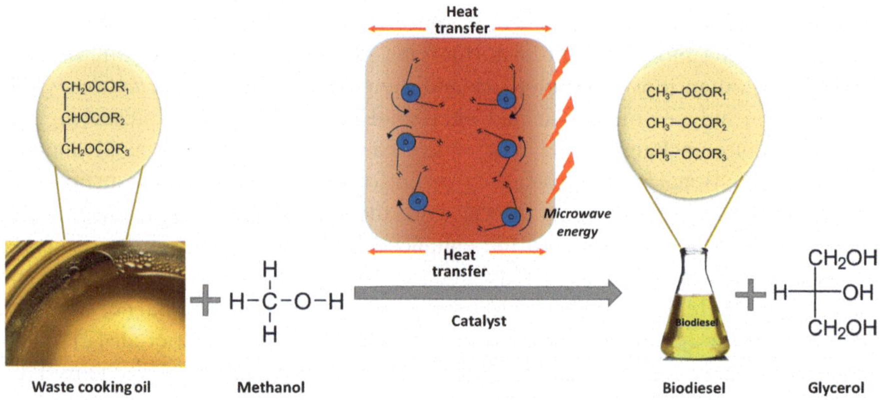

Fig. 5.1 Mode and mechanism of microwave-assisted transesterification of waste cooking oil

process of transesterification, a crucial step involves the transition through a tetrahedral state around the carboxyl carbon [45]. When triglyceride adopts a flattened structure with reduced dipole moment, it creates a larger region above it. This structural alteration makes it more favourable for methoxy group to initiate an attack on the carboxyl carbon. The mode and mechanism of microwave-assisted transesterification of WCO have been given in Fig. 5.1. In microwave-assisted conversion, biodiesel production can be affected by limitations in mass transfer leading to inefficient transport of reactants. These limitations may arise due to factors such as slow diffusion, inadequate mixing, or restricted contact between phases, and they often require optimization of reaction conditions, catalysts, and reactor design to overcome. To ensure efficient mass transfer inside the reaction mixture, it is important to optimize reaction parameters with proper mixing and stirring of the reactants, suitable solvents can be selected, and reactant concentrations can be adjusted with implementation of monitoring and control systems.

5.3 Factors Influencing Microwave-Assisted Transesterification of WCO

Various process parameters like alcohol to oil ratio, reaction time, temperature, catalyst concentration, and microwave power influence the transesterification reaction. Each of these factors is interlinked, and therefore, optimization using response surface methodology (RSM) based on full factorial design model [18], central composite design model [46], and Box–Behnken design model [47] have been carried out by researchers. The different factors that affect the efficiency of microwave-assisted transesterification of WCO either positively or negatively have been given in Fig. 5.2.

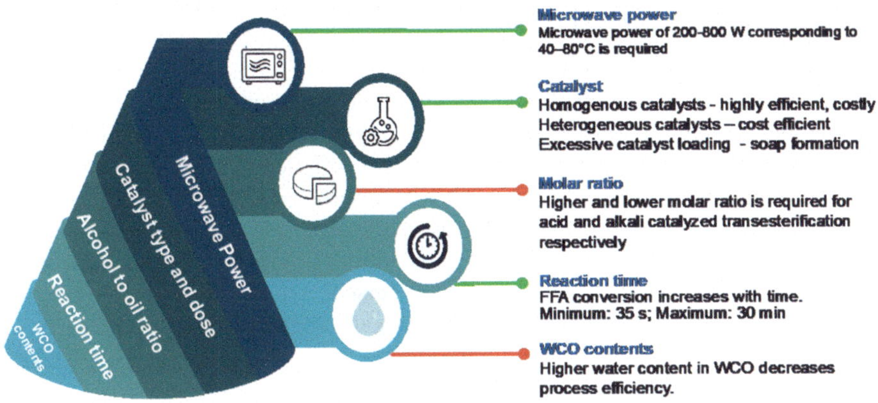

Fig. 5.2 Factors affecting the microwave-assisted transesterification of waste cooking oil

5.3.1 Effect of Microwave Power and Reaction Temperature

Microwave-assisted transesterification is an endothermic process and is influenced by reaction temperature. The viscosity of WCO reduces with increasing temperature, leading to sufficient mixing of reactants during the conversion process [48]. A microwave power of 200–800 W corresponding to a temperature of 40–80 °C is essential for transesterification depending on the FFA content of WCO [49]. Increasing microwave power above the optimal level can lead to the evaporation of alcohol content, thereby causing a reduction in biodiesel yield and increasing the by-product content. Hassan and Smith [42], developed a two-step microwave heating method, where the reactor is initially heated up to 300 W power for 1 min to obtain a temperature of 50 °C. Subsequently, the power is reduced to 100 W for 30 s to attain 60 °C, and then, the reaction time is recorded upon reaching this desired temperature. Although the reaction time is significantly reduced with microwave heating, a higher catalyst amount is needed to obtain a greater yield of biodiesel in short time [50]. With optimal conditions, methanol:oil ratio of 7.46:1, catalyst (calcium diglyceride) amount of 1.03 wt% at temperature of 62 °C, and highest biodiesel yield of 94.86% are obtained in a short time of 15 min using microwave heating at 850 W [46]. In contrast, using similar reaction parameters with conventional heating, the biodiesel yield is significantly low with 42.59%. The study also reported an activation energy of 26.56 kJ/mol for the biodiesel produced through microwave-assisted transesterification. It is essential to maintain an optimum temperature by adjusting the microwave power to facilitate the interaction of reactants, resulting in an increased yield of biodiesel.

5.3.2 *Effect of Catalyst*

Among all the factors, catalyst loading is identified as the most sensitive parameter affecting the conversion process. The use of catalysts with a dosage of 1–10 wt% in the transesterification process improves the process efficiency, reducing the reaction time and temperature required for the conversion process [36]. Elevating the catalyst concentration has the potential to enhance triglyceride conversion, leading to improved ester content. Inadequate catalyst quantities might result in the incomplete transformation of triglycerides into esters [51]. Based on its composition, the catalyst has been divided into homogeneous, heterogeneous, and enzymatic forms.

5.3.2.1 Homogenous Catalysts

Homogenous catalyst has high efficiency with lower reaction time and temperature, but this process is costly as it involves high purification costs for separating the catalyst from the reaction mixture [36]. Some of the acid catalysts like H_2SO_4, $ClSO_3H$, H_3CSO_3H, and HCl have been applied in transesterification process leading to conversion efficiency of up to 85% [52]. Alkali catalysts, including KOH, NaOH, and CH_3ONa, are soluble in alcohol as they can absorb microwave energy directly through the –OH group, leading to disruption of the two-tier arrangement of methanol and oil [53]. According to Ali et al. [47] and Babel et al. [52], the catalyst generated by activating limestone or coconut shell biochar with KOH exhibits insolubility in methanol or methyl esters, facilitating their easy separation from the reaction mixture after the transesterification process, resulting in biodiesel yield of 96% and 91%, respectively. Application of an acid or alkali catalyst in the transesterification process is based on the FFA% present in WCO [18]. The physical and chemical properties of WCO like viscosity, specific heat, surface tension, and %FFA content increase during the course of frying. This is due to the long-term heating at high temperatures (140–190 °C) and exposure to oxygen. Higher FFA content in WCO increases the soap formation with an alkali catalyst leading to deactivation of the catalyst. Hence, to reduce the FFA content, an acid catalyst is used in the conversion process as a pre-treatment step [36]. Soosai et al. [54] used 5% sulphuric acid-based esterification process to reduce FFA from 6.87 to 0.83%. For WCO with high %FFA content, two-step microwave-assisted transesterification process can be carried out with acid-catalysed esterification for minimizing the %FFA content, followed by alkali-catalysed reaction for production of biodiesel with an overall conversion efficiency of 95–97% [18, 54]. Some of the challenges associated with homogeneous catalysts include higher economic costs for separation of the catalyst from the reaction mixture and requirement of further purification steps to separate glycerol from the ester phase [50]. Due to these drawbacks, heterogeneous catalysts have been developed to reduce the process economics as well as to improve the conversion efficiency without requirement of additional purification steps for biodiesel production.

5.3.2.2 Heterogenous Catalysts

Heterogeneous catalysts are process-efficient, with easy separation, longer shelf life and reduced equipment corrosion [50]. These include various hetero-polyacids (HPA), metal oxides, acidic polymer resins, and natural substances like carbon, zeolites, or lignin-derived materials [36]. Nazir et al. [48] utilized a heterogeneous nano-catalyst with significantly high surface area of 30.31 m^2/g and pore size of 9.44 nm, with an acid density of 4.74 mmol/g. The catalyst had a good number of functional groups (sulphonic acid, carboxyl, and hydroxyl) and was reported to lose only 11% stability after 6 cycles of transesterification. Similarly, lignin-based catalysts have also been used by Yadav et al. [55, 56]. The use of biomass derived catalyst has been projected to make the microwave-assisted transesterification less energy intensive [57]. Because of high basicity, researchers have also focussed on strontium oxide (SrO) and zinc oxide (ZnO)-based bifunctional acid/base catalysts like Sr/ZrO_2, $ZnO\text{-}La_2O_3$, Sr-Ti, $SrO\text{-}ZnO/Al_2O_3$, and SrO-ZnO/MOF that showed a transesterification efficiency of 79–98% towards conversion of highly acidic WCO to biodiesel [34, 58]. Compared to others, SrO-based heterogeneous catalysts aid in better penetration of microwave irradiation, thereby enhancing the transesterification process [33]. 0.3% CaO-based catalysts, with 18:1 alcohol: oil ratio, at 270 W irradiation, after 114 s have been reported to yield 97.4% biodiesel [54]. Earlier studies by Khemthong et al. [59] have also shown the presence of 99.2% wt. of CaO in egg shells, with high strength basic sites and stability, that would result in increased catalytic heating during transesterification. Buasri et al. [60] projected the use of alkali-based catalysts prepared via calcination of animal bones for microwave-assisted transesterification, as an economic method. Heterogeneous catalysts are found to be more efficient than homogeneous catalysts considering the catalytic activity, energy efficiency, biodiesel yield, reaction time, and process economics.

5.3.2.3 Enzymatic Catalysts

Enzymatic catalysts are mainly fungi or bacteria-derived lipase proteins that offer easy catalyst separation from the reaction mixture, and are active at 40–60 °C [36]. However, there are limitations to this process as it takes greater reaction time for conversion, and there may be chances of thermal deactivation of the enzyme [61]. When solid catalysts are used in microwave-mediated reaction processes, they exhibit better results within shorter reaction times compared to conventional heating methods using the similar catalysts [37]. Additionally, the microwave heating mechanism enhances reaction kinetics by increasing the rate constant values, improving conversions.

5.3.3 Effect of Alcohol to Oil Ratio

Methanol and ethanol are widely used alcohols for the transesterification process because they can transform microwave energy into heat, leading to rapid boiling in a short time [35, 62]. Maintaining an appropriate molar ratio of methanol to oil is essential for ensuring a sufficient supply of methanol to effectively break the glycerol-fatty acid linkages [63]. Tangy et al. [33] reported that high methanol: oil ratio increases the contact area for triglyceride moieties and alcohol facilitating transesterification. Comparatively, higher molar ratio of methanol to oil is needed for acid-catalysed transesterification than for alkali-catalysed reaction under the same experimental conditions [53]. At a constant catalyst dosage of 1 wt% and fixed microwave power of 250 W, decreasing the methanol-to-oil molar ratio and prolonging the reaction time aims to favour a reversible reaction, consequently leading to a reduction in the overall yield of the process [18]. To move the transesterification reaction forward, a methanol-to-oil molar ratio of minimum 3:1 to a maximum of 18:1 is essential depending upon the %FFA content of WCO to obtain a biodiesel yield of 85–98% [42]. However, presence of excess methanol in the reaction mixture could lead to slower conversion process and greater settling time, while increasing the solubility of glycerol with the ester [29]. Enzymatic microwave transesterification processes are greatly affected by methanol leading to reduction in the enzymatic activity. Other solvents such as butanol or dimethyl carbonate at a molar ratio of 6:1 have been used for transesterification to achieve a biodiesel yield of 83–94% [61].

5.3.4 Effect of Reaction Time

Conventional transesterification processes based on reactor heating require a reaction time of 30 min to 180 min for biodiesel production [64]. Microwave irradiation is attributed to the direct absorption of radiation by the polar –OH group in alcohol. This absorption leads to immediate excitation of the –OH group by microwave radiation, causing a significantly higher local temperature than its surroundings [10]. Consequently, microwave-assisted conversion offers a reaction time of 3–30 min for complete conversion [37]. At an equivalent reaction time of 15 min, conventional heating resulted in a biodiesel yield of 42.59%, while microwave heating showed a significantly higher yield of 94.86% [46]. Reaction time also varies with the temperature and catalyst used during the process. Acid-catalysed conversion process with 2 wt% H_2SO_4, at a reaction temperature of 60 °C and reaction time of 200 min, resulted in a biodiesel yield of 24% only [42]. While for alkali-catalysed reaction at a temperature of 60 °C with 1 wt% KOH, a biodiesel yield of 93–98% is achieved within 60 min of reaction time. Often, an optimal reaction time provides an adequate contact time between irradiation and reactants, promoting interaction between glyceride moieties of oil and methanol [33]. Enzymatic microwave transesterification processes usually require a greater reaction time of 240 min for obtaining a biodiesel yield of up to 94%. Although the reaction time is very high, enzyme reusability and sustainability of the process make the process more economically viable [61]. These are the significant parameters influencing the transterification efficiency of WCO through microwave and the few related studies reported has been given in Table 5.1.

5.4 Enviro-Economic Feasibility of Microwave-Assisted Transesterification of WCO

Although biodiesel production from WCO is considered an acceptable option to produce alternative energy, the scale-up requirements often need validation of their economic feasibility and environmental impact assessment. One of the earliest studies by Marchetti et al. [69] reported that a significant fraction of biodiesel production costs (76–89%) are linked with feedstock/raw material prices. Thus, exploring WCO or recycled oil can bring down the overall costs. The study also emphasized using heterogeneous catalyst systems for transesterification since the homogenous catalyst systems involve more equipment and process costs. The scale of operation and the plant capacity play a significant role in deciding the selling price of WCO-based biodiesel, either in neat or blended forms. Hussain et al. [70] reported the price of biodiesel produced at a rate of 150 kt/year via transesterification of WCO with 15 years' lifetime to be higher than that of conventional diesel in the United Arab Emirates. Similar reports of biodiesel prices being 3.7 times higher than conventional diesel were also projected by Mohammadshirazi et al. [71] for Iran. Recent studies by Farid et al. [72] and Liu et al. [73] have conducted the techno-economic assessment for biodiesel production from WCO. Farid et al. [72] extended the process of alkaline transesterification resulting in 74.3% biodiesel yield with 96.9% purity, to energy and economic feasibility analysis using engineering principles and multipliers, considering a production capacity of 3.68 kt/annum. The authors have reported a net energy ratio (NER) of 1.15, with a net present value (NPV) of 1.43 million US $ at a 60% internal rate of returns (IRR) for B10 blends considering a lifetime of 10 years. Liu et al. [73] analysed the process feasibility through life-cycle cost analysis in Aspen Plus software via simulating a 2-step transesterification consisting of pre-treatment followed by alkali-catalysed transesterification of WCO, resulting in biodiesel with 99.6% purity. The cost of biodiesel was projected to be 65.28% higher than that of conventional diesel. Owing to the higher content of impurities and free fatty acids (FFA), the processing and pre-treatment costs account for 56.26% of total production costs. Thus, the use of a single-step process co-combined with advanced techniques like microwave-assisted transesterification under optimized conditions would help in making the technology feasible during scale-up.

Refaat et al. [74] compared the conventional and microwave-assisted transesterification processes of high FFA containing WCO and reported 97% and 94% reductions in reaction and separation time, respectively, with quality similar to the biodiesel produced during the conventional process. Reduced treatment and processing time

Table 5.1 Studies reported on microwave-assisted transesterification of waste cooking oil for biodiesel production

Feedstock	Alcohol: oil ratio	Reaction time	Catalyst dosage	Microwave power/ temperature	Reactor	Yield	References
WCO (8.17% FFA)	1st step MeOH: oil—19.57:1	35 s	H_2SO_4—1 wt%	250 W, 65 °C	Domestic microwave oven	94.6% (0.082% FFA)	Supraja et al. [18]
	2nd step MeOH: oil—5:1	30 min	NaOH—2 wt%				
WCO	MeOH: oil—12.21:1	55.26 min	KOH-activated limestone—5.47 wt%	900 W, 65 °C	Continuous microwave-assisted reactor (CMAR)	96.65% (0.38% FFA)	Ali et al. [47]
WCO (23% FFA)	MeOH: oil—0.3 v/v	60 min	KOH—1 wt%	400 W, 60 °C	Multimode microwave apparatus (CEM MARS)	97.4%	Hassan and Smith [42]
WCO	MeOH: oil—9:1	6 min	KOH—2 wt%	800 W, 80 °C	Domestic microwave oven	92%	Prafulla et al. [65]
WCO	MeOH: oil—12:1	6 min	BaO—2 wt%	800 W, 80 °C	Domestic microwave oven	96%	Prafulla et al. [65]
WCO	DMC: oil—6:1	240 min	Lipase 435—10 wt%	70 W	Domestic microwave oven	94%	Panadare and Rathod [61]
WCO (0.46% FFA)	Alcohol: oil—6:1	1 min	KOH—1 wt%	800 W	Domestic microwave oven	89.29%	Zare et al. [66]
WCO (0.68% FFA)	MeOH: oil—12:1	40 min	KOH-activated coconut shell biochar—5 wt%	80 °C	Domestic microwave oven	91.3%	Babel et al. [52]

(continued)

Table 5.1 (continued)

Feedstock	Alcohol: oil ratio	Reaction time	Catalyst dosage	Microwave power/ temperature	Reactor	Yield	References
Inedible olive oil (2.36% FFA)	MeOH: oil—9:1	9 min	KOH—1.2 wt%	500 W	Domestic microwave oven	93.52%	Dehghan et al. [29]
Waste frying palm oil	MeOH: oil—12:1	7 min	NaOH—1 wt%	312 W, 65 °C	Domestic microwave oven	88.63%	Yakoob et al. [67]
WCO	MeOH: oil—11.62:1	4.47 min	$NaOCH_3$—0.68 wt%	720 W, 65 °C	Continuous microwave-assisted reactor (CMAR)	97.13%	Ali et al. [68]
WCO	MeOH: oil—18:1	15 min	Lignin (–SO_3H, –COOH and –OH)—15 wt%	60 °C	Microwave reactor under continuous stirring	89.19%	Nazir et al. [48]
WCO (soyabean oil)	MeOH: oil—6:1	20 min	Corn cobs-derived sulfonated activated carbon—20 wt%	600 W, 75 °C	Microwave irradiated unit with a reflux condenser	88.7%	Rocha et al. [57]
WCO	MeOH: oil—12:1	10 min	Yellow horn oil utilizing hetero-polyacid catalyst $Cs_{2.5}H_{0.5}PW_{12}O_{40}$—1 wt%	500 W, 60 °C	MARS-II microwave accelerated reaction system	96.22%	Zhang et al. [24]
WCO	MeOH: oil—18:1	20 min	Guinea fowl bone-based heterogeneous catalyst- 4 wt%	800 W, 65 °C	Domestic microwave oven	95.82%	Singh and Sharma [49]
WCO	MeOH: oil—7.46:1	15 min	Calcium diglyceride—1.03 wt%	62 °C	Multi-wave PRO microwave reactor	94.86%	Gupta and Rathod [46]
WCO	MeOH: oil— 18:1	5 min	Natural Hydroxyapatite (NHAp) derived from pork bone—4 wt%	800 W	Domestic microwave oven	94%	Buasri et al. [60]

(continued)

Table 5.1 (continued)

Feedstock	Alcohol: oil ratio	Reaction time	Catalyst dosage	Microwave power/ temperature	Reactor	Yield	References
Palm oil	MeOH: oil—18:1	4 min	Industrial eggshell wastes containing CaO—15 wt%	900 W	Domestic microwave oven	96.7%	Khemthong et al. [59]
Non-edible silk-cotton seed oil	MeOH: oil—18:1	114 s	Calcium oxide (CaO)—0.3 wt%	270W	Domestic microwave oven	97.4%	Soosai et al. [54]
Oleic acid	MeOH: oleic acid—16:1	60 min	Lignin-rich biomass (WNS-SO_3H)—9 wt%	50 W, 85 °C	Discover SP microwave system	99.01%	Yadav et al. [56]
Oleic acid	MeOH: oleic acid—24:1	60 min	*Oryza sativa* catalyst—8 wt%	50 W, 80 °C	Discover SP microwave system	99.6%	Yadav et al. [55]
WCO	MeOH: oil—12:1	2 min	SrO—1.25 wt%	242 W, 65 °C	Flow SYNTH microwave system	94.5%	Tangy et al. [33]

are expected to have an economic advantage during the overall process. Patil et al. [19] reported lower energy requirements of 255 kJ during microwave-assisted extraction transesterification of *Nannochloropsis sp.* than the supercritical process requiring 600 kJ of energy. A promising aspect of using industrial-scale continuous microwave reactors for biodiesel production with positive energy returns has been reported by Muley and Bolder [75]. The use of single-step microwave extraction and acid-catalysed transesterification of *Rhodotorula glutinis* resulted in FAMEs, with the energy requirement of the microwave being 20% less than that of the energy content of biodiesel [76]. The study also projected the energy return on investment to be significantly influenced by reaction time than the process temperature. Waudby and Zein [77] projected that the feasibility of microwave-assisted technology is possible with the inclusion of recycled streams and an optimal combination of separation processes. Reviews by Manikandan et al. [78], Tibesigwa et al. [79], and Nayak et al. [36] have summarized the state-of-art linked with the techno-economic aspects of biodiesel production. To the best of the author's knowledge, a limited number of studies have been done with respect to the techno-economic feasibility of the transesterification of WCO. Nevertheless, the insights from the factors influencing industrial-scale microwave technology for different feedstocks provide its potential for scalability and could also be extrapolated for WCO.

Environmental assessment of WCO conversion into biodiesel is a crucial aspect to be considered for process scale-up. Chung et al. [80] reported reduced environmental impacts linked with using WCO as feedstock compared to that of the Jatropha due to the absence of agricultural processing involved. Further, using an eggshell-derived calcium oxide (CaO) based catalyst had a much lower overall environmental impact than homogenous KOH-based transesterification due to the processing of chemicals during the production and waste streams disposal involved thereof. A detailed LCA study by Foteinis et al. [81] compared the environmental impacts of biodiesel produced from WCO with those from the first and second generations. The biodiesel from WCO was found to be more environmentally sustainable with 0.55 tonnes CO_2 eq. and 58.37 Pt environmental footprints per tonne of biodiesel produced. Authors have projected these values to be approximately 40% lower than the first-generation biodiesel and almost an order of magnitude than the third-generation microalgal biofuel. Also, a threefold reduction in environmental impacts was reported by authors as compared to petroleum-based fuel. The non-renewable energy input during the processing, along with the use of chemicals, is the primary cause of environmental impacts. Iglesias et al. [82] projected that biodiesel production from WCO is environmentally sustainable when operated in a centralized territory rather than at a decentralized level. Though conversion of WCO is linked with a decrease in environmental impacts, most emissions are linked with the chemicals and energy involved during the transesterification step [83]. Alishahi et al. [84] compared the properties and the associated environmental impacts of biodiesel produced from WCO via magnetic stirrer esterification and microwave-assisted esterification, projecting their physicochemical properties and heating rate to be similar, however, the energy linked in the latter case was 4-times lower than the former. The energy ratio of 1.04, efficiency of 0.0178 kg/MJ, and the intensity of 56.07 MJ/kg were observed by Yari et al. [85]

for the biodiesel produced from waste fish oil. Vionery-Portillo et al. [86] reported that the B25 blend of biodiesel from WCO and conventional low-sulphur diesel used in a 33 kW power generator had almost 39% reductions in eutrophication potential, acidification potential, abiotic depletion, and potential for human toxicity. Also, the global warming potential declined by 36%. A decrease in NO_x by 42% and a 52% increase in CO was also observed in the exhaust emissions compared to the low-sulphur diesel. Limitations and challenges linked with the LCA of WCO into biodiesel, along with the potential research opportunities, have been summarized in detail by Hosseinzadeh-Bandbafha et al. [87].

5.5 Commercialization of WCO-Based Biodiesel: Challenges and Future Outlook

WCO is undoubtedly considered an ideal feedstock for conversion into second-generation biofuel owing to the issues linked with its subsequent use in cooking resulting in health hazards and environmental impacts associated with improper disposal. However, the commercialization process is hindered by several technical and socio-economic factors.

Over the years, even though technical advances have improved yield and brought down the price of biodiesel, still presently projected costs are higher than the existing counterparts, thereby questioning its commercial feasibility. Microwave-based reactors are often costly in terms of production, compared to the chemical synthesis reactors. Scaling-up is technically hindered by the development of spark or electric arc and sudden surge in temperature or localized hotspots due to thermal leakage [88]. Lack of information on dielectric properties of materials makes the process of optimization difficult, thereby declining the product yield and its subsequent properties. Shorter wavelength, low frequency with higher penetrative capacity, might result in potential health hazards when operated for prolonged interval of time. Also, supply chain management is another limiting factor hindering commercialization [15]. Smaller restaurants in several countries are seldom equipped with facilities to recover and separate the WCO for transport to adequate processing sites converting them into biofuel. Often insufficient daily discharge, concomitant high equipment costs, and lack of space are the primary cause of the unavailability of stringent supply chain, especially in developing countries. Thus, a significant fraction of WCO is disposed untreated due to the absence of stringent rules and policies. Few countries like the USA, Japan, and EU have enforced regulatory frameworks governing the disposal of WCO. The demand of WCO in developing countries like China, India, Indonesia, Japan, Republic of Korea and Malaysia has increased recently. Collectively, these countries collect 3700–5000 kT/year [89]. Liu et al. [90] during a survey in urban areas of Beijing reported 90.14 thousand tonnes produced in 2016, out of which 24% hotels do not submit it to the collectors, and thus 21.63 thousand tonnes are unaccounted. With economic incentives by government of China and subsidy value of 4

Yuan RMB/kg, the above problem can be avoided. There is an ardent need to extrapolate these laws and regulations and impose subsidies on the collection, recovery, and treatment facilities for WCO in developing countries. Appropriate recycling infrastructure and sales price of biofuel would definitely attract stakeholders and beneficiaries. The comparative analysis of standards of waste cooking oil in relation to biodiesel has been given in Table 5.2.

Another limiting feature hindering the commercialization of WCO-based biodiesel is its varying physicochemical properties in each batch. This variation mainly results from the differences in the FFA content, heat exposure time, complicated oil and food constituents, and their degraded counterparts. Thus, a stringent selective procedure must be used to pre-treat the raw WCO to bring its parameters under the appropriate limit before transesterification [94]. The emergence of advanced processes like microwave-assisted techniques under optimized conditions would aid in adequate sample preparation. Also, there is a need of better equipment for characterization and appropriate analysis of the physiochemical properties of samples for efficient conversion. The development of accurate temperature sensors is required to reduce the potential errors in transesterification.

Though the energetics for microwave-assisted transesterification have been investigated, the number of studies linked with the economics and environmental impacts of the aspects is minimal. Implementing technology to scale up WCO conversion into biodiesel requires a stringent scrutinization of the economics involved, from raw materials procurement and processing to its pre-treatment and microwave-based conversion into biodiesel. Thus, more techno-economic studies must be conducted considering microwave-assisted processes under optimized conditions at the semi-industrial scale. Environmental impact assessment studies must be carried out without zero-burden assumptions, as highlighted by [87], since the waste by-products generated during the transesterification process can be utilized as raw materials in other industrial processes. The process scale-up of microwave-assisted transesterification for substantial biodiesel production has to be extrapolated from lab-scale studies. These studies would aid in convincing policymakers about the sustainability aspects of microwave-based conversion of WCO into biodiesel.

5.6 Conclusions

WCO, as a second-generation feedstock for biofuel production, can curb the growing energy demands, avoiding the unwanted environmental pollution caused by their inadequate disposal. Microwave-based technology could facilitate the transesterification of WCO with much lower energy requirements and in lesser process time. Optimization of process conditions like the microwave power, alcohol to oil ratio, reaction time, and catalyst dosage for microwave-assisted transesterification is expected to improvise the process efficiency resulting in a better yield and quality of biodiesel with FAMEs suitable for engine applications. Better conversion efficiency could be obtained with the utilization of heterogeneous solid catalyst with capacity to adsorb

Table 5.2 Comparative analysis of biodiesel properties for waste cooking oil

WCO/ Biodiesel	Density (g/ cm^3)	Viscosity (mm^2/s)	% FFA	Calorific value (MJ/ kg)	Saponification value (mg of KOH/g oil)	Acid value (mg KOH/g oil)	Iodine value (mg KOH/g oil)	Cetane no	Flash point (°C)	Water content (%)	References
WCO	–	28.8	–	44.44	186.3	17.41	–	32.48	–	–	Prafulla et al. [65]
WCO	0.920	38.41	–	–	204.4	2.45	–	–	–	0.5	Gupta and Rathod [46]
WCO	0.910	48.10	8.17	–	193.9	16.24	193.04	–	–	–	Supraja et al. [18]
WCO	0.904	49.05	9.92	38.59	–	2.3	–	–	–	–	Milano et al. [91]
WCO	0.910	37.7	3.75	–	–	7.5	–	–	295	0.12	Nazir et al. [48]
WCO	0.9206	4.38	–	39.8	199.8	0.8	124.05	–	282	0.04	Azcan and Yilmaz [26]
WCO	–	4.2	1.47	–	181	2.57	103	–	–	0	Selvaraj et al. [92]
ASTM D6751 standard for biodiesel	0.87–0.89	1.9–6.0	< 0.45		< 370	< 0.5	< 115	> 47	> 130	< 0.05%	Motasemi and Ani [35]
EN 14,214 standard for biodiesel	0.86–0.90	3.5–5.0	–	–	–	< 0.5	< 120	> 51	> 101	< 0.05%	Motasemi and Ani [35]

(continued)

Table 5.2 (continued)

WCO/ Biodiesel	Density (g/ cm^3)	Viscosity (mm^2/s)	% FFA	Calorific value (MJ/ kg)	Saponification value (mg of KOH/g oil)	Acid value (mg KOH/g oil)	Iodine value (mg KOH/g oil)	Cetane no	Flash point (°C)	Water content (%)	References
IS 15607 standard for biodiesel	0.86–0.9	2.5–6.0	–	–	–	< 0.5	–	> 51	> 120	< 0.05%	Rao et al. [93]
Biodiesel	–	2.6	–	45	–	–	–	46	–	–	Prafulla et al. [65]
Biodiesel	0.87	4.85	–	–	–	0.3	–	–	165	–	Gupta and Rathod [46]
Biodiesel	0.892	4.68	0.02	–	110.6	0.04	112.02	–	–	–	Supraja et al. [18]
Biodiesel	0.862	5.01	–	38.766	–	0.13	–	–	154	–	Milano et al. [91]
Biodiesel	0.83	3.92	–	–	–	0.39	–	49.34	133	0.02	Nazir et al. [48]
Biodiesel	0.8883	2.35	–	38.529	–	–	–	–	185	0.03	Azcan and Yilmaz [26]
Biodiesel	–	5.2	0.48	–	186	0.96	111	–	–	0	Selvaraj et al. [92]
Biodiesel	0.87	4.2	–	36.6	–	0.34	–	–	183	–	Sharma et al. [64]

microwave energy, thereby converting them into heat. Economic and environmental assessment studies on microwave-assisted transesterification must be focussed on scale-up and implementation of semi-industrial or pilot-scale processes. More studies on combining microwave-assisted transesterification with supercritical and sub-critical methods, must be carried to achieve better diesel yield and quality. Stringent rules and policies on properly separating WCO from the source, transportation, and biochemical conversion would facilitate its real-time application. Attractive subsidies, market initiatives, and an appropriate sale price for biodiesel would encourage investors, thereby making upscaling and commercialization feasible.

Acknowledgements The authors acknowledge the Department of Biotechnology, Thapar Institute of Engineering and Technology, Patiala, the Department of Biochemical Engineering and Biotechnology, Indian Institute of Technology Delhi, the Department of Biotechnology and Medical Engineering, National Institute of Technology Rourkela and the Department of Biotechnology, and National Institute of Technology Jalandhar for providing the requisite facilities to carry out the study.

References

1. Behera, B., Acharya, A., Gargey, I. A., Aly, N., & Balasubramanian, P. (2019). Bioprocess engineering principles of microalgal cultivation for sustainable biofuel production. *Bioresource Technology Reports, 5*, 297–316.
2. Maity, J. P., Bundschuh, J., Chen, C. Y., & Bhattacharya, P. (2014). Microalgae for third generation biofuel production, mitigation of greenhouse gas emissions and wastewater treatment: Present and future perspectives—A mini review. *Energy, 78*, 104–113.
3. Behera, B., Unpaprom, Y., Ramaraj, R., Maniam, G. P., Govindan, N., & Paramasivan, B. (2021). Integrated biomolecular and bioprocess engineering strategies for enhancing the lipid yield from microalgae. *Renewable and Sustainable Energy Reviews, 148*, 111270.
4. Yacob, M. R., Kabir, I., & Radam, A. (2015). Households willingness to accept collection and recycling of waste cooking oil for biodiesel input in Petaling District, Selangor, Malaysia. *Procedia Environmental Sciences, 30*, 332–337.
5. Ali, C. H., Asif, A. H., Iqbal, T., Qureshi, A. S., Kazmi, M. A., Yasin, S., Danish, M., & Mu, B. Z. (2018). Improved transesterification of waste cooking oil into biodiesel using calcined goat bone as a catalyst. *Energy Sources, Part A: Recovery, Utilization, and Environmental Effects, 40*(9), 1076–1083.
6. Borah, M. J., Das, A., Das, V., Bhuyan, N., & Deka, D. (2019). Transesterification of waste cooking oil for biodiesel production catalyzed by Zn substituted waste egg shell derived CaO nanocatalyst. *Fuel, 242*, 345–354.
7. Maneerung, T., Kawi, S., Dai, Y., & Wang, C. H. (2016). Sustainable biodiesel production via transesterification of waste cooking oil by using CaO catalysts prepared from chicken manure. *Energy Conversion and Management, 123*, 487–497.
8. Rezania, S., Oryani, B., Park, J., Hashemi, B., Yadav, K. K., Kwon, E. E., Hur, J., & Cho, J. (2019). Review on transesterification of non-edible sources for biodiesel production with a focus on economic aspects, fuel properties and by-product applications. *Energy Conversion and Management, 201*, 112155.
9. Aghel, B., Mohadesi, M., Ansari, A., & Maleki, M. (2019). Pilot-scale production of biodiesel from waste cooking oil using kettle limescale as a heterogeneous catalyst. *Renewable Energy, 142*, 207–214.

10. Gaur, A., Mishra, S., Chowdhury, S., Baredar, P., & Verma, P. (2021). A review on factor affecting biodiesel production from waste cooking oil: An Indian perspective. *Materials Today: Proceedings, 46*, 5594–5600.
11. Mahesh, S. E., Ramanathan, A., Begum, K. M. S., & Narayanan, A. (2015). Biodiesel production from waste cooking oil using KBr impregnated CaO as catalyst. *Energy Conversion and Management, 91*, 442–450.
12. Martinez-Guerra, E., & Gude, V. G. (2014). Synergistic effect of simultaneous microwave and ultrasound irradiations on transesterification of waste vegetable oil. *Fuel, 137*, 100–108.
13. Talebian-Kiakalaieh, A., Amin, N. A. S., & Mazaheri, H. (2013). A review on novel processes of biodiesel production from waste cooking oil. *Applied Energy, 104*, 683–710.
14. Ho, K. C., Chen, C. L., Hsiao, P. X., Wu, M. S., Huang, C. C., & Chang, J. S. (2014). Biodiesel production from waste cooking oil by two-step catalytic conversion. *Energy Procedia, 61*, 1302–1305.
15. Goh, B. H. H., Chong, C. T., Ge, Y., Ong, H. C., Ng, J. H., Tian, B., Ashokkumar, V., Lim, S., Seljak, T., & Józsa, V. (2020). Progress in utilisation of waste cooking oil for sustainable biodiesel and biojet fuel production. *Energy Conversion and Management, 223*, 113296.
16. Gupta, A. R., Yadav, S. V., & Rathod, V. K. (2015). Enhancement in biodiesel production using waste cooking oil and calcium diglyceroxide as a heterogeneous catalyst in presence of ultrasound. *Fuel, 158*, 800–806.
17. Topare, N. S., & Patil, K. D. (2021). Biodiesel from waste cooking soybean oil under ultrasonication as an alternative fuel for diesel engine. *Materials Today: Proceedings, 43*, 510–513.
18. Supraja, K. V., Behera, B., & Paramasivan, B. (2020). Optimization of process variables on two-step microwave-assisted transesterification of waste cooking oil. *Environmental Science and Pollution Research, 27*, 27244–27255.
19. Patil, P. D., Gude, V. G., Mannarswamy, A., Cooke, P., Nirmalakhandan, N., Lammers, P., & Deng, S. (2012). Comparison of direct transesterification of algal biomass under supercritical methanol and microwave irradiation conditions. *Fuel, 97*, 822–831.
20. Cintas, P., Mantegna, S., Gaudino, E. C., & Cravotto, G. (2010). A new pilot flow reactor for high-intensity ultrasound irradiation. Application to the synthesis of biodiesel. *Ultrasonics Sonochemistry, 17*(6), 985–989.
21. Mohamad Aziz, N. A., Yunus, R., Kania, D., & Abd Hamid, H. (2021). Prospects and challenges of microwave-combined technology for biodiesel and biolubricant production through a transesterification: A review. *Molecules, 26*(4), 788.
22. Salaheldeen, M., Mariod, A. A., Aroua, M. K., Rahman, S. A., Soudagar, M. E. M., & Fattah, I. R. (2021). Current state and perspectives on transesterification of triglycerides for biodiesel production. *Catalysts, 11*(9), 1121.
23. Cheng, J., Huang, R., Li, T., Zhou, J., & Cen, K. (2014). Biodiesel from wet microalgae: Extraction with hexane after the microwave-assisted transesterification of lipids. *Bioresource Technology, 170*, 69–75.
24. Zhang, S., Zu, Y. G., Fu, Y. J., Luo, M., Zhang, D. Y., & Efferth, T. (2010). Rapid microwave-assisted transesterification of yellow horn oil to biodiesel using a heteropolyacid solid catalyst. *Bioresource Technology, 101*(3), 931–936.
25. Lokman, I. M., Rashid, U., & Taufiq-Yap, Y. H. (2015). Microwave-assisted methyl ester production from palm fatty acid distillate over a heterogeneous carbon-based solid acid catalyst. *Chemical Engineering and Technology, 38*(10), 1837–1844.
26. Azcan, N., & Yilmaz, O. (2013). Microwave assisted transesterification of waste frying oil and concentrate methyl ester content of biodiesel by molecular distillation. *Fuel, 104*, 614–619.
27. Gina, M. H., Valange, S., Barrault, J., Moreno, J. A., & López, D. P. (2014) Effect of microwave-assisted system on transesterification of castor oil with ethanol. *Universitas Scientiarum, 19*(3), 193–200.
28. Yadav, G. D., Hude, M. P., & Talpade, A. D. (2015). Microwave assisted process intensification of lipase catalyzed transesterification of 1, 2 propanediol with dimethyl carbonate for the green synthesis of propylene carbonate: Novelties of kinetics and mechanism of consecutive reactions. *Chemical Engineering Journal, 281*, 199–208.

29. Dehghan, L., Golmakani, M. T., & Hosseini, S. M. H. (2019). Optimization of microwave-assisted accelerated transesterification of inedible olive oil for biodiesel production. *Renewable Energy, 138*, 915–922.
30. Fatimah, I., Rubiyanto, D., Taushiyah, A., Najah, F. B., Azmi, U., & Sim, Y. L. (2019). Use of ZrO_2 supported on bamboo leaf ash as a heterogeneous catalyst in microwave-assisted biodiesel conversion. *Sustainable Chemistry and Pharmacy, 12*, 100129.
31. Nomanbhay, S., & Ong, M. Y. (2017). A review of microwave-assisted reactions for biodiesel production. *Bioengineering, 4*(2), 57.
32. Gude, V. G., Patil, P., Martinez-Guerra, E., Deng, S., & Nirmalakhandan, N. (2013). Microwave energy potential for biodiesel production. *Sustainable Chemical Processes, 1*, 1–31.
33. Tangy, A., Pulidindi, I. N., Perkas, N., & Gedanken, A. (2017). Continuous flow through a microwave oven for the large-scale production of biodiesel from waste cooking oil. *Bioresource Technology, 224*, 333–341.
34. Yang, J., Cong, W. J., Zhu, Z., Miao, Z. D., Wang, Y. T., Nelles, M., & Fang, Z. (2023). Microwave-assisted one-step production of biodiesel from waste cooking oil by magnetic bifunctional SrO–ZnO/MOF catalyst. *Journal of Cleaner Production, 395*, 136182.
35. Motasemi, F., & Ani, F. N. (2012). A review on microwave-assisted production of biodiesel. *Renewable and Sustainable Energy Reviews, 16*(7), 4719–4733.
36. Nayak, S. N., Bhasin, C. P., & Nayak, M. G. (2019). A review on microwave-assisted transesterification processes using various catalytic and non-catalytic systems. *Renewable Energy, 143*, 1366–1387.
37. Sajjadi, B., Aziz, A. A., & Ibrahim, S. (2014). Investigation, modelling and reviewing the effective parameters in microwave-assisted transesterification. *Renewable and Sustainable Energy Reviews, 37*, 762–777.
38. Tan, S. X., Lim, S., Ong, H. C., & Pang, Y. L. (2019). State of the art review on development of ultrasound-assisted catalytic transesterification process for biodiesel production. *Fuel, 235*, 886–907.
39. Usmani, Z., Sharma, M., Tripathi, M., Nizami, A. S., Gong, L., Nguyen, Q. D., Reddy, M. S., Thakur, V. K., & Gupta, V. K. (2022). Converting biowaste streams into energy–leveraging microwave assisted valorization technologies for enhanced conversion. *Journal of the Energy Institute, 107*, 101161.
40. Gao, Y., Remón, J., & Matharu, A. S. (2021). Microwave-assisted hydrothermal treatments for biomass valorisation: A critical review. *Green Chemistry, 23*(10), 3502–3525.
41. Refaat, A. A. (2010). Different techniques for the production of biodiesel from waste vegetable oil. *International Journal of Environmental Science and Technology, 7*, 183–213.
42. Hassan, A. A., & Smith, J. D. (2020). Investigation of microwave-assisted transesterification reactor of waste cooking oil. *Renewable Energy, 162*, 1735–1746.
43. Mazo, P. C., & Rios, L. A. (2010). Esterification and transesterification assisted by microwaves of crude palm oil: Homogeneous catalysis. *Latin American Applied Research, 40*(4), 337–342.
44. Saifuddin, N., Refal, H., & Kumaran, P. (2014). Research article rapid purification of glycerol by-product from biodiesel production through combined process of microwave assisted acidification and adsorption via chitosan immobilized with yeast. *Research Journal of Applied Sciences, Engineering and Technology, 7*(3), 593–602.
45. Asakuma, Y., Ogawa, Y., Maeda, K., Fukui, K., & Kuramochi, H. (2011). Effects of microwave irradiation on triglyceride transesterification: Experimental and theoretical studies. *Biochemical Engineering Journal, 58*, 20–24.
46. Gupta, A. R., & Rathod, V. K. (2018). Calcium diglyceroxide catalyzed biodiesel production from waste cooking oil in the presence of microwave: Optimization and kinetic studies. *Renewable Energy, 121*, 757–767.
47. Ali, M. A. M., Gimbun, J., Lau, K. L., Cheng, C. K., Vo, D. V. N., Lam, S. S., & Yunus, R. M. (2020). Biodiesel synthesized from waste cooking oil in a continuous microwave assisted reactor reduced PM and NO_x emissions. *Environmental Research, 185*, 109452.
48. Nazir, M. H., Ayoub, M., Zahid, I., Shamsuddin, R. B., Yusup, S., Ameen, M., & Qadeer, M. U. (2021). Development of lignin based heterogeneous solid acid catalyst derived from

sugarcane bagasse for microwave assisted-transesterification of waste cooking oil. *Biomass and Bioenergy, 146*, 105978.
49. Singh, V., & Sharma, Y. C. (2017). Low cost guinea fowl bone derived recyclable heterogeneous catalyst for microwave assisted transesterification of *Annona squamosa* L. seed oil. *Energy Conversion and Management, 138*, 627–637.
50. Khan, H. M., Iqbal, T., Mujtaba, M. A., Soudagar, M. E. M., Veza, I., & Fattah, I. R. (2021). Microwave assisted biodiesel production using heterogeneous catalysts. *Energies, 14*(23), 8135.
51. Leung, D. Y. C., & Guo, Y. (2006). Transesterification of neat and used frying oil: Optimization for biodiesel production. *Fuel Processing Technology, 87*(10), 883–890.
52. Babel, S., Arayawate, S., Faedsura, E., & Sudrajat, H. (2018). Microwave-assisted transesterification of waste cooking oil for biodiesel production. In: *Utilization and management of bioresources: Proceedings of 6th IconSWM 2016* (pp. 165–174). Springer Singapore.
53. Hsiao, M. C., Liao, P. H., Lan, N. V., & Hou, S. S. (2021). Enhancement of biodiesel production from high-acid-value waste cooking oil via a microwave reactor using a homogeneous alkaline catalyst. *Energies, 14*(2), 437.
54. Soosai, M. R., Moorthy, I. M. G., Varalakshmi, P., & Yonas, C. J. (2022). Integrated global optimization and process modelling for biodiesel production from non-edible silk-cotton seed oil by microwave-assisted transesterification with heterogeneous calcium oxide catalyst. *Journal of Cleaner Production, 367*, 132946.
55. Yadav, G., Yadav, N., & Ahmaruzzaman, M. (2023). Microwave-assisted sustainable synthesis of biodiesel on *Oryza sativa* catalyst derived from agricultural waste by esterification reaction. *Chemical Engineering and Processing: Process Intensification, 187*, 109327.
56. Yadav, N., Yadav, G., & Ahmaruzzaman, M. (2023). Microwave-assisted biodiesel production using-SO_3H functionalized heterogeneous catalyst derived from a lignin-rich biomass. *Science and Reports, 13*(1), 9074.
57. Rocha, P. D., Oliveira, L. S., & Franca, A. S. (2019). Sulfonated activated carbon from corn cobs as heterogeneous catalysts for biodiesel production using microwave-assisted transesterification. *Renewable Energy, 143*, 1710–1716.
58. Al-Saadi, A., Mathan, B., & He, Y. (2020). Esterification and transesterification over SrO–ZnO/Al_2O_3 as a novel bifunctional catalyst for biodiesel production. *Renewable Energy, 158*, 388–399.
59. Khemthong, P., Luadthong, C., Nualpaeng, W., Changsuwan, P., Tongprem, P., Viriya-Empikul, N., & Faungnawakij, K. (2012). Industrial eggshell wastes as the heterogeneous catalysts for microwave-assisted biodiesel production. *Catalysis Today, 190*(1), 112–116.
60. Buasri, A., Inkaew, T., Kodephun, L., Yenying, W., & Loryuenyong, V. (2015). Natural hydroxyapatite (NHAp) derived from pork bone as a renewable catalyst for biodiesel production via microwave irradiation. In *Key engineering materials* (Vol. 659, pp. 216-220). Trans Tech Publications Ltd.
61. Panadare, D. C., & Rathod, V. K. (2016). Microwave assisted enzymatic synthesis of biodiesel with waste cooking oil and dimethyl carbonate. *Journal of Molecular Catalysis B: Enzymatic, 133*, S518–S524.
62. Bunushree, B., Ravichandra, B., Rangabhashiyam, S., Jayabalan, R., & Balasubramanian, P. (2019). Qualitative analysis of biodiesel produced by alkali catalyzed transesterification of waste cooking oil using different alcohols. *Indian Journal of Chemical Technology, 25*(4), 330–336.
63. Predojević, Z. J. (2008). The production of biodiesel from waste frying oils: A comparison of different purification steps. *Fuel, 87*(17–18), 3522–3528.
64. Sharma, A., Kodgire, P., & Kachhwaha, S. S. (2019). Biodiesel production from waste cotton-seed cooking oil using microwave-assisted transesterification: Optimization and kinetic modeling. *Renewable and Sustainable Energy Reviews, 116*, 109394.
65. Prafulla, D. P., Veera Gnaneswar, G., Harvind, K. R., Tapaswy, M., & Shuguang, D. (2012). Biodiesel production from waste cooking oil using sulfuric acid and microwave irradiation processes. *Journal of Environmental Protection, 3*(1), 1–7.

66. Zare, M., Ghobadian, B., Fayyazi, E., Najafi, G., & Hosseinzadeh, B. (2013). Microwave-assisted biodiesel fuel production from waste cooking oil. *International Journal of Agriculture and Crop Sciences, 5*(12), 1314.
67. Yaakob, Z., Ong, B. H., Satheesh Kumar, M. N., & Kamarudin, S. K. (2009). Microwave-assisted transesterification of jatropha and waste frying palm oil. *International Journal of Sustainable Energy, 28*(4), 195–201.
68. Ali, M. M., Yunus, R. M., Cheng, C. K., & Gimbun, J. (2015). Successive optimisation of waste cooking oil transesterification in a continuous microwave assisted reactor. *RSC Advances, 5*(94), 76743–76751.
69. Marchetti, J. M., Miguel, V. U., & Errazu, A. F. (2008). Techno-economic study of different alternatives for biodiesel production. *Fuel Processing Technology, 89*(8), 740–748.
70. Hussain, M. N., Al Samad, T., & Janajreh, I. (2016). Economic feasibility of biodiesel production from waste cooking oil in the UAE. *Sustainable Cities and Society, 26*, 217–226.
71. Mohammadshirazi, A., Akram, A., Rafiee, S., & Kalhor, E. B. (2014). Energy and cost analyses of biodiesel production from waste cooking oil. *Renewable and Sustainable Energy Reviews, 33*, 44–49.
72. Farid, M. A. A., Roslan, A. M., Hassan, M. A., Hasan, M. Y., Othman, M. R., & Shirai, Y. (2020). Net energy and techno-economic assessment of biodiesel production from waste cooking oil using a semi-industrial plant: A Malaysia perspective. *Sustainable Energy Technology Assessment, 39*, 100700.
73. Liu, Y., Yang, X., Adamu, A., & Zhu, Z. (2021). Economic evaluation and production process simulation of biodiesel production from waste cooking oil. *Current Research in Green and Sustainable Chemistry, 4*, 100091.
74. Refaat, A. A., El Sheltawy, S. T., & Sadek, K. U. (2008). Optimum reaction time, performance and exhaust emissions of biodiesel produced by microwave irradiation. *International Journal of Environmental Science and Technology, 5*, 315–322.
75. Muley, P. D., & Boldor, D. (2013). Scale-up of a continuous microwave-assisted transesterification process of soybean oil for biodiesel production. *Transactions of the ASABE, 56*(5), 1847–1854.
76. Chuck, C. J., Lou-Hing, D., Dean, R., Sargeant, L. A., Scott, R. J., & Jenkins, R. W. (2014). Simultaneous microwave extraction and synthesis of fatty acid methyl ester from the oleaginous yeast *Rhodotorula glutinis*. *Energy, 69*, 446–454.
77. Waudby, H., & Zein, S. H. (2021). A circular economy approach for industrial scale biodiesel production from palm oil mill effluent using microwave heating: Design, simulation, techno-economic analysis and location comparison. *Process Safety and Environmental Protection, 148*, 1006–1018.
78. Manikandan, G., Kanna, P. R., Taler, D., & Sobota, T. (2023). Review of waste cooking oil (WCO) as a feedstock for biofuel—Indian perspective. *Energies, 16*(4), 1739.
79. Tibesigwa, T., Olupot, P. W., & Kirabira, J. B. (2022). The critical techno-economic aspects for production of B10 biodiesel from second generation feedstocks: A review. *International Journal of Sustainable Energy, 41*(7), 751–771.
80. Chung, Z. L., Tan, Y. H., San Chan, Y., Kansedo, J., Mubarak, N. M., Ghasemi, M., & Abdullah, M. O. (2019). Life cycle assessment of waste cooking oil for biodiesel production using waste chicken eggshell derived CaO as catalyst via transesterification. *Biocatalysis and Agricultural Biotechnology, 21*, 101317.
81. Foteinis, S., Chatzisymeon, E., Litinas, A., & Tsoutsos, T. (2020). Used-cooking-oil biodiesel: Life cycle assessment and comparison with first-and third-generation biofuel. *Renewable Energy, 153*, 588–600.
82. Iglesias, L., Laca, A., Herrero, M., & Díaz, M. (2012). A life cycle assessment comparison between centralized and decentralized biodiesel production from raw sunflower oil and waste cooking oils. *Journal of Cleaner Production, 37*, 162–171.
83. Ripa, M., Buonaurio, C., Mellino, S., Fiorentino, G., & Ulgiati, S. (2014). Recycling waste cooking oil into biodiesel: A life cycle assessment. *International Journal of Performability Engineering, 10*(4), 347.

84. Alishahi, A., Golmakani, M. T., & Niakousari, M. (2021). Feasibility study of microwave-assisted biodiesel production from vegetable oil refinery waste. *European Journal of Lipid Science and Technology, 123*(9), 2000377.
85. Yari, N., Mostafaei, M., Naderloo, L., & Safieddin Ardebili, S. M. (2022). Energy indicators for microwave-assisted biodiesel production from waste fish oil. *Energy Sources, Part A: Recovery, Utilization, and Environmental Effects, 44*(1), 2208–2219.
86. Viornery-Portillo, E. A., Bravo-Díaz, B., & Mena-Cervantes, V. Y. (2020). Life cycle assessment and emission analysis of waste cooking oil biodiesel blend and fossil diesel used in a power generator. *Fuel, 281*, 118739.
87. Hosseinzadeh-Bandbafha, H., Nizami, A. S., Kalogirou, S. A., Gupta, V. K., Park, Y. K., Fallahi, A., Sulaiman, A., Ranjbari, M., Rahnama, H., Aghbashlo, M., & Peng, W. (2022). Environmental life cycle assessment of biodiesel production from waste cooking oil: A systematic review. *Renewable and Sustainable Energy Reviews, 161*, 112411.
88. Zhu, H., He, J., Hong, T., Yang, Q., Wu, Y., Yang, Y., & Huang, K. (2018). A rotary radiation structure for microwave heating uniformity improvement. *Applied Thermal Engineering, 141*, 648–658.
89. Kristiana, T., Baldino, C., & Searle, S. (2022). An estimate of current collection and potential collection of used cooking oil from major Asian exporting countries. Work. Pap, 21.
90. Liu, T., Liu, Y., Wu, S., Xue, J., Wu, Y., Li, Y., & Kang, X. (2018). Restaurants' behaviour, awareness, and willingness to submit waste cooking oil for biofuel production in Beijing. *Journal of Cleaner Production, 204*, 636–642.
91. Milano, J., Ong, H. C., Masjuki, H. H., Silitonga, A. S., Chen, W. H., Kusumo, F., Dharma, S., & Sebayang, A. H. (2018). Optimization of biodiesel production by microwave irradiation-assisted transesterification for waste cooking oil-*Calophyllum inophyllum* oil via response surface methodology. *Energy Conversion and Management, 158*, 400–415.
92. Selvaraj, R., Moorthy, I. G., Kumar, R. V., & Sivasubramanian, V. (2019). Microwave mediated production of FAME from waste cooking oil: Modelling and optimization of process parameters by RSM and ANN approach. *Fuel, 237*, 40–49.
93. Rao, G. L. N., Ramadhas, A. S., Nallusamy, N., & Sakthivel, P. (2010). Relationships among the physical properties of biodiesel and engine fuel system design requirement. *International Journal of Energy and Environmental, 1*(5), 919–926.
94. Mazubert, A., Poux, M., & Aubin, J. (2013). Intensified processes for FAME production from waste cooking oil: A technological review. *Chemical Engineering Journal, 233*, 201–223.

Chapter 6
Green Catalysts Synthesized from Biomass for Biodiesel Production

Amirthavalli Velmurugan, Anita R. Warrier, and Gurunathan Baskar

Abstract The impact of any processing technology on sustainable development requires a zero-waste approach of reusing, recycling, and recovering. Biodiesel serves as an alternative source of renewable energy to fossil fuels, and it can be produced by transesterification of feedstock with alcohol in the presence of a catalyst. Nanocatalysts are preferred in the transesterification reaction when compared to acid, base, and enzymatic catalysts due to their high surface-to-volume ratio, high catalytic activity, easy recoverability, simple synthesis procedure, and low toxicity. Calcium and carbon-based green catalysts that offer a large surface area, improved porosity, stability, and inertness can be derived from plant, agricultural, and animal wastes. This chapter discusses the role of various green catalysts that are derived from biomass, along with their merits and demerits for large-scale production of biodiesel, and the possible ways of improving the yield to make biodiesel viable for commercial applications.

Keywords Biomass · Transesterification · Biodiesel · Green catalyst

A. Velmurugan
Department of Petroleum Engineering, Academy of Maritime Education and Training, Chennai, Tamil Nadu, India
e-mail: amirthavalli.v@ametuniv.ac.in

A. R. Warrier (✉)
Nanophotonics Research Laboratory, Department of Physics, Academy of Maritime Education and Training, Chennai, Tamil Nadu, India
e-mail: anitawarrier2@gmail.com

G. Baskar
Department of Biotechnology, St. Joseph's College of Engineering, Chennai, Tamil Nadu, India

G. Baskar et al. (eds.), *Circular Bioeconomy Perspectives in Sustainable Bioenergy Production*, Energy, Environment, and Sustainability,
https://doi.org/10.1007/978-981-97-2523-6_6

6.1 Introduction to Green Catalyst

Green catalysts are those that are synthesized from the pyrolysis of biomass which can be from agricultural waste, plant waste, and animal waste as shown in Fig. 6.1, and green catalysts are considered to be a good source of calcium and carbon. The advantages of green catalysts are non-toxicity, reusability, simple catalyst synthesis procedure, thermal stability, enhanced biodiesel yield when used in transesterification reaction, and it does not create any harm to society as well as to the environment [1, 2]. Moreover, there are no disposal issues since the catalyst itself is biodegradable and excludes the production of wastewater. Catalyst derived from biomass possesses an increased surface-to-volume ratio and finds use as a carbonaceous catalyst or as support in biodiesel production. Agricultural waste biomass especially lignocellulosic, i.e., rice husk, palm kernel shell, empty fruit bunch, and walnut shell, is gaining importance as an inexpensive and renewable feedstock for the synthesis of green catalysts [3]. Moreover, green catalysts obtained from these biomasses possess increased porosity and large surface area which is ideal for adsorption–desorption mechanism [4]. It has been reported that bio-based catalyst derived from agricultural waste and food waste shows high conversion efficiency in transesterification [5]. The presence of reduced ash content in activated carbon promotes transesterification, and also, it finds application in high pressure and temperature. Moreover, the surface properties do not alter at increased temperature or pressure [6].

Nanoparticles synthesized from plant sources are inexpensive, eco-friendly, and efficient and can be used for large-scale production without the requirement of any sophisticated equipment and complex procedures when compared to physical or chemical methods. The synthesis procedure can be carried out in a shorter duration

Fig. 6.1 Shows the various feedstocks for obtaining green catalysts from biomass

of time and requires less energy [7]. The production cost of nanoparticles synthesized using local origin plant sources is comparatively less [8].

Plants consist of alkaloids, flavonoids, terpenoids, and steroids which serve as reducing agents in nanoparticle synthesis. It has been reported that calcium oxide nanoparticles were synthesized from the phytochemical component of watermelon extract by reducing and stabilizing the precursor (calcium nitrate) [9]. Synthesized calcium oxide nanoparticles when used in the transesterification of soybean oil give a biodiesel yield of 96.2% at 20:1 methanol-to-oil molar ratio, 8 wt% of the amount of catalyst, and 65 °C of the reaction temperature in 2.5 h. Cholapandian et al. have reported on the synthesis of calcium oxide nanocatalyst from *Acalypha indica.* A biodiesel yield of 94.74% was obtained when the synthesized calcium oxide nanocatalyst was used in the transesterification of waste cooking oil at 11.8:1 methanol-to-oil molar ratio, 2.4 wt% of the amount of catalyst, and 63.7 °C of reaction temperature in 70 min [10]. Sumera et al. have reported on the green synthesis of cadmium oxide nanoparticles from the leaf extract of *Buxus papillosa* using in situ wet impregnation method. It has been reported that cadmium-based nanoparticle finds application in polymerization and transesterification [11]. When the synthesized cadmium oxide is used in the transesterification of *Diospyros malabarica* (Malabar Ebony) gives a biodiesel yield of 94 wt% at 1:9 oil-to-methanol molar ratio, 0.5 wt% of the amount of catalyst, and 90 °C of reaction temperature in 180 min [12].

Animal wastes in the form of chicken eggshell waste, waste fish scale, waste cockle shell, waste mud scrub shell, shells of mollusk, golden apple snail shells, ostrich eggshells, bones, oyster shells, and river snail shell ash can also serve as suitable feedstock in the synthesis of CaO nanocatalyst for biodiesel production [13, 14]. Animal waste is biodegradable, recyclable, and non-toxic to the environment and humans with improved catalytic efficiency [15, 16]. Animal waste excludes waste management costs but helps in obtaining inexpensive nanocatalysts that can be used in biodiesel production [17, 18].

6.2 Agricultural Waste

6.2.1 Rice Husk

Rice husk is an agricultural waste obtained from the rice milling process. During the milling process, 78% of the weight is obtained as rice, broken rice, and bran, and the remaining 22% as rice husk. Rice husk contains 15 wt% of silica, cellulose 28.7–35.6%, hemicellulose 12.0–29.3%, and lignin 15.4–20.0% [19], and it is not recommended as fodder for animals due to its low cellulose and sugar content. Rice husk is the hard outer covering that is developed due to the absorption of silica from the soil to shield the rice grains within the structure. It has been reported that one ton of rice gives 200 kg of rice husks which when combusted gives 40 kg of rice husk ash [20]. The disposal of large quantities of rice husk in developing countries is very tedious and expensive, and the majority of the rice husk is unused and heaped as waste material or burned in the agricultural fields which further creates environmental issues. Due to the presence of carbon in the rice husk, it serves as an important raw material in the production of activated carbon of amorphous nature with increased surface area after the removal of silica [21, 22]. Silica present in the rice husk is removed in the form of sodium silicate solution after it is made to react with dilute sodium hydroxide solution at increased temperature for a particular duration [23].

The pyrolysis of the rice husk generates bio-oil, syngas (carbon monoxide and hydrogen), and also by-product biochar (rice husk carbon and rice husk ash) [24]. Bio-oil and syngas are used in biofuel production with a heating value of (20 MJ/kg). Synthesis of activated carbon is carried out either directly by calcination or indirectly by burning the rice husk into rice husk carbon and then calcinating at a known temperature, which further undergoes chemical or physical activation followed by drying. Rice husk ash contains amorphous silica which can be extracted by alkali treatment [25, 26]. Rice husk ash serves as a raw material for producing silicates and silica due to its increased silica content. Othman Ali et al. synthesized ZSM-5 zeolite using rice husk ash [27] and Jan-Jezreel F. Saceda et al. synthesized zeolite NaY using rice husk ash. The extracted silica from rice husk ash can be used as SBA-15 materials which are used in the production of ordered mesoporous carbon like CMK-3 (carbon manufactured from korea-3) [28]. The ordered mesoporous carbon finds use as catalyst supports [29], adsorbents [30], electrode materials [31], hydrogen storage [32], sensors [33], and supercapacitors [34].

The porosity of activated carbon obtained from the carbonization step can be further improved by using activation steps. The activation methods enhance the pore volume and expand the diameter of the pores and surface area of activated carbon. In the activation method, improvement in the microporous structure can be attained by stripping out the disorganized carbon. The pores swell due to the heating of walls among the pores. Hence, the provisional pores and microporosity are enhanced after the activation methods. The activation step plays an important role in the preparation of activated carbon with good porosity [35]. The pores obtained on the activated carbon surface are distinguished into macropores greater than 25 nm, mesopores between 1 and 25 nm, and micropores less than 1 nm [36].

The calcination of rice husk carbon needs to be carried out in a muffle furnace with inert media (nitrogen) at a high temperature of 600–900 °C [37]. It is important to carry out calcination in an inert atmosphere to decrease the production of ash content and flue gas, especially carbon dioxide. The most commonly used physical activation methods are steam and gas activation. Steam activation is normally carried out after the thermal carbonization of biomass. Porosity developed in the biochar after thermal carbonization further undergoes steam activation which will enhance the porosity. The chemical reactions that occur during steam activation are [38]:

$$C + H_2O \rightarrow CO + H_2 \tag{6.1}$$

$$CO + H_2O \rightarrow CO_2 + H_2 \tag{6.2}$$

The reaction between water and carbon results in the removal of volatile matter, the growth of new micropores as well as expansion of existing pores. Gas activation is also carried out to enhance the surface area and pore volume by the reaction between the gas (CO_2, N_2, NH_3, air, and O_2) and the surface of the biochar and also yields microporous- and mesoporous-activated carbon structure. In the chemical activation methods, acids such as sulfuric acid (H_2SO_4), hydrochloric acid (HCl), phosphoric acid (H_3PO_4), and nitric acid (HNO_3) or bases such as sodium hydroxide (NaOH), potassium hydroxide (KOH), potassium carbonate (K_2CO_3), and calcium oxide (CaO) or salts like zinc chloride ($ZnCl_2$) or oxidants like hydrogen peroxide (H_2O_2) and potassium permanganate ($KMnO_4$) are used. The chemical activation method gives activated carbon in reduced time with improved surface area and mesoporous structure when compared to the physical activation.

Carbon present in the rice husk combines with oxygen molecules at high temperatures to form carbon dioxide which decreases the adsorbent property and also actives sites in the obtained biochar. Hence, the synthesis of activated carbon should be carried out under an inert atmosphere [39] which further increases the production cost and also needs a complicated experimental setup. Hence, a proportional integral derivative control muffle furnace is recommended.

Rice husk char consists of 50% of ash which suppresses the pore development, and hence, it has a low surface area. The ash present in the rice husk char is removed by reacting it with chemicals such as zinc chloride and alkali solution (potassium hydroxide or sodium hydroxide). The surface area of raw rice husk char varies from 25 to 28 m^2g^{-1}, whereas the surface area of rice husk char after removing ash by treating it with an alkali solution (sodium hydroxide) is reported to be 162, 174, and 331 m^2g^{-1}. It has been reported that rice husk char on further treatment with zinc chloride leads to a more porous-activated carbon with an enhanced surface area of 365–645 m^2g^{-1} [40].

Zhaoe et al. have reported that a biodiesel yield of 93.4% is obtained using highly porous rice husk biochar loaded with 30% of CaO as a catalyst with a surface area of 28.3 m^2g^{-1} under optimum conditions of 9:1 methanol-to-oil ratio, 8 wt% of catalyst in 180 min [41]. It has also been reported that rice husk biochar acts as a good carrier and the presence of Ca–O–Si bond on the catalyst surface aids in improving the catalytic activity and stability of the catalyst. Balkis Hazmi et al. have reported a biodiesel yield of 98.2% using a bifunctional magnetic nanocatalyst derived from rice husk char doped with potassium oxide and nickel oxide at 4 wt% of catalyst, 12:1 methanol-to-oil molar ratio, 65 °C of reaction temperature within a reaction duration of 2 h [42]. Chen et al. have reported that Li-modified rice husk ash is synthesized using rice husk ash and Li_2CO_3. The prepared catalyst with basic strength (OH-) above 15 shows good catalytic activity and when used in transesterification of soybean oil shows 99.5% of biodiesel under optimized conditions of 24:1 methanol-to-oil molar ratio, 4% of the amount of catalyst, and 65 °C for a reaction duration of 3 h [20].

The usage of inexpensive rice husk-derived biomass as the catalyst raw material can solve leftover problems of rice husk in the open atmosphere. Activated carbon obtained from rice husks can serve as a catalyst for large-scale biodiesel production

[43], photocatalytic degradation of dye [44], wastewater treatment, and enhancing power generation from fuel cells.

6.2.2 Palm Kernel Shell

It was reported by the Malaysian Palm Oil Board (MPOB) that one ton of palm fresh fruit bunch in the milling process yields 5–7% palm kern shell (PKS), 21–22% empty fruit bunch (EFB), and 12–16% mesocarp fiber (MF). Palm kernel shell consists of 48 wt% of lignin, 30 wt% of cellulose, and 22 wt% of hemicellulose. It mostly has 51 wt% of carbon, 39 wt% of oxygen, and minimum content of hydrogen and nitrogen approximately 3 and 7 wt% [45].

It has been reported that biochar derived from biomass can act as the catalyst for combining suitable functional active groups. Moreover, converting palm waste into a heterogeneous catalyst can decrease the production cost of biodiesel as well as decrease the health issues that can be caused due to dumped wastes. Waste generated from palm oil industries can give a high-value activated carbon that acts as the catalyst support for biodiesel production. Activated carbon obtained from waste palm oil, mainly palm kern shell, is used due to its high calorific value, reduced sulfur content and ash content, no variation in the species, enhanced shelf life, and availability of palm waste throughout the year. Activated carbon obtained from palm kernel shell is a highly porous carbon-rich content with an amorphous carbon form. Activated carbon obtained from palm kernel shells is inexpensive when compared to other agricultural byproducts. Activated carbon can be used as a catalyst provided it should have a minimum amount of ash content and minerals.

It has been reported that palm kernel shell possesses 50 wt% of lignin which is considered to be the maximum amount when compared with different palm waste materials such as empty fruit bunch, mesocarp fiber, oil palm trunk, and frond (21–35 wt% of lignin) [46–49]. Enhanced yield of activated carbon can be obtained from palm kernel shell due to the presence of complicated polymeric lignin structure [50]. Lignin structure has benzene rings within it which provides increased chemical stability to the decomposition reaction in comparison with the polysaccharides that are available in cellulose and hemicellulose. Hence, lignin can be converted into activated carbon in stable form with greater aromaticity in the carbonization method [51], instead of being converted into bio-oil or gases like methane, carbon dioxide, and carbon monoxide by decomposition reaction [52]. Palm kernel shell shows maximum content of carbon which makes it a potential resource for obtaining activated carbon with the help of pyrolysis. It also possesses increased wt% of oxygen which needs to be removed to obtain a carbon-rich product.

Abdulla et al. have reported on the synthesis of bifunctional nanocatalyst attained from waste palm kernel shell and synthesized catalyst (palm kernel shell-activated carbon-$K_2CO_{3(30\%)}CuO_{(5\%)}$, PKSAC-$KOH_{(30\%)}CuO_{(5\%)}$, PKSAC-$K_3PO_{4(30\%)}CuO_{(5\%)}$, and PKSAC-$NaOH_{(30\%)}CuO_{(5\%)}$) along with the impregnation of K_2CO_3, KOH, K_3PO_4, $Cu(NO_3)_2$, and NaOH. PKSAC-$K_2CO_{3(30\%)}CuO_{(5\%)}$ has high surface area of 438.08 m^2g^{-1}, pore volume of 0.367 cm^3 g^{-1}, and pore width of 3.8 nm. It has been reported that 95% of biodiesel yield is obtained from waste cooking oil under optimal conditions of 5 wt% catalyst loading, 12:1 methanol-to-oil molar ratio at 80 °C for 4 h using the mixture of potassium and copper on activated carbon surface (PKSAC-$K_2CO_{3(30\%)}CuO_{(5\%)}$) [53]. An enhanced amount of basicity in the range of 8.866 $mmolg^{-1}$ and acidity (27.016 $mmol^{-1}$) has favored the simultaneous transesterification and esterification reaction. Celine et al. have reported on the synthesis of magnetic palm kernel shell-derived catalysts (Fe-KOH-PKS) using impregnation of potassium hydroxide followed by pyrolysis and magnetic palm kernel shell catalyst (Fe-PKS) [54]. The synthesized catalyst shows increased catalytic properties due to the increased magnetic saturation of 22.155 emu/g, a basicity of 6.68 mmol/g when compared to the Fe-PKS catalyst. Fe-KOH-PKS when used in transesterification of palm oil gives a maximum biodiesel yield of 99.43% under optimal operating conditions of 7.5 wt% catalyst loading and 9:1 methanol-to-oil ratio at 65 °C after 2 h. The synthesized catalyst derived from biomass helps to obtain a huge amount of inexpensive green catalyst for large-scale biodiesel production. Kostic et al. obtained a biodiesel yield of 99% from sunflower oil using palm kernel shell biochar under the optimal reaction conditions of 9:1 methanol-to-oil molar ratio, 3 wt% of catalyst, and 65 °C in 4 h [55].

The disposal of these biomass residue has become a challenging task. The palm biomass waste is widely used in fertilizer and biogas production and also has a great scope to get converted into a biodegradable heterogeneous catalyst for biodiesel production and wastewater treatment due to its huge availability.

6.2.3 Sugarcane Bagasse

Sugarcane belongs to a grass family, and it is cultivated annually for its juice which in turn is used for the production of sugar. Sugarcane bagasse the leftover part after obtaining the juice is considered as a waste, and it consists of cellulose, hemicellulose, and lignin. It has been reported that one metric ton of sugarcane gives 280 kg of bagasse which finds application in the bioenergy industry after burning to supply energy and for the production of bioethanol [56]. Sugarcane bagasse consists of 40–45% of cellulose, 30–35% of hemicellulose, and 20–30% of lignin and ash. The cellulose consists of long chains of D-glucose units joined by glycosidic bonds, hemicellulose consists of xylose, mannose, and glucose saccharide units, and lignin is obtained by polymerization of aromatic alcohols [57, 58]. Mostly sugarcane bagasse consists of 60–80% of carbohydrates which predominantly consists of carbon and hydrogen that can be formed into activated carbon [59].

Chausali et al. have reported that the advantages of nanotechnology and biochar are displayed together when bulk biochar is decreased to the nano biochar level [60]. Research is being widely carried out for the synthesis of carbon nanotubes, and activated carbon monoliths derived from crop residue biochar and synthesized catalysts possess excellent properties such as good porosity, adsorption capacity, and low cost.

Akinfalabi et al. have synthesized bio-based sulfonated sugarcane bagasse and when used in transesterification of palm fatty acid distillate gives a biodiesel yield of 98.6% under optimal conditions of 10:1 methanol-to-oil molar ratio, 2 wt% of catalyst, and 60 °C reaction temperature for a duration of 1.5 h [61]. Subramaniapillai et al. synthesized biochar from sugarcane bagasse using an activation method followed by carbonization. Synthesized heterogeneous catalyst gives a biodiesel yield of 98.94% when used in the transesterification of waste cooking oil at a reaction temperature of 65 °C, 1:2 methanol-to-oil molar ratio, and 10 wt% of catalyst in 120 min [62]. Abdul Mutalib has synthesized SiO_2-rich sugarcane bagasse ash/CaO inexpensive heterogeneous catalyst and when used in the transesterification of palm olein oil gives a yield of 93.8% under optimum conditions of 20:1 methanol-to-oil molar ratio, 3 h reaction time, 6 wt% catalyst amount, and 65 °C reaction temperature [63]. It has also been reported that the presence of SiO_2 in the sugarcane bagasse ash plays an important role in improving the catalytic performance and decreasing the deactivation of the catalyst by reducing the leaching of active sites of the catalyst.

Activated carbon finds application as an absorbent material for various industrial applications such as the removal of organic and inorganic impurities from wastewater, and biodiesel production due to its increased thermal stability, porosity, and simple synthesis procedure with low cost [64–67].

6.2.4 Coconut Shell

Coconut shell consists of 14 wt% of cellulose, 32 wt% of hemicellulose, and 46 wt% of lignin [68]. It has been reported that coconut shell consists of reduced ash content, more carbon content, and possesses increased strength and hardness which makes it a suitable candidate for the development of catalyst.

Activated carbon can be produced from different raw carbon resources like lignite, peat, coal, and biomass resources such as wood, sawdust, bagasse, and coconut shells [69]. However, the abundant supply of coconut shells as a waste product from the coconut oil and desiccated coconut industry makes the production of activated carbon from this material more financially viable since using grain or coal as raw materials for activated carbon will require manufacturers extra amount of money for procurement. Furthermore, besides being an amorphous form of carbon that can absorb many gases, vapor, and colloidal solids, coconut shell-activated carbons are advantageous over carbons made from other materials because of their high density, high purity, and virtually dust-free nature. These carbons are harder and more resistant to attrition.

Shobhana et al. have reported on the synthesis of Ce-supported sulfated activated carbon obtained from coconut shell. It has been reported that free fatty acid conversion of 93% is obtained by transesterification of chicken fat and skin oil using 5Ce/ACcs-S at 90 °C of reaction temperature, 12:1 methanol-to-oil molar ratio with 3 wt% of catalyst loading in 1 h [70]. Enhanced physicochemical properties and stability of the catalyst are attained due to the rare-earth metal oxides such as CeO_2 and La_2O_3. Sulfated activated carbon possesses an increased volume of micropores and wide pore size distribution which supports adsorption when used in the production of biodiesel. Moreover, oxygen groups that are present on the surface serve as an attaching site for metallic precursors which further helps in the dispersion of dopants that are added during the reaction. Increased catalytic activity is due to the enhanced pore diameters in the catalyst which enables diffusion of reactants such as free fatty acid and triglycerides. Activated carbon obtained from coconut shells possesses uniform pore structure, increased density, and purity due to its hardness and increased resistance to attrition. The maximum number of pores falls in the category of micropores which are efficient in the removal of volatile organic compounds and residue of pesticides [71].

6.3 Plant Waste

6.3.1 Potato Waste

It is considered to be one of the crops which are grown worldwide in the majority and consumed by people in different forms such as potato chips, crisp, puree, and French fries. The usage of potatoes in different forms increases the peel waste which ranges from 15 to 40% of its original weight. China is the main producer of potatoes with 91.8 million tonnes annually followed by India with an annual production of 50.2 million tonnes. If the potato peels are not used properly and dumped as waste, it contributes to environmental pollution. Potato peels serve as suitable feedstock for the production of bio-oil and biochar through the pyrolysis process. Biochar obtained as the product of the pyrolysis process contains a biogenic potassium source which is used for the synthesis of heterogeneous catalysts. Biodiesel conversion of 97.5% was obtained from soybean waste cooking oil using biochar from potato peels under optimum conditions of 9:1 methanol-to-oil molar ratio, 3 wt% of catalyst, and 60 °C in 2 h [72].

6.3.2 Banana Waste

The banana is considered to be the earliest fruit that is familiar to humankind. Banana is the second greatest tropical fruit cultivated in the world, and huge amount of banana peel is obtained as waste by-product. India is considered to be the biggest producer and consumer of bananas in the world. Indian states like Maharashtra, Assam, Tamil Nadu, Arunachal Pradesh, Karnataka, Madhya Pradesh, and Gujarat are the major producers of 15–20 varieties of bananas. Banana peel finds application in bioethanol production and energy-related activities as well as acts as a biosorbent and biochar. The usage of biochar obtained from the banana peel, trunk, and rhizome as a catalyst for the transesterification process is gaining momentum among researchers since banana peel consists of carbohydrates, minerals, and nutrients, including cations of calcium, sodium, potassium, iron, magnesium, and anions of sulfur.

Niran et al. have synthesized banana peel pyrolyzed calcined biochar catalyst from banana peel which is rich in potassium, calcium, magnesium, sodium, and silicon. The obtained biochar acts as support material for the catalyst by using acid (H_2SO_4) or base (KOH, CaOH). Biodiesel yield of around 98% is obtained by transesterification of soybean waste cooking oil in the presence of banana peel pyrolyzed calcined biochar catalyst under optimal conditions of 9:1 methanol-to-oil molar ratio, 1.5 wt% of catalyst, and 60 °C reaction temperature in 2 h [73]. The synthesized catalyst displays improved catalytic activity due to the presence of more quantity of potassium in oxide and carbonate form. Jitjamnong et al. have reported a biodiesel yield of 99.15% from the transesterification of palm oil using a heterogeneous AC/KOH catalyst derived from the calcination of banana peel biochar-supported K_2CO_3.The optimal conditions for transesterification were found to be a 15:1 methanol-to-oil molar ratio, 4 wt% of catalyst, and 65 °C reaction temperature in 120 min [74].

Sujata Brahma et al. have reported on the synthesis of biochar derived from *Musa chinensis* peel, trunk, and rhizome. Biochar derived from *M. chinensis* peel has the highest basicity of 0.93 $mmolg^{-1}$ when compared to biochar derived from trunk and rhizome with a basicity of 0.15 and 0.2 $mmolg^{-1}$ and is micro-mesoporous. Synthesized biochar derived from *M. chinensis* peel when used in transesterification of a quinary oil mixture (soybean oil, sunflower oil, canola oil, jatropha oil, and pongamia oil) gives a biodiesel yield of 95.82% under optimal conditions of 9:1 methanol-to-oil molar ratio, 5 wt% of catalyst, and 65 °C of reaction temperature in 11 min [75]. Laskar has reported on the synthesis of *Mangifera indica* peel ash obtained by burning the mango peel in the open air. Synthesized catalyst when used in the transesterification of soybean oil gives a biodiesel yield of 98% under optimum conditions of 6:1 methanol-to-oil molar ratio and 6 wt% of catalyst in 4 h under room temperature. Enhanced catalytic activity of the synthesized catalyst with increased surface area and porosity is due to the presence of metal oxides such as K_2O, MgO, and CaO.

Recently ash catalyst derived from plant waste has gained momentum in biodiesel production due to its simple preparation methods, abundance, inexpensive, and does not harm the environment. Waste tucuma peels, Musa acuminate peduncle, *Musa acuminata* peel, *Musa balbisiana* Colla, rice husk ash, coconut waste, rubber seed shells, cocoa shells, and *Lemna perpusilla* Torrey ash are reported to be efficient feedstocks for the synthesis of catalyst used in biodiesel production.

6.3.3 Mango Waste

Mango is grown in more than 90 countries in the world, and it is obtained in 160 varieties. India is the main producer of mango in the world. Mango is considered to be a tropical fruit that is consumed in large quantities after banana, and residue of mango processing (peel, kernel, and seed) consists of 35–60% of total fruit weight, hence describing high volume resource of usable bio-based chemicals and materials. The mango peel provides 7–24% of the total weight of the fruit, which is heaped as a waste product. Laskar et al. have reported the synthesis of a catalyst derived from waste peels of mango (*M. indica*) with high basicity. The synthesized catalyst when used in transesterification of soybean oil gives a biodiesel yield of 98% at 6:1 methanol-to-oil molar ratio and 6 wt% of catalyst at 4 h when carried out under room temperature. Shah et al. have reported on the plasma synthesis of graphene using mango peels [76].

6.3.4 Pomelo Waste

Pomelo is a fruit widely grown in China, and huge quantities of waste are disposed in landfills. It has been reported that pomelo peel has been used to produce activated carbon which finds application in the removal of organic pollutants or as anode materials for batteries. The usage of biochar as catalyst support helps in the production of biodiesel on a large scale and also promotes the pyrolysis process to reduce the environmental impact. Che Zhao et al. have reported on the synthesis of biochar loaded with 25 wt% of potassium carbonate which possesses high basic sites. The biochar was derived from derived from pomelo peel. The synthesized catalyst when used in the transesterification of palm oil gives a biodiesel yield of 98% under optimal conditions of 8:1 methanol-to-oil molar ratio, 6 wt% of catalyst, and 65 °C of the reaction temperature in 2.5 h [77].

6.3.5 Mustard Plant Waste

The mustard plant is cultivated in all places of India, central Africa, and southern Russia, and it is considered to be one of the cultivated species back to a period of 5000 years. The oil obtained from mustard seeds is mainly used in cooking. The delicate leaves of the immature plants are consumed as vegetables and salads. After harvesting the mustard seeds, waste *Brassica nigra* plant is burnt as ash, and it is considered to be highly basic. Biswajit Nath et al. have reported that the catalyst consists of 56.13 wt% of potassium and 26.04 wt% of calcium in the form of oxide, carbonate, and chloride. A biodiesel yield of 98.79% is obtained while converting soybean oil using a catalyst derived from waste *B. nigra* plant under optimal conditions of 12:1 methanol-to-oil molar ratio, 7 wt% of catalyst, and 320 C reaction temperature in 75 min [78].

6.3.6 Moringa Waste

Moringa leaves consist of micro and macronutrients such as nitrogen, phosphorus, potassium, calcium, boron, zinc, sodium, manganese, and iron, and polyphenols, alkaloids, tannins, ascorbic acid, and carotenoid flavonoids. The composition of minerals and proteins varies in the plant based on their source. The leaves with alkaline materials are considered suitable feedstock for the synthesis of catalysts. Aleman Ramirez et al. have reported on the synthesis of moringa leaves ash and synthesized catalyst after calcination showing the presence of inorganic carbonate minerals of dolomite, calcite, and $K_2Ca(CO_3)_2$) which helps in the transesterification reaction. 86.7% of biodiesel yield is obtained when the synthesized catalyst is used in the transesterification of soybean oil under optimal operating conditions of 6:1 methanol/oil molar ratio, 6 wt% catalyst, and 65 °C of reaction temperature in 120 min [79].

Activated carbon can act as suitable catalyst support for CaO, KOH, or lipase-supported catalysts with large surface area and porosity. It can go through surface functionalization with acid or base groups. It shows thermal, acidic, or environmental stability as well as mechanical durability. It possesses a considerable volume of micropores with a good pore size distribution and adsorption surface area (800–1500 m^2/g), and it is considered to be inexpensive with inert carbon skeleton. It shows increased catalytic activity due to its surface properties and surface oxides. Surface oxygen groups form fastening sites for metallic precursors and metals which characterize the properties of activated carbon as a catalyst support material.

Biochar catalysts embedded with ferrite ions are known as magnetic biochar catalysts. It is synthesized by embedding ferrite ions into biomass accompanied by carbonization and functionalization of the magnetic support. It has been reported that the usage of biochar as a catalyst or catalyst support due to its reduced porosity and adsorptive properties. Hence, biochar is sulfonated with sulfuric acid to improve the surface area and porosity [80].

Beata et al. have reported on the transesterification of corn oil using KOH supported on activated carbon. The author has obtained a biodiesel yield of 92% under optimal reaction conditions of 3:1 methanol-to-oil molar ratio, 0.75 wt% of catalyst, and 62.5 °C in 1 h of reaction time [81]. Activated carbon was synthesized from beech tree biomass and has a low surface area with 12.4% of oxygen groups. Ayesha Hameed et al. have reported on the synthesis of calcium-loaded activated carbon catalyst by pyrolysis of peach shell succeeded by chemical activation with KOH followed by calcium loading with the help of wet impregnation method. A biodiesel yield of 96% is obtained by transesterification of waste cooking oil under optimal reaction conditions of 8:1 methanol-to-oil molar ratio, 5 wt% of catalyst, and 65 °C temperature in 160 min of reaction time [82].

6.4 Animal Waste

6.4.1 Calcium Oxide

Waste eggshells consist of a network of protein fibers along with crystals of calcium carbonate ($CaCO_3$), magnesium carbonate, calcium phosphate, organic substances, and water. $CaCO_3$ on calcination at high temperatures gets converted into calcium oxide (CaO). Eggshell consists of more than more than 90% of $CaCO_3$ which forms a nanoporous structure and occurs in the form of calcite. CaO catalyst synthesized from waste eggshell, shells of mullock is non-toxic, cheap, biodegradable, non-corrosive, highly basic, readily available, and shows good recyclability and reusability. CaO possesses a long shelf life as a catalyst, and it can be used in moderate reaction conditions [83]. It has been reported that basic sites of pure CaO get poisoned and converted to calcium carbonate and calcium hydroxide when it is kept in atmospheric air.

Catalytic activity, shelf life, and stability of CaO can be increased by doping with rare-earth elements such as lanthanum and cerium or forming mixed metal oxides. Improved catalytic performance of CaO was reported to be good when it is mixed with La_2O_3 but when it is exposed to air, there is a change in the structure of the catalyst. Mixed metal oxides can also be synthesized by impregnating CaO with other metals. Eu_2O_3/c-Al_2O_3 was considered to be costlier for large-scale applications. In the lanthanide series, praseodymium oxide is considered to be a basic catalyst with the highest oxygen ion mobility due to which various stable phases help in the quick changes in the oxygen state of praseodymium. Hence, praseodymium oxides find applications as catalysts or act as promoters or stabilizers in forming a catalyst. Sana Gohar Khan et al. have reported that 87.42% of biodiesel yield is obtained when transesterification of castor oil is carried out using Pr doped CaO derived from waste mussel shell at 8:1 methanol-to-oil molar ratio, 2.5 wt% of catalyst, and 60–65 °C of the reaction temperature in a duration of 4 h.

Rahman et al. reported the synthesis of mixed metal oxides (Zn/CaO, Cu/CaO) by incipient wet impregnation method, and the surface area of mixed metal oxides was higher when compared to pure CaO. Synthesized mixed metal oxide Zn/CaO when used in transesterification of eucalyptus oil gives a maximum yield of 93.8% when compared to Cu/CaO catalyst at 6:1 methanol-to-oil ratio, 5 wt% of catalyst, and 65 °C for a duration of 2.5 h [84]. Jiadi et al. synthesized an eggshell-derived CaO/Au composite in which eggshell was used to adsorb Au^{3+} in chloroauric acid and calcinated further. He has reported a biodiesel yield of 88.9% when eggshell-derived CaO/Au composite was used in the transesterification of soybean oil at 12:1 methanol-to-oil molar ratio, 1 wt% of catalyst, and 70 °C for 3 h. Eggshell-derived CaO/Au composite exhibits good catalytic activity due to the promising joining of CaO with Au nanoparticles [85]. Kavitha et al. obtained a biodiesel yield of 96% from dairy scum waste using a CaO catalyst under optimum conditions of 6:1 methanol-to-oil molar ratio, 2.4 wt% of catalyst, and 65 °C of reaction temperature for a reaction time of 3 h [86]. Chen et al. have reported on the synthesis of mixed metal oxide

$CaO–SiO_2$ using CaO and Na_2SiO_3. Synthesized $CaO–SiO_2$ catalyst when used in the transesterification of palm oil gives FAME yield of 90.2% [87]. Shan et al. have obtained a biodiesel yield of 92.4% from vegetable oil using a diatomite-supported CaO catalyst with high basicity. It shows enhanced catalytic activity and reusability when used as a catalyst in the transesterification reaction. It has also been reported that diatomite consists of SiO_2 which reacts with CaO to form a Ca–O–Si bond. It has also been reported that Ca/Si catalyst shows improved stability due to the well scattering of Ca compounds on the support and a stable Ca–O–Si bond is formed. The presence of the Ca–O–Si bond improves the stability of the Ca/Si composite catalyst and also decreases the loss of Ca^{2+} ions.

Krishnamurthy et al. have reported on the synthesis of CaO from snail shell waste and obtained a biodiesel yield of 96.92% when used in the transesterification of dairy waste scum oil under optimized conditions of 12.7:1 methanol-to-oil molar ratio, 0.866 wt% of catalyst, and 58.56 °C reaction temperature for 119.684 min. Kaolin a natural substance rich in calcium can also catalyze biodiesel production, but the catalytic activity will be less when it is used alone. Liu et al. have reported that a solid base catalyst with improved catalytic activity is synthesized by mixing different proportions of kaolin and snail shells followed by immersion in different KBr concentrations since K+ concentrations show improved catalytic performance. A biodiesel yield of 98.5% was attained when the synthesized catalyst was used in the transesterification of soybean oil under optimal conditions of 6:1 methanol-to-oil molar ratio, 2 wt% of catalyst, and 65 °C in a duration of 2 h.

6.4.2 *Chitosan*

Chitosan ($C_{56}H_{103}N_9O_{39}$) is obtained from chitin by deacetylation reaction and is widely found in aquatic organisms such as shrimp/crab shells, algal cell membranes, and bones of molluscus. It is widely used as a material membrane since it is biodegradable, hydrophilic with enhanced chemical stability, and naturally available, inexpensive, and non-toxic polymer. Chitosan consists of hydroxyl, C–O–C, and amino groups and is insoluble in water. Amino groups that are present in chitosan act as basic sites that support the production of biodiesel. Moreover, amino groups can also act as cations in an acidic medium. Chitosan can be used in catalysis in various forms such as solids, powder, films, hydrogels, fibers, flakes, membranes, or sponges. The drawbacks of polymer-based membranes are reduced mechanical strength and enlargement after the long-term operation, which in turn decreases the separation performance. The ways of improving the physical properties of chitosan-based polymers are inserting inorganic nanoparticles within a polymeric phase such as silica, calcium aluminosilicate, titanium oxide, silver, graphene oxide, and titanium dioxide.

Graphene oxide is considered to be inexpensive and also possesses a large surface area, improved mechanical strength, and consists of oxygen functional groups, e.g., hydroxyl, epoxide, diol, ketone, and carboxyl which react to the amine group of chitosan to form a bond between graphene oxide and biopolymer.

Maryam Helmi et al. have reported on the synthesis of NaOH/Chitosan-Fe_3O_4 catalyst with magnetic properties. A biodiesel yield of 92% is obtained under optimal reaction conditions of 6:1 methanol-to-oil molar ratio, 0.5 wt% of catalyst, and 25 °C reaction temperature in 4.5 h [88].

Leila et al. have reported on the synthesis of zeolite-chitosan biocomposite modified using Na^+. The catalytic activity of Na^+ is enhanced due to the reaction of chitosan's amino groups and Al–O–Na groups on the zeolite surface as well as with the physical process of Na^+ in pores and cages of the composite. Biodiesel yield of 96.5% is attained by electrocatalytic transesterification of waste fried oil using the synthesized catalyst under optimal reaction conditions of 8:1 methanol-to-oil molar ratio, 1 wt% of catalyst, and 10 wt% of acetone as a co-solvent at room temperature in 30 min. Wang et al. have reported on the synthesis of magnetic carbonaceous acid (Zr–CMC–SO_3H@3Fe–C_{400}) having Bronsted and Lewis sites. On esterification of oleic acid in the presence of the catalyst gives 97% biodiesel yield, whereas when used in the transesterification of soybean oil gives 95% biodiesel yield in the presence of the catalyst. Wang et al. synthesized a double-shell mesoporous magnetic solid acid catalyst in which Fe_3O_4 metal nanoparticles were coated with chitosan and silica two times and then sulfonated with p-toluene sulfonic acid. A biodiesel yield of 96.7% is obtained in the esterification of oleic acid using the synthesized catalyst at 15:1 methanol-to-oil molar ratio and 80 °C of reaction temperature. The improved catalytic performance of the catalyst along with chitosan is due to the ammonia NH_3 sites on the catalyst surface that are derived from the amine groups of chitosan (Table 6.1).

6.5 Nanocatalyst Derived from Plant Pigment

Flowers, leaves, bark, and fruits of plants contain pigments and dyes like chlorophylls and anthocyanin which can be removed using easy procedures, and these plant sources are considered to be inexpensive, non-toxic, eco-friendly, and widely available. These natural dyes and pigments like anthocyanin and chlorophyll from China rose, anthocyanins from cumin, anthocyanin from red radish, curcumin from turmeric, and also chlorophyll from cyanobacteria like Spirulina find application as a sensitizer in photocatalysis for the breaking down of organic pollutants. The presence of a porphyrin ring in chlorophylls serves as a light-harvesting antenna for solar energy collection. Chlorophyll can be used for photo-driven synthesis because of its affinity for oxygen. It has been reported that photocatalysts like titanium dioxide along with porphyrin can be synthesized using incipient wetness impregnation procedure, and synthesized photocatalyst serves as an efficient sensitizer for attaining visible light photocatalysis when compared to synthetic dyes. It has been reported that 85% of methylene blue degradation was attained at optimum conditions of amount of catalyst of 0.1 wt% chlorophyll, initial pH of 6, and 20 ppm of initial methylene blue concentration for 2 h of visible light exposure duration [89]. Moreover, chlorophyll pigment loaded on titanium dioxide helped in improving the photocatalytic

Table 6.1 Summarizes green catalyst derived from biomass for biodiesel production along with the conversion yield and operating conditions

S. No.	Type of biomass	Type of feedstock	Catalyst	Transesterification conditions					Biodiesel yield (%)	References
				FS	*T* (°C)	*t* (min)	MTOR	AC (wt%)		
1	Agricultural waste	Rice husk	Rice husk biochar loaded with 30% of CaO	Vegetable oil	65	180	9:1	8	93.4	[41]
2	Agricultural waste	Rice husk	Rice husk char doped withpotassium oxideandnickel oxide	Used cooking oil	65	120	12:1	4	98.2	[42]
3	Agricultural waste	Sugarcane bagasse	Sulfonated sugarcane bagasse	Palm fatty acid distillate	60	90	10:1	2	98.6	[61]
4	Agricultural waste	Coconut shell	Ce-supported sulfated activated carbon	Chicken fat and skin oil	90	60	12:1	3	93	[70]
5	Plant waste	Potato peels	Biochar	Soybean waste cooking oil	60	120	9:1	3	97.5	[72]
6	Plant waste	Banana waste	Heterogeneous AC/KOH catalyst-supported K_2CO_3	Palm oil	65	120	15:1	4	99.15	[74]
7	Plant waste	Pomelo waste	Biochar loaded with 25 wt% of potassium carbonate	Palm oil	65	150	8:1	6	98	[77]

(continued)

Table 6.1 (continued)

S. No.	Type of biomass	Type of feedstock	Catalyst	Transesterification conditions					Biodiesel yield (%)	References
				FS	*T* (°C)	*t* (min)	MTOR	AC (wt%)		
8	Animal waste	Eggshell	Eggshell-derived CaO/Au composite	Soybean oil	70	180	12:1	1	88.9	[85]
9	Animal waste	Chitosan	NaOH/ Chitosan-Fe_3O_4	Waste cooking oil	25	270	6:1	0.5	92	[87]
10	Plant waste	Kraft lignin	Potassium carbonate/kraft lignin-activated carbon	Rapeseed oil	65	120	15:1	3	99.6	[97]

FS feedstock, *T* reaction temperature, *t* reaction time, *MTOR* methanol-to-oil molar ratio, *AC* amount of catalyst efficiency for the degradation of pollutant molecules. Chlorophyll is not stable in organic solutions. Hence, doping on a desirable substrate, and silica materials can improve the stability. Vahhab et al. have reported that titanium oxide nanoparticles and chlorophyll were doped to change polyvinyl alcohol/Chitosan and can serve as a photocatalyst for the degradation of methylene blue dye. Methylene blue degradation efficiency of (96%) was attained under visible light irradiation (LED lamp 70 W by λ is 425 nm) in 60 min using bio nanocomposite film (poly(vinyl alcohol)/TiO_2/ chitosan/chlorophyll) [90].

6.6 Kraft Lignin as a Catalyst

Lignin is considered to be an inexpensive biopolymer with various links of carbon in the form of C–O–C and C–C. The active sites present in the lignin help in the adsorption of contaminated molecules, and the presence of hydroxyl groups helps in hydrophilicity and functionality [91–93]. Kraft pulping method supplies oxygen groups for lignin which enhances the properties of adsorption [94]. The structure and composition of kraft lignin are completely different from that of lignin [95, 96]. Hence, kraft lignin finds application in wastewater treatment as well as in biodiesel production. Zahra Nezafat has synthesized kraft lignin-titanium oxide/ palladium nanocomposite and used it in the degradation of methylene blue and methylene orange. Titanium oxide nanoparticles were immobilized on kraft lignin which acts as a substrate. Metal oxides such as titanium oxide improve the attachment of other metal nanoparticles such as palladium and also provide oxygen functional groups which further enhance the hydrophilicity. Palladium nanoparticles help in the removal of organic dyes from water. Moreover, it has been reported that bare palladium nanoparticles give less catalytic efficiency when compared to the attached nanoparticles. The author has reported that the degradation of methylene blue and methyl orange was carried out using 1 mg of KL-T/Pd catalyst in 23 and 8 s [97].

Xian-fa Li et al. have reported that potassium carbonate/kraft lignin-activated carbon (K_2CO_3/KLC) was synthesized using an in situ method of mixing K_2CO_3 with technical kraft lignin (KL) followed by activating at 800 °C for a duration of 2 h in an inert atmosphere. The synthesized catalyst K_2CO_3/KLC when used in the transesterification of rapeseed oil gives a biodiesel yield of 99.6% at optimal 15:1 methanol-to-oil molar ratio, 3 wt% of catalyst at a reaction temperature of 65 °C in 2 h [98]. Kraft lignin is obtained from the kraft pulping process which is considered to be inexpensive, biodegradable, and easily procurable. The synthesized catalyst has improved structural morphology and improved scattering of active components for biodiesel production at moderate conditions.

6.7 Important Factors that Affect the Production of Activated Carbon

1. Activation temperature
2. Activation duration
3. Chemical impregnation ratio.

6.8 Activation Temperature

Supply of heat to an impregnated material helps in improving the thermal degradation and volatilization process which helps in the development of pores and enhancement in surface area. The activation temperature is selected based on aspects such as the type of precursor and the chemical agent that is used in the method. Activation temperature for various biomass feedstock varies from 400 to 800 °C, whereas coal-based feedstock can progress as high as 900 °C. Huiping et al. have reported on the effect of varying activation temperature for 60 min and an impregnation ratio of 100%. An increase in activation temperature reduces the yield of activated carbon due to the burn-off of carbon [99]. Shenghui et al. have studied the effect of carbon dioxide activation on porous structures of coconut shell-based activated carbon. Micropore volume, BET surface area, and total pore volume of the activated carbon were lower at an activation temperature of 750 °C since there is a slow reaction between char and carbon dioxide. Whereas at an activation temperature of 900 °C, there is a substantial increase in micropore volume, surface area, and total pore volume due to the emission of volatiles as well as an enhanced reaction between carbon and carbon dioxide. He has obtained a micropore percentage of 87.1 at 900 °C of activation temperature and 600 cm^3/min of CO_2 flow rate [100]. Moreover, it is also reported that when the activation temperature is increased further to 950 °C, the widening of pores occurs rapidly when compared to the formation of new pores which results in enhancement in pore diameter and mesopore formation which in turn further reduces the micropore volume, micropore percentage and BET surface area, and the average pore diameter and total volume enhances. The widening of pores is due to the rapid reaction between carbon and carbon dioxide.

6.9 Activation Duration

The duration of the activation plays an important role in the progress of the carbon's porous networks. The duration should be sufficient to remove all the moisture and volatile components present in the precursor to create pores to develop. As soon as the volatile escape starts to end, the creation of pore structures starts to develop, and the activation should be stopped up to this point. Increased duration results in an increase of pores at the cost of the surface area. Control of activation time is important because a shorter duration reduces energy consumption and hence is considered to be more desirable [101].

Bin Liu et al. have found out the surface area of rice husk activated carbon formulated at a fixed concentration of zinc chloride and copper chloride at an activation temperature of 500 °C for various activation time [102]. The surface area of the micropore and the surface area obtained from BET analysis show an increasing trend as the activation time enhances from 0.5 to 1.5 h. Reduction in surface area was observed when activation time was extended beyond 1.5 h. The activation time can favor the pyrolysis of rice husk and the evaporation of volatiles, supporting incomplete combustion of carbon and creating new micropores. There was an enhancement in surface area due to the increase in the amount of micropores and mesopores. When the activation time is increased above a certain limit, there is an increase in the amount of evaporation of the activator. Transverse activation plays a predominant role and damages the pore structure of activated carbon. Hence, more micropores and macropores developed which caused a reduction in the surface area.

6.10 Chemical/Impregnation Ratio

In the process of chemical activation, the impregnation ratio plays an important role which has an impact on the quality of the obtained carbon. The impregnation ratio is the ratio of the weights of the chemical agent and the dry precursor. The chemical agents that are used in the activation method pierce deeper into the structure of the carbon which in turn results in the growth of tiny pores. The surface area of the small pores is considered to be larger [101]. K. Y. Foo et al. have reported that on enhancing the impregnation ratio (KOH/char) from 0.25 to 1, there is an increase in the yield of carbon; whereas a reduction in the yield of carbon is noticed when the impregnation ratio is increased beyond 1. It has been reported that metallic potassium structured during the gasification process would penetrate the internal structure of the char matrix broadening the available pores and generating new porosities. The activation method improved by increasing the impregnation ratio. Beyond the maximum value, there is an expansion of pores and burn off which decreases the yield of carbon. Hence, the impregnation ratio was set at 1 for efficient activation at the lowest consumption of the activating agent (Table 6.2).

6.11 Chemical Activation Mechanism of the Biomass-Derived Catalyst During Pyrolysis

The chemical activation mechanism of the biomass-derived catalyst during pyrolysis can be explained by the following steps [108]:

Table 6.2 Shows the activation effects due to several activation methods

Raw material	Chemical agent	Activation temperature	Activation effect	References
Coconut shell	NaOH	700	Mesoporous structure was observed when there was an increase in NaOH: biochar ratio and mesopores that are present in activated carbon act as an adsorbent for the removal of methylene blue from aqueous solutions	[103]
Sour cherry stone	$ZnCl_2$	700	Surface area and pore volume of activated carbon treated with $ZnCl_2$ is 1704 m^2g^{-1} and 1.566 cm^3g^{-1} when compared to the sour cherry stones (41.54 m^2g^{-1}) 0.0975 cm^3g^{-1}. $ZnCl_2$ helps in the formation of new micropores and also helps in the aggregation of micropores into mesopores	[104]
Banana leaf	K_2CO_3	750	The surface area and pore volume of activated carbon nanosheets are 1459 m^2g^{-1} and 0.652 cm^3 g^{-1} when compared to the surface area and pore volume of 23 m^2g^{-1} and a pore volume of 0.035 cm^3 g^{-1} for non-activated carbon Activation helps in the formation of micropores	[105]

(continued)

Table 6.2 (continued)

Raw material	Chemical agent	Activation temperature	Activation effect	References
Lotus leaves	KOH	800	Lotus leaf porous carbon after activation with KOH has a surface area of 2440 m^2g^{-1} when compared to the surface area of 2.131 for lotus leaf. The pore size of lotus leaf porous carbon is < 4 nm. KOH develops micro and mesopores on the lotus leaves. LLPC is used in the adsorption of RhB	[106]
Reedy grass leaves	H_3PO_4	500	BET surface area and pore volume of activated carbon is 1474 m^2/g, 0.560 cm^3/g H_3PO_4 acts as an activating reagent, which will increase the yield but will also change the thermal degradation of the precursor, leading to a subsequent change in the evolution of porosity	[107]

1. During the pyrolysis process, KOH reacts with the oxygen-containing groups like C=O, –OH, O–C=O, C–O, and –COOH that are present in the biomass to create more vacancies in the biochar due to the liberation of C=O, –OH, O–C=O, C–O, and –COOH groups from the biomass.
2. KOH also reacts with the C–C and C–H groups that are obtained from broken carbon particles to release hydrogen protons that form vacancies.
3. OH^- ions from KOH enter these vacancies and form an abundant number of new oxygen-containing groups in the biochar given in Eq. (6.3).

$$OH^- + C(\text{vacancies}) \rightarrow (C-O) + (-OH) + C-O + (O-C=O) + (-COOH) \tag{6.3}$$

4. Conversion of KOH into K_2CO_3 occurs due to the reactions taking place between KOH and oxygen-containing groups/broken carbon particles at a temperature of 400–700 °C shown in Eqs. (6.4)–(6.9) which further converts into K_2O at 800 °C given in Eqs. (6.10)–(6.13), followed by the evolution of potassium and gaseous products.

$$6KOH + 2C \rightarrow 2K_2CO_3 + 2K + 3H_2 \tag{6.4}$$

$$KOH + (-COOH)/(-O-C{=}O) \rightarrow K + K_2CO_3 + H_2 + CO_2 \tag{6.5}$$

$$KOH + (-C{=}O)/(-C-O-C) \rightarrow K + K_2CO_3 + H_2 + CO \tag{6.6}$$

$$KOH + (-O-CH_3) \rightarrow K + K_2CO_3 + H_2 + CH_4 \tag{6.7}$$

$$KOH + (C-OH) \rightarrow K_2CO_3 + K + H_2O + H_2 \tag{6.8}$$

$$KOH + (C-H) \rightarrow K_2CO_3 + K + H_2 \tag{6.9}$$

$$K_2CO_3 + C \rightarrow K_2O + 2CO \tag{6.10}$$

$$K_2CO_3 \rightarrow K_2O + CO_2 \tag{6.11}$$

$$2K + CO_2 \rightarrow K_2O + CO \tag{6.12}$$

$$K_2O + C \rightarrow 2K + CO \tag{6.13}$$

5. These steps favored enhanced porosity and oxygen content in the biochar.
6. Pyrolytic intermediates formed during pyrolysis on reacting with free radicals and KOH yield phenol, aromatics, and gaseous products such as hydrogen, carbon dioxide, carbon monoxide, and methane.
7. Some of the free radicals also convert directly into gaseous products.

6.12 Future Prospects

Green catalysts derived from various biomass sources help to reduce environmental pollution caused by the dumping of this waste. Inexpensive green catalysts synthesized from various biomass will be a boon to researchers in the field of wastewater treatment, biodiesel production, fuel cells, and CO_2 capture. Research on the synthesis of green catalysts derived from biomass in the form of films is important since the reusability of the catalyst is comparatively easier when compared to that in its powder form.

6.13 Conclusions

Green catalyst synthesized from various biomass serves as an inexpensive and efficient catalyst in biodiesel production. This chapter discusses various biomass from which green catalysts can be synthesized along with the composition of biomass, ways by which activation of biomass can be carried out to improve the porosity of the obtained green catalyst, factors that affect the production of activated carbon, and chemical activation mechanism of the biomass-derived catalyst during pyrolysis. The selection of the catalyst plays an important role in enhancing the conversion of biodiesel. Biodiesel with high yield and high quality can be produced on an industrial scale using inexpensive green catalysts derived from biomass such as activated carbon, NaOH/Chitosan-Fe_3O_4 catalyst, eggshell-derived CaO/Au composite, and biochar loaded with potassium carbonate which can serve as a substitute for the fossil fuels.

References

1. Ratan, J. K., Kaur, M., & Adiraju, B. (2018). Synthesis of activated carbon from agricultural waste using a simple method: Characterization, parametric and isotherms study. *Materials Today: Proceedings, 5*(2), 3334–3345. https://doi.org/10.1016/j.matpr.2017.11.576
2. Adegoke, K. A., & Bello, O. S. (2015). Dye sequestration using agricultural wastes as adsorbents. *Water Resources and Industry, 12*, 8–24. https://doi.org/10.1016/j.wri.2015.09.002
3. Kumar, R. J., Manjeet, K., & Bharadwaj, A. (2018). Synthesis of activated carbon from agricultural waste using a simple method: Characterization, parametric and isotherms study. *Materials Today: Proceedings, 5*(2), 3334–3345. https://doi.org/10.1016/j.matpr.2017.11.576
4. Cazetta, A. L., Vargas, A. M. M., Nogami, E. M., Kunita, M. H., Guilherme, M. R., Martins, A. C., Silva, T. L., Moraes, J. C. G., & Almeida, V. C. (2011). NaOH-activated carbon of high surface area produced from coconut shell: Kinetics and equilibrium studies from the methylene blue adsorption. *Chemical Engineering Journal, 174*, 117–125.
5. Daimary, N., Boruah, P., Eldiehy, KhalifaSH., Pegu, T., Bardhan, P., Bora, U., Mandal, M., & Deka, D. (2022). *Musa acuminata* peel: A bioresource for bio-oil and by-product utilization as a sustainable source of renewable green catalyst for biodiesel production. *Renewable Energy, 187*, 450–462.
6. Baroutian, S., Aroua, M. K., Raman, A. A. A., & Sulaiman, N. M. N. (2010). Potassium hydroxide catalyst supported on palm shell activated carbon for transesterification of palm oil. *Fuel Processing Technology, 91*, 1378–1385.
7. Irshad, M. A., Nawaz, R., Rehman, M. Z., Adrees, M., Rizwan, M., Ali, S., Ahmad, S., & Tasleem, S. (2021). Synthesis, characterization and advanced sustainable applications of titanium dioxide nanoparticles: a review. *Ecotoxicology and Environmental Safety, 212*, 111978.

8. Sivanesh, S., Aswin, K. N., Antony, A., SuryaVarma, M., Lakshmi, A., Kamalesh, K., Naageshwaran, M., Soundarya, S., & Subramanian, S. (2022). Biodiesel production from custard apple seeds and *Euglena sanguinea* using CaO nano-catalyst. *Bioresource Technology., 344*, 126418.
9. Sahu, S., Saikia, K., Gurunathan, B., Dhakshinamoorthy, A., & Rokhum, S. L. (2023). Green synthesis of CaO nanocatalyst using watermelon peels for biodiesel production, *547*, 113342.
10. Cholapandian, K., Gurunathan, B., Rajendran, N. (2022). Investigation of CaO nanocatalyst synthesized from *Acalypha indica* leaves and its application in biodiesel production using waste cooking oil, *312*, 122958.
11. Yusuff, A. S., Gbadamosi, A. O., & Popoola, L. T. (2021). Biodiesel production from transesterified waste cooking oil by zinc-modified anthill catalyst: Parametric optimization and biodiesel properties improvement. *Journal of Environmental Chemical Engineering, 9*(2), 104955.
12. Arshad, S., Ahmad, M., Munir, M., Sultana, S., Zafar, M., Dawood, S., Rozina, AhmadM., Alghamdi, S. A., Bokhari, A., Mubashir, M., Chuah, L. F., & Show, P. L. (2023). Assessing the potential of green CdO_2 nano-catalyst for the synthesis of biodiesel using non-edible seed oil of Malabar Ebony. *Fuel, 333*, 126492.
13. Yaşar, F. (2019). Biodiesel production via waste eggshell as a low-cost heterogeneous catalyst: Its effects on some critical fuel properties and comparison with CaO. *Fuel, 255*, 115828. https://doi.org/10.1016/j.fuel.2019.115828
14. Mamo, T. T., & Mekonnen, Y. S. (2019). Microwave-assisted biodiesel production from microalgae, scenedesmus species, using goat bone–made nano-catalyst. *Applied Biochemistry and Biotechnology, 190*, 1147–1162. https://doi.org/10.1007/s12010-019-03149-0
15. Mosaddegh, E., & Hassankhani, A. (2014). Preparation and characterization of nano-CaO based on eggshell waste: Novel and green catalytic approach to highly efficient synthesis of pyrano[4,3-b] pyrans. *Chinese Journal of Catalysis, 35*(2), 351–356.
16. Gao, Y., & Xu, C. (2012). Synthesis of dimethyl carbonate over waste eggshell catalyst. *Catalysis Today, 190*(1), 107–111. https://doi.org/10.1016/j.cattod.2011.12.004
17. Ahmad, A. L., Yasin, N. H. M., Derek, C. J. C., & Lim, J. K. (2011). Microalgae as a sustainable energy source for biodiesel production: A review. *Renewable and Sustainable Energy Reviews, 15*(1), 584–593.
18. Jazie, A. A., Sinha, H., & Pramanik, A. S. K. (2013). Transesterification of peanut and rapeseed oils using the waste of animal bone as a cost-effective catalyst. *Materials for Renewable and Sustainable Energy, 2*(11), 1–10.
19. Isikgor, F. H., & Becker, C. R. (2015). Lignocellulosic biomass: A sustainable platform for the production of bio-based chemicals and polymers. *Polymer Chemistry, 6*, 4497–4559.
20. Chen, K.-T., Wang, J.-X., Dai, Y.-M., Wang, P.-H., Liou, C.-Y., Nien, C.-W., Wu, J.-S., & Chen, C.-C. (2013). Rice husk ash as a catalyst precursor for biodiesel production. *Journal of the Taiwan Institute of Chemical Engineers, 44*(4), 622–629. https://doi.org/10.1016/j.jtice.2013.01.006
21. Liou, T.-H., & Wu, S.-J. (2009). Characteristics of microporous/mesoporous carbons prepared from rice husk under base- and acid-treated conditions. *Journal of Hazardous Materials, 171*, 693–703.
22. Lin, L., Zhai, S.-R., Xiao, Z.-Y., Song, Y., An, Q.-D., & Song, X.-W. (2013). Dye adsorption of mesoporous activated carbons produced from NaOH-pretreated rice husks. *Bioresource Technology, 136*, 437–443.
23. Muniandy, L., Adam, F., Mohamed, A. R., & Ng, E.-P. (2014). The synthesis and characterization of high purity mixed microporous/mesoporous activated carbon from rice husk using chemical activation with NaOH and KOH. *Microporous and Mesoporous Materials, 197*, 316–323. https://doi.org/10.1016/j.micromeso.2014.06.020
24. Quispe, I., Navia, R., & Kahhat, R. (2017). Energy potential from rice husk through direct combustion and fast pyrolysis: A review. *Waste Management, 59*, 200–210. https://doi.org/10.1016/j.wasman.2016.10.001

25. Krishnarao, R. V., Subrahmanyam, J., & Jagadish Kumar, T. (2001). Studies on the formation of black particles in rice husk silica ash. *Journal of European Ceramic Society, 21*, 99–104.
26. Prasetyoko, D., Ramli, Z., Endud, S., Hamdan, H., & Sulikowski, B. (2006). Conversion of rice husk ash to zeolite beta. *Waste Management, 26*, 1173–1179.
27. Othman Ali, I., Hassan, A. M., Shaaban, S. M., & Soliman, K. S. (2011). Synthesis and characterization of ZSM-5 zeolite from rice husk ash and their adsorption of Pb^{2+} onto unmodified and surfactant-modified zeolite. *Separation and Purification Technology, 83*, 38–44. https://doi.org/10.1016/j.seppur.2011.08.034
28. Pimprom, S., Sriboonkham, K., Dittanet, P., Föttinger, K., Rupprechter, G., & Kongkachuichay, P. (2015). Synthesis of copper–nickel/SBA-15 from rice husk ash catalyst for dimethyl carbonate production from methanol and carbon dioxide. *Journal of Industrial Engineering Chemistry, 31*, 156–166. https://doi.org/10.1016/j.jiec.2015.06.019
29. López, M., Palacio, R., Royer, S., Mamede, A. S., & Fernández, J. J. (2020). Mesostructured CMK-3 carbon supported Ni–ZrO_2 as catalysts for the hydrodeoxygenation of guaiacol. *Microporous Mesoporous Material, 292*, 109694. https://doi.org/10.1016/j.micromeso.2019.109694109694
30. Jeong, Y., Cui, M., Choi, J., Lee, Y., Kim, J., Son, Y., & Khim, J. (2020). Development of modified mesoporous carbon (CMK-3) for improved adsorption of bisphenol-A. *Chemosphere, 238*, 124559. https://doi.org/10.1016/j.chemosphere.2019.124559124559
31. Zhang, T., Qiu, H., Zhang, M., Fang, Z., Zhao, X., Wang, L., Chen, G., Wei, Y., Yue, H., Wang, C., & Zhang, D. (2017). A unique 2D-on-3D architecture developed from $ZnMn_2O_4$ and CMK-3 with excellent performance for lithium-ion batteries. *Carbon, 123*, 717–725. https://doi.org/10.1016/j.carbon.2017.08.013
32. Juárez, J. M., Ledesma, B. C., Costa, M. G., Beltramone, A. R., & Anunziata, O. A. (2017). Novel preparation of CMK-3 nanostructured material modified with titania applied in hydrogen uptake and storage. *Microporous Mesoporous Materials, 254*, 146–152. https://doi.org/10.1016/j.micromeso.2017.03.056
33. Regiart, M., Fernández-Baldo, M. A., Villarroel-Rocha, J., Messina, G. A., Bertolino, F. A., Sapag, K., Timperman, A. T., & Raba, J. (2017). Microfluidic immunosensor based on mesoporous silica platform and CMK-3/poly-acrylamide-co-methacrylate of dihydrolipoic acid modified gold electrode for cancer biomarker detection. *Analytica Chimica Acta, 963*, 83–92. https://doi.org/10.1016/j.aca.2017.01.029
34. Zhang, G., Chen, Y., Huang, K., Chen, Y., & Guo, H. (2018). CMK-3/$NiCo_2S_4$ nanostructures for high-performance asymmetric supercapacitors. *Materials Chemistry and Physics, 220*, 270–277. https://doi.org/10.1016/j.matchemphys.2018.09.008
35. Ahmed, M. J. (2016). Preparation of activated carbons from date (*Phoenix dactylifera* L.) palm stones and application for wastewater treatments. *Review Process Safety and Environmental Protection, 102*, 168–182. https://doi.org/10.1016/j.psep.2016.03.010
36. Huang, F. C., Lee, C. K., Han, Y. L., Chao, W. C., & Chao, H. P. (2014). Preparation of activated carbon using micro-nano carbon spheres through chemical activation. *Journal of Taiwan Institute of Chemical Engineers, 45*(5), 2805–2812.
37. Naji, S. Z., & Tye, C. T. (2022). A review of the synthesis of activated carbon for biodiesel production: Precursor, preparation, and modification. *Energy Conversion and Management: X, 13*, 100152.
38. Mazlan, M. A. F., Uemura, Y., Yusup, S., Elhassan, F., Uddin, A., & Hiwada, A. I. (2016). Activated carbon from rubber wood sawdust by carbon dioxide activation. *Procedia Engineering, 148*, 530–537.
39. Liu, D., Zhang, W., Lin, H., Li, Y., Lu, H., & Wang, Y. (2016). A green technology for the preparation of high capacitance rice husk-based activated carbon. *Journal of Cleaner Production, 112*, 1190–1198.
40. Ahiduzzaman, M., & Sadrul Islam, A. K. M. (2016). Preparation of porous bio-char and activated carbon from rice husk by leaching ash and chemical activation. *Springerplus, 5*, 1248.

41. Zhao, C., Yang, L., Xing, S., Luo, W., Wang, Z., & Lv, P. (2018). Biodiesel production by a highly effective renewable catalyst from pyrolytic rice husk. *Journal of Cleaner Production, 199*, 772–780.
42. Hazmi, B., Rashid, U., Ibrahim, M. L., Nehdi, I. A., Azam, M., & Al-Resayes, S. I. (2021). Synthesis and characterization of bifunctional magnetic nano-catalyst from rice husk for production of biodiesel. *Environmental Technology & Innovation, 21*, 101296.
43. Zhao, C., Yang, L., Xing, S., Luo, W., Wang, Z., & Lv, P. (2018). Biodiesel production by a highly effective renewable catalyst from pyrolytic rice husk. *Journal of Cleaner Production, 199*, 772–780. https://doi.org/10.1016/j.jclepro.2018.07.242
44. Vinayagam, M., Ramachandran, S., Ramya, V., & Sivasamy, A. (2018). Photocatalytic degradation of orange G dye using ZnO/biomass activated carbon nanocomposite. *Journal of Environmental Chemical Engineering, 6*(3), 3726–3734.
45. Liew, R. K., Chong, M. Y., Osazuwa, O. U., Nam, W. L., Phang, X. Y., Su Huan, M. H., Cheng, C. K., Chong, C. T., & Lam, S. S. (2018). Production of activated carbon as catalyst support by microwave pyrolysis of palm kernel shell: A comparative study of chemical versus physical activation. *Research on Chemical Intermediates, 44*(6), 3849–3865. https://doi.org/10.1007/s11164-018-3388-y
46. Srimachai, T., Thonglimp, V., & Sompong, O. (2014). *Ethanol and methane production from oil palm frond by two-stage SSF*. *Energy Procedia, 52*, 352.
47. Ang, S. K., Shaza, E. M., Adibah, Y., Suraini, A. A., & Madihah, M. S. (2013). Production of cellulases and xylanase by *Aspergillus fumigatus* SK1 using untreated oil palm trunk through solid state fermentation. *Process Biochemistry, 48*(9), 1293–1302.
48. Ishola, M. M., Isroi, T., & Mohammad, J. (2014). Effect of fungal and phosphoric acid pretreatment on ethanol production from oil palm empty fruit bunches (OPEFB). *Bioresource Technology, 165*, 9–12. https://doi.org/10.1016/j.biortech.2014.02.053
49. Saidu, M., Yuzir, A., Salim, M. R., Salmiati, S. A., & Abdullah, N. (2014). Biological pre-treated oil palm mesocarp fiber with cattle manure for biogas production by anaerobic digestion during acclimatization phase. *International Biodeterioration & Biodegradation, 95*, 189–194. https://doi.org/10.1016/j.ibiod.2014.06.014
50. Neutelings, G. (2011). Lignin variability in plant cell walls: Contribution of new models. *Plant Science, 181*(4), 379–386.
51. Cao, J., Xiao, G., Xu, X., Shen, D. K., & Jin, B. (2013). Study on carbonization of lignin by TG-FTIR and high-temperature carbonization reactor. *Fuel Processing Technology, 106*, 41–47. https://doi.org/10.1016/j.fuproc.2012.06.016
52. Collard, F.-X., & Blin, J. (2014). A review on pyrolysis of biomass constituents: Mechanisms and composition of the products obtained from the conversion of cellulose, hemicelluloses, and lignin. *Renewable and Sustainable Energy Reviews, 38*, 594–608. https://doi.org/10.1016/j.rser.2014.06.013
53. Abdullah, R. F., Rashid, U., Taufiq-Yap, Y. H., Ibrahim, M. L., Ngamcharussrivichai, C., & Azam, M. (2020). Synthesis of bifunctional nanocatalyst from waste palm kernel shell and its application for biodiesel production. *Royal Society of Chemistry, 10*, 27183–27193.
54. Goh, C. M. H., Tan, Y. H., Mubarak, N. M., Kansedo, J., Rashid, U., Khalid, M., & Walvekar, R. (2021). Synthesis of magnetic basic palm kernel shell catalyst for biodiesel production and characterization and optimization by Taguchi method. *Applied Nanoscience, 12*, 3721–33. https://doi.org/10.1007/s13204-021-01815-6
55. Kostić, M. D., Bazargan, A., Stamenković, O. S., Veljković, V. B., & McKay, G. (2016). Optimization and kinetics of sunflower oil methanolysis catalyzed by calcium oxide-based catalyst derived from palm kernel shell biochar. *Fuel, 163*, 304–313. https://doi.org/10.1016/j.fuel.2015.09.042
56. Cheah, K. Y., Toh, T. S., & Koh, P. M. (2010). Palm fatty acid distillate biodiesel: Next generation palm biodiesel. *International News on Fats, Oils, and Related Materials, 21*, 264–266.
57. Betancur, G. J. V., & Pereira, N. (2010). Sugar cane bagasse as feedstock for second generation ethanol production. Part I: Diluted acid pretreatment optimization. *Electronic Journal of Biotechnology, 13*(3), 1–9.

58. Kumar, A., Negi, Y. S., Choudhary, V., & Bhardwaj, N. K. (2020). Characterization of cellulose nanocrystals produced by acid-hydrolysis from sugarcane bagasse as agro-waste. *Journal of Materials Physics and Chemistry, 2*(1), 1–8.
59. Rezende, C. A., De Lima, M., Maziero, P., Deazevedo, E., Garcia, W., & Polikarpov, I. (2011). Chemical and morphological characterization of sugarcane bagasse submitted to a delignification process for enhanced enzymatic digestibility. *Biotechnology for Biofuels, 4*(1), 1–19.
60. Chausali, N., Saxena, J., & Prasad, R. (2021). Nanobiochar and biochar based nanocomposites: Advances and applications. *Journal of Agriculture and Food Research, 5*, 100191.
61. Akinfalabi, S.-I., Rashid, U., Ngamcharussrivichai, C., & Nehdi, I. A. (2020). Synthesis of reusable biobased nano-catalyst from waste sugarcane bagasse for biodiesel production. *Environmental Technology & Innovation, 18*, 100788. https://doi.org/10.1016/j.eti.2020.100788
62. Niju, S., Kanna, S. K. A., Ramalingam, V., Kumar, M. S., & Balaji, M. (2019). Sugarcane bagasse derived biochar—A potential heterogeneous catalyst for transesterification process. *Energy Sources, Part A: Recovery, Utilization, and Environmental Effects, 45*(4), 9815–9826. https://doi.org/10.1080/15567036.2019.1680771
63. Mutalib, A. A. A., Ibrahim, M. L., Matmin, J., Kassim, M. F., Mastuli, M. S., Taufiq-Yap, Y. H., Shohaimi, N. A. M., Islam, A., Tan, Y. H., & Kaus, N. H. M. (2020). SiO_2-rich sugar cane bagasse ash catalyst for transesterification of palm oil. *BioEnergy Research, 13*, 986–997. https://doi.org/10.1007/s12155-020-10119-6
64. El Nemr, A., El-Sikaily, A., & Khaled, A. (2010). Modeling of adsorption isotherms of methylene blue onto rice husk activated carbon. *Egyptian Journal of Aquatic Research, 36*(3), 403–425.
65. El Nemr, A., El Sadaawy, M. M., Khaled, A., & El Sikaily, A. (2014). Adsorption of the anionic dye direct red 23 onto new activated carbons developed from *Cynara cardunculus*: Kinetics, equilibrium and thermodynamics. *Blue Biotechnology Journal, 3*(1), 121–142.
66. Shoaib, A. G. M., El-Sikaily, A., El Nemr, A., Mohamed, A.E.-D.A., & Hassan, A. A. (2020). Preparation and characterization of highly surface area activated carbons followed type IV from marine red alga (*Pterocladia capillacea*) by zinc chloride activation. *Biomass Convers Biorefinery, 12*, 2253–2265. https://doi.org/10.1007/s13399-020-00760-8
67. Shoaib, A. G. M., El-Sikaily, A., El Nemr, A., Mohamed, A.E.-D.A., & Hassan, A. A. (2020). Testing the carbonization condition for high surface area preparation of activated carbon followed type-IV from green alga *Ulva lactuca*. *Biomass Conversion and Biorefinery, 12*, 3303–3318. https://doi.org/10.1007/s13399-020-00823-w
68. Sekhon, S. S., Kaur, P., & Park, J.-S. (2021). From coconut shell biomass to oxygen reduction reaction catalyst: Tuning porosity and nitrogen doping. *Renewable and Sustainable Energy Reviews, 147*, 111173. https://doi.org/10.1016/j.rser.2021.111173
69. Ioannidou, O., & Zabaniotou, A. (2006). Agricultural residues as precursors for activated carbon production—A review. *Renewable and Sustainable Energy Reviews, 11*(9), 1966–2005.
70. Shobhana-Gnanaserkhar, A.-M., AbdulKareem-Alsultan, G., Sivasangar-Seenivasagam, S. M., & Izham, T.-Y. (2020). Biodiesel production via simultaneous esterification and transesterification of chicken fat oil by mesoporous sulfated Ce supported activated carbon. *Biomass and Bioenergy, 141*, 105714.
71. Mozammel, H. M., Masahiro, O., & Bhattacharya, S. C. (2002). Activated charcoal from coconut shell using $ZnCl_2$ activation. *Biomass and Bioenergy, 22*(5), 397–400. https://doi.org/10.1016/s0961-9534(02)00015-6
72. Daimary, N., Eldiehy, K. S. H., Boruah, P., Deka, D., Bora, U., & Kakati, B. K. (2022). Potato peels as a sustainable source for biochar, bio-oil and a green heterogeneous catalyst for biodiesel production. *Journal of Environmental Chemical Engineering, 10*(1), 107108.
73. Daimary, N., Boruah, P., KhalifSHEldiehy, T. P., Bardhan, P., Bora, U., Mandal, M., & Deka, D. (2022). *Musa acuminata* peel: A bioresource for bio-oil and by-product utilization as a sustainable source of renewable green catalyst for biodiesel production. *Renewable Energy, 187*, 450–462.

74. Jitjamnong, J., Thunyaratchatanon, C., Luengnaruemitchai, A., Kongrit, N., Kasetsomboon, N., Sopajarn, A., & Khantikulanon, N. (2020). Response surface optimization of biodiesel synthesis over a novel biochar-based heterogeneous catalyst from cultivated (*Musa sapientum*) banana peels. *Biomass Conversion and Biorefinery, 11*, 2795–2811. https://doi.org/10.1007/s13399-020-00655-8
75. Brahma, S., Basumatary, B., Basumatary, S. F., Das, B., Brahma, S., Rokhum, S. L., & Basumatary, S. (2023). Biodiesel production from quinary oil mixture using highly efficient *Musa chinensis* based heterogeneous catalyst. *Fuel, 336*, 127150.
76. Shah, J., Janneth Loez-Mercado, M., Carreon, G., Lopez-Miranda, A., & Carreon, M. L. (2018). Plasma synthesis of graphene from mango peel. *ACS Omega, 3*(1), 455–463. https://doi.org/10.1021/acsomega.7b01825
77. Zhao, C., Lv, P., Yang, L., Xing, S., Luo, W., & Wang, Z. (2018). Biodiesel synthesis over biochar-based catalyst from biomass waste pomelo peel. *Energy Conversion and Management, 160*, 477–485. https://doi.org/10.1016/j.enconman.2018.01.059
78. Nath, B., Das, B., Kalita, P., & Basumatary, S. (2019). Waste to value addition: Utilization of waste *Brassica nigra* plant derived novel green heterogeneous base catalyst for effective synthesis of biodiesel. *Journal of Cleaner Production, 239*, 118112.
79. Aleman-Ramireza, J. L., Moreiraa, J., Torres-Arellanoa, S., Adriana Longoria, C., Okoyeb, P. U., & Sebastian, P. J. (2021). Preparation of a heterogeneous catalyst from Moringa leaves as a sustainable precursor for biodiesel production. *Fuel, 284*, 118983.
80. Rakopoulos, C. D., Antonopoulos, K. A., Rakopoulos, D. C., Hountalas, D. T., & Giakoumis, E. G. Comparative performance and emissions study of a direct injection diesel engine using blends of diesel fuel with vegetable oils or bio-diesels of various origins. *Energy Conversion and Management, 47*(18–19), 3272–3287.
81. Narowska, B., Kułażyński, M., Łukaszewicz, M., & Burchacka, E. (2019). Use of activated carbons as catalyst supports for biodiesel production. *Renewable Energy, 135*, 176–185.
82. Hameed, A., Naqvi, S. R., Sikandar, U., & Chen, W.-H. (2022). One-step biodiesel production from waste cooking oil using cao promoted activated carbon catalyst from *Prunus persica* seeds. *Catalysts, 12*, 592. https://doi.org/10.3390/catal12060592
83. Kouzu, M., Kasuno, T., Tajika, M., Sugimoto, Y., Yamanaka, S., & Hidaka, J. (2008). Calcium oxide as a solid base catalyst for transesterification of soybean oil and its application to biodiesel production. *Fuel, 87*, 2798–2806.
84. Rahman, W. U., Fatima, A., Anwer, A. H., Athar, M., Khan, M. Z., Khan, N. A., & Halder, G. (2018). Biodiesel synthesis from eucalyptus oil by utilizing waste egg shell derived calcium-based metal oxide catalyst. *Process Safety and Environmental Protection, 122*, 313–319. https://doi.org/10.1016/j.psep.2018.12.015
85. Liu, J., Liu, M., Chen, S., Wang, B., Chen, J., Yang, D.-P., Zhang, S., & Du, W. (2020). Conversion of Au(III)-polluted waste eggshell into functional CaO/Au nanocatalyst for biodiesel production. *Green Energy & Environment, 7*(2), 352–359. https://doi.org/10.1016/j.gee.2020.07.019
86. Kavitha, V., Geetha, V., Jacqueline, P., & Jennita. (2019). Production of biodiesel from dairy waste scum using eggshell waste. *Process Safety and Environmental Protection, 125*, 279–287.https://doi.org/10.1016/j.psep.2019.03.021
87. Chen, G., Shan, R., Li, S., & Shi, J. (2015). A biomimetic silicification approach to synthesize CaO–SiO_2 catalyst for the transesterification of palm oil into biodiesel. *Fuel, 153*, 48–55. https://doi.org/10.1016/j.fuel.2015.02.109
88. Helmi, M., & Hemmati, A. (2021). Synthesis of magnetically solid base catalyst of NaOH/Chitosan-Fe_3O_4 for biodiesel production from waste cooking oil: Optimization, kinetics and thermodynamic studies. *Energy Conversion and Management, 248*, 114807.
89. Krishnan, S., & Shriwastav, A. (2020). Application of TiO_2 nanoparticles sensitized with natural chlorophyll pigments as catalyst for visible light photocatalytic degradation of methylene blue. *Journal of Environmental Chemical Engineering, S2213-3437*(20), 31048-4.

90. Soltaninejad, V., & Maleki, A. (2021). A green, and eco-friendly bionanocomposite film (poly(vinyl alcohol)/TiO_2/chitosan/chlorophyll) by photocatalytic ability, and antibacterial activity under visible-light irradiation. *Journal of Photochemistry & Photobiology, A: Chemistry, 404*, 112906.
91. Ge, Y., & Li, Z. (2018). Application of lignin and its derivatives in adsorption of heavy metal ions in water: A review. *ACS Sustainable. Chemistry and Engineering, 6*, 7181–7192.
92. Kajita, S., Hishiyama, S., Tomimura, Y., Katayama, Y., & Omori, S. (1997). Structural characterization of modified lignin in transgenic tobacco plants in which the activity of 4-coumarate: Coenzyme A ligase is depressed. *Plant Physiology, 114*, 871–879.
93. Li, Z., Zhang, J., Qin, L., & Ge, Y. (2018). Enhancing the antioxidant performance of lignin by enzymatic treatment with laccase. *ACS Sustainable Chemistry and Engineering, 6*, 2591–2595.
94. Mohan, D., Pittman, C. U., Jr., & Steele, P. H. (2006). Single, binary and multi-component adsorption of copper and cadmium from aqueous solutions on Kraft lignin-a biosorbent. *Journal of Colloid and Interface Science, 297*, 489–504.
95. Schlee, P., Hosseinaei, O., Baker, D., Landmér, A., Tomani, P., Mostazo-López, M. J., Cazorla-Amorós, D., Herou, S., & Titirici, M. M. (2019). From waste to wealth: from kraft lignin to free-standing supercapacitors. *Carbon, 145*, 470–480.
96. Chatterjee, S., & Saito, T. (2015). Lignin-derived advanced carbon materials. *ChemSusChem-Chemistry Europe, 8*, 3941–3958.
97. Nezafat, Z., Mohazzab, B. F., Jaleh, B., Nasrollahzadeh, M., Baran, T., & Shokouhimehr, M. (2021). A promising nanocatalyst: Upgraded kraft lignin by titania and palladium nanoparticles for organic dyes reduction. *Inorganic Chemistry Communications, 130*, 108746
98. Li, X.-f, Zuo, Y., Zhang, Y., Fu, Y., & Guo, Q.-X. (2013). In situ preparation of K_2CO_3 supported Kraft lignin activated carbon as a solid base catalyst for biodiesel production. *Fuel, 113*, 435–442.
99. Zhang, H., Yan, Y., & Yang, L. (2010). Preparation of activated carbon from sawdust by zinc chloride activation. *Adsorption, 16*(3), 161–166. https://doi.org/10.1007/s10450-010-9214-5
100. Guo, S., Peng, J., Li, W., Yang, K., Zhang, L., Zhang, S., & Xia, H. (2009). Effects of CO_2 activation on porous structures of coconut shell-based activated carbons. *Applied Surface Science, 255*(20), 8443–8449. https://doi.org/10.1016/j.apsusc.2009.05.150
101. Gratuito, M. K. B., Panyathanmaporn, T., Chumnanklang, R. A., Sirinuntawittaya, N., & Dutta, A. (2008). Production of activated carbon from coconut shell: Optimization using response surface methodology, *99*(11), 4887–4895. https://doi.org/10.1016/j.biortech.2007.09.042
102. Liu, B., Gu, J., & Zhou, J. (2016). High surface area rice husk-based activated carbon prepared by chemical activation with $ZnCl_2$–$CuCl_2$ composite activator. *Environmental Progress & Sustainable Energy, 35*(1), 133–140. https://doi.org/10.1002/ep.12215
103. Cazetta, A. L., Vargas, A. M. M., Nogami, E. M., Kunita, M. H., Guilhermea, M. R., Martins, A. C., Silva, T. L., Moraes, J. C. G., & Almeida, V. C. (2011). NaOH-activated carbon of high surface area produced from coconut shell: Kinetics and equilibrium studies from the methylene blue adsorption. *Chemical Engineering Journal, 174*, 117–125.
104. Angin, D. (2014). Production and characterization of activated carbon from sour cherry stones by zinc chloride. *Fuel, 115*, 804–811.
105. Roy, C. K., Shah, S. S., Reaz, A. H., Sultana, S., Chowdhury, A.-N., Firoz, S. H., Zahir, M. H., Qasem, M. A. A., & Aziz, M. A. (2020). Preparation of hierarchical porous activated carbon from banana leaves for high-performance supercapacitor: Effect of type of electrolytes on performance. *Chemistry—An Asian Journal, 16*(4), 296–308. https://doi.org/10.1002/asia.202001342
106. Li, A., Huang, W., Qiu, N., Mou, F., & Wang, F. (2020). Porous carbon prepared from lotus leaves as potential adsorbent for efficient removal of rhodamine B. *Materials Research Express, 7*, 055505.
107. Xu, J., Chen, L., Qu, H., Jiao, Y., Xie, J., & Xing, G. (2014). Preparation and characterization of activated carbon from reedy grass leaves by chemical activation with H_3PO_4. *Applied Surface Science, 320*, 674–680.

108. Chen, W., Gong, M., Li, K., Xia, M., Chen, Z., Xiao, H., Fang, Y., Chen, Y., Yang, H., & Chen, H. (2020). Insight into KOH activation mechanism during biomass pyrolysis: Chemical reactions between O-containing groups and KOH. *Applied Energy, 278*, 115730. https://doi.org/10.1016/j.apenergy.2020.115730

Chapter 7
Sustainable Solutions for Bioenergy Production from Hospital-Based Plastic Waste—Thinking Beyond Landfills

Patitapaban Dash, Chirasmita Mohanty, Pratyush Kumar Das, Joseph M Anto Simon, Debasish Sahoo, and Gurunathan Baskar

Abstract The use of plastics and the growth of modern plastic industries began in the early twentieth century, thus providing the human populace with a very useful tool that found widespread applications in various sectors. Mostly synthesized from petrochemicals, plastic's high durability makes it one of the most widely used synthetic materials. Its characteristic properties like lightweight, ability to be molded, insulating ability, and resistance against degradation have made it an ideal choice for being used to manufacture different consumer goods, medical equipment, and safety gears. The hospital sector is one of the largest users of plastic products owing to their non-reactive and durable nature simultaneously producing a proportionate amount of disposable plastic waste. The COVID pandemic saw a huge rise in the use of plastic-based materials in the healthcare sector, thus increasing the waste load in the environment. The long-term environmental persistence of plastics due to their

P. Dash · C. Mohanty
Centre for Biotechnology, Siksha 'O' Anusandhan (Deemed to be University), Bhubaneswar, Odisha 751003, India

P. K. Das (✉)
Department of Phytopharmaceuticals, School of Agricultural Engineering and Bioprocessing (SoABE), Centurion University of Technology and Management, Paralakhemundi, Odisha 761200, India
e-mail: pratyush.das@cutm.ac.in

M. A. S. Joseph
Department of Biotechnology, Sri Krishna Arts and Science College, Bharathiar University, Coimbatore, Tamil Nadu 641008, India

D. Sahoo
BioInnovale Lifescience (P) Ltd, Bengaluru, Karnataka, India

G. Baskar
Department of Biotechnology, St. Joseph's College of Engineering, Chennai, Tamil Nadu, India

C. Mohanty
Department of Biotechnology, School of Bio Sciences and Technology, Vellore Institute of Technology, Vellore 632014, Tamil Nadu, India

G. Baskar et al. (eds.), *Circular Bioeconomy Perspectives in Sustainable Bioenergy Production*, Energy, Environment, and Sustainability,
https://doi.org/10.1007/978-981-97-2523-6_7

non-biodegradable nature coupled with improper disposal and conventional waste management systems have led to the accumulation of these synthetic polymers in different environmental matrices, thereby causing pollution. It is essential to devise sustainable solutions to reduce the load of plastic waste in the environment, thereby nullifying its toxicity. In this context, the conversion of plastic waste into suitable energy forms is suggested as a means to sustainably reduce the plastic contaminants in the environment and parallelly supply energy alternatives to the dwindling conventional energy resources.

Keywords Bioenergy · Environmental toxicity · Marine pollutants · Pyrolysis of plastic · Sustainable solutions · Valorization of plastic

7.1 Introduction

Plastics are one of the most important inventions ever come across by human civilization. Plastics are generally made up of synthetic polymers, with several fascinating properties that make them essential for domestic and industrial applications. Diverse groups of polymers contribute to plastics manufacturing, making them cheap, lighter, durable, resistant to corrosion, and with high insulating (thermal and electrical) properties [1]. These properties have made plastics a cost-effective option for manufacturing a variety of products [2]. There are several hundreds of plastic materials that are commercially available but only a few of them qualify as thermoplastics for commodity level, mainly attributed to their high volume and cheaper price. The commodity plastics are mainly comprised of low-density polyethylene (LDPE), high-density polyethylene (HDPE), polyvinylchloride (PVC), polystyrene (PS), polypropylene (PP), and polyethylene terephthalate (PET). These plastics contribute to 90% of global plastic production.

Plastics due to their sterility and lightweight are the foremost choice for the production of a maximum of hospital-related equipment including personal protective equipment (PPEs). The COVID-19 pandemic situation witnessed a steep rise in the demand for PPEs and thus that of plastics. The large-scale use and disposal of PPEs and other COVID-19 wastes has created real havoc in the environment [3], thus piling up a huge number of plastic wastes. The highly recalcitrant nature of plastics has made it difficult to manage the waste load as they persist in the environment for several years [4], thus leading to environmental pollution [5]. The slow degradation of plastics in the environment under the influence of abiotic factors like temperature, pressure, light, moisture, and UV further compounds the problem, thereby endangering flora and marine fauna along with raising risks to human health. Conventional methods of plastic waste management fail to provide a sustainable solution and thus need further attention and research in this regard. Valorization of plastic wastes into bioenergy is a sustainable way to handle the waste load in the environment. Conversions of plastic waste into different forms of chemicals or fuels can be a good option.

Conversion of plastic wastes into useful energy forms can be possible by utilizing several techniques [6]. Heat treatment and oxidation technologies have been found useful in the conversion of plastic wastes into fuels. Plastic waste can be easily converted into various energy-usable forms like liquid fuels [7] and combustible gases. Organic by-products like carboxylic and sulfonic acids can also be used as a source of carbon in the chemical industries [8].

Pyrolysis is an advanced technique that seems promising as far as sustainable energy production is concerned. The chapter discusses the importance of plastics in the healthcare sector and the burden created due to the disposal of such plastics on the environment. The conventional plastic waste management methods are also discussed along with their various shortcomings. Pyrolysis of plastic wastes into energy/fuel sources has been emphasized.

7.2 The Intangible Relationship Between the Healthcare Sector and Plastics

The human race has been using medical devices for thousands of years. Metal was the primary material used to create medical equipment during the early periods of history. Ceramics began to be employed in the late 1920s, and subsequently, glass entered the picture. However, some of the problems with these materials paved the door for the use of plastics in medical care. Despite the fact that scientists have been experimenting with plastics since the early nineteenth century, the medical business was not completely transformed by plastics until the middle of the twentieth century. According to a report by Market and Markets, the global market for medical plastics was valued at roughly USD 22.8 billion in 2019 and is projected to reach USD 31.7 billion, at a CAGR of 6.8% by 2024 (Market and Markets 2021). Plastics are utilized in many different applications, such as MRI casings, surgical equipment that replace ceramic and other metals, bedpans, and implants (Table 7.1).

Numerous medical manufacturers treat plastic with antimicrobial coatings that can resist or kill bacteria, lowering the risk of infection and preventing cross-contamination. The following is a list of some of the polymers used in medical applications: polyethylene (PE), polyvinylchloride (PVC), polycarbonate (PC), polypropylene (PP), polystyrene (PS), polyethylene terephthalate (PET), polyamide (PA), acrylonitrile butadiene styrene (ABS), and polyurethane (PU).

In the modern world, plastics and synthetic polymers have assimilated into every aspect of our existence. Every type of plastic that has been manufactured up to this point is still in use, posing a continuing threat to the environment [4]. The COVID-19 pandemic has exacerbated the problems associated with managing plastic garbage, which were already a problem [10]. There is a consistent rise in COVID-19 cases, particularly with the emergence of the second wave in India. This has highlighted the pressing requirement for significant amounts of personal protective equipment (PPE).

Table 7.1 Plastics used in medical applications

Property	Applications in medical devices (percentage usage)	Types of plastics	Medical device applications
Commodity plastics	70% of all plastics	Polyethylene Polypropylene Polystyrene Polyvinylchloride	Tubing Films, Packaging Connectors Labware IV bags Catheters Face masks Drug-delivery components Housings Luers Membranes Sutures
Engineering thermoplastics	20% of all plastics	Polyamides, Nylons Polyesters Polycarbonates Polyurethanes Acrylics Acetals	Surgical instruments, Balloons Blood set components, Blood bowls Blood oxygenators, Syringes Moving parts and components Luers Catheters
High-temperature engineering thermoplastics and other polymers	10% of all plastics	Polyimides Polyetherimides Polysulfones Polyether ether ketone Polyphenylene sulfide Fluoropolymers Liquid crystalline polymers Biopolymers Thermosets and adhesives	Surgical instruments, Surgical trays Syringes Implants Dentals implants Bone implants Moving parts and components High precision parts Electronic components Luers Bioresorbable sutures

Source Sastri [9]

Despite having a significant amount of plastic in PPE, they appear to be the most reliable and cost-effective way to prevent virus transmission. PPE has become essential for healthcare and frontline workers due to the nature of virus transmission from person to person, which is accomplished by aerosols. More asymptomatic instances have led to a surge in demand for PPE kits, particularly single-use disposable face masks, face shields, and gloves, without adequate attention to proper disposal mechanisms. Consequently, there has been a sharp increase in global manufacturing and production of these items [11, 12].

7.3 The Burden of Hospital-Generated Plastic Waste

Plastics due to their high durability and sterility have been the most preferred substrate for use in the field of healthcare. With the onset of COVID-19, plastics have become more important as a day-to-day material in hospitals. The frequent and large-scale use of face masks during the COVID regime has been found to largely contribute to global plastic pollution. The problem is even aggravated due to the improper disposal of face masks by human communities. According to Shiferie [13], improper disposal of 1% of face masks will approximately lead to 10 million masks accumulating in the environment on a monthly basis, thus leading to pollution. These 10 million masks roughly account for 40,000 kg of plastic waste in the environment assuming the weight of a single face mask to be 4 g [14].

The healthcare sector is one of the largest consumers of durable plastics and thus produces significantly larger amounts of plastic waste. To make it more worrisome, the onset of COVID-19 even has compounded the problem with the generation of huge volumes of plastic waste. Global plastic waste generation was estimated to be 380 million tons in 2018 [15]. Considering the rate of plastic consumption by the biomedical sector during the pandemic, the generation of waste thereof was estimated to double in 2020 [16]. Figure 7.1 provides a detailed overview of plastic waste generation from 2010 to 2018 and estimated values for 2019–20 and 2050.

India itself accounted for 20,400 kg of plastic waste generation from its hospitals in 2005 which further increased to 198,766 kg in 2019, an increase of approximately 89% (Fig. 7.2). The onset of COVID-19 even worsened the problem.

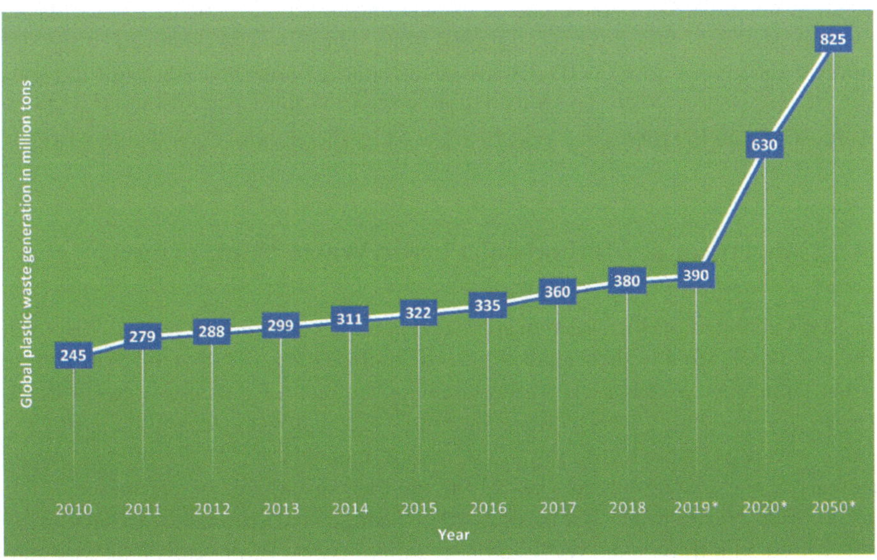

Fig. 7.1 Global plastic waste generation from 2010 to 2018, and estimation for 2019–20, and 2050. *Source* Khoo et al. [17]

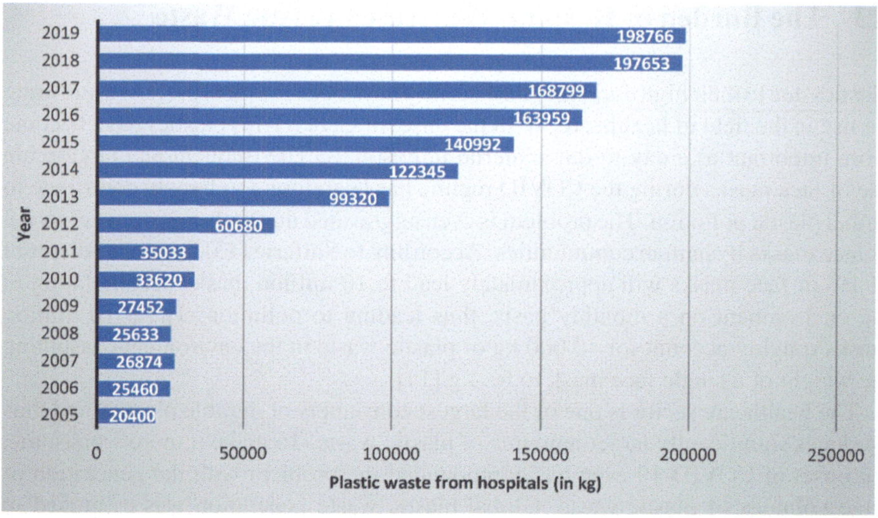

Fig. 7.2 Plastic waste generation from hospitals in India (2005–2019). *Source* Rai et al. [18]

Wuhan in China, the epicenter of COVID-19 witnessed the generation of high amounts of PPE plastics on a daily basis which was almost six times higher than that of the pre-pandemic period [11]. However, recent estimates suggest the generation of 685 kg of medical waste generated per day for every 100 confirmed COVID-19 cases [16, 19, 20]. Peng et al., estimated that two-thirds of the global plastic waste was from hospitals with Asian countries accounting for 72% of the global plastic waste burden [21]. Moreover, on a global scale, plastic waste disposed of by 193 countries in relation to COVID-19 was also found to breach the 8.4 million tons mark by August 23, 2021. The hospital-generated plastic waste during COVID-19 was mostly from the protective gear being used by the frontline healthcare workers.

7.4 Conventional Methods of Plastic Waste Management

Hospital-generated plastics need proper segregation and treatment prior to disposal. However, humans out of their unawareness or negligence generally tend to mix these wastes with other garbage items, thus making the task of disposal quite difficult. Plastics in general play a very important role in the healthcare sector, and replacing them seems impossible as of now. However, the management of plastic waste must be taken up with the utmost necessity to maintain a sustainable environment.

The three conventional methods for the management of plastic waste include incineration, landfilling, and recycling. The process of recycling can be divided into four types (primary, secondary, tertiary, and quaternary) and thus lead to different products (Fig. 7.3).

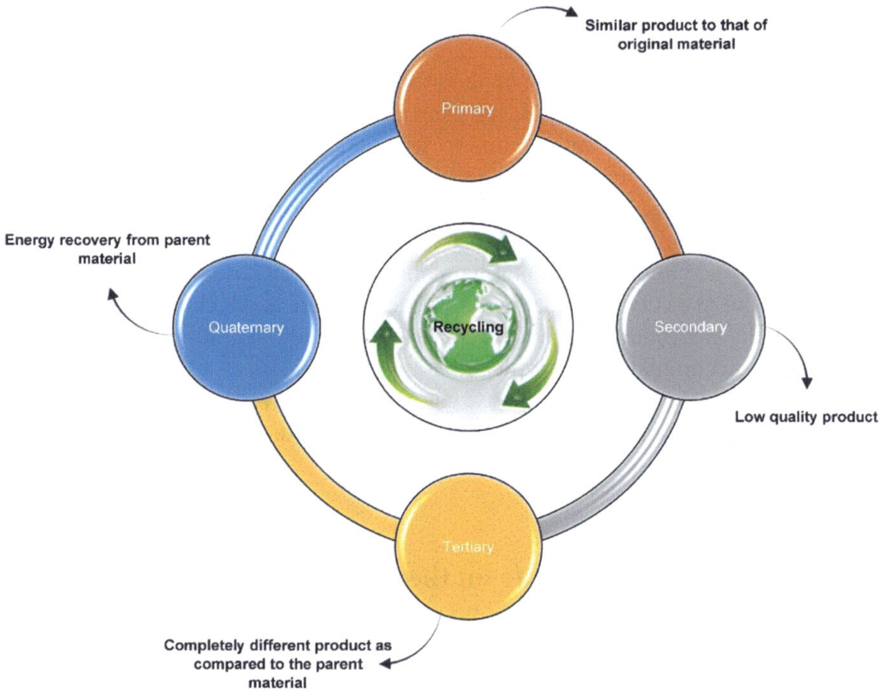

Fig. 7.3 Different recycling processes and the nature of their corresponding product obtained. *Source* Sastri [9]

However, plastic waste from the healthcare sector is mostly contaminated and requires prior proper sterilization. The purity of plastics is also a major concern and one of the bottlenecks in the recycling process. Plastics composed of mixed polymeric structures or certain other additives are difficult to recycle [10]. Moreover, mixed polymeric plastics have a very low recyclable value due to their low-quality post recycling [22]. Several limitations of the recycling processes leave out only the options of incineration and landfilling.

A thermal treatment process such as incineration is another method that utilizes high temperatures to decontaminate as well as reduce the volume of plastic waste generated by the healthcare sector. The process results in the complete combustion of the plastics resulting in the formation of water and carbon dioxide (CO_2) [23]. Moreover, this process also helps in the conversion of these wastes into energy particularly in the form of heat [24]. The incineration process leads to the release of a high amount of CO_2, thus leading to atmospheric pollution and global warming [25]. Not only CO_2 but incineration of plastic wastes is also involved in the release of toxic pollutants into the atmosphere such as particulate matter (PM), volatile organic compounds (VOC), carbon monoxide (CO), and nitrogen oxides (NOx). The incineration processes no doubt reduce the load of plastic waste to a much greater extent but simultaneously also contribute toward environmental pollution [26], thus raising the risk level on public health [2].

Landfilling is the most common method followed for the disposal of plastic waste generated from hospitals and healthcare centers. It is widely followed in most under-developed and developing nations due to its low cost of operation. The onset of COVID-19 saw a large-scale generation of plastic waste from the healthcare sector which was mostly disposed of in landfills due to a lack of proper incineration facilities with high capacities. Plastic wastes at landfill sites tend to degrade naturally under the influence of environmental temperature, atmospheric pressure, moisture, and sunlight but in a very slow manner. However, the rate at which plastic wastes are produced in the environment superexceeds the rate of their degradation, thus leading to large-scale accumulation. Slow degradation of plastic wastes from landfill sites also contributes to the global greenhouse gas pool in the form of CO_2, CO, and others [27]. The long persistence of these wastes also results in these toxic gases remaining in the surrounding air for a quite long period [28].

7.5 Impact of Plastic Waste on the Environment

Plastic waste being persistent in the environment for a longer period of time poses several environmental concerns. Plastics degrade very slowly under the influence of factors like temperature, pressure, light, reaction with chemicals, and some microbial actions [20]. Plastics being a complex mixture of several chemical compounds mainly contribute to chemical pollution of the environment. Chemicals from plastics due to weathering can leach into the soil, thereby affecting the soil profile as well as groundwater. Certain volatile chemicals present in plastic wastes pose the chances of being released into the atmosphere, thus reducing the air quality. Plastics because of their intrinsic properties like high surface area and presence of certain functional groups can also adsorb certain toxic pollutants from the environment, thus increasing the environmental toxicity load [29]. Chemical pollution due to plastic waste is more pronounced in aquatic bodies as compared to terrestrial ecosystems. This may be attributed to the lower temperature of the water as compared to the land system. A lower temperature ensures slow degradation of the waste, thus making it persistent for a longer period [30]. Plastics can further undergo degradation to form micro and nano plastics which adsorb other chemical pollutants from the environment [31].

Management of plastic waste can also lead to deterioration of the air quality, thus causing severe pollution of the atmosphere. The incineration of plastic wastes leads to the generation of several toxic pollutants including greenhouse gases (GHGs) and particulate matter (PM). Besides this, particular waste management process also releases several volatile toxins into the atmosphere [12]. Incineration of PPE wastes during COVID-19 has been found to generate particulate matter in the dimensions of 10 μm (PM_{10}) and 2.5 μm ($PM_{2.5}$).

Plastic wastes are also a major source of marine pollutants with recent data suggesting approximately 25.9 kilotons (kt) of these being released into the marine environment during the pandemic [21]. The initial phases of COVID-19 during 2020 witnessed the disposal of 1.56 billion face masks in the environment which roughly weighed around 5.66 million tons (Mt) and were disposed of offshore [32]. Adsorption of various heavy metals onto the disposed PPEs in water bodies can create a disturbance in aquatic environmental health [33]. Plastic particles generated from disposable hospital-generated plastic wastes can also be consumed by marine organisms, thus leading to higher mortality rates [34]. Plastic particles are known to exert ecotoxic, cytotoxic, and genotoxic effects on marine organisms. Plastic wastes result in the generation of reactive oxygen species (ROS), thus creating oxidative stress and disturbance in the DNA repair mechanism [35]. Plastic particles in the water also prevent the interaction of oxygen with the aquatic bodies, thus creating an oxygen-depleted environment and subsequent death of the aquatic flora and fauna.

Plastic pollution also leads to microbial pollution which may be attributed to the environmental interaction between the microbes and the plastics. Plastics provide a solid surface for the formation of microbial biofilms and can facilitate horizontal gene transfer among microbial species, thus leading to increased pathogenicity [36].

7.6 Valorization of Plastic Wastes into Bioenergy

The approach of using landfills appeared inadequate for managing the accumulation of contagious COVID-19-related medical waste. Budiman and Ardiansyah, stated that in May 2020, the stacking of COVID-19 medical garbage collapsed the landfill's wall, letting tons of waste enter the Cisadane River [37]. This improper waste handling led to contamination of the main river, affecting the public water source. According to Corburn et al. [38], the improper disposal of plastic and medical waste through landfills and local burning exacerbated waste problems in Indian municipalities during the pandemic and contributed to the spread of the virus. As a result, there exists an issue in managing unconventional waste sustainably while reducing carbon emissions, minimizing secondary virus transmission, and mitigating potential health risks. Moreover, in the absence of conventional waste management and emergency strategies to address the epidemic, the world could face devastating consequences. Given the adverse environmental effects of plastic waste and its disposal methods, innovative environmentally friendly approaches for plastic waste treatment need to be developed [39]. The dwindling energy scenario which is responsible for the creation of an energy crisis on a global basis needs to be countered by devising alternative sources of energy generation [40–43]. In this regard, the valorization of plastic medical waste will not only solve the environmental problem but also suffice for the ever-increasing energy needs. Pyrolysis is one of the most advanced technologies currently being researched and poses the ability to sustainably convert medical-based plastic wastes into suitable bioenergy forms. The current section highlights the role of pyrolysis in the conversion of these plastic wastes into bioenergy and also discusses several advancements in the field.

7.6.1 Pyrolysis

The pyrolysis approach piqued the researchers' interest because it offers high waste decomposition, a simple process, and promotes green waste management over current incineration practices. Pyrolysis involves a thermochemical transformation of plastic waste at high temperatures in an inert deoxygenated environment, resulting in the conversion of the waste into lower molecular weight compounds [44].

In a prior study by Xue et al. [45], pyrolysis was recognized as a viable and efficient technique for managing industrial and medical solid waste. It serves as an alternative method for thermal disinfection of COVID-19-related waste. The heat and pressure used in waste pyrolysis result in combustible lower molecular weight molecules, primarily gases like hydrogen, methane, and carbon monoxide [46]. Liquids are predominantly formed at high operating temperatures, including methanol, acetone, acetic acid, acetaldehyde, tar, solvent oil, and other organic materials [44]. Additionally, low-temperature pyrolysis yields solid end-products like coke, char, and carbon black (Fig. 7.4). The collected end-products are repurposed, thus reducing pollution. This approach is highlighted as a great solution for transforming plastic waste into usable by-products [47]. Vivero et al. [48] emphasized that pyrolysis is optimal for converting plastic waste into hydrocarbon blends, aiding in recycling and reducing PPE plastic trash.

It aids in the reduction of plastic waste, the recovery of chemicals, and the substitution of fuels and other virgin materials. Table 7.2 compares the solid yields of pyrolysis of polypropylene, including char, coke, and waxy residue. According to the published study, polypropylene plastic waste showed slow pyrolysis with a 5 °C/min heating rate, resulting in over 2.67% char product yield [49]. Ahmad et al. [50] studied pyrolysis temperature effects on PP waste, ranging from 250 to 400 °C, and achieved the highest conversion at 250 °C. In terms of char yields, the yield of solid residue dropped with increasing temperature, from 13.68 to 5.7%. Witkowski et al. [51] suggested that this decrease in char output at higher temperatures is linked to the primary depolymerization of plastics and the secondary depolymerization of produced char.

7.6.1.1 Cold Plasma Pyrolysis

Recent research has explored plasma technology's application in pyrolysis systems. Cold plasma pyrolysis presents an effective approach to managing infectious medical plastic waste, aided by high pyrolytic temperatures and a sophisticated cooling system. This method utilizes excited hot electrons generated from electrodes to efficiently break down plastic chemical bonds under controlled conditions and in less time than traditional pyrolysis [58]. However, this approach is energy-intensive due to the advanced cooling system and higher process temperatures required to thermally break down plastic waste into lighter hydrocarbons. Globally, it is crucial to

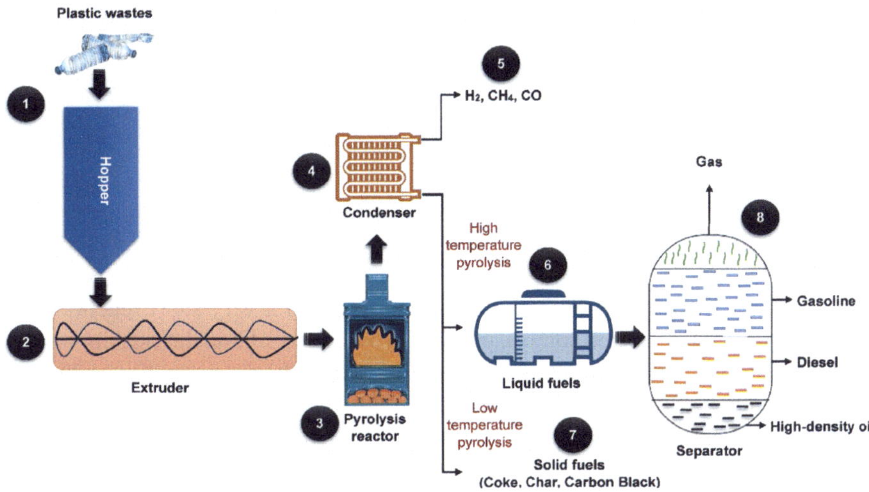

Fig. 7.4 Steps in pyrolytic conversion of plastic wastes into valuable energy forms. (Step-1: Collected plastic wastes are fed into a hopper which is then passed onto an extruder. Step-2: The plastic wastes are converted into smaller fragments or particles in the extruder and pushed forward into the pyrolysis reactor. Step-3: The pyrolysis reactor degrades the plastic wastes under high temperatures and in the absence of oxygen. The vapors from the pyrolysis reactor move forward to the condenser. Step-4: The condenser primarily produces gases like H_2, CH_4, and CO. Step-6: Moreover, condensation also produces a mixture of liquid fuels if the pyrolysis process involves very high temperatures. Step-7: While, a low-temperature pyrolysis yields coke, char, and carbon black that can be used as a solid fuel. Step-8: The final step involves the use of a separation chamber or separator that separates different liquid fuels along with the release of any gases)

Table 7.2 Pyrolysis of polypropylene under different conditions and respective char yield

Reactor	Temperature (°C)	Other conditions	Char yield (wt%)	References
Batch reactor	380	Pressure—1 atm Heating rate—3 °C/min	13.30	[52]
Batch reactor	740	Heating rate—10 °C/min	1.60	[53]
Fluidized bed reactor	703	Feed rate—7 g/min	6.90	[54]
Semi-batch reactor	450	Pressure—1 atm Heating rate—25 °C/min	3.60	[55]
Fixed bed and batch reactor	400	Heating rate—25 °C/min Residence time—45 min	2.50	[49]
Fixed bed reactor	900	Residence time—21 min	2.80	[56]
Microwave-assisted semi-batch reactor	250	HZSM-5 catalyst	1.54	[57]

adopt economically simple processes and energy-saving technologies for handling extensive plastic waste production.

7.6.1.2 Microwave-Assisted Pyrolysis

This process involves the thermal decomposition of plastic wastes at higher temperatures of 573–1073 K with the assistance of microwave irradiation, thus producing biogas and bio-oils [56, 57]. Microwave irradiation-based pyrolysis is characterized by faster heating rates, low operating temperatures, and a very short reaction time [59]. Materials like activated carbon, graphene oxide, and silicon dioxide are used as absorbents during microwave irradiation to improve the heating rate, thus achieving high temperatures within a very short time as compared to conventional techniques [60]. Co-pyrolysis of waste plastics and waste cooking oil at 500 °C carried out through microwave irradiation has been found to produce a high yield of liquid fuel within 10 min. The liquid fuel thus obtained demonstrated characteristic similarity with that of diesel used in transportation. The fuel besides exhibiting higher stability also remained free from nitrogen and sulfur and had a very low oxygen content. Moreover, the liquid fuel had much high energy content of 42–46 MJ/kg [61].

7.6.1.3 Plasma Pyrolysis

The plasma pyrolysis or plasma gasification technique for managing plastic wastes is mostly being followed in the countries of the European Union [62]. In this process (Fig. 7.5), gasification of the plastic wastes is mostly carried out by thermal plasma-based heating generated by DC current-operated arc plasma torches. This operates at a very high reaction temperature of 2000–14,000 °C and a short residence time of less than 30 min. Any carbonaceous materials including plastics when fed into the plasma reactor are heated by the arcs, thus liberating volatile matter which when cracked leads to the formation of syngas and methane [63].

A recent study explored a novel hydrothermal processing technology, which effectively addresses medical plastic waste using a high-pressure system. This method employs a hot compressed aqueous medium at elevated pressures to yield monomers, chemicals, hydrochar, and crude oil, which can be used in plastic production and other applications [64]. Moreover, hydrothermal processing contributes to lower oxygen content, improved energy content in liquid crude oil, cost-effective processing, and efficient storage. This alternative approach holds promise for reducing plastic pollution and addressing environmental concerns.

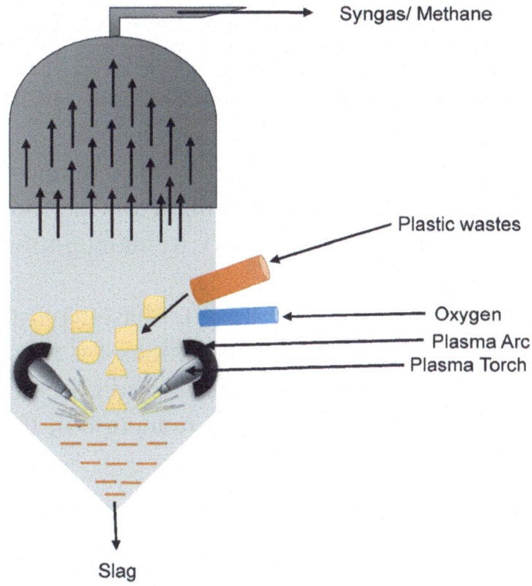

Fig. 7.5 Plasma pyrolysis of plastic waste leads to the generation of syngas and methane

7.7 Challenges and Future Prospects

Taking into consideration the vast amount of plastic waste generated by the hospital and healthcare sector and the pollution caused by them, an advanced technique like pyrolysis offers a sustainable alternative to the problem. Pyrolysis not only converts these plastic wastes into energy-usable and other valuable forms, but also reduces the risk of pollution as observed in the case of conventional treatment methods. Pyrolysis of plastic wastes leads to the formation of liquid, gaseous, and solid fuels of high importance and energy yield. However, valorization of the hospital-based plastic waste mostly comes with several challenges.

The selection of feedstock is a major challenge in the pyrolysis process. Heterogeneous plastic materials create a problem in the pyrolysis process. Thermal degradation of materials made of PET and PVC hinders the conversion process. Chlorinated hydrocarbons result due to thermal degradation of PVC along with the formation of HCl. These by-products in turn corrode the reactor and make the oil halogenated. Phthalic acid is formed as a result of PET degradation leading to the clogging of the reactor pipes and deterioration in the oil quality [65].

Plastic waste undergoing pyrolysis needs prior pretreatment to remove other impurities like glass, metals, and wood. Moreover, plastic waste generated from hospital settings needs to be pretreated to remove any microbial contamination. Hospital-generated plastic wastes pose the threat of infecting persons employed in the collection, sorting, and processing of the wastes and must be taken utmost care. Plastic wastes obtained from hospital settings also vary in size as per their utilities. Pretreatment is also required to bring them to a uniform size, thus adding extra cost to the entire process.

Another major problem is the formation of wax as a final product in the pyrolysis process. Reactors with inefficient recovery systems can lead to the scaling of wax on the inner surface of the condensing systems, thus making the recovery process too tedious [66].

Pyrolysis liquids are quite unstable and often tend to repolymerize, thus making their storage difficult for a longer time period. Post-treatment techniques like blending and dewaxing are necessary to maintain the quality of the liquid fuels for a longer period [67, 68].

The conversion of hospital-based plastic wastes into useful bioenergy forms provides a lot of research options for the overall advancement of the technology as well as to reduce its operational costs. Further, research in this regard is highly essential for the sustainable reduction of these wastes in the environment in a more efficient manner.

7.8 Conclusions

Plastics being recalcitrant in nature are a hazard to the environment as well as their biotic components. However, looking at their usefulness, social, and economic relevance, it is quite impossible to limit or discontinue their uses as of now. Improper disposal and management of plastic load in the environment can be really troublesome and pose several health hazards to the living components including humans. Proper management of plastic waste by means of converting it into sustainable energy forms can be a win–win situation. Pyrolysis of plastic wastes into energy-usable forms can not only contemplate the energy crisis situation but also lead to the formation of a sustainable environment. At present, the management of plastic waste involves a lot of expenditures. These expenditures could be reversed by the conversion of these wastes into usable forms such as bioenergy. The pyrolysis technique despite its several advantages also faces numerous challenges which include the selection of feedstocks, pretreatment processes involved, and stability concerns. As such, it opens up wide avenues for research in the field with the sole aim of making the technique more sustainable and cost-effective.

References

1. Kumar, B., Pundir, A., Mehta, V., Singh, B., & Solanki, R. (2020). A review paper on plastic, its variety, current scenario and its waste management. *Plant Arch, 20*, 53–56.
2. Er, C. T., Sen, L. Z., Srinophakun, P., & Wei, O. C. (2023). Recent advances and challenges in sustainable management of plastic waste using biodegradation approach. *Bioresource Technology,* 128772. https://doi.org/10.1016/j.biortech.2023.128772
3. Zarocostas, J. (2022). WHO concerned over COVID-19 health-care waste. *The Lancet, 399*(10324), 507. https://doi.org/10.1016/S0140-6736(22)00225-2

4. Gewert, B., Plassmann, M. M., & MacLeod, M. (2015). Pathways for degradation of plastic polymers floating in the marine environment. *Environmental Science: Processes & Impacts, 17*(9), 1513–1521. https://doi.org/10.1039/C5EM00207A
5. Lee, J., Park, H. J., Moon, M., Lee, J. S., & Min, K. (2021). Recent progress and challenges in microbial polyhydroxybutyrate (PHB) production from CO_2 as a sustainable feedstock: A state-of-the-art review. *Bioresource Technology, 339*, 125616. https://doi.org/10.1016/j.biortech.2021.125616
6. Chen, X., Wang, Y., & Zhang, L. (2021). Recent progress in the chemical upcycling of plastic wastes. *Chemsuschem, 14*(19), 4137–4151.
7. Papari, S., Bamdad, H., & Berruti, F. (2021). Pyrolytic conversion of plastic waste to value-added products and fuels: A review. *Materials, 14*(10), 2586.
8. Li, N., Liu, H., Cheng, Z., Yan, B., Chen, G., & Wang, S. (2022). Conversion of plastic waste into fuels: A critical review. *Journal of Hazardous Materials, 15*(424), 127460. https://doi.org/10.1016/j.jhazmat.2021.127460
9. Sastri, V. R. (2021). *Plastics in medical devices: Properties, requirements, and applications*. William Andrew.
10. Inamdar, I. (2022). Recycling of plastic wastes generated from COVID-19: A comprehensive illustration of type and properties of plastics with remedial options. *Science of The Total Environment*, 155895. https://doi.org/10.1016/j.scitotenv.2022.155895
11. Adyel, T. M. (2020). Accumulation of plastic waste during COVID-19. *Science, 369*(6509), 1314–1315. https://doi.org/10.1126/science.abd9925
12. Parashar, N., & Hait, S. (2021). Plastics in the time of COVID-19 pandemic: Protector or polluter? *Science of the Total Environment, 759*, 144274. https://doi.org/10.1016/j.scitotenv.2020.144274
13. Shiferie, F. (2021). Improper disposal of face masks during COVID-19: Unheeded public health threat. *The Pan African Medical Journal, 38*. https://doi.org/10.11604/pamj.2021.38.366.29063
14. Patrício Silva, A. L., Prata, J. C., Walker, T. R., Duarte, A. C., Ouyang, W., Barcelo, D., & Rocha-Santos, T. (2021). Increased plastic pollution due to COVID-19 pandemic: Challenges and recommendations. *Chemical Engineering Journal, 405*, 126683. https://doi.org/10.1016/j.cej.2020.126683
15. Shams, M., Alam, I., & Mahbub, M. S. (2021). Plastic pollution during COVID-19: Plastic waste directives and its long-term impact on the environment. *Environmental advances, 5*, 100119. https://doi.org/10.1016/j.envadv.2021.100119
16. Khoo, K. S., Ho, L. Y., Lim, H. R., Leong, H. Y., & Chew, K. W. (2021). Plastic waste associated with the COVID-19 pandemic: Crisis or opportunity? *Journal of hazardous materials, 417*, 126108. https://doi.org/10.1016/j.jhazmat.2021.126108
17. Rai, P. K., Sonne, C., Song, H., & Kim, K. H. (2023). Plastic wastes in the time of COVID-19: Their environmental hazards and implications for sustainable energy resilience and circular bio-economies. *Science of The Total Environment, 858*, 159880. https://doi.org/10.1016/j.scitotenv.2022.159880
18. Achra, A., Mahajan, R. K., & Sahoo, S. (2021). The changing pattern of the quantum of biomedical waste generated from a tertiary care hospital in Delhi. *Journal of Laboratory Physicians, 13*(01), 080–083. https://doi.org/10.1055/s-0041-1723056
19. Patricio Silva, A. L., Prata, J. C., Duarte, A. C., Soares, A. M., Barceló, D., & Rocha-Santos, T. (2021). Microplastics in landfill leachates: The need for reconnaissance studies and remediation technologies. *Case Studies in Chemical and Environmental Engineering, 3*, 100072.
20. Patricio Silva, A. L., Prata, J. C., Duarte, A. C., Barcelò, D., & Rocha-Santos, T. (2021). An urgent call to think globally and act locally on landfill disposable plastics under and after covid-19 pandemic: Pollution prevention and technological (Bio) remediation solutions. *Chemical Engineering Journal, 426*, 131201. https://doi.org/10.1016/j.cej.2021.131201
21. Peng, Y., Wu, P., Schartup, A. T., & Zhang, Y. (2021). Plastic waste release caused by COVID-19 and its fate in the global ocean. *Proceedings of the National Academy of Sciences, 118*(47), e2111530118. https://doi.org/10.1073/pnas.2111530118

22. Lange, J. P. (2021). Managing plastic waste\\sorting, recycling, disposal, and product redesign. *ACS Sustainable Chemistry & Engineering, 9*(47), 15722–15738. https://doi.org/10.1021/acssuschemeng.1c05013
23. Evode, N., Qamar, S. A., Bilal, M., Barceló, D., & Iqbal, H. M. (2021). Plastic waste and its management strategies for environmental sustainability. *Case Studies in Chemical and Environmental Engineering, 4*, 100142. https://doi.org/10.1016/j.cscee.2021.100142
24. Klemeš, J. J., Van Fan, Y., Tan, R. R., & Jiang, P. (2020). Minimising the present and future plastic waste, energy and environmental footprints related to COVID-19. *Renewable and Sustainable Energy Reviews, 127*, 109883. https://doi.org/10.1016/j.rser.2020.109883
25. Devasahayam, S., Raju, G. B., & Hussain, C. M. (2019). Utilization and recycling of end of life plastics for sustainable and clean industrial processes including the iron and steel industry. *Materials Science for Energy Technologies, 2*(3), 634–646. https://doi.org/10.1016/j.mset.2019.08.002
26. Jeswani, H., Krüger, C., Russ, M., Horlacher, M., Antony, F., Hann, S., & Azapagic, A. (2021). Life cycle environmental impacts of chemical recycling via pyrolysis of mixed plastic waste in comparison with mechanical recycling and energy recovery. *Science of the Total Environment, 769*, 144483. https://doi.org/10.1016/j.scitotenv.2020.144483
27. Royer, S. J., Ferrón, S., Wilson, S. T., & Karl, D. M. (2018). Production of methane and ethylene from plastic in the environment. *PLoS ONE, 13*(8), e0200574. https://doi.org/10.1371/journal.pone.0200574
28. Prata, J. C., Silva, A. L., Walker, T. R., Duarte, A. C., & Rocha-Santos, T. (2020). COVID-19 pandemic repercussions on the use and management of plastics. *Environmental Science & Technology, 54*(13), 7760–7765. https://doi.org/10.1021/acs.est.0c02178
29. Kumar, H., Azad, A., Gupta, A., Sharma, J., Bherwani, H., Labhsetwar, N. K., & Kumar, R. (2021). COVID-19 creating another problem? Sustainable solution for PPE disposal through LCA approach. *Environment, Development and Sustainability, 23*, 9418–9432. https://doi.org/10.1007/s10668-020-01033-0
30. Duan, J., Bolan, N., Li, Y., Ding, S., Atugoda, T., Vithanage, M., Sarkar, B., Tsang, D. C., & Kirkham, M. B. (2021). Weathering of microplastics and interaction with other coexisting constituents in terrestrial and aquatic environments. *Water Research, 196*, 117011. https://doi.org/10.1016/j.watres.2021.117011
31. Kutralam-Muniasamy, G., Pérez-Guevara, F., & Shruti, V. C. (2021). A critical synthesis of current peer-reviewed literature on the environmental and human health impacts of COVID-19 PPE litter: New findings and next steps. *Journal of Hazardous Materials, 422*, 126945. https://doi.org/10.1016/j.jhazmat.2021.126945
32. Yuan, X., Wang, X., Sarkar, B., & Ok, Y. S. (2021). The COVID-19 pandemic necessitates a shift to a plastic circular economy. *Nature Reviews Earth & Environment, 2*(10), 659–660. https://doi.org/10.1038/s43017-021-00223-2
33. De-la-Torre, G. E., Dioses-Salinas, D. C., Pizarro-Ortega, C. I., Severini, M. D., López, A. D., Mansilla, R., Ayala, F., Castillo, L. M., Castillo-Paico, E., Torres, D. A., & Mendoza-Castilla, L. M. (2022). Binational survey of personal protective equipment (PPE) pollution driven by the COVID-19 pandemic in coastal environments: Abundance, distribution, and analytical characterization. *Journal of Hazardous Materials, 426*, 128070. https://doi.org/10.1016/j.jhazmat.2021.128070
34. Ray, S. S., Lee, H. K., Huyen, D. T., Chen, S. S., & Kwon, Y. N. (2022). Microplastics waste in environment: A perspective on recycling issues from PPE kits and face masks during the COVID-19 pandemic. *Environmental Technology & Innovation, 26*, 102290. https://doi.org/10.1016/j.eti.2022.102290
35. Tagorti, G., & Kaya, B. (2022). Genotoxic effect of microplastics and COVID-19: The hidden threat. *Chemosphere, 286*, 131898. https://doi.org/10.1016/j.chemosphere.2021.131898
36. Rogers, K. L., Carreres-Calabuig, J. A., Gorokhova, E., & Posth, N. R. (2020). Micro-by-micro interactions: How microorganisms influence the fate of marine microplastics. *Limnology and Oceanography Letters, 5*(1), 18–36. https://doi.org/10.1002/lol2.10136

37. Budiman, Y. C., & Ardiansyah, T. (2020). In Indonesia, coronavirus floods Cisadane River with extra hazard: Medical waste. *TheJakartaPost*. https://www.thejakartapost.com/news/2020/09/01/in-indonesia-coronavirus-floods-cisadane-river-with-extra-hazardmedical-waste.html
38. Corburn, J., Vlahov, D., Mberu, B., Riley, L., Caiaffa, W. T., Rashid, S. F., Ko, A., Patel, S., Jukur, S., Martínez-Herrera, E., & Jayasinghe, S. (2020). Slum health: Arresting COVID-19 and improving well-being in urban informal settlements. *Journal of Urban Health, 97*, 348–357. https://doi.org/10.1007/s11524-020-00438-6
39. Harussani, M. M., Sapuan, S. M., Rashid, U., Khalina, A., & Ilyas, R. A. (2022). Pyrolysis of polypropylene plastic waste into carbonaceous char: Priority of plastic waste management amidst COVID-19 pandemic. *Science of The Total Environment, 803*, 149911. https://doi.org/10.1016/j.scitotenv.2021.149911
40. Das, P. K., Das, B. P., Dash, P., & Gurunathan, B. (2022). Production of biofuel from genetically modified microalgal biomass and its effects on environment and public health. In *Biofuels and bioenergy* (pp. 505–519). Elsevier. https://doi.org/10.1016/B978-0-323-85269-2.00004-6
41. Das, P. K., Das, B. P., & Dash, P. (2021). Application of nanotechnology in the production of bioenergy from algal biomass: Opportunities and challenges. *Nanomaterials*, 355–377. https://doi.org/10.1016/B978-0-12-822401-4.00024-6
42. Das, P. K., Das, B. P., & Dash, P. (2020). Potentials of postharvest rice crop residues as a source of biofuel. In *Refining biomass residues for sustainable energy and bioproducts* (pp. 275–301). Academic Press. https://doi.org/10.1016/B978-0-12-818996-2.00013-2
43. Das, P. K., Das, B. P., & Dash, P. (2019). Role of energy crops to meet the rural energy needs: An overview. *Biomass Valorization to Bioenergy, 12*, 11–30. https://doi.org/10.1007/978-981-15-0410-5_2
44. Verma, A., Budiyal, L., Sanjay, M. R., & Siengchin, S. (2019). Processing and characterization analysis of pyrolyzed oil rubber (from waste tires)-epoxy polymer blend composite for lightweight structures and coatings applications. *Polymer Engineering & Science, 59*(10), 2041–2051. https://doi.org/10.1002/pen.25204
45. Xue, Y., Kelkar, A., & Bai, X. (2016). Catalytic co-pyrolysis of biomass and polyethylene in a tandem micropyrolyzer. *Fuel, 166*, 227–236. https://doi.org/10.1016/j.fuel.2015.10.125
46. Kairytė, A., Kremensas, A., Vaitkus, S., Członka, S., & Strąkowska, A. (2020). Fire suppression and thermal behavior of biobased rigid polyurethane foam filled with biomass incineration waste ash. *Polymers, 12*(3), 683. https://doi.org/10.3390/polym12030683
47. Worrell, W. A., & Vesilind, P. A. (2012). Integrated solid waste management. In *Solid waste engineering* (2nd ed.). Cengage Learning.
48. Vivero, L., Barriocanal, C., Alvarez, R., & Diez, M. A. (2005). Effects of plastic wastes on coal pyrolysis behaviour and the structure of semicokes. *Journal of Analytical and Applied Pyrolysis, 74*(1–2), 327–336. https://doi.org/10.1016/j.jaap.2004.08.006
49. Sogancioglu, M., Yel, E., & Ahmetli, G. (2020). Behaviour of waste polypropylene pyrolysis char-based epoxy composite materials. *Environmental Science and Pollution Research, 27*, 3871–3884. https://doi.org/10.1007/s11356-019-07028-3
50. Ahmad, I., Khan, M. I., Khan, H., Ishaq, M., Tariq, R., Gul, K., & Ahmad, W. (2015). Pyrolysis study of polypropylene and polyethylene into premium oil products. *International Journal of Green Energy, 12*(7), 663–671. https://doi.org/10.1080/15435075.2014.880146
51. Witkowski, A., Stec, A. A., & Hull, T. R. (2016). Thermal decomposition of polymeric materials. In *SFPE handbook of fire protection engineering* (pp. 167–254).
52. Sakata, Y., Uddin, M. A., & Muto, A. (1999). Degradation of polyethylene and polypropylene into fuel oil by using solid acid and non-acid catalysts. *Journal of analytical and Applied Pyrolysis, 51*(1–2), 135–155. https://doi.org/10.1016/S0165-2370(99)00013-3
53. Demirbas, A. (2004). Pyrolysis of municipal plastic wastes for recovery of gasoline-range hydrocarbons. *Journal of Analytical and Applied Pyrolysis, 72*(1), 97–102. https://doi.org/10.1016/j.jaap.2004.03.001
54. Jung, S. H., Cho, M. H., Kang, B. S., & Kim, J. S. (2010). Pyrolysis of a fraction of waste polypropylene and polyethylene for the recovery of BTX aromatics using a fluidized bed reactor. *Fuel Processing Technology, 91*(3), 277–284. https://doi.org/10.1016/j.fuproc.2009.10.009

55. Abbas-Abadi, M. S., Haghighi, M. N., Yeganeh, H., & McDonald, A. G. (2014). Evaluation of pyrolysis process parameters on polypropylene degradation products. *Journal of Analytical and Applied Pyrolysis, 109*, 272–277. https://doi.org/10.1016/j.jaap.2014.05.023
56. Liu, X., Burra, K. G., Wang, Z., Li, J., Che, D., & Gupta, A. K. (2020). On deconvolution for understanding synergistic effects in co-pyrolysis of pinewood and polypropylene. *Applied Energy, 279*, 115811. https://doi.org/10.1016/j.apenergy.2020.115811
57. Duan, D., Wang, Y., Dai, L., Ruan, R., Zhao, Y., Fan, L., Tayier, M., & Liu, Y. (2017). Ex-situ catalytic co-pyrolysis of lignin and polypropylene to upgrade bio-oil quality by microwave heating. *Bioresource Technology, 241*, 207–213. https://doi.org/10.1016/j.biortech.2017.04.104
58. Yao, L., King, J., Wu, D., Chuang, S. S., & Peng, Z. (2021). Non-thermal plasma-assisted hydrogenolysis of polyethylene to light hydrocarbons. *Catalysis Communications, 150*, 106274. https://doi.org/10.1016/j.catcom.2020.106274
59. Loy, A. C., Quitain, A. T., Lam, M. K., Yusup, S., Sasaki, M., & Kida, T. (2019). Development of high microwave-absorptive bifunctional graphene oxide-based catalyst for biodiesel production. *Energy Conversion and Management, 180*, 1013–1025. https://doi.org/10.1016/j.enconman.2018.11.043
60. Bu, Q., Chen, K., Xie, W., Liu, Y., Cao, M., Kong, X., Chu, Q., & Mao, H. (2019). Hydrocarbon rich bio-oil production, thermal behavior analysis and kinetic study of microwave-assisted co-pyrolysis of microwave-torrefied lignin with low density polyethylene. *Bioresource Technology, 291*, 121860. https://doi.org/10.1016/j.biortech.2019.121860
61. Mahari, W. A., Chong, C. T., Cheng, C. K., Lee, C. L., Hendrata, K., Yek, P. N., Ma, N. L., & Lam, S. S. (2018). Production of value-added liquid fuel via microwave co-pyrolysis of used frying oil and plastic waste. *Energy, 162*, 309–317. https://doi.org/10.1016/j.energy.2018.08.002
62. Mączka, T., Śliwka, E., & Wnukowski, M. (2013). Plasma gasification of waste plastics. *Journal of Ecological Engineering, 14*(1), 33–9. https://doi.org/10.5604/2081139X.1031534
63. Al Rayaan, M. B. (2021). Recent advancements of thermochemical conversion of plastic waste to biofuel-A review. *Cleaner Engineering and Technology, 2*, 100062. https://doi.org/10.1016/j.clet.2021.100062
64. Hongthong, S., Leese, H. S., & Chuck, C. J. (2020). Valorizing plastic-contaminated waste streams through the catalytic hydrothermal processing of polypropylene with lignocellulose. *ACS Omega, 5*(32), 20586–20598.
65. Qureshi, M. S., Oasmaa, A., Pihkola, H., Deviatkin, I., Tenhunen, A., Mannila, J., Minkkinen, H., Pohjakallio, M., & Laine-Ylijoki, J. (2020). Pyrolysis of plastic waste: Opportunities and challenges. *Journal of Analytical and Applied Pyrolysis, 152*, 104804. https://doi.org/10.1016/j.jaap.2020.104804
66. Scheirs, J., & Kaminsky, W. (2006). *Feedstock recycling and pyrolysis of waste plastics*. Wiley.
67. Butler, E., Devlin, G., & McDonnell, K. (2011). Waste polyolefins to liquid fuels via pyrolysis: Review of commercial state-of-the-art and recent laboratory research. *Waste and Biomass Valorization., 2*, 227–255.
68. Lam, S. S., Mahari, W. A., Ok, Y. S., Peng, W., Chong, C. T., Ma, N. L., Chase, H. A., Liew, Z., Yusup, S., Kwon, E. E., & Tsang, D. C. (2019). Microwave vacuum pyrolysis of waste plastic and used cooking oil for simultaneous waste reduction and sustainable energy conversion: Recovery of cleaner liquid fuel and techno-economic analysis. *Renewable and Sustainable Energy Reviews, 115*, 109359. https://doi.org/10.1016/j.rser.2019.109359

Chapter 8
Functionalized Biochar for Green and Sustainable Production of Biodiesel

Hlawncheu Zohmingliana, Joseph V. L. Ruatpuia, and Samuel Lalthazuala Rokhum

Abstract Biomass is a renewable energy source generated from agricultural waste, solid waste, etc. In addition to being utilized extensively for storing energy and gas, biomass is a major source of catalysts for biodiesel production. Biochar is a by-product of thermochemically degrading biomass through processes like pyrolysis, hydrothermal conversion, carbonization, etc., which has great potential as an energy storage material, a precursor to nanotubes and graphene, and as a catalyst for various chemical reactions. The high porosity, large surface area, low cost, and ease of generation make biochar a great contender for a catalyst. The post modification includes the physical or chemical activation which makes the surface area larger by increasing the pores. It also includes functionalization by adding a specific functional group for a particular chemical reaction. Acid-functionalized and base-functionalized biochar can be used for the transesterification of triglycerides and the esterification of free fatty acids. This production of biodiesel using functionalized biochar as a catalyst can be enhanced by optimizing different reaction parameters like time, temperature, etc. Overall, functionalized biochar is a strong candidate as a catalyst for the production of biodiesel.

Keywords Biomass · Biodiesel · Pyrolysis · Hydrothermal · Trans/esterification · Biochar

8.1 Introduction

With the depletion of fossil fuel stock and regard for the environmental impact like rising temperature, air pollution, melting of ice, etc., there has been research for finding an alternative energy source for more than a decade [1]. Although renewable energy like solar, wind, geothermal, and hydropower were studied and have been used as a source of energy in many places, the inconsistent energy output, the impact on

H. Zohmingliana · J. V. L. Ruatpuia · S. L. Rokhum (✉)
Department of Chemistry, National Institute of Technology Silchar, Silchar 788010, India
e-mail: rokhum@che.nits.ac.in

G. Baskar et al. (eds.), *Circular Bioeconomy Perspectives in Sustainable Bioenergy Production*, Energy, Environment, and Sustainability,
https://doi.org/10.1007/978-981-97-2523-6_8

the environment, and the irregular pattern posed an issue on the availability through the year [2]. Because biomass is readily available and provides energy in all states of physical existence, it is thought to be a good replacement for fossil fuels in light of all these challenges and problems [3]. Biomass is a naturally occurring organic resource that is renewable and derived from plant and animal waste [4]. This biomass can be transformed into fuel, energy storage material, a catalyst for different chemical reactions, and some compounds with added value [5]. The different possible conversion method includes pyrolysis, gasification, hydrothermal conversion, etc. The products from the degradation of biomass can be biochar (solid), crude (liquid), and syngas (gas) [6].

When biomass is broken down, a carbon-rich material called biochar results. There are two main ways to obtain it: pyrolysis, which includes heating the biomass in the absence of oxygen, or hydrothermal conversion, which requires employing high temperatures and pressure in an environment based on water [7]. Although pyrolysis is one of the common methods for biochar production, the necessity for the raw material to be moisture-free and should be done in the absence of oxygen makes the hydrothermal process more convenient where the process does not need moisture removal [8]. In comparison with other carbonaceous material from other chemical process, biochar is much more inexpensive and easy to produce [9]. The biochar still needs to be activated to increase the surface area by physical or chemical methods which will enhance the activity of the biochar [10]. The functionalization approach has also been investigated for further use of the biochar as a catalyst. The biochar produced differs significantly in its physicochemical properties depending on the raw biomass, activation process, conversion method, and functionalization [11]. The activated and functionalized biochar has been widely used for a variety of applications, including gas storage, energy storage, supercapacitors, wastewater treatment, and, more importantly, as a versatile catalyst and catalyst support due to its large surface area and abundance of functional groups [12].

Biodiesel are produced from plants and vegetable oil and as well as animal oil as a feedstock [13]. This oil mainly consists of mixtures of triglycerides and free fatty acids depending on the quality of the oil. This oil reacts with low molecular weight alcohol such as methanol, ethanol, and propanol [14]. For practical purposes, methanol is commonly used as it is the least expensive. The reaction is known as transesterification which produces a mixture of fatty acids methyl ester and glycerol [15] (Fig. 8.1).

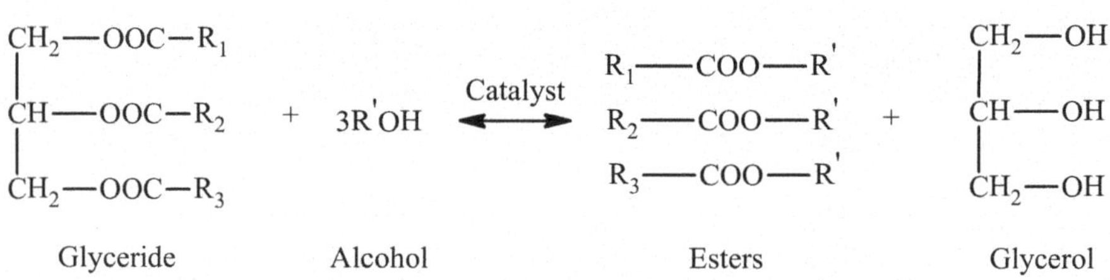

Fig. 8.1 Transesterification of glyceride

$$\underset{\text{Fatty acid}}{R_1\text{—COOH}} + \underset{\text{Alcohol}}{R'OH} \xrightleftharpoons{\text{Catalyst}} \underset{\text{Ester}}{R_1\text{—COO—}R'} + \underset{\text{Water}}{H_2O}$$

Fig. 8.2 Esterification of fatty acids

Free fatty acids (FFA) and alcohol can be used to create biodiesel by catalyzing an esterification reaction that yields both biodiesel and water [16] (Fig. 8.2).

The transesterification process can be catalyzed by various different catalysts through different processes: (1) homogeneous alkali, (2) homogeneous acid, (3) heterogeneous basic catalyst, and (4) heterogeneous acid catalyst [17–20]. For the industrial purposes, the homogeneous alkali-catalyzed method is widely used. It produced a high yield within a very short period, but it is the most expensive process since the separation of the homogeneous catalyst is very hard and needs a lot of washing to produce high purity [21]. Due to their limited tolerance for water and free fatty acids (FFA), basic catalysts provide difficulties in the manufacture of biodiesel. Due to the need for pure vegetable oil as a feedstock, the procedure is, therefore, more expensive overall [17]. The homogeneous acid catalyst improves the transesterification with low quality oil and high FFA content of up to 5 wt.% [22].

However, to get a conversion of more than 95%, the reaction took more than 48 h in mild condition [23]. Both the homogeneous alkali and acidic processes require a significant amount of water to separate the catalyst, which increases the amount of waste produced and calls for a more involved procedure. Heterogeneous catalysts are the most economical since they can be easily separated from the product and can be reused without much loss of activity [24, 25]. It does not require water for separation and neutralization and hence has a much easier process. Between the basic and acidic heterogeneous catalysts, the acid catalyst is significantly more effective because it can concurrently catalyze the esterification and transesterification reactions, enabling the processing of feedstocks with lower quality and cheaper cost [26].

8.2 Biochar Production

The conversion of biomass to biochar generally includes methods such as pyrolysis, carbonization, hydrothermal process, liquefaction, torrefaction, and liquefaction [28]. They are divided into two major types depending on whether water is added as a solvent or catalyst and whether it is absent in the process. The amount and properties of the biochar produced depend on the feedstock biomass as well as the type of conversion process used. In this chapter, we will discuss only pyrolysis and hydrothermal conversion (Fig. 8.3).

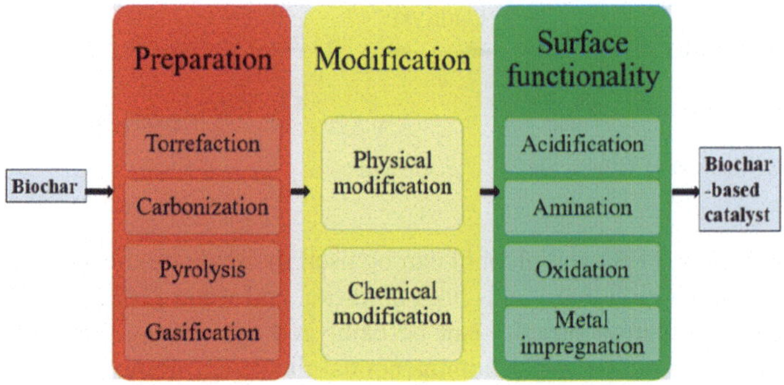

Fig. 8.3 Methods involved in producing biochar-based catalyst. Reproduced from Ref. [27]

8.2.1 *Pyrolysis*

Pyrolysis, which is the heating of organic material waste to some high temperature in the absence of oxygen, is one of the method to produce biochar [29, 30]. Co-pyrolysis is a method that involves simultaneously pyrolyzing multiple feedstocks, such as plastic waste and biomass, to create a variety of useful products. This method enables the synthesis of solid carbon, hydrogen, value-added compounds, and fuels [31]. The pyrolysis is divided into two categories based on the heating rate: slow pyrolysis (heating rates of 0.01–1 °C/s) [32] and rapid pyrolysis (heating rates of 10–1000 °C/s) [33]. The average heating range is 400–600 °C, and the residence duration in an atmosphere without oxygen ranges from a few minutes to several hours [32, 33]. The slow pyrolysis maximizes the biochar [34], while the fast pyrolysis increases the bio-oil yield [35].

The biochar produced by pyrolysis are termed as pyrochar [36]. Biomass pyrolysis involves free radical reaction which is very reactive, and hence, the overall reaction is a very complex process. Free radicals produced during the pyrolysis process can participate in several chemical processes, such as free radical substitution, free radical addition, and coupling reactions. These interactions can take place between the free radicals themselves or with other substances already in the system. As a result, a variety of products can be created, including syngas, biochar, and bio-oil [37]. The pyrolysis parameters, including temperature, heating rate, residence time, and the kind of feedstock, can have a big impact on the properties of the biochar that is produced, including its stability, carbon content, surface area, capacity to retain nutrients, pH, and cation exchange capacity [32].

Mechanism of Biochar Formation by Pyrolysis

The mechanism of biochar synthesis during pyrolysis involves several complex thermal and chemical processes. The three main stages of the process are drying, pyrolysis, and carbonization [38]. Following is a general description of each phase:

I. Drying: In the beginning, the feedstock is heated to remove moisture. At the relatively low temperature, which is typically below 100 °C, the main goal is to eliminate the water content from the material [39].

II. Pyrolysis: The temperature is raised to start the pyrolysis process once the drying stage is finished, typically between 200 and 500 °C. When there is little or no oxygen available, pyrolysis takes place. This phase sees the feedstock's organic molecules decompose as a result of intricate chemical interactions [35].

   (a) Volatile Release: Volatile substances from the feedstock, including water, gases (methane, ethylene, carbon monoxide, etc.), and tars, are released as the temperature rises. This process is frequently known as devolatilization [40].
   (b) Fragmentation and Rearrangement: Thermal decomposition is used to further break down the solid residue. Due to reactions like cracking, dehydrogenation, decarboxylation, and depolymerization, bigger organic molecules become fragmented at high temperatures. As a result, numerous intermediate chemicals are created [37].
   (c) Char Formation: Carbonaceous elements start to build and turn into a solid residue known as biochar as the thermal decomposition process advances. Rearranging and condensing the residual carbon-rich pieces during the carbonization process results in the production of a stable carbon matrix [41].

III. Carbonization: The maturation and stabilization of the carbonaceous structure continue during the production of biochar. As the temperature rises, usually above 500 °C, and the thermal breakdown reactions become more constrained, this happens [42]. The carbonization procedure increases the biochar's carbon content, lowers volatile matter, and improves its stability [43].

Based on variables including feedstock content, heating rate, temperature, and residence time, the precise reaction routes and mechanisms during pyrolysis can change. These factors may affect the amount of volatile emission, the level of carbonization, and the ultimate characteristics of the biochar produced [43]. It is crucial to remember that pyrolysis is a complicated process, and current research is currently being done to determine the precise mechanisms and reaction routes. To make biochar with desirable qualities for uses like soil amendment, carbon sequestration, and energy production, researchers are still looking into and adjusting pyrolysis settings (Fig. 8.4).

8.2.2 *Hydrothermal Conversion*

A technique known as hydrothermal conversion uses high-temperature and high-pressure water to transform biomass or other organic materials into valuable goods [45]. It is a type of thermochemical conversion in which several chemical processes

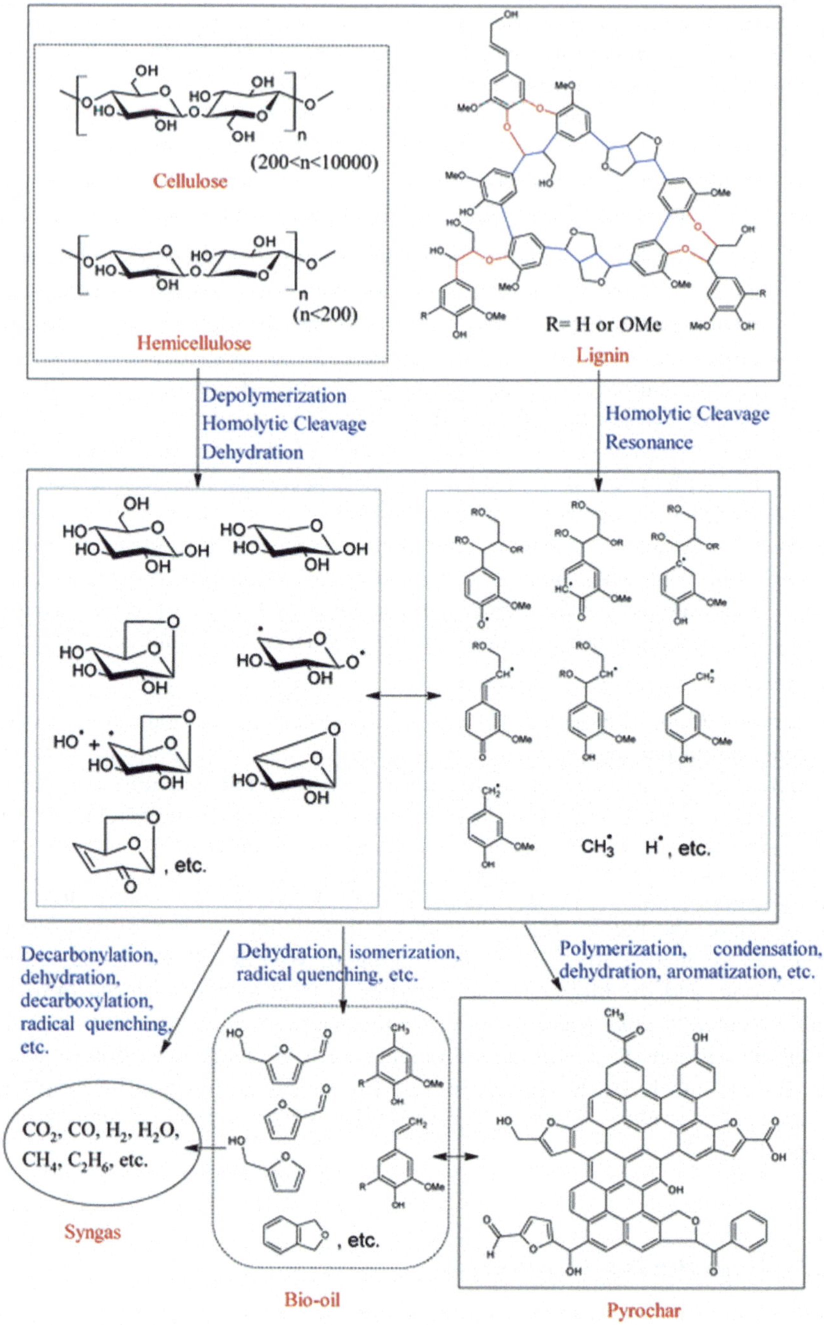

Fig. 8.4 Possible formation pathway of pyrochar. Reproduced from Ref. [44] with permission from the Royal Society of Chemistry

are carried out using water as a medium. Hydrothermal conversion is commonly selected as the preferred method when the feedstock cannot be pyrolyzed or has a high moisture content that would require a considerable energy input for drying [46].

Biomass or organic materials are frequently heated to temperatures exceeding 180 ºC and compressed above atmospheric pressure during hydrothermal conversion. High-pressure water works as a catalyst and solvent to help break down complex chemical compounds into smaller ones. Three different processes can occur: gasification (gas), liquefaction (liquid), and carbonization (solid) [47].

8.2.2.1 Hydrothermal Carbonization

Biomass or organic waste is transformed into hydrochar or biocarbon through the thermochemical process of hydrothermal carbonization (HTC) [48]. Because it occurs in the presence of water under intense pressure and heat, it is frequently referred to as "wet carbonization." Biomass or organic waste is often combined with water and heated to temperatures between 180 and 250 °C under pressures between 10 and 25 bar during the hydrothermal carbonization process. The presence of water and the high temperature and pressure conditions encourage the breakdown of complex organic molecules in the feedstock and start several chemical processes [49].

Several concurrent reactions take place during hydrothermal carbonization. The biomass is initially partially hydrolyzed, which results in the creation of water-soluble molecules including sugars and organic acids. These soluble molecules go through additional reactions as the process progresses, such as dehydration, decarboxylation, and polymerization. This causes the production of solid hydrochar, a carbon-rich substance that resembles char [50]. The hydrochar created by HTC has several advantageous qualities. Its high carbon content, which normally ranges from 50 to 90%, is what makes it unique. Similar to coal, it has a high energy density and can be utilized as a solid fuel for the production of heat and power [51]. In addition, hydrochar is stable and has a low oxygen content, which makes it more resistant to biodegradation and less prone to combustion than raw biomass [52].

8.2.2.2 Hydrothermal Liquefaction (HTL)

HTL entails converting biomass into a liquid substance known as bio-oil or crude [53]. In this procedure, the complex organic molecules in the biomass are broken down by the high-temperature and high-pressure water, creating a mixture of liquid organic chemicals. By further refining the created bio-oil, important items including transportation fuels, chemicals, and bio-based materials can be obtained [54].

8.2.2.3 Hydrothermal Gasification (HTG)

Biomass is transformed into a mixture of gases, chiefly hydrogen (H_2), carbon monoxide (CO), and methane (CH_4), in the process of hydrogen-to-gas (HTG) [55]. Syngas, also known as synthesis gas, is created as a result of the organic material's reaction with the high-pressure water. Syngas is a feedstock that can be used to make chemicals, fertilizers, and biofuels.

Compared to other methods of biomass conversion, hydrothermal conversion has several benefits. Agricultural waste, energy crops, algae, and organic waste are just a few of the many feedstocks it can process. When compared to traditional pyrolysis or gasification, it is a comparatively quick process and may be carried out at lower temperatures. Additionally, the breakdown of biomass components is aided by the high-pressure water conditions, allowing for a more thorough conversion and minimizing the production of char or solid leftovers [48].

Mechanism of Hydrochar Formation by Hydrothermal Carbonization

The process of hydrothermal carbonization (HTC), which produces hydrochar, comprises several intricate processes that take place in high-temperature, high-pressure water environments [56]. Numerous crucial processes are thought to play a role in the creation of hydrochar, even though the precise mechanisms are not entirely known. Here are a few of the suggested mechanisms:

I. Hydrolysis: Complex polysaccharides, lignin, and other organic components are hydrolyzed in the early stages of HTC when biomass or organic waste is subjected to hot water. These large molecules undergo hydrolysis, which reduces them to smaller, water-soluble substances like sugars and organic acids. These soluble molecules are capable of additional reactions to produce other products [57].

II. Dehydration: Water removal from the hydrolysis products causes dehydration processes to take place. Furans, aldehydes, ketones, and other reactive intermediates are created as a result of the elimination of water molecules. The carbonization process is aided by these chemicals' ability to generate more durable aromatic structures through subsequent reactions [57].

III. Decarboxylation: Decarboxylation reactions include the removal of carboxyl groups (–COOH) from organic acids and other carboxyl-containing substances. This process results in the production of aromatic chemicals and the release of carbon dioxide (CO_2). Decarboxylation increases the carbon content of the final hydrochar product, which helps to carbonize the biomass [57].

IV. Polymerization: Hydrochar is created primarily by polymerization processes. When the reactive intermediates generated by hydrolysis, dehydration, and decarboxylation are subjected to polymerization, they interact with one another to create bigger, more complex molecules. Cross-linked structures are created as a result of polymerization events, which also help the hydrochar to become more aromatic [57].

V. Condensation: In condensation reactions, smaller molecules create chemical interactions with one another to generate larger, more condensed structures. These reactions aid in the carbonization process and the synthesis of molecules rich in aromatic carbon. Several reactive intermediates, including furans, phenols, and aldehydes, can undergo condensation reactions with one another [57].

VI. Carbonization and Solidification: The organic compounds continue to change as the reactions advance, and eventually, the system reaches a point where the hydrochar starts to harden. A carbon-rich matrix with a porous structure is created during this solidification process. Temperature, pressure, reaction time, and the makeup of the biomass fuel are all elements that affect the solidification process [56].

The precise reaction pathways and processes can differ depending on the type of biomass feedstock, the reaction environment, and the experimental setup, which is vital to highlight. To optimize the HTC process and enhance the quality and attributes of the hydrochar produced, researchers are still investigating the intricate principles of hydrochar creation.

8.3 Feedstock for Biochar Production

Wood chips and sawdust, which include lignocellulosic components, are the most frequently utilized feedstock. Different lignin, hemicellulose, and cellulose types can be found in different types of agricultural and forestry waste [58]. These leftovers, which are frequently not used after harvesting, might be used to make biochar. Nut shells like those from almonds, walnut, and coconut can be utilized as raw materials to make biochar. These shells can be used to create biochar, even though they are generally dumped as waste [59]. Livestock waste, such as poultry litter, cow dung, and pig manure, can be used as feedstocks for biochar production [60]. This not only helps in waste management but also produces biochar with added nutrient content. Fast-growing energy crops like bamboo, miscanthus, and switchgrass can be used as feedstocks for biochar production. These crops are specifically grown for biomass production and can be sustainably harvested for biochar [61].

Organic waste from households, restaurants, and other urban sources can be used as feedstocks for biochar production. This helps in waste diversion and reduces the amount of waste sent to landfills [62]. Certain types of algae and aquatic plants, such as water hyacinth and duckweed, can be used as feedstocks for biochar production. These plants are often considered invasive species and can be harvested to produce biochar [63]. However, animal waste and aquatic materials that have low lignin and cellulose content have not been studied extensively due to their difficult handling condition. Different biomass with different chemical compositions (lignin, hemicellulose, and cellulose) are thought to have a significant impact on the properties of the biochar produced, surface area, reactivity with chemical activating agent, and yields.

The choice of feedstock depends on factors such as availability, cost, and desired biochar characteristics.

8.4 Activation and Functionalization of Biochar

The biochar produced after just pyrolysis and hydrothermal carbonization may not have enough surface area and large porosity and hence may not serve well for application. Biochar activation refers to a process that enhances the physical and chemical properties of biochar, making it more effective for specific applications [42]. Activated biochar has increased surface area, pore volume, and reactivity, which improves its ability to adsorb and retain nutrients, pollutants, and other substances [64]. In addition to its core function, activated biochar has a wide range of uses due to its improved surface and adsorption qualities. It can be used in a variety of fields, including environmental adsorbents, soil amendment, carbon sequestration agents, and activated carbon.

8.4.1 Types of Biochar Activation Method

8.4.1.1 Physical Activation

Physical activation involves subjecting biochar to high temperatures above 700 °C in the presence of gases, typically steam or carbon dioxide. This process, known as steam or CO_2 activation, causes the carbon matrix to expand, creating additional pores and increasing surface area [8]. Removing the unstructured parts, this disintegration only leaves the carbon structure with its tiny pores. As a result, the interior surface area of the activated biochar increases. The next stage of the activation process involves the depletion of crystalloid carbon, which is defined as carbonized material or carbon with small pores. Because of this depletion, the activated biochar develops bigger pores. The pores in the biochar that are loaded with carbon volatiles are opened under precise circumstances during the activation process. These volatiles can diffuse with nearby pores thanks to this opening. These occurrences result in a large increase in the biochar's specific surface area, the production of micropores, and a noticeable decrease in the concentration of mesopores [42] (Fig. 8.5).

The yield of activated carbon may decrease as a result of the increased danger of combustion from air activation processes that take longer to complete. The likelihood that the carbonaceous material will reach its ignition point and undergo combustion is increased by the lengthier exposure to air. As a result, it is essential to maintain the temperature and exposure time to keep the pore volume and specific surface area. In general, activation by air is less expensive than activation by CO_2, which produces more pores and eventually increases the biochar's surface area [8].

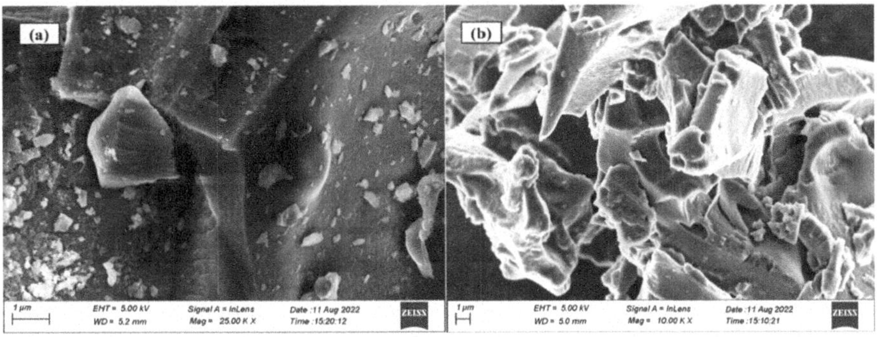

Fig. 8.5 SEM image of **a** steam-activated wood apple **b** H_2SO_4 *impregnated wood* apple. Reproduced from Ref. [65]

The type of biomass used, the kind of gas used during activation, and the particular activation circumstances used can all have a major impact on the features of activated biochar. According to Nabais et al. [11], the use of steam increased the activation rate of biochar produced from the pyrolysis of coffee endocarp, whereas CO_2 increased the specific area and pore size. Reinoso et al. [66] used steam and CO_2 to activate biochar from olive seed and compared the results. According to their studies, activated carbon made with pure steam as the activation agent has smaller micropore volume than activated carbon made with CO_2. Additionally, it has been found that steam activation produces meso- and macro-porosity more efficiently than CO_2 activation, resulting in larger pores. A study was done by Koltowski et al. [10] that looked at activating biochar made from willow that had undergone slow pyrolysis. The study's goal was to look at how CO_2 and steam as activation agents affected the properties of the activated biochar that resulted. They have reported that the steam activation and CO_2 activation have improved the surface area, and the steam activation have higher surface areas (840.6 $m^2\ g^{-1}$) than CO_2 activation (512.0 $m^2\ g^{-1}$). Additionally, according to the research results, steam-activated biochar showed larger pore diameters and a greater specific surface area than CO_2-activated biochar under similar activation settings. This suggests that the steam activation procedure resulted in the formation of a more porous and structurally varied biochar material.

8.4.1.2 Chemical Activation

Chemical activation involves first impregnating the biochar made from biomass with an activating chemical, such as KOH, NaOH, $ZnCl_2$, K_2CO_3, H_2SO_4, and H_3PO_4, before heating it at a high temperature while being surrounded by inert gas [8, 67, 68]. Although chemical activation has drawbacks like corrosion of the apparatus, difficult recovery of the chemical, and cost of chemicals, it has a lot of advantages like high efficiency, a larger surface area of more than 4000 m^2g-1, controlled porosity, low-temperature range, and higher carbon yield. The temperature utilized during chemical activation, the kind of chemical agent used, the kind of biomass from which the biochar is formed, and the concentration of the chemical agent are all variables that affect the physical qualities of biochar that have undergone chemical activation [42, 69].

The production of biochar with a significant surface area and porosity has been seen when potassium hydroxide (KOH) is used as an activating agent in the chemical activation process. This is due to the cooperative actions of numerous mechanisms, such as metallic K^+ intercalation, physical activation, chemical erosion, and lattice expansion of carbon [42]. According to research done by Ros et al. [70], using phosphoric acid (H_3PO_4) to sludge char as an activating agent did not significantly improve the specific surface area or pore volume. Similar to how alkali metal activating agents increased specific surface area and pore volume, activation with carbon dioxide (CO_2) had a much lower effect. Two activators, zinc chloride ($ZnCl_2$) and ammonium chloride (NH_4Cl), were used in a different investigation to activate biochar made from mandarin peel. A tubular furnace heated to 700 °C was used for the activation procedure [71]. It was found out that although NH_4Cl increased the surface area to 181 $m^2\ g^{-1}$, it was not comparable to the surface area increased by using $ZnCl_2$ to 1085 $m^2\ g^{-1}$.

In a study by Park et al. [72], they used different chemical activators, such as HCl, H_2SO_4, H_3PO_4, KOH, MgO, $ZnCl_2$, and K_2SO_4, to activate sesame straw biochar. Their study's goal was to find out how the various activation agents affected the amount of phosphorus that could be absorbed by the resulting activated biochar. The research results showed that $ZnCl_2$ and MgO were more effective than the other chemical activating agents at increasing the phosphorus adsorption capacity of sesame straw biochar than the other activating agents evaluated. Particularly, the $ZnCl_2$-activated biochar had the highest phosphorus adsorption capability, with a staggering maximum of 15 g per kilogram (g kg^{-1}). It is therefore a highly effective phosphorus adsorbent, capable of adsorbing up to 15 g of phosphorus for every kilogram of $ZnCl_2$-activated biochar [72].

8.4.1.3 Biological Activation

When microorganisms are inoculated or added to biochar to encourage biological activity and nutrient cycling in the soil, this process is referred to as biological activation. In this procedure, the biochar is "primed" or "pre-charged" with microbes that can colonize its surface and foster microbial activity. Biological activation may involve the employment of advantageous bacteria, fungi, and other soil microbes. They can be obtained from vermicompost, compost, or other organic materials rich in microorganisms. The activation process typically involves mixing biochar with the microbial inoculant and allowing it to interact and colonize the biochar surface.

The removal of lignin from biowaste and the hydrolysis of cellulose using cellulose enzymes created highly microporous biochar with a larger surface area. The addition of paraffin through vacuum impregnation then created a 3D porous structure with a high energy capacity of 141.47 J/g, demonstrating that bio activation of biowaste can create a potential carbon source for energy applications [73]. A highly porous carbon material with a remarkable surface area of 1831 m^2/g, mostly made up of micro- and mesopores, was produced when lignin was activated utilizing green bacterial processes. For the material to function well in many applications, its broad porosity is essential. The fungus Trichoderma was grown in a mixture of wheat straw and food scraps. The obtained biomass underwent a carbonization procedure before being heated to high temperatures between 800 and 900 °C for activation. This procedure produced a carbon substance with a remarkable surface area of about 4000 m^2/g. Its volume of 2.37 cm^3/g further demonstrated the material's porosity and potential for effective charge storage [74].

8.4.2 Biochar Functionalization

The biochar obtained from biomass usually has low surface functionality with only limited C–O, C=O, and OH groups present and possessing very limited porosity and surface area (usually < 150 m^2/g) [75]. The surface functionality, however, can be tuned very easily which offers a promising platform for synthesizing various functional materials. In general, biochar can gain surface activity via surface modification or active substance deposition [64].

Biochar modification is a flexible method for adjusting its characteristics for certain uses. One efficient technique involves adding particular functional groups, especially acid groups, to the surface of the organic biochar. Sulfonation with concentrated sulfuric acid (H_2SO_4) or its derivatives is one of the most often used processes for modifying biochar. Using strong sulfuric acid or an acid derivative, such as fuming sulfuric acid (oleum), the biochar is subjected to this process to add sulfonic acid functional groups ($-SO_3H$) onto the surface of the material [42]. Other weak acid functional groups, such as carboxylic acid groups (–COOH), can also be introduced to the biochar surface during alteration in addition to sulfonic acid groups ($-SO_3H$). The addition of –COOH groups can increase the modified biochar's catalytic activity and reactivity, making it a flexible solid acid catalyst for a variety of chemical reactions [76].

Solid acids derived from biochar that have functional groups added, including –SO_3H and –COOH, demonstrate a number of key benefits that make them attractive substitutes for conventional mineral acid catalysts in a variety of catalytic reactions [77]. By subjecting biochar to gaseous sulfur trioxide (SO_3) concentrations greater than 20%, a process known as "gaseous sulfonation" is used to alter the material. Between 25 and 150 °C, relatively low temperatures are commonly used for this operation. The biochar interacts with the SO_3 gas during the sulfonation process, which results in the insertion of sulfonic acid functional groups ($-SO_3H$) onto its surface [78]. Another way to alter biochar is by treating it with a liquid sulfonating agent, usually concentrated sulfuric acid (H_2SO_4) or its derivatives. Liquid sulfonation typically makes it easier for the surface area of the biochar to expand more than gaseous sulfonation does.

There are several methods for chemical functionalization of biochar, and here are a few commonly employed techniques:

I. Acid Treatment: Acids like nitric acid (HNO_3) and hydrochloric acid (HCl) can be used to treat biochar. By adding carboxylic acid (–COOH) functional groups to the surface of the biochar during acid treatment, it becomes more hydrophilic and has a greater chance of interacting with other molecules.
II. Base Treatment: Similar to acid treatment, biochar can be treated with bases like sodium hydroxide (NaOH) or potassium hydroxide (KOH). Basic functional groups (–OH) are added to the surface of biochar during base treatment, altering its properties and reactivity.
III. Silanization: It is possible to silanize biochar by using organosilanes, which contain one silicon atom and organic functional groups. This process enables the addition of additional functional groups to the surface of the biochar, such as amino (–NH_2) or alkyl (–CH_3) groups. For adapting biochar's surface chemistry for specific applications, silanization offers a flexible technique.
IV. Ozonation: The treatment of charcoal with ozone (O_3) gas is known as ozonation. This process exposes the biochar surface to oxygen-containing functional groups, such as carbonyl (–C=O) or hydroxyl (–OH) groups. Ozonation, which raises the surface reactivity of biochar, can increase its ability to absorb substances.
V. Plasma Treatment: Plasma is used to treat biochar by exposing it to a low-temperature plasma discharge. Radicals, a class of extremely reactive species created by plasma, can interact with biochar's surface to introduce different functional groups. Plasma treatment is an efficient and well-controlled method of functionalizing biochar.

The chemical functionalization of biochar can alter its surface chemistry, hydrophobicity/hydrophilicity, porosity, and reactivity, making it suitable for diverse applications. Functionalized biochar has been used in environmental remediation, water treatment, soil improvement, energy storage, and as catalyst supports, among other areas. The specific functionalization method chosen depends on the desired properties and targeted applications of the biochar (Fig. 8.6).

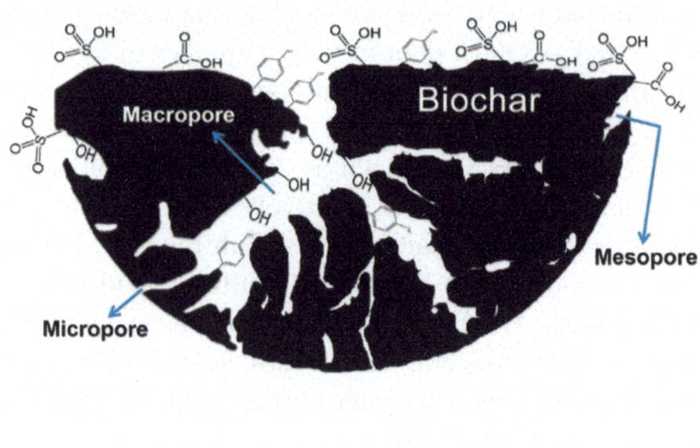

Fig. 8.6 Model of porous biochar containing different functional groups. Reproduced from Ref. [77]

8.5 Biodiesel Production by Biochar-Based Catalyst

A sustainable and eco-friendly substitute for petroleum diesel is biodiesel. It is a sustainable fuel made by esterification and transesterification of animal fats, vegetable oils, or used cooking oil with alcohols like methanol or ethanol. These processes produce a mixture of mono-alkyl esters, which combine to generate biodiesel [79]. Esterification and transesterification are the two primary types of processes used in the manufacturing of biodiesel to transform vegetable or animal fats into mono-alkyl esters. These procedures necessitate the use of a catalyst and require the conversion of triglycerides and free fatty acids into biodiesel. Both acidic and basic catalysts can aid in the transesterification and esterification procedures required for the manufacture of biodiesel. Numerous heterogeneous and homogeneous catalysts have been employed for this purpose [80].

In reality, homogeneous catalysts allow for rapid and complete conversion of feedstocks while also offering a high level of reaction efficiency for the production of biodiesel. But their major drawback is that it takes a lot of work to purify the product and is difficult to reuse, which raises production expenses [81]. Accordingly, several carbon-based acid and basic catalyst as well as heterogeneous acid and basic catalyst have been employed to produce biodiesel. These catalysts include CaO, MgO, Amberlyst-15, TiO_2/ZrO_2, Al_2O_3/ZrO_2, and WO_3/ZrO_2 [82]. The use of carbon-based biochar as a catalyst has many benefits. These advantages include its affordability, substantial surface area, wide accessibility, and capacity to tailor its surface functional groups to particular applications. The biochar is activated by some activating agent which increase the porosity and hence the surface area after which it is subjected for functionalization and make an acid or basic catalyst, which is used for biodiesel production [83].

Acid-functionalized biochar catalysts and base-functionalized biochar catalysts are two different types of biochar catalysts that can be useful in the generation of biodiesel. Sulfonated biochars have been the most popular heterogeneous catalysts for this use among them. Sulfonic acid groups ($-SO_3H$) are added to the surface of biochar during the sulfonation process. This alteration increases the biochar's catalytic activity and makes it a powerful solid acid catalyst for the synthesis of biodiesel. Sulfonation can be accomplished in a variety of ways. One method includes heating a concentrated solution of sulfuric acid (H_2SO_4) and soaking the charcoal in it. The sulfonic acid groups are immobilized on the surface of the biochar during this procedure. Another way to sulfonate biochar is to expose it to gaseous sulfur trioxide (SO_3). This exposure causes $-SO_3H$ groups to be incorporated onto the surface of the biochar, turning it into a powerful solid acid catalyst. These $-SO_3H$ groups can catalyze the esterification of free fatty acids (FFAs) or promote the transesterification reactions of triglycerides with alcohols to form biodiesel [79] (Fig. 8.7).

Rokhum et al. [84] synthesized a carbon-based solid acid catalyst from glucose via a an efficient one-pot hydrothermal carbonization-sulfonation without the need of high temperature. The catalyst is combined with aromatic structure of hydrophilic –OH and –COOH groups and a high density of $-SO_3H$ centers. In their study, the

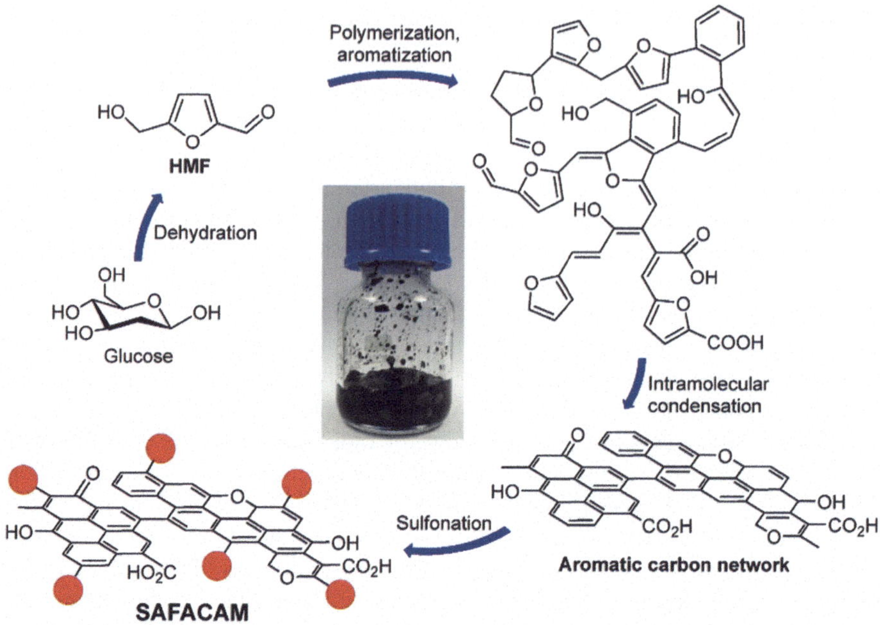

Fig. 8.7 Representation of the one-pot, multistep conversion of glucose to the mesoporous SAFACAM pictured in the reagent bottle (red circle = SO_3H). Reproduced from Ref. [84]

relevance of the catalyst was tested with oleic acid, where the level of SO_3H (0.81–1.29 $mmolg^{-1}$) proved synthetically tunable. In a study, Kasner et al. synthesized sulfonated biochar using two different techniques: (a) concentrated sulfuric acid (H_2SO_4) and (b) gaseous sulfur trioxide (SO_3). Both of these techniques were used to add sulfonic acid groups ($-SO_3H$) to the biochar's surface, creating sulfonated biochar catalysts. Their investigation proved that the biochar matrix's surface area and pore volume increased as a result of sulfonation using concentrated sulfuric acid (H_2SO_4). This indicates that there were more active sites and available area for catalytic reactions in the improved biochar. The biochar's surface was successfully increased throughout the H_2SO_4 sulfonation procedure, making catalytic activities easier [78]. In their studies, they found that the sulfonated biochar catalyst had impressive catalytic activity for the esterification of fatty acids during their investigations. The outcomes demonstrated a remarkable conversion rate of the fatty acids, which produced the matching fatty acid methyl esters (FAMEs) at a rate of about 90–100%. It only took between 30 and 60 min for this high conversion rate to be achieved. The esterification processes were carried out in the 55–60 °C temperature range.

González et al. [85] synthesized biochar using oat hull as a feedstock, and then, they used concentrated sulfuric acid (H_2SO_4) to sulfonate the finished product. In order to perfect the alteration, they tested with different temperatures throughout the sulfonation process in a microwave reactor. After getting sulfonated biochar catalysts with various levels of sulfonation, researchers looked into how well these substances worked as catalysts for the synthesis of biodiesel. They specifically employed discarded cooking oil as the feedstock for the synthesis of biodiesel. The same microwave reactor was used to produce biodiesel from start to finish, including the esterification of free fatty acids and the transesterification of triglycerides. Dehkhoda et al. [16] sulfonated hardwood char using concentrated sulfuric acid and fuming sulfuric acid, activated it chemically with 10 M KOH to increase the specific surface area, and applied it to the transesterification of vegetable oil and the esterification of FFA. Li et al. [83] sulfonated pyrolyzed rice hulk by concentrated sulfuric acid and used the prepared catalyst for simultaneous esterification and transesterification to produced biodiesel from waste cooking oil. In their study, they have showed that the yield of fatty acid methyl ester is 87.57% after 15 h.

In fact, base-functionalized biochar catalysts have been used in addition to acid-functionalized biochar catalysts to produce biodiesel. CaO/biochar and KOH/biochar are two examples of such catalysts [86, 87]. Using a CaO-based palm kernel shell biochar catalyst, McKay and colleagues created a new catalyst for their study. This base-functionalized biochar catalyst was created by impregnating biochar made from palm kernel shells in calcium oxide (CaO). They examined the effectiveness of the CaO-based biochar catalyst in the transesterification reaction for the creation of fatty acid methyl esters (FAMEs) from sunflower oil. In their investigation, it was found that under particular ideal conditions, they were able to produce an amazing 99% yield of fatty acid methyl esters (FAMEs). A 3 wt.% catalyst loading, 65 °C reaction temperature, and a 9:1 (MTOR) were all used to achieve this high yield. Wang et al. [81] also synthesized a series of peat biochar supported K_2CO_3 catalyst using wet impregnation method with different K_2CO_3 loading (20–40 wt.%) [88]. Table 8.1 presents the production of biodiesel from different feedstock using different biochar-based acid and basic catalyst.

8.5.1 Comparison of Biochar and Functionalized Biochar in Biodiesel Production

In summary, the comparison between biochar and functionalized biochar as catalysts for biodiesel production highlights the superior catalytic efficiency of the latter. As research progresses, optimizing reaction parameters, such as temperature and duration, promises to further enhance the efficiency of biodiesel production when functionalized biochar is utilized. This innovation holds great promise for the sustainable and cost-effective production of biodiesel, contributing to a more environmentally friendly and economically viable energy source. From entries 1 and 12, we can see that without sulfonation biochar the biodiesel yield were decreased compared to entries (2–11, 13, and 14) having sulfonated biochar. For entry 15, although it is non-sulfonated biochar, the yield of biodiesel was quite high; this can be due to very high reaction temperature for transesterification reaction.

Table 8.1 Biodiesel production under the catalysis of various biochar-based acid or basic catalyst

Sl. No.	Biochar	Functionalization condition	Feedstock	Reaction condition[a]	Biodiesel yield (%)	References
	Oat hull biochar	None	Waste cooking oils	140, 0.5 (microwave), 10, 10:1	0.5	[85]
	Oat hull biochar	Sulfonated with H_2SO_4 in microwave reactor at 100 °C for 30 min	Waste cooking oils	140, 0.5 (microwave), 10, 10:1	28	[85]
	Oat hull biochar	Sulfonated with H_2SO_4 in microwave reactor at 140 °C for 30 min	Waste cooking oils	140, 0.5 (microwave), 10, 10:1	72	[85]
	Biochar from fast pyrolysis of different feedstocks	20 g of biochar sulfonated with 200 mL of H_2SO_4 at 150 °C for 24 h	Vegetables oils	60, 3, 5, EtOH/oil (18:1)	77–89	[16]
	Biochar from peanut hulls, pine pellets, and wood chips	Sulfonated with H_2SO_4 at 100 °C for 12–18 h or gaseous SO_3 at room temperature for six days	5% palmitic and stearic acid	65, 3, 20:1	> 80 (FFA conversion)	[78]
	Wood biochar	Sulfonated with fuming H_2SO_4 at 150 °C for 15 h	Canola oil	65, 12, 5, 15:1	44.2	[89]
	Rice husk biochar	Sulfonated with H_2SO_4 at 90 °C for 0.5 h	Waste cooking oils	110, 15, 5, 20:1	87.6 (FAME yield) > 98 (FFA conversion)	[83]
	Rice husk biochar	Sulfonated with H_2SO_4 at 70–150 °C for 0.25–4 h	Oleic acid	110, 2, 5, 4:1	98.7	[90]
	Corn straw biochar	Sulfonated with fuming H_2SO_4 at 180 °C for 4 h	Oleic acid	60, 4, 7, 7:1	98	[91]
	Palm kernel shell biochar	Loaded with CaO at a total basicity of 0.516 mmol g^{-1}	Sunflower oil	60, 5, 3, 9:1	99	[86]

(continued)

Table 8.1 (continued)

Sl. No.	Biochar	Functionalization condition	Feedstock	Reaction condition[a]	Biodiesel yield (%)	References
	Palm kernel shell biochar	Loaded with 50.6% of CaO	Sunflower oil	60, 5, 3, 9:1	99	[87]
	Peat biochar	None	Palm oil	65, 1.5, 5, 8:1	10.5	[81]
	Peat biochar	Loaded with 20–40 wt.% of K_2CO_3	Palm oil	65, 1.5, 5, 8:1	95.2–98.6	[81]
	Flamboyant pods biochar	Impregnated in KOH solution	*Hevea brasiliensis* oil	60, 1, 3.5, 15:1	89.3	[88]
	Biochar from maize residue	None	Waste cooking oil	300, 1, 50:1	~ 90	[92]

[a]Temperature (°C), time (in hour), Catalyst loading (wt.%), methyl-to-oil ratio (unless otherwise stated)

8.5.2 Catalyst Reusability and Kinetics Study

Recycling poses a significant concern due to the limited availability of resources. In industrial settings, the imperative is to utilize catalysts that have a prolonged lifespan and can be easily retrieved. Heterogeneous catalysts enable straightforward recycling, potentially lowering the overall expenses of chemical processes and enhancing environmental conditions. According to Gonzales et al. [85] findings, when the catalyst was employed for successive cycles, both with and without hexane washing, there was a notable decline in catalytic activity. Specifically, the FAME yield dropped from approximately 86% during the initial cycle to approximately 14% by the third cycle, signifying an 84% reduction in catalytic efficiency. Conversely, when the catalyst was subjected to hexane washing, its performance exhibited a more modest decrease. The FAME yield began at around 90% during the first cycle and declined to approximately 60% after the third cycle, resulting in only a 33% decrease in activity. In another study by Wang et al. [81], the catalyst demonstrated exceptional catalytic performance, resulting in a remarkable 98.6% biodiesel yield, primarily attributed to its superior total basicity. Additionally, their investigation into the catalyst's reusability revealed that it maintained favorable catalytic activity, yielding above 81% biodiesel even after undergoing nine successive uses. This finding confirm that biochar catalyst has great potential for biodiesel production.

In the presence of excess methanol, the esterification and transesterification reactions used in biodiesel production exhibit pseudo-first-order kinetics, which permits the neglect of the reverse process [84, 86]. The activation energies reported in the literature vary for different biochar catalysts when using various feedstocks, including sunflower oil (108.8 kJ/mol), waste frying oil (79–84 kJ/mol), jatropha oil (29.5 kJ/mol), and canola oil (102–136 kJ/mol) [86]. Hence, the response rate (r) can be expressed using Eq. 8.1:

$$r = -\frac{d[O]}{dt} = k[O] \tag{8.1}$$

where $[O]$ stands for the oil concentration, t for the reaction time, and k for the rate constant. The methyl ester conversion at time t was measured from Eq. 8.2 to determine first-order rate constants. The Arrhenius equation (Eq. 8.3) was used to determine the activation energy (E_a) for the transesterification processes using the rate constants.

$$-\ln(1 - X) = kt \tag{8.2}$$

$$\ln k = -\frac{E_a}{RT} + \ln P \tag{8.3}$$

8.6 Future Prospect

The expansion of the utilization of biomass as a catalyst source in the production of biodiesel appears to be on the horizon. A multipurpose substance that can be further improved for catalytic uses is biochar, a by-product of biomass breakdown. As research advances, it is foreseeable that optimization of reaction parameters such as temperature and duration will enhance the efficiency of biodiesel production, utilizing functionalized biochar as a catalyst. This will result in processes that are both economically feasible and environmentally sustainable. Functionalized biochar, particularly acid-functionalized and basic-functionalized variants, shows potential as a catalyst for chemical reactions. These materials can find applications in various processes, including transesterification and esterification, contributing to the sustainable production of biodiesel and other chemicals. Further studies could uncover new ways to improve its catalytic capabilities.

8.7 Conclusions

Biochar emerges as a suitable low-cost and sustainable catalyst and catalyst support for chemical synthesis, biofuel production, energy storage, gas adsorption, etc., giving comparable or superior performance to the commercial one. In this review, we discuss the biochar produced by pyrolysis and hydrothermal processes. The physicochemical properties of the biochar depend on the feedstock, temperature, time, and the process of formation. The surface area of the biochar is not high enough, and due to the absence of a functional group, the biochar further needs to be activated and functionalized to be used as a catalyst. The surface area and porosity of the biochar can be enhanced to different extents depending on the activation methods employed. Furthermore, depending on the intended application of the biochar, it can be further modified with specific functional groups or substances to carry out particular tasks, such as selective adsorption and catalysis. The biochars that have been activated and functionalized demonstrate significant promise for serving as versatile catalysts or catalyst supports in biomass upgrading. These biochars have been effectively utilized in various applications such as carbohydrate hydrolysis and dehydration, biodiesel production, biomass pyrolysis, gasification, and bio-oil upgrading. Further investigation has to be done in developing the catalytic properties of biochar which will be important to designing active, selective, and stable biochar catalysts. Furthermore, for biochar to be a substitute for industrial heterogeneous catalysts, it has to be producible on an industrial scale.

References

1. Demirbaş, A. (2001). Biomass resource facilities and biomass conversion processing for fuels and chemicals. *Energy Conversion and Management, 42*, 1357–1378. https://doi.org/10.1016/S0196-8904(00)00137-0
2. Kunkes, E. L., Simonetti, D. A., West, R. M., Serrano-Ruiz, J. C., Gärtner, C. A., & Dumesic, J. A. (2008). Catalytic conversion of biomass to monofunctional hydrocarbons and targeted liquid-fuel classes. *Science (80-.), 322*, 417–421. https://doi.org/10.1126/SCIENCE.1159210/SUPPL_FILE/KUNKES.SOM.PDF
3. Ruatpuia, J. V. L., Changmai, B., Pathak, A., Alghamdi, L. A., Kress, T., Halder, G., Wheatley, A. E. H., & Rokhum, S. L. (2023). Green biodiesel production from Jatropha curcas oil using a carbon-based solid acid catalyst: A process optimization study. *Renewable Energy, 206*, 597–608. https://doi.org/10.1016/J.RENENE.2023.02.041
4. Soffian, M. S., Abdul Halim, F. Z., Aziz, F., Rahman, M. A., Mohamed Amin, M. A., & Awang Chee, D. N. (2022). Carbon-based material derived from biomass waste for wastewater treatment. *Environmental Advances, 9,* 100259. https://doi.org/10.1016/J.ENVADV.2022.100259
5. Zhou, C. H., Xia, X., Lin, C. X., Tong, D. S., & Beltramini, J. (2011). Catalytic conversion of lignocellulosic biomass to fine chemicals and fuels. *Chemical Society Reviews, 40*, 5588–5617. https://doi.org/10.1039/C1CS15124J
6. Ma, L., Wang, T., Liu, Q., Zhang, X., Ma, W., & Zhang, Q. (2012). A review of thermal–chemical conversion of lignocellulosic biomass in China. *Biotechnology Advances, 30*, 859–873. https://doi.org/10.1016/J.BIOTECHADV.2012.01.016
7. Sri Shalini, S., Palanivelu, K., Ramachandran, A., & Raghavan, V. (2020). Biochar from biomass waste as a renewable carbon material for climate change mitigation in reducing greenhouse gas emissions—A review. *Biomass Conversion and Biorefinery, 115*(11), 2247–2267. https://doi.org/10.1007/S13399-020-00604-5
8. Cha, J. S., Park, S. H., Jung, S. C., Ryu, C., Jeon, J. K., Shin, M. C., & Park, Y. K. (2016). Production and utilization of biochar: A review. *Journal of Industrial and Engineering Chemistry, 40*, 1–15. https://doi.org/10.1016/J.JIEC.2016.06.002

9. Maroušek, J., Vochozka, M., Plachý, J., & Žák, J. (2017). Glory and misery of biochar. *Clean Technologies and Environmental Policy, 19*, 311–317. https://doi.org/10.1007/S10098-016-1284-Y/METRICS
10. Kołtowski, M., Charmas, B., Skubiszewska-Zięba, J., & Oleszczuk, P. (2017). Effect of biochar activation by different methods on toxicity of soil contaminated by industrial activity. *Ecotoxicology and Environmental Safety, 136*, 119–125. https://doi.org/10.1016/J.ECOENV.2016.10.033
11. Nabais, J. M. V., Nunes, P., Carrott, P. J. M., Ribeiro Carrott, M. M. L., García, A. M., Díaz-Díez, M. A. (2008). Production of activated carbons from coffee endocarp by CO_2 and steam activation. *Fuel Processing Technology, 89,* 262–268. https://doi.org/10.1016/J.FUPROC.2007.11.030
12. Uchimiya, M., Pignatello, J. J., White, J. C., Hu, S. L., & Ferreira, P. J. (2017). Surface interactions between gold nanoparticles and biochar. *Science Reports, 71*(7), 1–9. https://doi.org/10.1038/s41598-017-03916-1
13. Bhuiya, M. M. K., Rasul, M. G., Khan, M. M. K., Ashwath, N., Azad, A. K., & Hazrat, M. A. (2014). Second generation biodiesel: potential alternative to-edible oil-derived biodiesel. *Energy Procedia, 61*, 1969–1972. https://doi.org/10.1016/J.EGYPRO.2014.12.054
14. Meher, L. C., Vidya Sagar, D., & Naik, S. N. (2006). Technical aspects of biodiesel production by transesterification—A review. *Renewable and Sustainable Energy Reviews, 10,* 248–268. https://doi.org/10.1016/J.RSER.2004.09.002
15. Demirbas, A. (2008). Comparison of transesterification methods for production of biodiesel from vegetable oils and fats. *Energy Conversion and Management, 49*, 125–130. https://doi.org/10.1016/J.ENCONMAN.2007.05.002
16. Dehkhoda, A. M., West, A. H., & Ellis, N. (2010). Biochar based solid acid catalyst for biodiesel production. *Applied Catalysis, A: General, 382*, 197–204. https://doi.org/10.1016/J.APCATA.2010.04.051
17. Freedman, B., Pryde, E. H., & Mounts, T. L. (1984). Variables affecting the yields of fatty esters from transesterified vegetable oils. *Journal of the American Oil Chemists Society, 61*, 1638–1643. https://doi.org/10.1007/BF02541649/METRICS
18. West, A. H., Posarac, D., & Ellis, N. (2008). Assessment of four biodiesel production processes using HYSYS plant. *Bioresource Technology, 99,* 6587–6601. https://doi.org/10.1016/J.BIORTECH.2007.11.046
19. Saka, S., & Kusdiana, D. (2001). Biodiesel fuel from rapeseed oil as prepared in supercritical methanol. *Fuel, 80*, 225–231. https://doi.org/10.1016/S0016-2361(00)00083-1
20. López, D. E., Goodwin, J. G., Bruce, D. A., & Lotero, E. (2005). Transesterification of triacetin with methanol on solid acid and base catalysts. *Applied Catalysis, A: General, 295*, 97–105. https://doi.org/10.1016/J.APCATA.2005.07.055
21. Zhang, Y., Dubé, M. A., McLean, D. D., & Kates, M. (2003). Biodiesel production from waste cooking oil: 1. Process design and technological assessment. *Bioresource Technology, 89*, 1–16. https://doi.org/10.1016/S0960-8524(03)00040-3
22. Lam, M. K., Lee, K. T., & Mohamed, A. R. (2010). Homogeneous, heterogeneous and enzymatic catalysis for transesterification of high free fatty acid oil (waste cooking oil) to biodiesel: A review. *Biotechnology Advances, 28*, 500–518. https://doi.org/10.1016/J.BIOTECHADV.2010.03.002
23. Canakci, M., & Van Gerpen, J. (1999). Biodiesel production via acid catalysis. *Transactions of the ASAE, 42,* 1203–1210. https://doi.org/10.13031/2013.13285
24. Warabi, Y., Kusdiana, D., & Saka, S. (2004). Reactivity of triglycerides and fatty acids of rapeseed oil in supercritical alcohols. *Bioresource Technology, 91*, 283–287. https://doi.org/10.1016/S0960-8524(03)00202-5
25. Ruatpuia, J. V. L., Halder, G., Mohan, S., Gurunathan, B., Li, H., Chai, F., Basumatary, S., & Lalthazuala, S. (2023). Microwave-assisted biodiesel production using ZIF-8 MOF-derived nanocatalyst : A process optimization, kinetics, thermodynamics and life cycle cost analysis. *Energy Conversion and Management, 292*, 117418. https://doi.org/10.1016/j.enconman.2023.117418

26. Furuta, S., Matsuhashi, H., & Arata, K. (2004). Biodiesel fuel production with solid superacid catalysis in fixed bed reactor under atmospheric pressure. *Catalysis Communications, 5*, 721–723. https://doi.org/10.1016/J.CATCOM.2004.09.001
27. Velusamy, K. Devanand, J., Senthil Kumar, P., Soundarajan, K., Sivasubramanian, V., Sindhu, J., & Vo, D. V. N. (2021). A review on nano-catalysts and biochar-based catalysts for biofuel production. *Fuel, 306*, 121632. https://doi.org/10.1016/J.FUEL.2021.121632
28. Jha, S., Nanda, S., Acharya, B., & Dalai, A. K. (2022). A review of thermochemical conversion of waste biomass to biofuels. *Energies, 15*, 6352. https://doi.org/10.3390/EN15176352
29. Han, T. U., Kim, Y. M., Watanabe, A., Teramae, N., Park, Y. K., & Kim, S. (2017). Pyrolysis kinetic analysis of poly(methyl methacrylate) using evolved gas analysis-mass spectrometry. *Korean Journal of Chemical Engineering, 34*, 1214–1221. https://doi.org/10.1007/S11814-016-0354-5/METRICS
30. Heidari, A., Stahl, R., Younesi, H., Rashidi, A., Troeger, N., & Ghoreyshi, A. A. (2014). Effect of process conditions on product yield and composition of fast pyrolysis of Eucalyptus grandis in fluidized bed reactor. *Journal of Industrial and Engineering Chemistry, 20*, 2594–2602. https://doi.org/10.1016/J.JIEC.2013.10.046
31. Wang, Z., Burra, K. G., Lei, T., & Gupta, A. K. (2021). Co-pyrolysis of waste plastic and solid biomass for synergistic production of biofuels and chemicals—A review. *Progress in Energy and Combustion Science, 84*, 100899. https://doi.org/10.1016/J.PECS.2020.100899
32. Ronsse, F., van Hecke, S., Dickinson, D., & Prins, W. (2013). Production and characterization of slow pyrolysis biochar: Influence of feedstock type and pyrolysis conditions. *GCB Bioenergy, 5*, 104–115. https://doi.org/10.1111/GCBB.12018
33. Bridgwater, A. V. (2012). Review of fast pyrolysis of biomass and product upgrading. *Biomass and Bioenergy, 38*, 68–94. https://doi.org/10.1016/J.BIOMBIOE.2011.01.048
34. Williams, P. T., & Besler, S. (1996). The influence of temperature and heating rate on the slow pyrolysis of biomass. *Renewable Energy, 7*, 233–250. https://doi.org/10.1016/0960-1481(96)00006-7
35. Bridgwater, A. V., Meier, D., & Radlein, D. (1999). An overview of fast pyrolysis of biomass. *Organic Geochemistry, 30*, 1479–1493. https://doi.org/10.1016/S0146-6380(99)00120-5
36. Wagner, A., & Kaupenjohann, M. (2014). Suitability of biochars (pyro- and hydrochars) for metal immobilization on former sewage-field soils. *European Journal of Soil Science, 65*, 139–148. https://doi.org/10.1111/EJSS.12090
37. Demirbaş, A. (2000). Mechanisms of liquefaction and pyrolysis reactions of biomass. *Energy Conversion and Management, 41*, 633–646. https://doi.org/10.1016/S0196-8904(99)00130-2
38. kyu Choi, M., Park, H. C., & Choi, H. S. (2018). Comprehensive evaluation of various pyrolysis reaction mechanisms for pyrolysis process simulation. *Chemical Engineering and Processing: Process Intensification, 130*, 19–35. https://doi.org/10.1016/J.CEP.2018.05.011
39. Lindfors, C., Kuoppala, E., Oasmaa, A., Solantausta, Y., & Arpiainen, V. (2014). Fractionation of bio-oil. *Energy and Fuels, 28*, 5785–5791. https://doi.org/10.1021/EF500754D
40. Yu, J., Lucas, J. A., & Wall, T. F. (2007). Formation of the structure of chars during devolatilization of pulverized coal and its thermoproperties: A review. *Progress in Energy and Combustion Science, 33*, 135–170. https://doi.org/10.1016/J.PECS.2006.07.003
41. Wang, T., Zhai, Y., Zhu, Y., Li, C., & Zeng, G. (2018). A review of the hydrothermal carbonization of biomass waste for hydrochar formation: Process conditions, fundamentals, and physicochemical properties. *Renewable and Sustainable Energy Reviews, 90*, 223–247. https://doi.org/10.1016/J.RSER.2018.03.071
42. Liu, W. J., Jiang, H., & Yu, H. Q. (2015). Development of biochar-based functional materials: Toward a sustainable platform carbon material. *Chemical Reviews, 115*, 12251–12285. https://doi.org/10.1021/acs.chemrev.5b00195
43. Tomczyk, A., Sokołowska, Z., & Boguta, P. (2020). Biochar physicochemical properties: Pyrolysis temperature and feedstock kind effects. *Reviews in Environmental Science & Biotechnology, 19*, 191–215. https://doi.org/10.1007/S11157-020-09523-3/TABLES/3
44. Cao, X., Sun, S., & Sun, R. (2017). Application of biochar-based catalysts in biomass upgrading: A review. *RSC Advances, 7*, 48793–48805. https://doi.org/10.1039/C7RA09307A

45. Kang, S., Li, X., Fan, J., & Chang, J. (2013). Hydrothermal conversion of lignin: A review. *Renewable and Sustainable Energy Reviews, 27*, 546–558. https://doi.org/10.1016/J.RSER.2013.07.013
46. Carpenter, D., Westover, T. L., Czernik, S., & Jablonski, W. (2014). Biomass feedstocks for renewable fuel production: A review of the impacts of feedstock and pretreatment on the yield and product distribution of fast pyrolysis bio-oils and vapors. *Green Chemistry, 16*, 384–406. https://doi.org/10.1039/C3GC41631C
47. Tekin, K., Karagöz, S., & Bektaş, S. (2014). A review of hydrothermal biomass processing. *Renewable and Sustainable Energy Reviews, 40*, 673–687. https://doi.org/10.1016/J.RSER.2014.07.216
48. Khan, T. A., Saud, A. S., Jamari, S. S., Rahim, M. H. A., Park, J. W., & Kim, H. J. (2019). Hydrothermal carbonization of lignocellulosic biomass for carbon rich material preparation: A review. *Biomass and Bioenergy, 130*, 105384. https://doi.org/10.1016/J.BIOMBIOE.2019.105384
49. Lucian, M., & Fiori, L. (2017). Hydrothermal carbonization of waste biomass: process design, modeling, energy efficiency and cost analysis. *Energies, 10*, 211. https://doi.org/10.3390/EN10020211
50. Cao, Y., He, M., Dutta, S., Luo, G., Zhang, S., & Tsang, D. C. W. (2021). Hydrothermal carbonization and liquefaction for sustainable production of hydrochar and aromatics. *Renewable and Sustainable Energy Reviews, 152*, 111722. https://doi.org/10.1016/J.RSER.2021.111722
51. Xiao, L. P., Shi, Z. J., Xu, F., & Sun, R. C. (2012). Hydrothermal carbonization of lignocellulosic biomass. *Bioresource Technology, 118*, 619–623. https://doi.org/10.1016/J.BIORTECH.2012.05.060
52. Román, S., Nabais, J. M. V., Laginhas, C., Ledesma, B., & González, J. F. (2012). Hydrothermal carbonization as an effective way of densifying the energy content of biomass. *Fuel Processing Technology, 103*, 78–83. https://doi.org/10.1016/J.FUPROC.2011.11.009
53. Chan, Y. H., Yusup, S., Quitain, A. T., Uemura, Y., & Sasaki, M. (2014). Bio-oil production from oil palm biomass via subcritical and supercritical hydrothermal liquefaction. *Journal of Supercritical Fluids, 95*, 407–412. https://doi.org/10.1016/J.SUPFLU.2014.10.014
54. Pinheiro Pires, A. P., Arauzo, J., Fonts, I., Domine, M. E., Fernández Arroyo, A., Garcia-Perez, M. E., Montoya, J., Chejne, F., Pfromm, P., Garcia-Perez, M. (2019). Challenges and opportunities for bio-oil refining: A review. *Energy & Fuels, 33*, 4683–4720. https://doi.org/10.1021/ACS.ENERGYFUELS.9B00039
55. Kruse, A. (2009). Hydrothermal biomass gasification. *Journal of Supercritical Fluids, 47*, 391–399. https://doi.org/10.1016/J.SUPFLU.2008.10.009
56. Funke, A., & Ziegler, F. (2010). Hydrothermal carbonization of biomass: A summary and discussion of chemical mechanisms for process engineering. *Biofuels, Bioproducts and Biorefining, 4*, 160–177. https://doi.org/10.1002/BBB.198
57. Coronella, C. J., Lynam, J. G., Reza, M. T., & Uddin, M. H. (2014). *Hydrothermal carbonization of lignocellulosic biomass* (pp. 275–311). https://doi.org/10.1007/978-3-642-54458-3_12
58. Yousuf, A., Pirozzi, D., & Sannino, F. (2020). Fundamentals of lignocellulosic biomass. *Lignocellulosic biomass to liquid biofuels* (pp. 1–15). https://doi.org/10.1016/B978-0-12-815936-1.00001-0
59. Alfattani, R., Shah, M. A., Siddiqui, M. I. H., Ali, M. A., & Alnaser, I. A. (2021). Bio-char characterization produced from walnut shell biomass through slow pyrolysis: sustainable for soil amendment and an alternate bio-fuel. *Energies, 15*, 1. https://doi.org/10.3390/EN15010001
60. Maj, I. (2022). Significance and challenges of poultry litter and cattle manure as sustainable fuels: A review. *Energies, 15*, 8981. https://doi.org/10.3390/EN15238981
61. Singh, A., Nanda, S., & Berruti, F. (2020). A review of thermochemical and biochemical conversion of *Miscanthus* to biofuels. In *Biorefinery of alternative resources: targeting green fuels and platform chemicals* (pp. 195–220). https://doi.org/10.1007/978-981-15-1804-1_9/COVER

62. O'Connor, J., Hoang, S. A., Bradney, L., Dutta, S., Xiong, X., Tsang, D. C. W., Ramadass, K., Vinu, A., Kirkham, M. B., & Bolan, N. S. (2021). A review on the valorisation of food waste as a nutrient source and soil amendment. *Environmental Pollution, 272*, 115985. https://doi.org/10.1016/J.ENVPOL.2020.115985
63. Singh, P., Sharma, S., & Dhanorkar, M. (2022). Aquatic plant biomass-derived porous carbon: Biomaterials for sustainable waste management and climate change mitigation. *International Journal of Environmental Science and Technology, 2022*, 1–16. https://doi.org/10.1007/S13762-022-04601-1
64. Ahmed, M. B., Zhou, J. L., Ngo, H. H., Guo, W., & Chen, M. (2016). Progress in the preparation and application of modified biochar for improved contaminant removal from water and wastewater. *Bioresource Technology, 214*, 836–851. https://doi.org/10.1016/J.BIORTECH.2016.05.057
65. Ghosh, N., Rhithuparna, D., Khatoon, R., Rokhum, S. L., & Halder, G. (2023). Sulphonic-acid functionalized novel *Limonia acidissima* carbonaceous catalyst for biodiesel synthesis from *Millettia pinnata* oil: optimization, kinetics, thermodynamics and cost analysis. *Journal of Cleaner Production, 394*, 136362. https://doi.org/10.1016/J.JCLEPRO.2023.136362
66. Rodríguez-Reinoso, F., Molina-Sabio, M., & González, M. T. (1995). The use of steam and CO_2 as activating agents in the preparation of activated carbons. *Carbon N. Y., 33*, 15–23. https://doi.org/10.1016/0008-6223(94)00100-E
67. Yorgun, S., Vural, N., & Demiral, H. (2009). Preparation of high-surface area activated carbons from Paulownia wood by $ZnCl_2$ activation. *Microporous and Mesoporous Materials, 122*, 189–194. https://doi.org/10.1016/J.MICROMESO.2009.02.032
68. Angin, D., Altintig, E., & Köse, T. E. (2013). Influence of process parameters on the surface and chemical properties of activated carbon obtained from biochar by chemical activation. *Bioresource Technology, 148*, 542–549. https://doi.org/10.1016/J.BIORTECH.2013.08.164
69. Azargohar, R., & Dalai, A. K. (2008). Steam and KOH activation of biochar: experimental and modeling studies. *Microporous and Mesoporous Materials, 110*, 413–421. https://doi.org/10.1016/J.MICROMESO.2007.06.047
70. Ros, A., Lillo-Ródenas, M. A., Fuente, E., Montes-Morán, M. A., Martín, M. J., & Linares-Solano, A. (2006). High surface area materials prepared from sewage sludge-based precursors. *Chemosphere, 65*, 132–140. https://doi.org/10.1016/J.CHEMOSPHERE.2006.02.017
71. Park, H., Kim, J., Lee, Y. G., & Chon, K. (2021). Enhanced adsorptive removal of dyes using mandarin peel biochars via chemical activation with NH_4Cl and $ZnCl_2$. *Water (Switzerland), 13*, 1495. https://doi.org/10.3390/W13111495/S1
72. Park, J. H., Ok, Y. S., Kim, S. H., Cho, J. S., Heo, J. S., Delaune, R. D., & Seo, D. C. (2015). Evaluation of phosphorus adsorption capacity of sesame straw biochar on aqueous solution: Influence of activation methods and pyrolysis temperatures. *Environmental Geochemistry and Health, 37*, 969–983. https://doi.org/10.1007/S10653-015-9709-9/TABLES/2
73. Tian, S., Yang, R., Pan, Z., Su, X., Li, S., Wang, P., & Huang, X. (2022). Anisotropic reed-stem-derived hierarchical porous biochars supported paraffin wax for efficient solar-thermal energy conversion and storage. *Journal of Energy Storage, 56*, 106153. https://doi.org/10.1016/J.EST.2022.106153
74. Venkatachalam, C. D., Sekar, S., Sengottian, M., Ravichandran, S. R., & Bhuvaneshwaran, P. (2023). A critical review of the production, activation, and morphological characteristic study on functionalized biochar. *Journal of Energy Storage, 67*, 107525. https://doi.org/10.1016/J.EST.2023.107525
75. Mullen, C. A., Boateng, A. A., Goldberg, N. M., Lima, I. M., Laird, D. A., & Hicks, K. B. (2010). Bio-oil and bio-char production from corn cobs and stover by fast pyrolysis. *Biomass and Bioenergy, 34*, 67–74. https://doi.org/10.1016/J.BIOMBIOE.2009.09.012
76. Deng, A., Lin, Q., Yan, Y., Li, H., Ren, J., Liu, C., & Sun, R. (2016). A feasible process for furfural production from the pre-hydrolysis liquor of corncob via biochar catalysts in a new biphasic system. *Bioresource Technology, 216*, 754–760. https://doi.org/10.1016/J.BIORTECH.2016.06.002

77. Lee, J., Kim, K. H., & Kwon, E. E. (2017). Biochar as a catalyst. *Renewable and Sustainable Energy Reviews, 77*, 70–79. https://doi.org/10.1016/J.RSER.2017.04.002
78. Kastner, J. R., Miller, J., Geller, D. P., Locklin, J., Keith, L. H., & Johnson, T. (2012). Catalytic esterification of fatty acids using solid acid catalysts generated from biochar and activated carbon. *Catalysis Today, 190*, 122–132. https://doi.org/10.1016/J.CATTOD.2012.02.006
79. Konwar, L. J., Boro, J., & Deka, D. (2014). Review on latest developments in biodiesel production using carbon-based catalysts. *Renewable and Sustainable Energy Reviews, 29*, 546–564. https://doi.org/10.1016/J.RSER.2013.09.003
80. Rafi, J. M., Rajashekar, A., Srinivas, M., Rao, B. V. S. K., Prasad, R. B. N., & Lingaiah, N. (2015). Esterification of glycerol over a solid acid biochar catalyst derived from waste biomass. *RSC Advances, 5*, 44550–44556. https://doi.org/10.1039/C5RA06613A
81. Wang, S., Zhao, C., Shan, R., Wang, Y., & Yuan, H. (2017). A novel peat biochar supported catalyst for the transesterification reaction. *Energy Conversion and Management, 139*, 89–96. https://doi.org/10.1016/J.ENCONMAN.2017.02.039
82. Balajii, M., & Niju, S. (2019). Biochar-derived heterogeneous catalysts for biodiesel production. *Environmental Chemistry Letters, 17*, 1447–1469. https://doi.org/10.1007/S10311-019-00885-X/TABLES/10
83. Li, M., Zheng, Y., Chen, Y., & Zhu, X. (2014). Biodiesel production from waste cooking oil using a heterogeneous catalyst from pyrolyzed rice husk. *Bioresource Technology, 154*, 345–348. https://doi.org/10.1016/J.BIORTECH.2013.12.070
84. Rokhum, S. L., Changmai, B., Kress, T., & Wheatley, A. E. H. (2022). A one-pot route to tunable sugar-derived sulfonated carbon catalysts for sustainable production of biodiesel by fatty acid esterification. *Renewable Energy, 184*. https://doi.org/10.1016/j.renene.2021.12.001
85. González, M. E., Cea, M., Reyes, D., Romero-Hermoso, L., Hidalgo, P., Meier, S., Benito, N., & Navia, R. (2017). Functionalization of biochar derived from lignocellulosic biomass using microwave technology for catalytic application in biodiesel production. *Energy Conversion and Management, 137*, 165–173. https://doi.org/10.1016/J.ENCONMAN.2017.01.063
86. Kostić, M. D., Bazargan, A., Stamenković, O. S., Veljković, V. B., & McKay, G. (2016). Optimization and kinetics of sunflower oil methanolysis catalyzed by calcium oxide-based catalyst derived from palm kernel shell biochar. *Fuel, 163*, 304–313. https://doi.org/10.1016/J.FUEL.2015.09.042
87. Bazargan, A., Kostić, M. D., Stamenković, O. S., Veljković, V. B., & McKay, G. (2015). A calcium oxide-based catalyst derived from palm kernel shell gasification residues for biodiesel production. *Fuel, 150*, 519–525. https://doi.org/10.1016/J.FUEL.2015.02.046
88. Dhawane, S. H., Kumar, T., & Halder, G. (2015). Central composite design approach towards optimization of flamboyant pods derived steam activated carbon for its use as heterogeneous catalyst in transesterification of *Hevea brasiliensis* oil. *Energy Conversion and Management, 100*, 277–287. https://doi.org/10.1016/j.enconman.2015.04.083
89. Yu, J. T., Dehkhoda, A. M., & Ellis, N. (2010). Development of biochar-based catalyst for transesterification of canola oil. *Energy and Fuels, 25*, 337–344. https://doi.org/10.1021/EF100977D
90. Li, M., Chen, D., & Zhu, X. (2013). Preparation of solid acid catalyst from rice husk char and its catalytic performance in esterification. *Chinese Journal of Catalysis, 34*, 1674–1682. https://doi.org/10.1016/S1872-2067(12)60634-2
91. Liu, T., Li, Z., Li, W., Shi, C., & Wang, Y. (2013). Preparation and characterization of biomass carbon-based solid acid catalyst for the esterification of oleic acid with methanol. *Bioresource Technology, 133*, 618–621. https://doi.org/10.1016/J.BIORTECH.2013.01.163
92. Lee, J., Jung, J. M., Oh, J. I., Ok, Y. S., Lee, S. R., & Kwon, E. E. (2017). Evaluating the effectiveness of various biochars as porous media for biodiesel synthesis via pseudo-catalytic transesterification. *Bioresource Technology, 231*, 59–64. https://doi.org/10.1016/J.BIORTECH.2017.01.067

Chapter 9
Versatile Pretreatment Approaches to Improve the Bioethanol Production from Various Biomass Feedstocks

R. Rajesh Kannan, V. Saravanan, M. Rajasimman, Panchamoorthy Saravanan, and Gurunathan Baskar

Abstract The result in climate change, fossil resource depletion, and transportation alternatives to petroleum are in great demand. The structural components belonging to lignocellulosic biomass like lignin, cellulose, hemicellulose and as well as technological unit phases like pretreatment and as well as hydrolysis of enzymes are discussed. The sole purpose of pretreatment stage is to increase the surface area of carbohydrate that is available for the enzymatic saccharification and also lower the inhibitor concentrations. The review focuses on bioethanol production from lignocellulosic biomass derived from plants. For synthesizing effective bioethanol, diverse lignocellulosic biomass like feedstocks of wood, agricultural wastes, and marine algae are used as substrates, and different enzyme techniques are used in hydrolysis protocols. This paper provides a detailed review in the synthesizing of lignocellulosic bioethanol via biochemical pathway while focusing on the widely used pretreatment technologies as well as important enzymatic hydrolysis and fermentation operational parameters in accordance to the yield of sugar and ethanol. Also, the importance of immobilization is discussed in this study as well as numerous detoxification processes for eliminating hazardous substances. According to this review, modified microbes could potentially be employed to boost fermentation of glucose and xylose, cellulolytic enzyme synthesis and also stress response.

Keywords Bioethanol · Pretreatment techniques · Lignocellulosic biomass · Sugar · Starch

R. Rajesh Kannan · V. Saravanan (✉) · M. Rajasimman
Department of Chemical Engineering, Annamalai University, Chidambaram 608002, Tamil Nadu, India
e-mail: sarav304@gmail.com

P. Saravanan
Department of Petrochemical Technology, UCE—BIT Campus, Anna University, Tiruchirappalli 620024, Tamil Nadu, India

G. Baskar
Department of Biotechnology, St. Joseph's College of Engineering, Chennai, Tamil Nadu, India

G. Baskar et al. (eds.), *Circular Bioeconomy Perspectives in Sustainable Bioenergy Production*, Energy, Environment, and Sustainability,
https://doi.org/10.1007/978-981-97-2523-6_9

9.1 Introduction

Renewable and sustainable energy sources which include bioethanol, biodiesel, biohydrogen, and methane are obtained from various raw resources to address the energy and environmental issues. Based on the raw materials used from a scientific standpoint, all biofuels are divided into four types [1]. Food-based agricultural waste namely starch (corn, wheat, cassava, sugarcane) yields biofuels which are categorized as first-generation biofuels since they are in competence with arable land as well as feed supply [2]. These generation biofuels will most certainly constitute a significant share of worldwide biofuel production in 2021. In the coming years, second-generation biofuels which are based on lignocellulosic-based materials could be a potential fuel supplement. Biofuels of the next generation include algal biofuels. Fourth generation of biofuels is made from genetically modified algae having high lipid content [3]. In 2023, global bioethanol output is predicted to reach 150 billion litres, from roughly 120 billion litres in 2019. The great majority of global bioethanol production is from the United States. Due to the biofuel policies of both the United States' and the European Union's which stimulate the development of cellulose-based biofuels worldwide, the cellulosic bioethanol commercialization from agricultural wastes is predicted to enhance second-involvement generations in the global bioethanol market [4–8].

9.1.1 First-Generation Bioethanol

First-generation bioethanol is a liquid biofuel generated from high-starch, high-sugar agricultural crops that are intended for use in automobiles [9]. Because *Saccharomyces cerevisiae* cannot break down complex carbohydrates, both starchy and lignocellulosic components must be hydrolyzed [10]. The synthesis of first-generation bioethanol and its raw materials are shown in Fig. 9.1. As a result, there are severe worries about large-scale production's social and environmental effects [11]. The production of first-generation biofuel demands for water and arable land with food production due to over fertilization and may lead to depletion of necessary resources which includes shortage of water availability and degradation of soil and water resources [12]. Also, food prices for grains, crops, vegetable oils, and livestock feed may rise as a result. Furthermore, emission of greenhouse gases is lower for biofuels than fossil fuel combustion [13]. Biodiesel which is created by transesterifying vegetable oils, aside from food-based bioethanol, is also another first-generation biofuel [9].

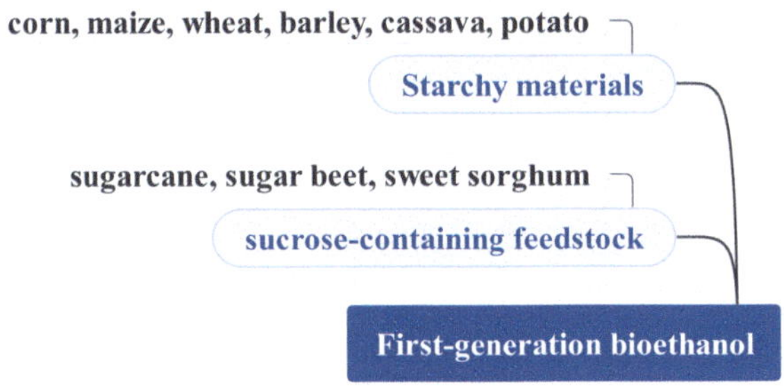

Fig. 9.1 Raw materials used for first generation of bioethanol production

9.1.2 Second-Generation Bioethanol

The biofuels of second and later generations including bioethanol do not compete with feed supply because they are created from non-food basic elements [14]. The primary feedstock used for the production of second-generation bioethanol is lignocellulosic biomass, however, industrial wastes like whey [10] or crude glycerol can also be employed. In most cases, such biomass is both cheap and plentiful in a given location [15–20]. The Lignocellulose is both a renewable and sustainable carbon source that can be found in a wide range of raw materials from plants [16, 21–28] as given in Fig. 9.2. The availability of amount of lignocellulosic biomass is predetermined by climate conditions. Converting the lignocellulose to reducing sugars is the most challenging compared to starch conversion [29–36]. Researchers have investigated the usage of many varieties of plant biomass in the production of biofuels (Fig. 9.3). The biofuels of second and later generations which includes bioethanol do not compete with the feed supply because they are created from non-feed basic elements [14].

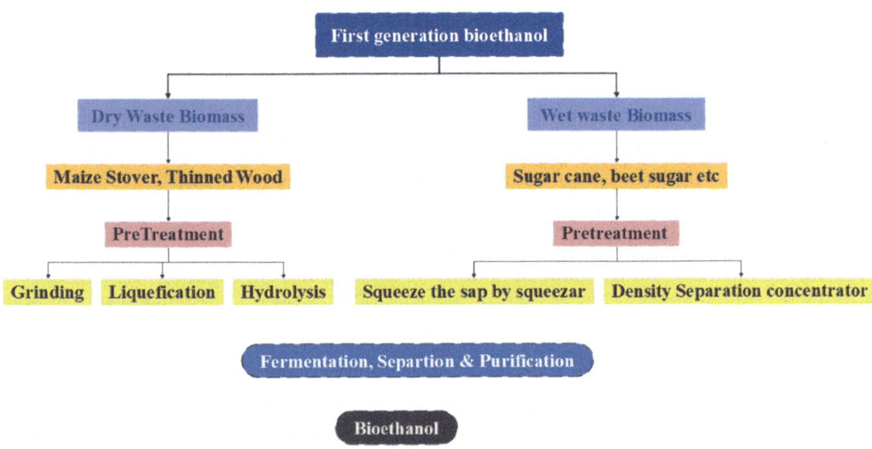

Fig. 9.2 The first-generation bioethanol production process

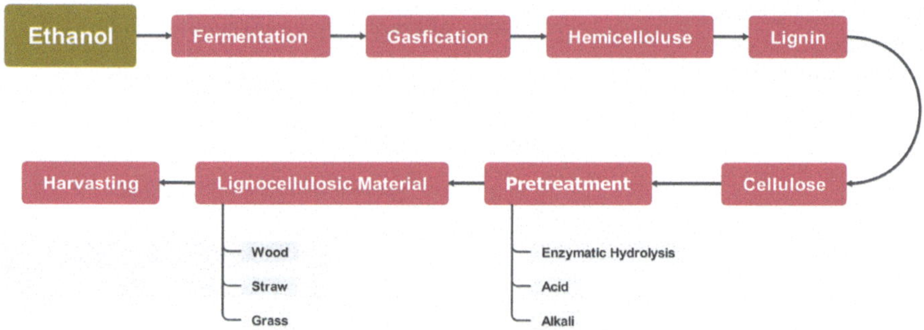

Fig. 9.3 Second-generation bioethanol manufacturing procedures

9.1.3 Third-Generation Bioethanol

The production of microalgae or unicellular microorganisms (such as *Cyanidium*, *Caldarium*, or *Synechococcus*) belonging to eukaryotes and prokaryotes (cyanobacteria) is the third generation of biofuels [36]. Active microalgal biomass such as live biocatalysts can use nutrients namely carbon, nitrogen, phosphate, or sulphur from industrial effluents as substrates to produce high biomass concentrations. Industrial power plant effluent gases, wastewater, organic waste hydrolysis products, and digestate are among the waste streams (waste from biogas production). As a result, manufacturing third-generation biofuels can aid in the reduction of waste streams from a wide range of industries. Algae, for example, are an excellent alternative fuel because of their high lipid and carbohydrate content. By alkaline transesterification with methanol, microalgae oils provide a possible raw material to vegetable oils commonly used in the production of biodiesel [34]. Technology developments such as enhanced oil extraction from cellular biomass are required for the manufacturing of biodiesel from microalgae. Microalgae residual biomass can be converted into environmentally friendly and energy carriers for a long-term (biofuels) like bioethanol and biogas as well as used for fermentation of methane and biohydrogen generation [36]. Anaerobic fermentation of starch as well as starch-like polysaccharides produces bioethanol from algal biomass. Starch is a conserved element in microalgae like *Cryptophyta* and *Chlorophyta* [36]. Despite the process's intricacy, yields are produced at a smaller rate for producing bioethanol from biomass. The by-product biohydrogen is a subsidiary of production of methane from manufactured organic acids during the acidogenic phase of anaerobic digestion from organic waste transformation [36]. Nevertheless, due to the biohydrogen amount manufactured by algae is currently restricted, the biohydrogen manufacturing process must be enhanced [36]. Algal cultivation is straightforward and can be done in various water environments [37].

The raw materials of bioethanol can be made from various origins of biomass. They usually fall into three categories as depicted in Fig. 9.4. The availability of feedstock containing sugar and starch (first generation) is, however, affected by their

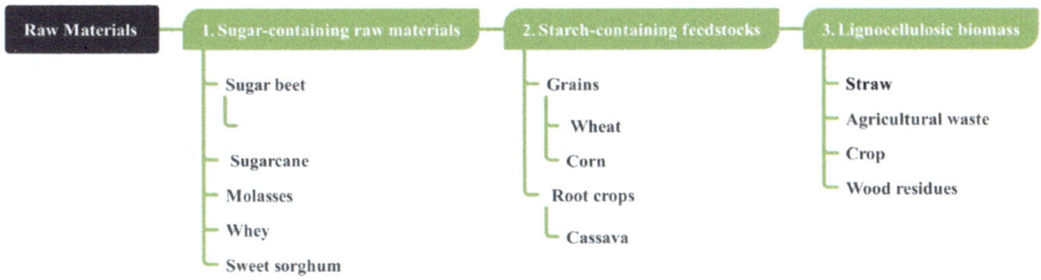

Fig. 9.4 Raw materials utilized in the manufacturing of bioethanol

usage as feed. A possible feedstock for the manufacture of bioethanol is Lignocellulosic biomass (second generation), due to its low capital, availability, and large dispersion. They are also not competitive with crops used for feed [27].

9.2 Raw Materials for Bioethanol Production

9.2.1 Sugar-Containing Raw Materials

Sugar cane and beet are the two most appropriate sugar-producing plants on earth. Sugar cane is responsible for global sugar output, while sugar beet is responsible for one-third of left [28] as given in Table 9.1. Most *Saccharomyces* species generate the enzyme invertase, which can efficiently hydrolyze them. As a result, bioethanol synthesis from sugar (sucrose) feedstocks does not require pretreatment, making it more practical than starch feedstocks [28]. For sugar extraction, sugar crops are needed to be milled and then transferred to a fermentation medium, where ethanol is extracted directly from the juice or molasses [29]. Table 9.1 shows the major achievement and investigation of ethanol production from various sources of sugar-containing raw materials and its composition.

9.2.2 Starch-Containing Raw Materials

The raw materials from starch for production of bioethanol are Cassava tubers comprised of a mass of 80% carbohydrate content and mass protein content of less than 1%. To process cassava tubers for bioethanol production, phases like dusting, flaking, fragmentation, and aeration are included. After that, the bioethanol is processed from dried cassava chips [5]. Linear (amylose) and branching (amylopectin) polyglucans both acting as a mixture makes up starch. -amylase is a starch hydrolase enzyme that works on the -1, 4 but not the -1, 6 links in amylopectin [37]. The hydrolysis of starch (mostly by -amylase and glucoamylase) into glucose

Table 9.1 Sugar-containing raw material for ethanol production and its chemical composition

Name of the crops	Major investigation	Major achievements	Freesugar (%)	Moisture (%)	References
Sugarcane (*Saccharum officinarum*)	Juices were studied (i) without adding supplement, (ii) with addition of 0.5% yeast extract, and (iii) with addition of yeast extract, thiamine, and micronutrients	The highest ethanol concentration (39.4–42.1 g/L) was found in (iii), while the lowest was 11.0 g/L and found in (i)	12–17.6	68.0–82.4	[31, 32]
	Enrichment technique was applied to isolate thermo tolerant yeast from cane juice for ethanol production at elevated temperature	Isolation and selection of thermo tolerant yeast strain that produced high concentrations of ethanol at both 40 and 45 °C from the juice			
	Juices were supplemented and cells were adapted to galactose medium	Higher (30.0%) ethanol was found with adapted cells than non-adapted cells			
Sugarbeet (*Beta vulgaris*)	Juices were ultrafiltered and were supplemented with mineral salt	Ethanol concentration was found to be 85.0–87.0 g/L	16.5	82.6	[8, 33]
	Flocculating and non-flocculating yeasts along with *Z. mobilis* were used through immobilization on loofa sponge	The highest ethanol yield (0.44 g/g) with the lowest productivity (0.08 g/h/L) was found by *Z. mobilis*			

(continued)

Table 9.1 (continued)

Name of the crops	Major investigation	Major achievements	Freesugar (%)	Moisture (%)	References
Sweet sorghum (*Sorghum bicolor*)	Five different genetic varieties (Keller, BJ 248, SSV84, Wray, and NSSH104) were investigated for ethanol production	Keller variety produced the highest ethanol (9.0%,w/v)	16–21.8	78.2	
Watermelon (*Citrullus lanatus*)	Juice was used as diluent, additive, and nitrogen source for processed sugar	As much as 25.0% w/v sugar was fermented at pH3.0 (ethanol yield was 0.41–0.46 g/g) or up to 35.0%w/v sugar at pH 5.0 with an ethanol yield of 0.36–0.41 g/g	70–10	90–93	

syrup is then converted to ethanol by yeast *Saccharomyces cerevisiae*, required for the production of bioethanol from starch-containing feedstocks. When compared with bioethanol synthesis from sugar-containing feedstocks, this step is more expensive [38]. Bacteria such as *Bacillus licheniformis* as well as genetically altered organisms produce α-amylase and glucoamylase [39, 40].

The *Saccharomyces cerevisiae* yeast transforms glucose to ethanol under anaerobic circumstances. In terms of mass conversion, the efficiency is about 51%, in the highest yield of glucose into ethanol. Glucose is needed for yeast for the development of cell and production of other metabolic products, and its efficiency in optimum conversion is also reduced. In practise, glucose is converted to ethanol at a mass conversion rate of 40–48% [41]. Starch-derived ethanol enhances enzyme application and yeast strains with high ethanol tolerance when compared to sugar-derived ethanol [42].

9.2.3 Lignocellulose-Containing Raw Materials

Due to being renewable and non-competitive with food crops, bioethanol production from lignocellulosic biomass is more appealing and sustainable. Additionally, the acquisition of usage of bioethanol from lignocellulosic biomass is associated with a large decrease in greenhouse gas emissions [44]. Unlike fossil fuels, lignocellulosic biomass is uniformly spread throughout the world, assuring supply security

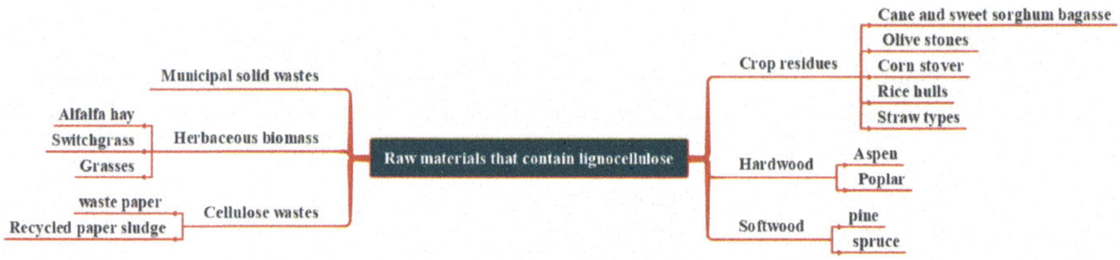

Fig. 9.5 Raw materials for bioethanol production comprise lignocellulosic materials

by utilizing local energy sources [38]. It can also be created using a wide variety of wastes or harvested from the forest and is often less expensive compared to feedstocks including sugar or starch, which necessitates a comprehensive agricultural breeding approach [3, 45]. Many different types of raw materials containing lignocellulosic components are depicted in Fig. 9.5.

Lignocellulosic biomass has 43% cellulose, 27% lignin, 20% hemicellulose, and 10% of other constituents when calculated for an average [3]. Figure 9.5 depicts the procedures where lignocellulosic biomass is required in the manufacture of bioethanol. Because lignocellulosic biomass has a more heterogeneous structure than the homogeneous and similar raw materials which is used in chemical industry, it needs more sophisticated chemical operations [46, 47]. Furthermore, lignocellulosic crop harvesting is often not possible all year, making biomass suppliers' duties more difficult [48–50]. As a result, this problem must be solved by the stabilization of biomass in such a way, especially for the availability of long-term storage and sustaining continuous biorefinery operation for the coming years [51–56].

9.3 Pretreatment Overview

Lignocellulosic biomass, such as wood, grass, agricultural, and forest residues, are potential resources for the production of bioethanol. The current biochemical process of converting biomass to bioethanol typically consists of three main steps: pretreatment, enzymatic hydrolysis, and fermentation. For this process, pretreatment is probably the most crucial step since it has a large impact on the efficiency of the overall bioconversion. The aim of pretreatment is to disrupt recalcitrant structures of cellulosic biomass to make cellulose more accessible to the enzymes that convert carbohydrate polymers into fermentable sugars [57–59]. Different pretreatment methods, including dilute acid pretreatment, steam explosion pretreatment, organosolv, liquid hot water, ammonia fibre expansion, soaking in aqueous ammonia, sodium hydroxide/lime pretreatments, and ozonolysis are intensively introduced and discussed. In this minireview, the key points are focused on the structural changes primarily in cellulose, hemicellulose, and lignin during the above-leading pretreatment technologies [60–68].

The pretreatment methods can be classified into different categories according to various criteria [69–75]. It is common to differentiate pretreatment methods that are efficient at high, low, or neutral pH. Basically, low pH methods require addition of acids to increase the hydrolytic capacity while higher pH methods need pH-adjusting agents such as sodium hydroxide or ammonia. The neutral pH methods, mainly liquid hot water (LHW) pretreatment, simply apply water in the process. However, the substrate medium in the neutral methods is weakly acidic due to the release of organic acids from the biomass during the pretreatments. Another way to do the grouping is based on the principal mechanism acting during pretreatment. As such, the methods can be classified as physical, physicochemical, and chemical pretreatments. In this review, mainly the first classification is used. The acidic pretreatments include dilute acid pretreatment, steam explosion pretreatment, and organosolv pretreatment. The neutral pretreatment is liquid hot water, and the sodium hydroxide/lime pretreatment and ammonia fibre expansion are classified into the alkaline pretreatments (Fig. 9.6). Besides the above methods, pretreatment of biomass with high-pressure ozone (ozonolysis) was recently developed in bioethanol production [70–79]. In this review, a brief process description is first given with recent developments for each pretreatment, followed by discussion of the technology's advantages and disadvantages. Different feedstocks, including herbaceous/agricultural residues, hardwood and softwood biomasses, applied in these technologies are highlighted (Figs. 9.7 and 9.8). The key points are focused on the structural changes primarily in cellulose, hemicellulose, and lignin during the above leading pretreatment technologies [6–10].

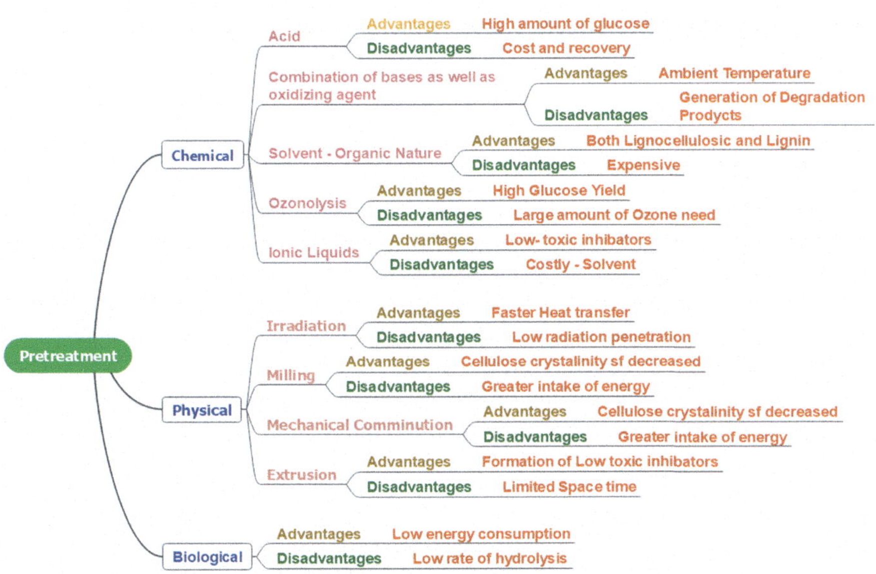

Fig. 9.6 The benefits and drawbacks of pretreatment of lignocellulose-containing raw materials for bioethanol production

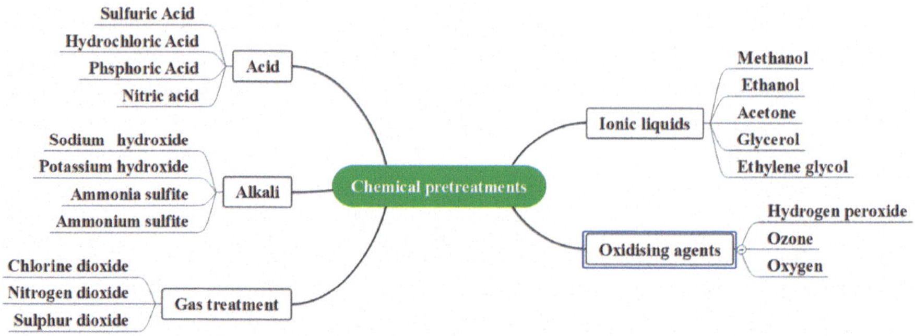

Fig. 9.7 Chemical treatment methods for bioethanol production

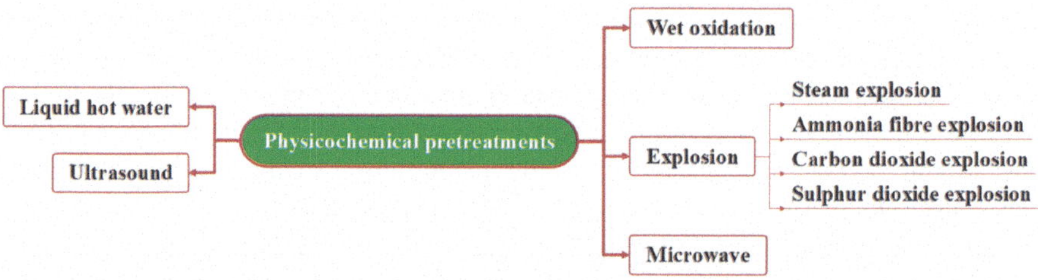

Fig. 9.8 Physicochemical processing types of lignocellulose-containing raw materials

9.3.1 Biological Processing of Lignocellulose-Containing Raw Materials

Comparing other existing pretreatments, biological pretreatments are regarded as illustrated in Fig. 9.9.

The employment of enzymes to hydrolyze raw lignocellulosic materials is also part of biological pretreatment. The rate at which lignocellulosic feedstock hydrolyzes into fermentable sugars determines the overall bioprocess efficiency [44]. Cellulose enzymatic hydrolysis by cellulase allows reducing sugars to be fermented into ethanol by the usage of yeasts or bacteria [57]. Endoglucanases, exoglucanases, and -glucosidases are the three major groups of cellulases involved in the hydrolysis of cellulose [9–12]. The two existing phases of enzymatic hydrolysis are primary and secondary. During the first hydrolysis stage, liberation of oligosacharides into liquid

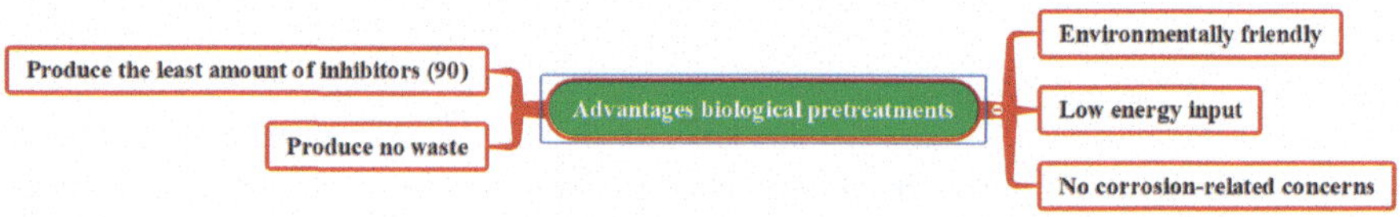

Fig. 9.9 Advantages of biological pretreatment for bioethanol production

phase are the work of both endoglucanases and exoglucanases on the solid substrate surface. During secondary hydrolysis step, Oligosacharides are further degraded to cellobiose and glucose [44].

9.4 Production of Bioethanol

9.4.1 Using Sugar-Containing Raw Materials

The microbe *Saccharomyces cerevisiae* is commonly used for the production of bioethanol from sugar-containing feedstocks because of its ability that break down sucrose to hexoses (glucose and fructose) (see Fig. 9.10). Because *S. cerevisiae* cells need oxygen of very small amounts for synthesizing fatty acid and sterol during bioethanol generation, important bioprocess parameter includes aeration (3–7). *S. cerevisiae* cannot bear increased concentrations of sugar and salts, as well as elevated temperatures in the medium. The media containing cane molasses exhibits high osmolarity, which limits ethanol formation, due to sugar and salt concentrations in the medium. A large quantity of tests had been done to develop strains of *S. cerevisiae* that are tolerant to high salt and temperature (7).

In the continuous bioprocesses for the production of bioethanol, concentration of immobilized cells is relatively high, and bioprocesses can be easily controlled at higher dilution rates, resulting in developed bioprocess productivities [41]. The immobilization approaches are classified by the following members as shown in Fig. 9.11.

Yeast cell trapping or mechanical separation is generally less effective when compared to surface adsorption immobilization. Although some of the system is

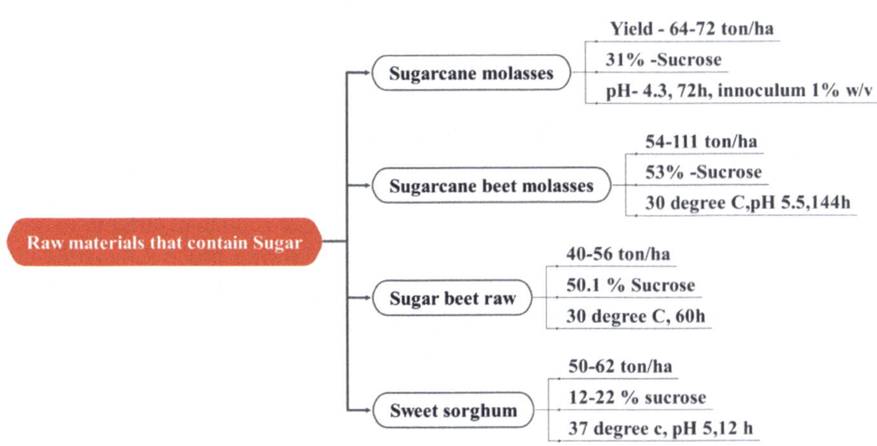

Fig. 9.10 The bioethanol production process using sugar as a starting material

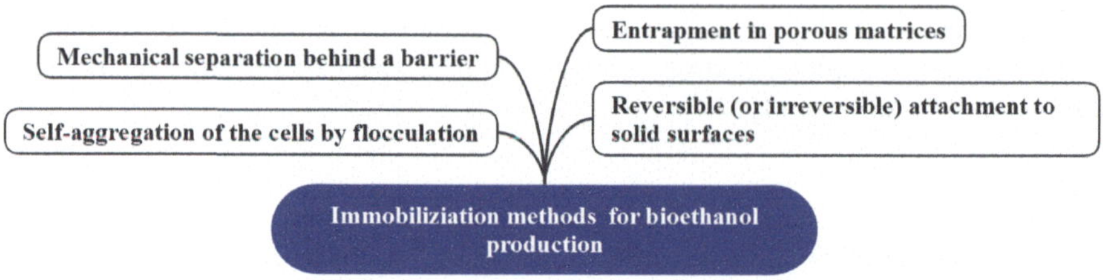

Fig. 9.11 Immobilization procedures employed in the synthesis of bioethanol

washed out of some yeast cells, surface adsorption studies have shown unaffected development in yeast cell [18]. Efficacy of self-flocculating yeast cells in producing bioethanol was comparable to yeast cells immobilized on supporting materials. Moreover, bioprocess is simpler as well as more cost-effective than when the yeast cell is immobilized on supporting materials due to no usage of supporting material.

9.4.2 Using Starch-Containing Raw Components

Yeasts (*Saccharomyces cerevisiae*, *Saccharomyces pastorianus*, *Schizosaccharomyces pombe*, and *Kluyveromyces* sp., etc.) that are competent in metabolizing starch hydrolysates are used in the dry-grind (67%) and wet-mill (33%) processes for bioethanol synthesis from maize starch, respectively [11]. Production of bioethanol in the United States often uses dry milling due to its operating costs and lower capital [11]. The whole corn is ground and formed into a mash by combining it with water. Furthermore, in a jet cooker which is set to 80–90 °C, the mash is cooked for 15–20 min. -amylase is used to enhance liquefaction during jet cooking (in very small doses). In secondary liquefaction, more -amylase is added, which takes 90 min at 95°.

A few adjustments to the dry-grind process were made before fermentation, to retrieve the maize germ and fibre [19]. A variety of value-added coproducts (such as fibre, germ, starch, and gluten) are produced using wet milling before fermentation, making the process energy-efficient and cost-effective [11]. The requirements of wet milling procedures are clean, steeped, and degermed maize to efficiently procure the extraction of oil from corn germ. The defibration of maize is done in order to extract the fibres, gluten, and starch. The bioethanol manufacturing process includes the same phases as the dry-grind method [12]. The yield of wet-mill is given as 0.2919, while the dry-grind yield is 0.3235 per kg of corn [11]. For bioethanol production, the technology which includes the growth of yeast (from active dry yeasts) in the bioreactor during early saccharification is also used. Simultaneous saccharification, yeast proliferation, and fermentation are some of the names used for this method (SSYPF) [12].

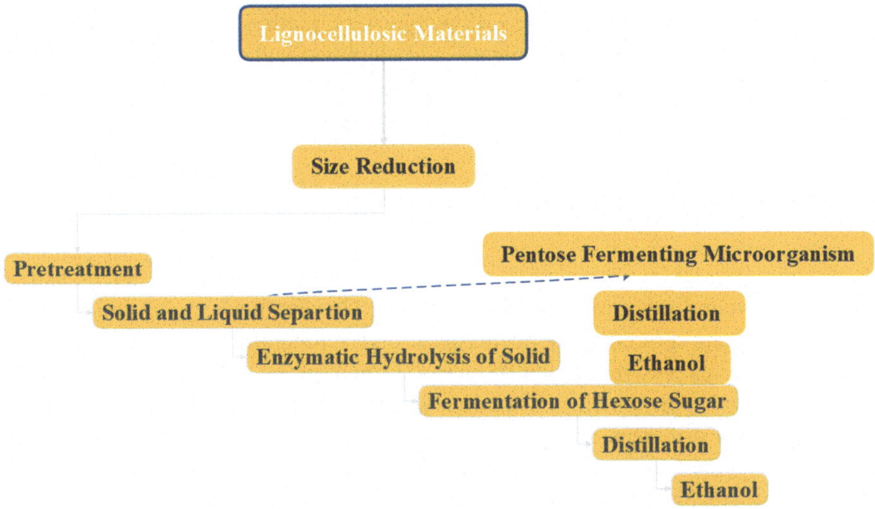

Fig. 9.12 The steps required in producing bioethanol from lignocellulosic raw materials

9.4.3 Using Lignocellulose-Containing Raw Materials

The conversion of lignocellulosic raw materials to fermentable sugars has been an extensive research subject in the biorefining field. Despite the high level of interest and advancement in lignocellulosic bioethanol research and development, several hurdles remain [26, 56]. Figure 9.12 depicts the utilized processes in the manufacture of bioethanol using raw materials which contains lignocellulose.

The lignocellulosic raw material pretreatment for the production of bioethanol was covered in the preceding section. As a result, the fermentation of lignocellulosic hydrolysates will be the emphasis of this section. During fermentation of glucose to ethanol, a number of microorganisms are used, the most common of which being the *Saccharomyces cerevisiae*. Unlike monosaccharides and disaccharides, pentoses are incapable of being broken down. Furthermore, *S. cerevisiae* is unable to digest cellulose and hemicellulose directly [23]. Recombinant DNA technology or pentose-fermenting bacteria (e.g. *Pichia stipitis*, *Pachysolen tannophilus*, and *Candida shehatae*) is a certain procedure utilized to circumvent this constraint [22]. These pentose-fermenting bacteria produce ethanol at a rate, at least five times slower than *S. cerevisiae* from glucose. Furthermore, the tolerance is 2–4 times lower for these microbes in the presence oxygen and ethanol [21–23].

Arabinose-metabolizing genes obtained from yeasts like *Candida aurigiensis* have been inserted into *S. cerevisiae* and, and as a result, there have recently been a lot of efforts to find an optimal microbe which produces ethanol directly from any form of carbohydrate [11].

The efficiency of ethanol extraction using vegetable oils, such as coconut, olive, safflower, and castor oil, and their derivatives, alcohols, and esters was also examined.

These oils were compared with the following esters: methyl laurate, methyl oleate, methyl linoleate, and methyl ricinoleate, and alcohols: lauryl (1-dodecanol), oleyl, and ricinoleyl.

9.5 Future Direction and Perspective

Ethanol is a fundamental chemical product, an important solvent, an industrial building block, and a promising renewable fuel. The chemical equilibrium conversion of the hydration of ethylene decreases at high temperatures to increase the rate of reaction. The excessive amount of energy to heat gases generates high pressure and utilizes crude oil, which is a nonrenewable resource. Ethanol synthesis from biomass residue produces a huge amount of CO_2 with a high cost of operation because of expensive machinery and fuel. The ethanol production from total sugars in lignocellulosic materials is inhibited by the action of pentose sugars, which are not fermentable by the brewer's yeast, i.e. *S. cerevisiae*, during the hydrolysis of hemicellulose. Although the fermentation route has been commercially realized, the cost of operation for this process is expensive due to the energy-intensive distillation steps at high dilution rates, which allow for high productivities, and the incomplete utilization of substrate, resulting in low yields to meet the market demand. Creating highly operative and selective nanocatalysts for the electrochemical CO_2 reduction reaction remains a major issue in the electrochemical CO_2 reduction process. The reaction mechanism is also closely associated with multiple catalytic active sites controlling every elementary reaction over catalysts. Thus, catalysts for optimum active sites should be designed and prepared to improve CO_2 conversion and ethanol selectivity. The formation rate of ethanol increases with increasing reaction temperature because thermodynamic equilibrium is still not reached. However, many challenges, e.g. the control of the ratio of multiple active sites, regulation of interface sites, and device of catalyst structures need to be explored in future research to improve the conversion and selectivity in ethanol production.

9.6 Conclusions

Bioethanol appears to be a viable solution to the current fuel crisis based on the information presented. Bioethanol separation and purification, as well as renewable biomass pretreatment, cellulase production, and sugar co-fermentation (pentose and hexose), have all made significant progress in recent years. But bioethanol is still not cost-competitive with the fossil fuels (the only exception being bioethanol produced from sugar cane in Brazil). Furthermore, the difficult issue at hand is the lowering of cost of bioethanol production. As a result, the concept of biorefinery is required to better utilize renewable feedstocks and to produce additional value-added coproducts (such as bio-based materials from lignin) that minimize the bioethanol production cost. As a result, the comparison of bioethanol to fossil fuels proves that bioethanol will be a more cost-effective one. Based on the presented data, it is obvious that bioethanol can be an alternative solution for the current fuel issue. The biggest challenge remains how to reduce the production cost of bioethanol. Therefore, the biorefinery concept is needed to utilize renewable feedstocks more comprehensively and to manufacture more value-added coproducts (*e.g.* bio-based materials from the lignin) that would reduce the cost of bioethanol production. This will make bioethanol more economically competitive than fossil fuels.

References

1. Bhaskar, T., Bhavya, B., Singh, R., Naik, D. V., Kumar, A., & Goyal, H. B. (2011). Thermochemical conversion of biomass to biofuels. In A. Pandey, C. Larroche, S. C. Ricke, C. G. Dussap, & E. Gnansounou (Eds.), *Biofuels—Alternative feedstocks and conversion processes* (pp. 51–77). Academic Press. https://doi.org/10.1016/B978-0-12-385099-7.00003-6
2. Forster, P., Ramaswamy, V., Artaxo, P., Berntsen, T., Betts, R., & Fahey, D. W. (2007). Changes in atmospheric constituents and in radiative forcing. In S. Solomon, D. Qin, M. Manning, Z. Chen, M. Marquis, & K. B. Averyt (Eds.), *Climate change 2007: The physical science basis* (pp. 129–234). Cambridge University Press.
3. Cherubini, F., & Strømman, A. H. (2011). Principles of biorefining. In A. Pandey, C. Larroche, S. C. Ricke, C. G. Dussap, & E. Gnansounou (Eds.), *Biofuels—Alternative feedstocks and conversion processes* (pp. 3–24). Academic Press. https://doi.org/10.1016/B978-0-12-385099-7.00001-2
4. Lange, J. P. (2007). Lignocellulose conversion: An introduction to chemistry, process and economics. *Biofuels, Bioproducts and Biorefining, 1*(1), 39–48. https://doi.org/10.1002/bbb
5. Khanal, S. K. (2008). *Anaerobic biotechnology for bioenergy production: Principles and applications*. Wiley-Blackwell. https://doi.org/10.1002/9780813804545
6. Yüksel, F., & Yüksel, B. (2004). The use of ethanol–gasoline blend as a fuel in an SI engine. *Renewable Energy, 29*(7), 1181–1191. https://doi.org/10.1016/j.renene.2003.11.012
7. World Bioenergy Association. (2017). *WBA Global Bioenergy Statistics*. www.worldbioenergy.org/global-bioenergy-statistics
8. Roadmap, B. (2012). *German Federal Government action plans for the material and energetic utilisation of renewable raw materials*. https://www.bmbf.de/pub/Roadmap_Biorefineries_eng.pdf
9. Kamm, B., Kamm, M., Gruber, P. R., & Kromus, S. (2008). *Biorefinery—Industrial processes and products: Status quo and future directions* (pp. 1–40). Weinheim: Wiley-VCH Verlag GmbH. https://doi.org/10.1002/9783527619849
10. Cherubini, F. (2010). The biorefinery concept: Using biomass instead of oil for producing energy and chemicals. *Energy Conversion and Management, 51*(7), 1412–1421. https://doi.org/10.1016/j.enconman.2010.01.015
11. Grilc, M., & Likozar, B. (2017). Levulinic acid hydrodeoxygenation, decarboxylation and oligmerization over NiMo/Al_2O_3 catalyst to bio-based value-added chemicals: Modelling of mass transfer, thermodynamics and micro-kinetics. *ChemEng J, 330*, 383–397. https://doi.org/10.1016/j.cej.2017.07.145
12. Huš, M., Bjelić, A., Grilc, M., & Likozar, B. (2018). First-principles mechanistic study of ring hydrogenation and deoxygenation reactions of eugenol over Ru(0001) catalysts. *Journal of Catalysis, 358*, 8–18. https://doi.org/10.1016/j.jcat.2017.11.020
13. Bjelić, A., Grilc, M., & Likozar, B. (2018). Catalytic hydrogenation and hydrodeoxygenation of lignin-derived model compound eugenol over Ru/C: Intrinsic microkinetics and transport

phenomena. *Chemical Engineering Journal, 333*, 240–259. https://doi.org/10.1016/j.cej.2017.09.135
14. Kromus, S., Wachter, B., Koschuh, W., Mandl, M., Krotscheck, C., & Narodoslawsky, M. (2004). The green biorefinery Austria—Development of an integrated system for green biomass utilization. *Chemical Biochemical Engineering Quartely, 18*(1), 7–12.
15. Koller, M., Bona, R., Hermann, C., Horvat, P., Martinz, J., & Neto, J. (2005). Biotechnological production of poly(3-hydroxybutyrate) with Wautersia eutropha by application of green grass juice and silage juice as additional complex substrates. *Biocatalysis Biotransformation, 23*(5), 329–337. https://doi.org/10.1080/10242420500292252
16. Khuong, L. S., Masjuki, H. H., Zulkifli, N. W. M., NizaMohamad, E., Kalam, M. A., & Alabdulkarem, A. (2017). Effect of gasoline-bioethanol blends on the properties and lubrication characteristics of commercial engine oil. *RSC Advances, 7*(25), 15005–15019. https://doi.org/10.1039/C7RA00357A
17. OECD/Food and Agriculture Organization of the United Nations. (2015). *OECD-FAO Agricultural Outlook 2015*. OECD Publishing.
18. European Biofuels Technology Platform. (2010). *Strategic research agenda 2010 update: Innovation driving sustainable biofuels*. http://www.biofuelstp.eu/srasdd/SRA_2010_update_web.pdf
19. Van Thuijl, E., & Deurwaarder, E. P. (2006). *European biofuels policies in retrospect. Petten, the Netherlands: Energy Research Centre of the Netherlands (ECN)*. https://www.ecn.nl/publications/PdfTetch.aspx?nr=ECN-C-06-016
20. Prieur-Vernat, A., & His, S. (2007). *Biofuels worldwide*. Panorama 2007. IFP-Innovation Energy Environment. http://www.firp.ula.ve/archivos/material_web_4xx/07_IFP_Biofuels.pdf
21. Li, K., Liu, S., & Liu, X. (2014). An overview of algae bioethanol production. *International Journal of Energy Research, 38*(8), 965–977. https://doi.org/10.1002/er.3164
22. Balat, M., & Balat, H. (2009). Recent trends in global production and utilization of bio-ethanol fuel. *Applied Energy, 86*(11), 2273–2282. https://doi.org/10.1016/j.apenergy.2009.03.015
23. Festel, G., Würmseher, M., Rammer, C., Boles, E., & Bellof, M. (2013) *Modelling production cost scenarios for biofuels and fossil fuels in Europe*. Discussion paper No. 13–075. ZEW—Centre for European Economic Research. http://ftp.zew.de/pub/zew-docs/dp/dp13075.pdf
24. Zhao, L., Zhang, X., Xu, J., Ou, X., Chang, S., & Wu, M. (2015). Techno-economic analysis of bioethanol production from lignocellulosic biomass in China: Dilute-acid pretreatment and enzymatic hydrolysis of corn stover. *Energies, 8*(5), 4096–4117. https://doi.org/10.3390/en8054096
25. What determines retail prices for gasoline and diesel? Volta Oil (2018). https://www.voltaoil.com/what-makes-up-retail-price-for-gasoline/
26. Mussatto, S. I., Dragone, G., Guimarães, P. M. R., Silva, J. P. A., Carneiro, L. M., & Roberto, I. C. (2010). Technological trends, global market, and challenges of bio-ethanol production. *Biotechnology Advances, 28*(6), 817–830. https://doi.org/10.1016/j.biotechadv.2010.07.001
27. Tomás-Pejó, E., Alvira, P., Ballesteros, M., & Negro, M. J. (2011). Pretreatment technologies for lignocellulose-to-bioethanol conversion. In A. Pandey, C. Larroche, S. C. Ricke, C. G. Dussap, & E. Gnansounou (Eds.), *Biofuels—Alternative feedstocks and conversion processes* (pp. 149–176). Academic Press. https://doi.org/10.1016/B978-0-12-385099-7.00007-3
28. Linoj, K. N. V., Dhavala, P., Goswami, A., & Maithel, S. (2006). Liquid biofuels in South Asia: Resources and technologies. *Asian Biotechnology and Development Review, 8*(2), 31–49.
29. İçöz, E., Tuğrul, K. M., Saral, A., & İçöz, E. (2008). Research on ethanol production and use from sugar beet in Turkey. *Biomass and Bioenergy, 33*(1), 1–7. https://doi.org/10.1016/j.biombioe.2008.05.005
30. Tabak, J. (2009). *Biofuels*. Infobase Publishing.
31. Pavlečić, M., Vrana, I., Vibovec, K., IvančićŠantek, M., Horvat, P., & Šantek, B. (2010). Ethanol production from different intermediates of sugar beet processing. *Food Technology and Biotechnology, 48*(3), 362–367.

32. Pavlečić, M., Rezić, T., IvančićŠantek, M., Horvat, P., & Šantek, B. (2017). Bioethanol production from raw sugar beet cossettes in horizontal rotating tubular bioreactor. *Bioprocess and Biosystems Engineering, 40*(11), 1679–1688. https://doi.org/10.1007/s00449-017-1823-x
33. Senthilkumar, V., & Gunasekaran, P. (2009). Bioethanol from biomass production of ethanol from molasses. In A. Pandey (Ed.), *Handbook of plant-based biofuels* (pp. 73–86). CRC Press.
34. Manikandan, K., Saravanan, V., & Viruthagiri, T. (2008). Kinetics studies on ethanol production from banana peel waste using mutant strain of *Saccharomyces cerevisiae. Indian Journal of Biotechnology, 1*(7), 83–88.
35. Jobling, S. (2004). Improving starch for food and industrial applications. *Current Opinion in Plant Biology, 7*(2), 210–218. https://doi.org/10.1016/j.pbi.2003.12.001
36. Solomon, B. D., Barnes, J. R., & Halvorsen, K. E. (2007). Grain and cellulosic ethanol: History, economics, and energy policy. *Biomass and Bioenergy, 31*(6), 416–425. https://doi.org/10.1016/j.biombioe.2007.01.023
37. Mousdale, D. M. (2008). *Biofuels. Biotechnology, chemistry and sustainable development.* CRC Press.
38. Soccol, C. R., Faraco, V., Karp, S., Vandenberghe, L. P. S., Thomaz-Soccol, V., & Woiciechowski, A. (2011). Lignocellulosic bioethanol: Current status and future perspectives. In A. Pandey, C. Larroche, S. C. Ricke, C. G. Dussap, & E. Gnansounou (Eds.), *Biofuels—Alternative feedstocks and conversion processes* (pp. 101–122). Academic Press. https://doi.org/10.1016/B978-0-12-385099-7.00005-X
39. Pandey, A., Nigam, P., Soccol, C. R., Soccol, V. T., Singh, D., & Mohan, R. (2000). Advances in microbial amylases. *Biotechnology and Applied Biochemistry, 31*(Pt 2), 135–152. https://doi.org/10.1042/BA19990073
40. Shigechi, H., Koh, J., Fujita, Y., Matsumoto, T., Bito, Y., & Ueda, M. (2004). Direct production of ethanol from raw corn starch via fermentation by use of a novel surface-engineered yeast strain codisplaying glucoamylase and α-amylase. *Applied and Environment Microbiology, 70*(8), 5037–5040. https://doi.org/10.1128/AEM.70.8.5037-5040.2004
41. Lee, S., Speight, J. G., & Loyalka, S. K. (Eds.) (2007). *Handbook of alternative fuel technologies.* CRC Press. https://doi.org/10.1201/9781420014518
42. Schubert, C. (2006). Can biofuels finally take center stage? *Nature Biotechnology, 24*, 777–784. https://doi.org/10.1038/nbt0706-777
43. Tanadul, O. U., VanderGheynst, J. S., Beckles, D. M., Powell, A. L. T., & Labavitch, J. M. (2014). The impact of elevated CO_2 concentration on the quality of algal starch as a potential biofuel feedstock. *Biotechnology and Bioengineering, 111*(7), 1323–1331. https://doi.org/10.1002/bit.25203
44. Binod, P., Sindhu, R., Singhania, R. R., Vikram, S., Devi, L., & Nagalakshmi, S. (2010). Bioethanol production from rice straw: An overview. *Bioresource Technology, 101*(13), 4767–4774. https://doi.org/10.1016/j.biortech.2009.10.079
45. Quintero, J. A., Rincón, L. E., & Cardona, C. A. (2011). Production of bioethanol from agroindustrial residues as feedstocks. In A. Pandey, C. Larroche, S. C. Ricke, C. G. Dussap, & E. Gnansounou (Eds.), *Biofuels—Alternative feedstocks and conversion processes* (pp. 251–285). Academic Press. https://doi.org/10.1016/B978-0-12-385099-7.00011-5
46. Dale, B. E., & Kim, S. (2010). Biomass refining global impact—The biobased economy of the 21st century. In B. Kamm, P. R. Gruber, & M. Kamm (Eds.), *Biorefineries—Industrial processes and products (Status quo and future directions)* (pp. 41–66). Wiley-VCH.
47. Hatti-Kaul, R. (2010). Biorefineries—A path to sustainability? *Crop Science* 50(Suppl_1):S152–S156. https://doi.org/10.2135/cropsci2009.10.0563
48. Mansfield, S. D., Mooney, C., & Saddler, J. N. (1999). Substrate and enzyme characteristics that limit cellulose hydrolysis. *Biotechnology Progress, 15*(5), 804–816. https://doi.org/10.1021/bp9900864
49. Kumar, R., & Wyman, C. E. (2010). Key features of pretreated lignocellulosic biomass solids and their impact on hydrolysis. In K. Waldron (Ed.), *Bioalcohol production—Biochemical conversion of lignocellulosic biomass. Series in Energy: Number 3* (pp. 73–121). Woodhead Publishing.

50. Laureano-Perez, L., Teymouri, F., Alizadeh, H., & Dale, B. E. (2005). Understanding factors that limit enzymatic hydrolysis of biomass. *Applied Biochemistry Biotechnology, 121–124*(1–3), 1081–1099. https://doi.org/10.1385/ABAB:124:1-3:1081
51. Mosier, N., Wyman, C., Dale, B. D., Elander, R., Lee, Y. Y., & Holtzapple, M. (2005). Features of promising technologies for pretreatment of lignocellulosic biomass. *Bioresource Technology, 96*(6), 673–686. https://doi.org/10.1016/j.biortech.2004.06.025
52. Pan, X., Xie, D., Gilkes, N., Gregg, D. J., & Saddler, J. N. (2005). Strategies to enhance the enzymatic hydrolysis of pretreated softwood with high residual lignin content. *Applied Biochemistry Biotechnology, 121–124*, 1069–1079. https://doi.org/10.1385/ABAB:124:1-3:1069
53. Chandra, R. P., Bura, R., Mabee, W. E., Berlin, A., Pan, X., & Saddler, J. N. (2007). Substrate pretreatment: The key to effective enzymatic hydrolysis of lignocellulosics? In: L. Olsson (Ed.) *Biofuels: Advances in biochemical engineering/biotechnology* (vol. 108, pp. 67–93). Springer. https://doi.org/10.1007/10_2007_064
54. Huang, H. J., Ramaswamy, S., Tschirner, U. W., & Ramarao, B. V. (2008). A review of separation technologies in current and future biorefineries. *Separation and Purification Technology, 62*(1), 1–21. https://doi.org/10.1016/j.seppur.2007.12.011
55. Kucera, D., Benesova, P., Ladicky, P., Pekar, M., Sedlacek, P., & Obruca, S. (2017). Production of polyhydroxyalkanoates using hydrolyzates of spruce sawdust: Comparison of hydrolyzates detoxification by application of overliming, active carbon, and lignite. *Bioengineering (Basel), 4*(2), 53. https://doi.org/10.3390/bioengineering4020053
56. Taherzadeh, M. J., & Karimi, K. (2008). Pretreatment of lignocellulosic wastes to improve ethanol and biogas production: A review. *Int J MolSci, 9*(9), 1621–1651. https://doi.org/10.3390/ijms9091621
57. Sun, Y., & Cheng, J. (2002). Hydrolysis of lignocellulosic materials for ethanol production: A review. *Bioresource Technology, 83*(1), 1–11. https://doi.org/10.1016/S0960-8524(01)00212-7
58. Talebnia, F., Karakashev, D., & Angelidaki, I. (2010). Production of bioethanol from wheat straw: An overview on pretreatment, hydrolysis and fermentation. *Bioresource Technology, 101*(13), 4744–4753. https://doi.org/10.1016/j.biortech.2009.11.080
59. Karunanithy, C. (2010). Effect of extruder parameters and moisture content of switchgrass, prairie cord grass on sugar recovery from enzymatic hydrolysis. *Applied Biochemistry Biotechnology, 162*(6), 1785–1803. https://doi.org/10.1007/s12010-010-8959-3
60. Yoo, J., Alavi, S., Vadlani, P., & Amanor-Boadu, V. (2011). Thermo-mechanical extrusion pretreatment for conversion of soybean hulls to fermentable sugars. *Bioresource Technology, 102*(16), 7583–7590. https://doi.org/10.1016/j.biortech.2011.04.092
61. Kootstra, A. M. J., Beeftink, H. H., Scott, E. L., & Sanders, J. P. M. (2010). Comparison of dilute mineral and organic acid pretreatment for enzymatic hydrolysis of wheat straw. *Biochemical Engineering Journal, 46*(2), 126–131. https://doi.org/10.1016/j.bej.2009.04.020
62. Carvalheiro, F., Duarte, L. C., & Gírio, F. M. (2008). Hemicellulose biorefineries: A review on biomass pretreatments. *Journal of Scientific and Industrial Research (India), 67*, 849–864.
63. Cheng, Y. S., Zheng, Y., Yu, C. W., Dooley, T. M., Jenkins, B. M., & VanderGheynst, J. S. (2010). Evaluation of high solids alkaline pretreatment of rice straw. *Applied Biochemistry and Biotechnology, 162*(6), 1768–1784. https://doi.org/10.1007/s12010-010-8958-4
64. Kumar, P., Barrett, D. M., Delwiche, M. J., & Stroeve, P. (2009). Methods for pretreatment of lignocellulosic biomass for efficient hydrolysis and biofuel production. *Industrial and Engineering Chemistry Research, 48*(8), 3713–3729. https://doi.org/10.1021/ie801542g
65. García-Cubero, M. A., González-Benito, G., Indacoechea, I., Coca, M., & Bolado, S. (2009). Effect of ozonolysis pretreatment on enzymatic digestibility of wheat and rye straw. *BioresourceTechnology, 100*(4), 1608–1613. https://doi.org/10.1016/j.biortech.2008.09.012
66. Silverstein, R. A., Chen, Y., Sharma-Shivappa, R. R., Boyette, M. D., & Osborne, J. (2007). A comparison of chemical pretreatment methods for improving saccharification of cotton stalks. *Bioresource Technology, 98*(16), 3000–3011. https://doi.org/10.1016/j.biortech.2006.10.022
67. Barros, R. R., Paredes, R. D. S., Endo, T., da Silva Bon, E. P., & Lee, S. H. (2013). Association of wet disk milling and ozonolysis as pretreatment for enzymatic saccharification of sugarcane bagasse and straw. *Bioresource Technology, 136*, 288–294. https://doi.org/10.1016/j.biortech.2013.03.009

68. Olivier-Bourbigou, H., Magna, L., & Morvan, D. (2010). Ionic liquids and catalysis: Recent progress from knowledge to applications. *Applied Catalysis A, 373*(1–2), 1–56. https://doi.org/10.1016/j.apcata.2009.10.008
69. Hayes, D. J. (2009). An examination of biorefining processes, catalysts and challenges. *Catalysis Today, 145*(1–2), 138–151. https://doi.org/10.1016/j.cattod.2008.04.017
70. Banerjee, C., Mandal, S., Ghosh, S., Kuchlyan, J., Kundu, N., & Sarkar, N. (2013). Unique characteristics of ionic liquids comprised of long-chain cations and anions: A new physical insight. *Journal of Physics Chemistry B, 117*(14), 3927–3934. https://doi.org/10.1021/jp4015405
71. Latif, M. N., Wan Isahak, W. N. R., Samsuri, A., Hasan, S. Z., Manan, W. N., & Yaakob, Z. (2023). Recent advances in the technologies and catalytic processes of ethanol production. *Catalysts, 13*(7), 1093. https://doi.org/10.3390/catal13071093
72. Shukla, A., Kumar, D., Girdhar, M., Kumar, A., Goyal, A., Malik, T., & Mohan, A. (2023). Strategies of pretreatment of feedstocks for optimized bioethanol production: distinct and integrated approaches. *Biotechnology for Biofuels and Bioproducts, 16*, 44.
73. Klongklaew, A., Unban, K., Kalaimurugan, D., Kanpiengjai, A., Azaizeh, H., Schroedter, L., Venus, J., & Khanongnuch, C. (2023). Bioconversion of dilute acid pretreated corn stover to L-lactic acid using co-culture of furfural tolerant *Enterococcus mundtii* WX1 and *Lactobacillus rhamnosus* SCJ9. *Fermentation, 9*, 112.
74. Gong, J.-S.-Q., Su, J.-E., Cai, J.-Y., Zou, L., Chen, Y., Jiang, Y.-L., & Hu, B.-B. (2023). Enhanced enzymolysis and bioethanol yield from tobacco stem waste based on mild synergistic pretreatment. *Frontiers in Energy Research, 10*, 989393.
75. Usino, D. O., Sar, T., Ylitervo, P., & Richards, T. (2023). Effect of acid pretreatment on the primary products of biomass fast pyrolysis. *Energies, 16*, 2377.
76. Abdurrahman, A., Richard, P. I., Galadima, A. U., Muhammad, A., & Adamu, M. (2022). Dilute sulphuric acid pre-treatment for efficient production of bioethanol from sugarcane bagasse using *Saccharomyces cerevisiae. Journal of Biotechnology, 1*, 56–65.
77. Zhao, H., Baker, G. A., & Cowins, J. V. (2010). Fast enzymatic saccharification of switchgrass after pretreatment with ionic liquids. *Biotechnology Progress, 26*(1), 127–133. https://doi.org/10.1002/btpr.331
78. Nguyen, T. A. D., Kim, K. R., Han, S. J., Cho, H. Y., Kim, J. W., & Park, S. M. (2010). Pretreatment of rice straw with ammonia and ionic liquid for lignocellulose conversion to fermentable sugars. *Bioresource Technology, 101*(19), 7432–7438. https://doi.org/10.1016/j.biortech.2010.04.053
79. da Costa Lopes, A. M., João, K. G., Rubik, D. F., Bogel-Łukasik, E., Duarte, L. C., & Andreaus, J. (2013). Pre-treatment of lignocellulosic biomass using ionic liquids: Wheat straw fractionation. *Bioresource Technology, 142*, 198–208. https://doi.org/10.1016/j.biortech.2013.05.032
80. Park, N., Kim, H. Y., Koo, B. W., Yeo, H., & Choi, I. G. (2010). Organosolv pretreatment with various catalysts for enhancing enzymatic hydrolysis of pitch pine (*Pinus rigida*). *Bioresource Technology, 101*(18), 7057–7064. https://doi.org/10.1016/j.biortech.2010.04.020
81. Mesa, L., González, E., Cara, C., González, M., Castro, E., & Mussatto, S. I. (2011). The effect of organosolv pretreatment variables on enzymatic hydrolysis of sugarcane bagasse. *Chemical Engineering Journal, 168*(3), 1157–1162. https://doi.org/10.1016/j.cej.2011.02.003

Chapter 10
Biorefinery Avenues for Processing Urban Solid Waste: Potential for Value-Added Chemicals and Energy

Swapna Gade, Yuvraj Patil, and Bhalchandra Bhanage

Abstract Global urbanization against rural settlements has been on a steady rise over the past 6 decades. Nearly one-third of the entire population of India residing in urban locations (roughly 5.5% of available residential land) has placed an unprecedented burden on the municipal economy, energy utilization, and waste management systems. Sustainability as regards resources and municipal solid waste (MSW) management remains a challenge. The focus of this chapter concerns these challenges, the various avenues for processing MSW, biorefinery approaches for de-novo generation of commercial products, energy, etc. Additionally, the sustainability of these approaches, cost to the taxpayer, and the conversion of MSW to value-added resources or energy are also discussed. Harnessing bioenergy from MSW biomass may be an attractive route for addressing the ongoing global energy crisis. Carbon dioxide can be trapped via reactive amines to reduce emissions and generate value-added compounds such as urea. Certain wastes within MSW can be processed to obtain glycerol-based compounds which have wide applications in industry. Segregated waste management strategies will help in selectively processing organic wastes using the biorefinery route, which entails broad-spectrum cellular/enzymatic tools to break down the organic material. Systematic management of MSW can be expected to be a major resource salvage route. Harvesting MSW is thus proposed here to be a vital mechanism to generate a circular bioeconomy aimed at sustainability and economic viability.

S. Gade
Chemical Engineering and Process Development Division, National Chemical Laboratory, Homi Bhaba Road, Pashan, Pune 411008, India

Y. Patil
School of Health Sciences and Technology, Dr. Vishwanath Karad MIT World Peace University, Pune 411038, India

B. Bhanage (✉)
Department of Chemistry, Institute of Chemical Technology, Nathalal Parekh Marg, Matunga, Mumbai 400019, India
e-mail: bm.bhanage@gmail.com

G. Baskar et al. (eds.), *Circular Bioeconomy Perspectives in Sustainable Bioenergy Production*, Energy, Environment, and Sustainability,
https://doi.org/10.1007/978-981-97-2523-6_10

Keywords Value-added chemicals · Biorefinery-enriched Economy · Municipal solid waste processing

10.1 Introduction

Global fuel crisis, resource restrictions, growing human population, and the burden put on urban economies and administration are unprecedented in human history. Over the last 3 decades, attempts have been made to look at alternatives for resources, especially those that are finite, such as fossil fuel. Resources for powering society and commerce in general may stem from potentially renewable origins, provided these are handled appropriately. An example of such a resource is plant-based fiber or oil which has several applications in fabrics or biodiesel domain, for instance. Population burden-wise, the United Nations World Urbanization Prospects report indicates that nearly one-third of the population of India has come to reside in urban locations which constitute about 5.5% of available residential (non-agricultural land) land. The phenomenon has placed a burden on the municipal economy and waste management systems leading to several challenges for municipal solid waste (MSW) solutions.

10.1.1 Challenges of a Post-covid World: Energy Crisis, Lower CO_2 Emission Milestones, and Poor Resource Availability

The past decade has seen multiple conflicts in the Eurasian region particularly mapped as a major crude oil and natural gas resource. As a direct consequence, Europe and other regions have seen fuel restrictions and inflation in fuel prices [1, 2]. As the global commerce network continues to be intimately linked, such regional disturbances translate to global upsets that sustain long-drawn economic and fuel crises. With the ever-looming global warming and climate change scenario, developed nations are also pushing the envelope on curbing carbon footprints across the globe [3, 4]. The Paris agreement outlines a 43% decrease in greenhouse gas (GHG) emissions by 2030 if we are to prevent global warming of the planet exceeding 1.5 °C, to avoid far-reaching climatic disasters. For instance, with a 2 °C increase in global temperature, coral reefs are in imminent threat of annihilation [4]. Thus, with two major crises and an active international policy for reduced use of carbon fuel, significant work and discussion have been carried out that covers alternative and sustainable approaches to energy, and resource development as well as the framework that these approaches are based upon. A practical approach to a more sustainable future lies in managing the spent resources or waste, or in other words, to implement a circular economy [5]. The current chapter discusses avenues of managing a crucial aspect of waste generated and its potential processing in biorefineries to produce value-added products. This circular approach or, a circular economy, aims at reutilizing the waste to resupply value-added products back into the economy and manage resource restriction as well as salvage compounds that would normally be lost [5]. In a review by Ferreira et al., a circular bioeconomy idea is proposed to rely on the utilization of waste(s) as a feedstock for different biotechnological processes to efficiently obtain a broad-spectrum yield of biochemical components through a biorefinery approach [6].

10.1.2 Urban Ecosystems, Municipal Waste, and Sustainability

MSW, in the context of this chapter, is fundamentally different from other anthropological wastes. It does not bear the same constituents as agricultural or forest waste. The diverse nature of waste presents multiple issues that require attention in the following areas: Waste collection/segregation, Waste Processing or Recycling, and Material viability for Biorefinery [5, 6]. The composition of MSW is dynamic in nature and varies by region and culture. The changing nature also extends to the temporal dimension and therefore requires periodic review for optimal processing of MSW. The typical composition of MSW has six recurring waste matter groups; food, paper, wood, metals, rubber/plastics/fabric, glass, and a miscellaneous umbrella group [7, 8]. Figure 10.1 depicts the mean proportions of the waste categories across different global locations. As a general rule, the amount of waste generated increases in proportion to the national GDP, but local factors may drive the proportion of the waste. As observed, a few peculiarities emerge from the composition study, namely, (a) irregular wastage reported for wood and paper, (b) reported data for food/kitchen waste, and (c) the broad miscellaneous category which may contain key elements that need attention.

China and by extension, Asian nations such as India appear to have a large proportion of food wastage which could be considered as biomass. While paper waste appears in all regions, wood waste is limited or virtually absent in European/Asian data. Observations such as these impact the broader consideration of biorefinery approaches because of the implied availability or limitation of the feedstock.

Interestingly, according to a report by Ministry of Housing and Urban Affairs (2021), India demonstrates a food waste of about 45%, like the EU model. A similar profile for glass, metals, wood (construction), etc., is also noted. The rubber/plastic/fabric waste is however the highest and on par or greater than that demonstrated by China. It may be speculated that being Asian super-centers of consumer goods production, India and China may show similar tendencies of resource wasting when it comes to plastic, rubber, or fabrics.

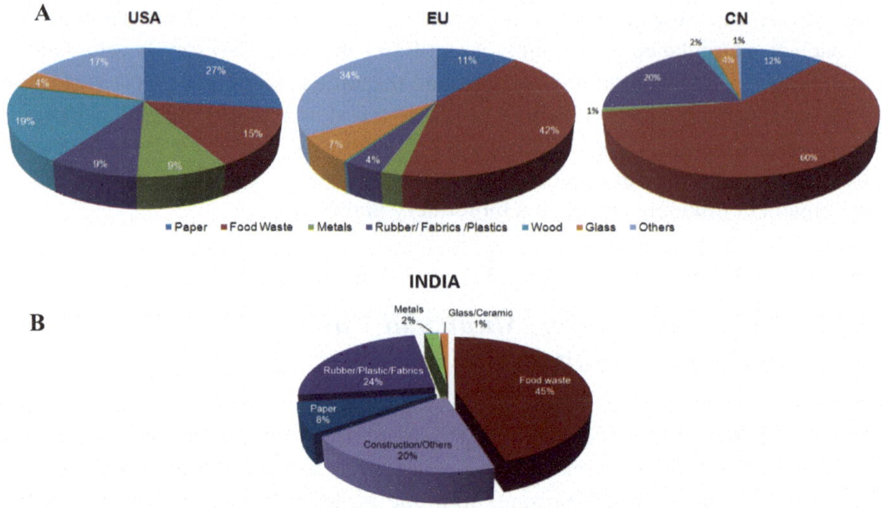

Fig. 10.1 Typical composition of MSW and regional variation. Data Adapted from [7, 8] and MoHUA Report on Circular Economy and Waste Management (2021). **a** Typical MSW composition as observed across developed nations and in China, **b** Denotes break-away Pie chart data depicting India's average MSW composition

10.1.3 Municipal Waste Management and Challenges with Solid Wastes

Management of MSW is a challenging task, especially when it is associated with different waste categories. The task is even more tedious in the developing countries on account of randomization of waste collection and non-adherence to segregation rules [7, 8]. Problems that are frequently linked with poor management of MSW are given below:

- Poor solid waste management and treatment,
- Unregulated waste dumping,
- Insufficient operator skills,
- Lack of responsiveness of health hazards,
- Unmanaged waste assortment system,
- Lack of awareness about recycling.

These mentioned difficulties are typically found in middle/poor-income countries. Urbanization in such countries is frequently without appropriate planning. Rapid proliferation of industries without proper planning is also one of the reasons leading to poor waste management [3]. Occurrences such as these are critical in importance in low-income to mid-income nations, where unregulated migration from rural zones to urban locales is normal, urban expansion sans adequate city-planning, and hastened industrialization with lacking infrastructure and supporting networks. Uncontrolled population upticks, as mentioned earlier and also noted by the United Nations World Urbanization Prospects report (2018), are also the cause of some of the ill effects.

10.2 Conventional Approaches to MSW Management

MSW is an urban planning problem of great proportions and one which grows exponentially with a greater standard of living and growing economy. While the nature of MSW may not be altered significantly across economic statures, the volume of MSW generated can and will certainly increase [5, 7]. Typical approaches toward managing MSW or garbage in general include the following.

A. Landfill

The earliest findings of landfills from the Roman era highlight the traditional municipal approach to disposing of garbage by accumulating wastes and burying or covering it [9]. As noted in Fig. 10.2, many countries prefer to use landfill as a viable option for MSW disposal. With relatively available non-agricultural land space landfill is opted over any of the other MSW management alternatives. The preference for landfill also loosely correlates with the economic status and political bent of the nation. It is interesting to note that economic analysis of landfill utility also includes cost payable as landfill gate fees, the latter being of an incremental nature. This implies that not only is landfill space limited, but that it comes at a premium, a cost which is undoubtedly passed on the taxpayer [5].

Nations such as Japan have virtually no space for landfill, leading the charge for alternative MSW processing. One such avenue is MSW incineration, which is considered as energy retrieval [7, 10, 11].

2 Incineration

As noted in Fig. 10.2, Japan as a first world nation with a large economy shows a surprising dependence on incineration of MSW. While incineration undoubtedly dramatically reduces the volume of MSW, it is associated with thermal inputs for the municipal system. MSW incineration has been attributed to significant electricity generation and supply for district heating [10]. Incineration is an advanced thermal

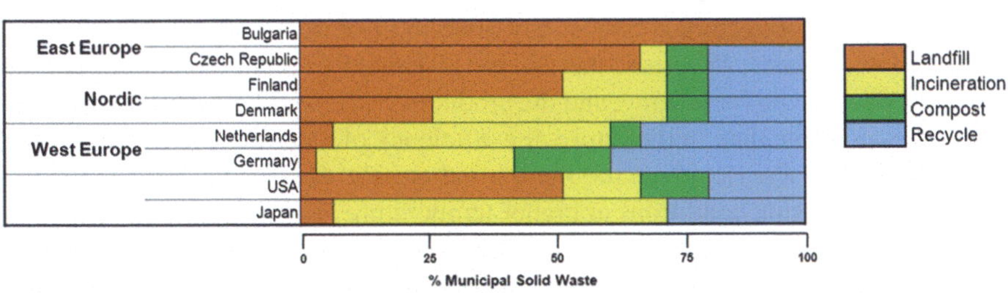

Fig. 10.2 Fate of municipal solid waste across different nations

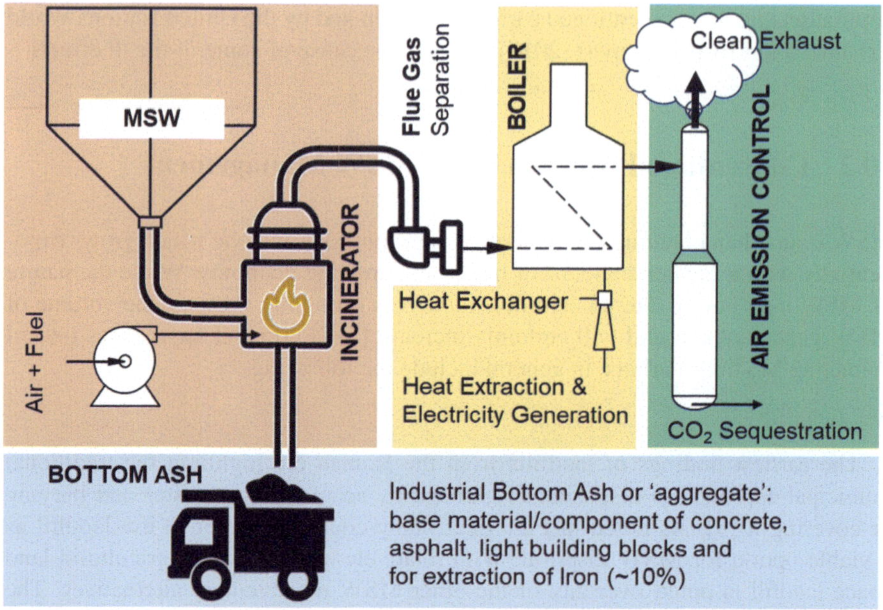

Fig. 10.3 Schematic map of MSW incineration

process carried out at 850 °C in presence of oxygen. Further waste is then transformed to flue gas i.e., carbon dioxide (CO_2), water, and non-combustible materials named bottom ash. Flue gas is useful in generating electricity and heat via boiler and after dusting the flue gas, which is emitted to the atmosphere. Figure 10.3 shows the schematic map of MSW Incineration. As an island nation, Japan can optimize MSW disposal while addressing its energy demands. Incineration has potent drawbacks, such as release of toxic metal and similar material particles which may lead to cancer [11].

An incineration process producing heat which has a thermal efficacy of around 80–90% can be used for electricity generation at around 17–30% crude effectiveness. The competence of an incinerator for energy corporations is inferior to gas-fired power station or a large coal. Normally, the efficiency of coal fire power station is in between 33 and 38% while gas turbine power station electric power efficiency could be ~ 50%. However, technical issues may arise in the case of larger-scale coal and gas-fired power plants. Whereas, the energy generated from an incinerator can be used to produce steam for heating, and the productivity is like a boiler fired by natural oil/gas. It is possible that, in the future, the efficiency of electricity generation using incineration will increase, given the trend in other solid fuel applications involving more severe operating conditions at higher temperature [7, 10]. It is conceivable, in our opinion that, the future may hold greater upward mobility of the electricity generating potential, based on the rising trend of solid fuel applications which frequently involve more severe operating conditions at higher temperatures [7].

3 Composting

Composting is biological decomposition of the organic fraction which can be done under precise conditions. Typically, end-to-end composting is executed as—compost preparation—decomposition-post processing and marketing [12]. It is the most common waste treatment process for organic waste segment or for garden waste, and as such is suitable for organic and biodegradable waste. Routinely, large swaths of land area are needed for composting and are subsequently locked up in the activity. Furthermore, it suffers a critical drawback that, with this process, we cannot control emission of gases. The implication here is that green routes are not followed making this a less satisfactory and a potentially unsustainable method for waste disposal.

MSW composting is a relatively new practice for conversion of MSW into a value-added soil additive. As would be expected from compost, MSW composting has been shown to enhance growth characteristics of plants such as wheat, Swiss chard and basil, among others [12, 13]. Interestingly, it is not noted to affect plant growth negatively by potential addition of heavy metals. Furthermore, heavy metals may possibly remain bound to the organic fraction of MSWC. A study conducted in Turkey has also shown MSW composting to have remarkable properties of sequestration of metals like nickel and cadmium in contaminated soil samples [14]. Unfortunately, despite definitive advantages to composting of MSW, relatively few nations have leaned toward adoption of the approach (see Fig. 10.2).

4 Recycling

While recycling has been promoted for a long time, prior to the MSW management crisis, recent government initiatives and incentives are a strong driver for MSW recycling of plastics, etc. Despite initiatives, though, without public involvement, recycling may not broadly benefit MSW management [15, 16]. As discussed earlier, developed nations such as the United States demonstrate a greater paper waste compared to other nations. Consequently, recycling has indeed emerged as a mitigation remedy for MSW management [7]. On the other hand, countries such as Bulgaria with a missing axis of recycling, composting, or incineration (energy recovery) are poor candidates for implementing circular economy (see Fig. 10.2).

It has been observed that the lower-income families produce comparatively more organic food waste. Large amounts of glass, plastics, wood, and metals-based waste are frequently noted as the waste output of higher-income families. The ability to reprocess waste depends on the MSW composition. For example, in low-income nations, organic waste can be utilized for making cheap fertilizer by composting it [17]. Effective collection, separation of waste, and further processing are directly proportional to the wealth of the community. Countries that are aware about the effective circular economy are therefore paying a major premium on waste management, waste collection, and transportation. A few examples of waste collection percentage in different countries as portrayed by the World Bank report on Wastes (2012), highlight the dependency of economy and MSW management: Philippines and Sri Lanka (40%), Paraguay and Vietnam (50%), India (70%), Malaysia, Switzerland, Germany, Japan, and United States (100%).

As Indian commerce catches up to the Chinese phenomenon of cheap, mass production of plastic (and other materials) commodities, there is an inclement rise in plastic wastes to the tune of nearly 15 K tons per day. Plastics add up to nearly 8% of all solid wastes, especially in cities such as Ahmedabad, Kolkata, and Delhi. Plastics are considered to be process-competent for recycling and to some extent for production of alternative fuel via pyrolysis.

Like Rishikesh, other Indian municipalities have invested in and explored the potential of MSW management with the view of creating circular economy. Indore, as reported by TERI (2018), which has recently gained the unique standing of being the cleanest city in India, also manages its MSW by segregation of waste at the source, followed by composting and fermentation to make manure and biofuel in the form of CNG.

Rawat and Daverey have reported on Classification of household solid waste and existing position of MSW management in Rishikesh and Uttarakhand, in India [18]. Based on broader review of literature concerning MSW, Table 10.1 displays various components that are grouped under the varied waste categories.

The presence of paper-based items in the grey zone is a result of product contamination, either with chemicals, potentially toxic dyes, inks, etc. They have associated the kitchen and other waste generated from houses, and the rates of production of waste are diverse worldwide. From Table 10.2, it can be seen that the per head left-over production rate in Rishikesh is almost identical to other metro cities of evolving countries such as Suzhou, Beijing. The waste production is very large in Mexicali city (0.981 kg/c/d), as compared to metropolitan city [18].

Ding et al. have published a review on management and technologies used for the treatment and utilization of resources of MSW. They have also made a comparative

Table 10.1 Classification of MSW

Types	MSW composition	Collection color code
Kitchen	Vegetable stems, animal skeleton viscera, leftovers, plant waste, waste kitchen oil, fruit waste, etc.	**Green**
Hazardous waste products	Waste tube bulb, scrap printer cartridges, batteries, waste electronics, hair dye, overdue drugs and cosmetics, waste paint can, pesticide container, etc.	**Red**
Recyclable	**Metal**—pop-tin cans waste, toothpaste cover, etc. **Paper**—newspapers, cardboard waste, book pages, cartons etc. **Glass**—various colored/non-colored glass bottles, mirror glass, etc. **Textile**—school bags, waste clothes, etc. **Plastics**—disposable plastic tableware, plastic bottles, plastic cup, plastic foam, plastic packaging, etc.	**Blue**
Other waste	Contaminated plastic bags, cigarette, dirty papers, shell covers, dust, non-recyclable glass, waste/broken ceramics, contaminated diapers, etc.	**Gray/ Black**

Bold text: Indicate material categories which form the basis of composition of MSW. The categorical proportions vary from region to region

Table 10.2 Household waste production rate and influence of food waste (%) from diverse cities of world

City (country)	Household waste generation (kg/c/d)	Food waste (%)	References
Wales (England)	–	26.79	[19]
Dublin (Ireland)	–	40.48	[20]
Takoradi (Ghana)	0.7	61	[21]
Mexicali (Mexico)	0.98	–	[22]
Bolgatanga (Ghana)	0.21	61	[21]
Suzhou (China)	0.28	65.7	[23]
Beijing (China)	0.23	69.3	[24]
Rishikesh (India)	0.26	57.3	[18]

study of international regions waste management [8]. The authors have presented different aspects of eight different cities in China, mainly situated in coastal regions.

The study is based on gross domestic production, MSW generation, MSW segregation (collection, transport, separation, composition), and acts and regulations related to the same. The study also includes advancement in technologies in MSW source consumption for waste-to-material and waste-to-energy. MSW is mainly separated into household food waste, recyclable waste, harmful waste, perishable wasteland, and other waste. The review includes different types of MSW separation trends that have been followed in China, mainly they are of four types which are collected in different containers (Table 10.2) [8].

In other examples of countries/cities with MSW management systems, we have Indonesia, where 60–70% of the produced waste is moved to landfills, whereas the remaining 30–40% of waste generated is independently managed by the community, burned it, or dropped in rivers [25]. The municipal waste in Berlin is mostly separated into five types: trade waste (10.06%), domestic waste (73.80%), bulky waste (6.95%), commercial waste (2.30%), and road sweepings (3.78%) [8].

Tokyo has an extremely organized waste collection strategy which is the most important part of waste collection that has been done using the slogan "separation is resource and mixing is waste" [26]. Waste management in Tokyo has four large categories: flammable waste (3.3 million tons), non-flammable left-over (0.1 million tons), recyclable matter (0.5 million tons), and bulky waste (0.08 million tons) [8, 26].

In the case of Singapore, MSW is separated in two main systems, general waste collection system and public waste collection scheme. General waste collection system attends industrial and commercial locations. These locations solid waste includes organic waste, inorganic waste, sludge, and also grease. All these wastes are collected by the licensed, general waste collection system is maintained by private waste collectors. The public waste collection scheme mainly targets the MSW produced by general household waste, garden waste, furnishings and domestic infrastructure waste, and remnants of animals. The accumulated waste is moved to different processing sites for reprocessing and handling. Organic/inorganic wastes can be sorted and further transferred to incinerators or transfer stations for handling post-contracting. Advanced technologies are also included in the waste collection systems such as pneumatic waste collection instrumentation, applied in novel buildings and private housing lands in Singapore. To transport the waste, underground pipes are used with vacuum suction to reach the assembly point of waste. This method is more hygienic and productive as compared to the above two methods [27, 28].

Biorefineries are increasingly being viewed as a viable commercial portal for mass valorization of waste products [29]. The chief reason for this change of perspective is not just resource limitation, but environmental concerns as well as central policies on sustainability and economy. The latter frequently dictates scientific approaches and broad adoption of the same, for instance, Fig. 10.1 depicts various approaches toward urban waste management in different countries. It is interesting to note a geographical disparity in policy approaches as noted in the depiction [29]. Eastern European nations rely heavily on the basic waste management solution of landfill while western nations may employ more sustainable approaches of composting and recycling waste. Although some of these western nations may also resort to waste negation by incineration, this approach may be driven by lack of space for either landfill or recycling/composting activities. Greece and Portugal display differences in MSW management from other western European nations, possibly attributed to national economic priorities and GDP. The United States and Japan as non-continental comparators are interesting in that they show opposing preferences for land-use while at the same time favor sustainable MSW management. Japan's surprising reliance on incineration may be directly related to restriction of space availability in the island nation [29].

Ghumra and colleagues discuss the ramifications of MSW management, in particular the valorization of MSW attributed to organic origin. The organic fraction or food-sourced waste is a large potential source of ethanol, butanol, organic acids, and gases. The biorefinery here is facilitated by pre-treatment of waste followed by thermochemical or biochemical mass conversion. Subsequently, fermentation of the retrieved sugars, amino acids, and other compounds results in value-added products. Organic components of MSW are typically a complex mixture of cellulose, carbohydrates, proteins, and lipids. With regard to thermochemical mass conversion, Plasma gasification is one of the advanced technologies being tested for altering MSW into circular economy. This advanced technology needs much attention, training, and need to spread awareness among research and Municipal bodies who are working on waste-to-circular economy program. Need to do much attention and research for successful commercialization [28].

Plasma gasification can be an effective technique for the disposal of MSW and could be much more beneficial for the waste-to-value circular economy. Successful industrial application of this process is needed. Since last decade, few research groups have been studying on application of plasma gasification, the problem associated with this process, etc. [28, 30–33].

10.3 Biorefinery to Bioeconomy

To improve circular economy, we need to reconsider organic/bio-waste and its potential role in circular bioeconomy. We need to improve eco-friendly utilization of resources. It may well influence the reduction in municipal waste, carbon demand, and greenhouse gases.

Bioeconomy structures surpass circular economic points that contain product or facility zones. It is mainly determined by vast number of biological processes, animal waste, fisheries, forestry, food waste, and animal feedstuff. These biomass wastes are used in various applications. In chemical processes and aquaculture (aquarium), etc. biomass waste is used. Furthermore, decaying crops and foodstuff can go back into the biological and nutrient cycle. Recycling of animal fats, bioplastic, and vegetable waste can be converted into oleo-chemicals which are also expected as a part of circular economy [34, 35] (Fig. 10.4). Biorefinery approaches take advantage of enzymes such as proteases, cellulases, amylases, and lipases to breakdown the biomass to yield fermentable sugars, peptides, amino acids, and free fatty acids as building blocks that are vital for valorization [36].

Though the idea of a similarity of biorefinery and petroleum refineries, the transformations are vital (Fig. 10.5). Throughout the petrochemical refinery procedure, fossil routes are converted into energy (e.g., fuel) also the list of the other products like Plastic, chemicals, etc., In case of a biorefinery, a widespread range of bio-based

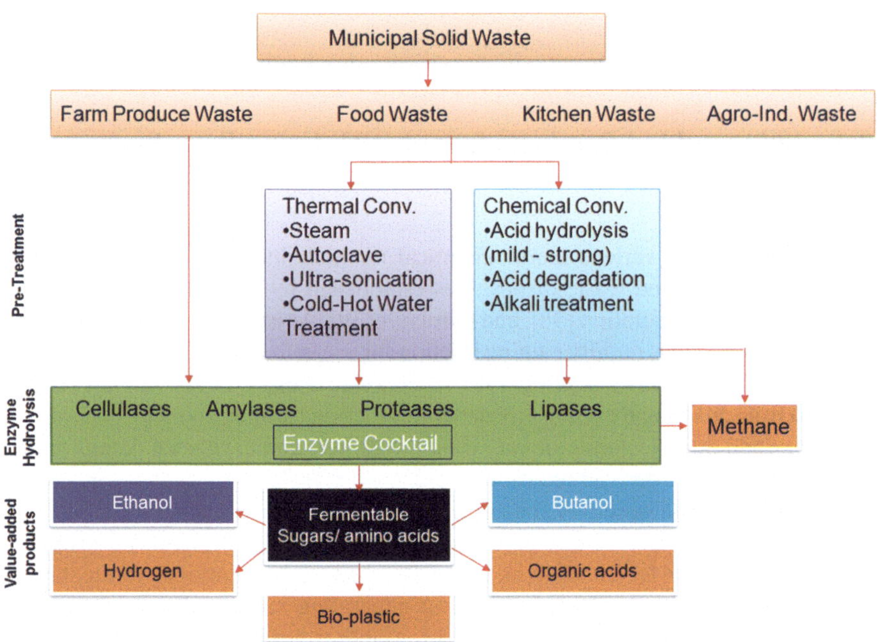

Fig. 10.4 Biorefinery approach to process municipal solid waste

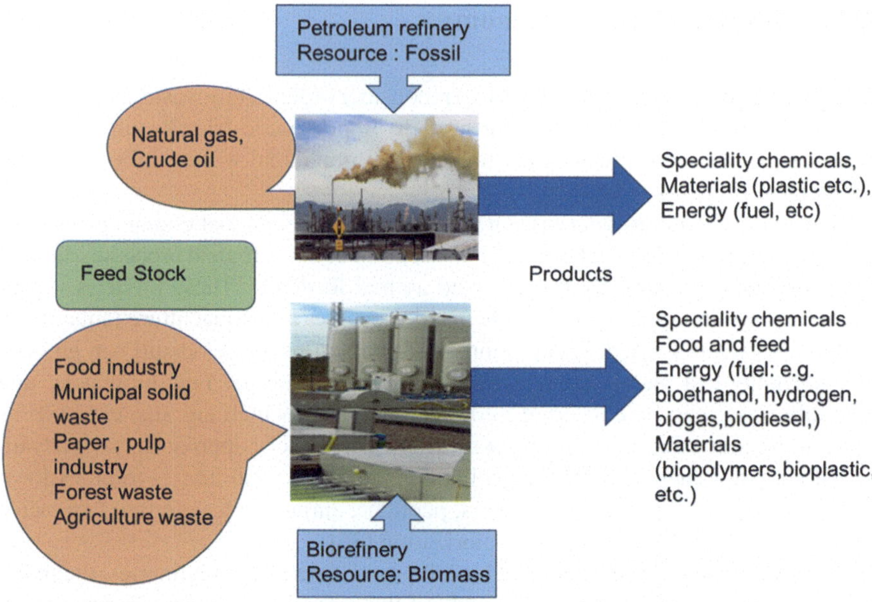

Fig. 10.5 Comparison of a biorefinery and a petroleum refinery; schematic based on [29]

goods like biochemicals, food and feed, bioenergy, biomaterials, etc. are formed from the biomass feedstock.

10.4 Value-Added Products from MSW Using Biorefinery Approaches

As discussed earlier, various thermochemical methods have been used for waste-to-value processing and waste disposal, as established by several studies on steam gasification [37], incineration [38], and pyrolysis [39]. As a distinct methodology, the biorefinery approach is rapidly gaining traction for use in managing MSW and wastes classified as wet waste, such as algal waste, etc. Under this process, biomass is first separated from MSW and used in bioenergy production by biorefinery [40]. Gasification [28] or and other technologies are established [41] and have been demonstrated for use with biomass-derived feedstocks.

Biorefinery can be generally regarded as a complex of approaches that, (a) separate biomass/materials that can be subject to processing, (b) primary processing that helps derivatize feedstock which can be used in conventional fermentation, (c) advanced processing which is aimed at generating value-added products, and (d) processing to maximize extraction of usable compounds with commercial value. Consequently, biorefinery is fundamentally distinct from the above-mentioned avenues of processing municipal solid waste. An interesting example of use of biorefinery in conjunction with an indispensable industry is the integration of biorefinery concept with paper processing plants. Wood pulping and paper production involve manipulation of ligno-cellulosic contents of the starting material. Post lignin removal, cellulose, and hemi-cellulosic content is capable of being harnessed for fermentation yielding fermentable sugars along with other value-added compounds including pharmaceutical ingredients such as mannitol.

Another instance of biorefinery approach in managing MSW is the process engineering for using a myriad methodological technique such as a microwave or ultrasonication assisted extraction, supercritical or hydrothermal methods for extraction of compounds from food waste. Food waste from markets, domestic, and public eatery waste, etc., is a veritable resource for not just starches, sugars, lipids, cellulose, and lignin but also organic constituents, biomass for fermentation and potentially salvaging of vitamins, co-factors, and biopolymers. Pretreatment with supercritical aqueous media and subsequent enzymatic or catalytic treatment help to extract fermentable sugars to further yield bioethanol, methane, lactic acid etc. Lipids are already being utilized to generate biodiesel, which is primarily an ester-based fuel and subsequently glycerol as well.

Various thermochemical methods have been used for waste-to-value processing and waste disposal, as established by several studies on steam gasification [37], incineration [38], and pyrolysis [39]. Biomass is first separated from MSW, it can be used in bio-energy production by biorefinery [40]. Gasification [28] and other technologies are established [41]. As mentioned earlier, within the process of MSW disposal under the biorefinery umbrella, biomass is first parted from MSW. Further, it can be utilized to produce bioenergy through biorefinery [40], gasification [28], or other technologies [41]. The bioenergy produced from such a source can be in the form of biofuel, biogas, crude oil, and solid fuel. These alternatives are substitutes for fossil fuels, which could help in reducing emission of greenhouse gas [42]. Furthermore, bio ethanol, bio-methanol, methane, and even hydrogen can be produced from biogas [43].

Utilization of waste food (organic waste) into a source of renewable energy has converted the food sector to a possibly practicable carbon economy [44]. To achieve circular bioeconomy, removal of bioactive and economically vital compounds from the MSW and application of the wastes to generate biofuels are substitutions to the remaining fossil fuels.

As described earlier, biomass is essentially any organic matter that is available as a renewable source, including certain organic industrial wastes, forest wastes, agricultural crop waste, and municipal waste. Specifically, referring to agricultural crops, plants such as switchgrass, jatropha like plants are known for providing energy sources, via fermentation and biorefinery. Primary biomass components vary by type and source of biomass. Mainly lignocellulose, sugar, triglyceride, and starch are known for biorefinery applications. Chemical structures of those compounds are shown in Fig. 10.6 [45].

Food waste includes lipids, proteins, and carbohydrates. These compounds can be used as a reactant for the manufacture of various bioactive compounds. For example,

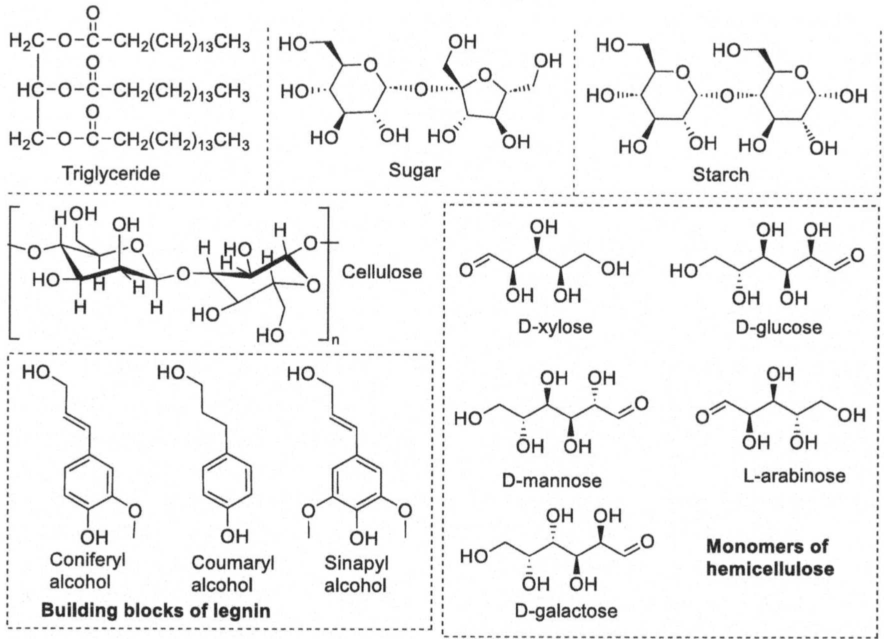

Fig. 10.6 Biorefinery value-added products

animal wastes can be the source of gelatin, polypeptides, and collagen. Dairy waste contains lactose and casein which can be used after extraction treatment on source [46]. Seafood waste is a source of protein isolates and hydrolysates. Kitchen waste includes cellulose, flavonoids, anthocyanins, etc. [46, 47].

10.5 Recent Trends in Biomass Conversion

MSW compositions from Asian nations like India and China reveal a large segment for food wastes, in addition to other organic wastes. As described earlier, a more appropriate resource-recouping ideology is to adopt the bio-process/biorefinery route. Literature review reveals certain insights into the contemporary trends in this domain. While the area of municipal waste management is a stand-alone challenge, the proposed integration with biorefinery and circular economy will set the stage for a sustainable industry and civil management program.

Inclusivity and Fundamental Rethinking: As with a radical reimagining of a concept, broad-sweeping changes need to be enacted for promoting and enabling circular bioeconomy. To this effect, floating an educational outline for the propagation of the concept is needed. To achieve the goal, it is vital to extend and venture beyond any single domain and implement design thinking for innovations that may bring commercial viability. True multi-disciplinary blending may be required to improve success in the MSW-Biorefinery field. Experts from areas such as law, policy-making, industry, STEM fields, and even social sciences would be required to work together to perceive and purview issues in the field and to devise custom solutions for the same.

Global Partnerships: Unsurprisingly, the key to success in complex, globally relevant problems is collaboration. Due to the unique geographical, cultural, or scientific dispositions, some nations can be better situated for economic and timely solutions to specific issues. For instance, if Bulgaria were to exploit composting instead of relying exclusively on landfill strategies for MSW management, an alliance with Germany as a regional partner would be highly effective. Beyond practical alliances, the collaborations are envisioned to educate all stakeholders to the nature of natural challenges occurring, which are the primary drivers of this sea change [48]. Climate change and efforts to mitigate the same should be the prime focus of such collaborations.

Goal Setting: Not unlike personal development strategies, nations need to categorize goals for achieving sustainability and circular bioeconomy. The establishment of bioeconomy linked to sustainable development goals (SDG) can vary substantially, depending on the strategic objectives adopted by a given nation for its development and goal. It follows that the nation's SDG program is intricately linked to the choice for enabling bioeconomy [49].

Administrative Accountability: As with inclusivity, lawmakers and administrations are deeply involved with enabling and maintaining a circular bioeconomy. The government must take initiatives in MSW regulation and drive nation-wide responsibility in MSW collection, classification, and disposal route.

Green Route: Beyond cliché, the process of bioeconomy should follow a green route, at least in part. The green pathway is primarily to promote biomass conversions such as waste-to-material and waste-to-economy. While the latter is broad in scope, the idea of enacting value-added-services/products is central to this theme. The accompanying theme is hence to minimize waste in the first place, judicious use of resources, re-use and recycling of resources as a natural pathway in civil decisions [8].

The summarized snapshot of the insights gained from this review is depicted in Fig. 10.7, showing country/region-specific MSW generation pattern. Depending on the proportions of specific wastes, the regional MSW management mechanism can be optimized to serve, for instance, higher organic waste by the biorefinery pathway. Organic wastes can be sorted for maximum efficiency of enzymatic pre-processing and degradation, prior to fermentation or bio-esterification (biodiesel synthesis). In an ideal example of multi-modal integration, heat released from incineration of wastes can be used to drive boilers/distillation unit for separation of high-grade biodiesel. Resulting gaseous wastes can be systematically processed to reduce CO_2 emissions as well as produce value-added products (VAPs) such as urea, glycidol, and glycerol carbonate. Most VAPs are circularly re-employed in civil and agricultural domains which feedback into the loop. A well-regulated MSW management pathway and auxiliary domains, as indicated by several initiatives and studies, will ultimately reduce waste production and enable sustainability [50–58].

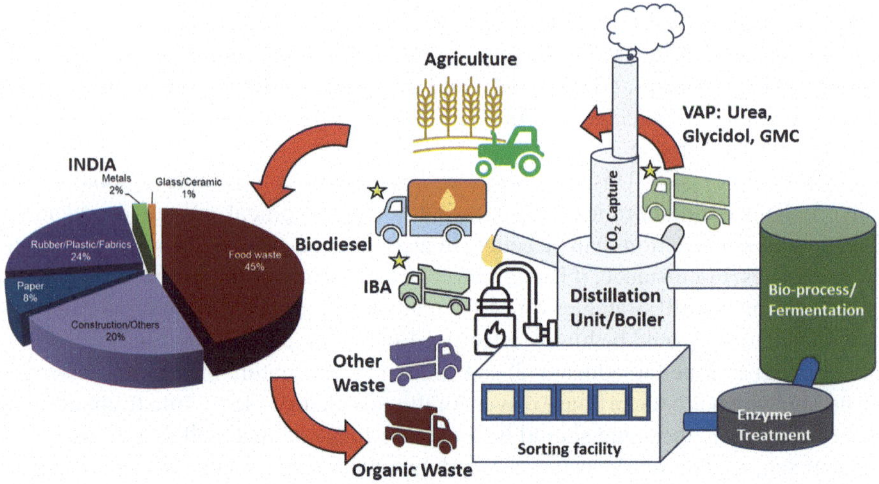

Fig. 10.7 Municipal solid waste management with value-added products in a circular bioeconomy

10.6 Conclusions

The global economy and resource pool situation may have long passed their peak and with the specter of fossil fuel exhaustion, urban localization of population, and unmanageable municipal waste scenarios, serious attention needs to be devoted to managing the municipal solid wastes. This chapter highlights and emphasizes the need for a circular economy revolving around urban waste and specifically a circular bioeconomy, owing to various value-added products that may be reaped from judicious management of certain MSW materials. Various MSW management models with the outlook of circular bioeconomy have been broadly studied. Biorefinery plays a pivotal role in circular economy. With proper methodology, technology biorefinery with a zero-waste goal can be achieved. It should be implemented in various industrial areas, societies, municipal garbage collection areas, etc. Sustainability of the mechanism of MSW management and the establishment of the circular economy will reduce the impact of MSW pressures on the taxpayer as well as the municipalities in question. The successful management of MSW in Indian cities like Rishikesh and Indore lays a roadmap for the rest of the country to implement best MSW management practices and maximize utilization of the biorefinery pathway to a sustainable and better economy. Beyond India, a broader study of nations like China, Japan, the United States, and countries within the European Union provide an insight into the attitudes of their citizens as well as fundamental differences which result in differing patterns of consumption and therefore wastage.

As proposed in this work, multiple overlapping approaches may be entertained to institutionalize the circular bioeconomy, for instance, adoption of the carbon sequestration technologies as illustrated in Fig. 10.2 in combination with not just MSW incineration but also biodiesel fueled industries to minimize the impact of greenhouse gases while safely managing anthropogenic organic wastes. Biodiesel production in turn generates glycerol and other compounds which can be processed to obtain higher value-added products.

A detailed study of the existing data and literature show that the widespread adoption of biorefineries to enable circular bioeconomy is still largely affected by socio-economic perceptions and social "acceptability," virtually creating a barrier for the path ahead. Such a stalemate requires intervention at the highest levels and hence support from policymakers and scientists alike. The potential value provided by organic MSW and associated biomass has taken the biorefinery role past its boundaries, well on the way to installing the much-desired circular bioeconomy and adding value to even secondary products.

References

1. Guan, Y., Yan, J., Shan, Y., Zhou, Y., Hang, Y., et al. (2023). *Nature Energy, 8*, 304–316.
2. Tollefson, J. (2022). What the war in Ukraine means for energy, climate and food. In *Nature news feature* (pp. 232–233). Nature Publishing Group.
3. Edenhofer, O. (2015). *Climate change 2014: Mitigation of climate change*. Cambridge University Press.
4. Change, I. C. (2014). *Contribution of working group III to the fifth assessment report of the intergovernmental panel on climate change*, 1454, 147.
5. Peter, D., Rathinam, J., & Vasudevan, R. T. (2022). *Biotechnology for zero waste* (pp. 409–420).
6. Ferreira, A., Ribeiro, B., Ferreira, A. F., Tavares, M. L. A., & Vladic, J., et al. (2019). *Biofuels, Bioproducts and Biorefining, 13*, 1169–1186
7. Azapagic, A. (2010). *Sustainable development in practice* (pp. 261–325).
8. Ding, Y., Zhao, J., Liu, J.-W., Zhou, J., Cheng, L., et al. (2021). *Journal of Cleaner Production, 293*, 126144.
9. Gadot, Y. (2022). *The landfill of early Roman Jerusalem: The 2013–2014 excavations in area D3*. Penn State University Press.
10. Ozturk, M., & Dincer, I. (2020). *Greenhouse Gases: Science and Technology, 10*, 855–864.
11. Denison, R. A., & Silbergeld, E. K. (1988). *Risk Analysis, 8*, 343–355.
12. Lakhdar, A., Iannelli, M. A., Debez, A., Massacci, A., Jedidi, N., & Abdelly, C. (2010). *Journal of the Science of Food and Agriculture, 90*, 965–971.
13. Zheljazkov, V. D., & Warman, P. R. (2004). *Journal of Environmental Quality, 33*, 542–552.
14. Elmaslar, Ö. E. (2015). *Environmental Progress & Sustainable Energy, 34*, 1372–1378.
15. Hamburg, K. T., Haque, C. E., & Everitt, J. C. (1997). *Canadian Geographies/Géographies canadiennes, 41*, 149–165.
16. Giordano, N. (1993). *European Environment, 3*, 14–17.
17. Periathamby, A. (2011). *Waste*. In T. M. Letcher & D. A. Vallero (Eds.) (pp. 109–125). Academic Press.
18. Rawat, S., & Daverey, A. (2018). *Environmental Engineering Research, 23*, 323–329.
19. Burnley, S. J., Ellis, J. C., Flowerdew, R., Poll, A. J., & Prosser, H. (2007). *Resources Conservation and Recycling, 49*, 264–283.
20. Dennison, G. J., Dodd, V. A., & Whelan, B. (1996). *Resources Conservation and Recycling, 17*, 227–244.
21. Miezah, K., Obiri-Danso, K., Kádár, Z., Fei-Baffoe, B., & Mensah, M. Y. (2015). *Waste Management, 46*, 15–27.
22. Ojeda-Benítez, S., Vega, C. A.-D., & Marquez-Montenegro, M. Y. (2008). *Resources, Conservation and Recycling, 52*, 992–999.

23. Gu, B., Wang, H., Chen, Z., Jiang, S., Zhu, W., et al. (2015). *Resources Conservation and Recycling, 98*, 67–75.
24. Qu, X.-Y., Li, Z.-S., Xie, X.-Y., Sui, Y.-M., Yang, L., & Chen, Y. (2009). *Waste Management, 29*, 2618–2624.
25. Damanhuri, E., Wahyu, I. M., Ramang, R., & Padmi, T. (2009). *Journal of Material Cycles and Waste Management, 11*, 270–276.
26. Matsumoto, S. (2011). *Resources Conservation and Recycling, 55*, 325–334.
27. Zhang, D., Keat, T. S., & Gersberg, R. M. (2010). *Waste Management, 30*, 921–933.
28. Munir, M. T., Mardon, I., Al-Zuhair, S., Shawabkeh, A., & Saqib, N. U. (2019). *Renewable and Sustainable Energy Reviews, 116*, 109461.
29. Clark, J., & Deswarte, F. (2015). *Introduction to chemicals from biomass* (pp. 1–29).
30. Gomez, E., Rani, D. A., Cheeseman, C. R., Deegan, D., Wise, M., & Boccaccini, A. R. (2009). *Journal of Hazardous Materials, 161*, 614–626.
31. Sanlisoy, A., & Carpinlioglu, M. O. (2017). *International Journal of Hydrogen Energy, 42*, 1361–1365.
32. Changming, D., Chao, S., Gong, X., Ting, W., & Xiange, W. (2018). *Waste Management, 77*, 373–387.
33. Fabry, F., Rehmet, C., Rohani, V., & Fulcheri, L. (2013). *Waste and Biomass Valorization, 4*, 421–439.
34. Giampietro, M. (2019). *Ecological Economics, 162*, 143–156.
35. Stegmann, P., Londo, M., & Junginger, M. (2020). *Resources Conservation & Recycling: X, 6*, 100029.
36. Ghumra, D. P., Rathi, O., Mule, T. A., Khadye, V. S., Chavan, A., et al. (2022). Biofuels. *Bioproducts and Biorefining, 16*, 877–890.
37. Arena, U. (2012). *Waste Management, 32*, 625–639.
38. Abd Kadir, S. A. S., Yin, C.-Y., Rosli Sulaiman, M., Chen, X., & El-Harbawi, M. (2013). *Renewable and Sustainable Energy Reviews, 24*, 181–186.
39. Velghe, I., Carleer, R., Yperman, J., & Schreurs, S. (2011). *Journal of Analytical and Applied Pyrolysis, 92*, 366–375.
40. Bolan, N. S., Thangarajan, R., Seshadri, B., Jena, U., Das, K. C., et al. (2013). *Bioresource Technology, 135*, 578–587.
41. Dehkordi, S., Taghipour, A., Ferdowsi, A., Shumal, M., & Dehnavi, A. (2019). *Renewable and Sustainable Energy Reviews, 119*, 109586.
42. You, S., Tong, H., Armin-Hoiland, J., Tong, Y. W., & Wang, C.-H. (2017). *Applied Energy, 208*, 495–510.
43. Furtado Amaral, A., Previtali, D., Bassani, A., Italiano, C., Palella, A., et al. (2020). *Energy, 203*, 117820.
44. Girotto, F., Alibardi, L., & Cossu, R. (2015). *Waste Management, 45*, 32–41.
45. Mailaram, S., Kumar, P., Kunamalla, A., Saklecha, P., & Maity, S. K. (2021). *Sustainable fuel technologies handbook* (51–87). https://doi.org/10.1016/B978-0-12-822989-7.00003-2
46. Anal, A. K. (2017). *Food processing by-products and their utilization* (pp. 1–10).
47. Tsouko, E., Alexandri, M., Fernandes, K. V., Guimarães Freire, D. M., Mallouchos, A., & Koutinas, A. A. (2019). *Food Technol Biotechnol, 57*, 29–38.
48. Aguilar, A., & Patermann, C. (2020). *New Biotechnology, 59*, 20–25.
49. Calicioglu, Ö., & Bogdanski, A. (2021). *New Biotechnology, 61*, 40–49.
50. Sacchi, S., Lotti, M., & Branduardi, P. (2021). *New Biotechnology, 60*, 72–75.
51. Bhattacharya, R. R. N., Chandrasekhar, K., Deepthi, M. V., Roy, P., & Khan, A. (2018). *TERI-challenges and opportunities in plastic waste management in India* (p. 20).
52. Mongkhonsiri, G., Charoensuppanimit, P., Anantpinijwatna, A., Gani, R., & Assabumrungrat, S. (2020). *Journal of Cleaner Production, 255*, 120278.
53. Sarangi, P. K., Singh, A. K., Sonkar, S., Shadangi, K. P., Srivastava, R. K., Gupta, V. K., Parikh, J., Sahoo, U. K., & Govarthanan, M. (2023). *Industrial Crops and Products, 205*, 117488.
54. Department for Environment Food and Rural Affairs. (2013). *Waste technology briefs* (p. 56). www.defra.gov.uk

55. IPPC reference document on best available technologies for waste incineration, August 2006.
56. United Nations, Department of Economic and Social Affairs, Population Division. (2019). *World urbanization prospects: The 2018 revision (ST/ESA/SER.A/420)*. United Nations.
57. Hoornweg, D., & Bhada-Tata, P. (2012). *What a waste: A global review of solid waste management*. World Bank—Urban Development Series.
58. Ministry of Housing and Urban Affairs, Government of India. (2021). *Circular economy in municipal solid and liquid waste* (p. 112). MoHUA/2021/June/30.

Chapter 11
Pyrolytic Conversion of Heterogenic Natural Waste Biomass from Rural Communities with Concomitant Valorization

M. Anil Kumar, Pareshkumar G. Moradeeya, K. Manikanda Bharath, P. Jakulin Divya Mary, and K. S. Giridharan

Abstract The challenge of waste disposal and management has become more prominent with increasing awareness. In rural areas, natural waste management is carried out within available space constraints. The conventional methods used for solid waste disposal include burning, landfilling, and composting, often regarded as the most effective and widely used practices. There is a significant concern over the adverse environmental consequences of implementing non-sustainable practices in solid waste management. This chapter aims to comprehensively analyse the many sources of solid waste, focusing on lignocellulosic biomasses originating from rural areas, mainly within the Indian sub-continent. Pyrolytic conversion is a thermal process that transforms waste materials into three distinct forms: solid residue known as char, liquid fuel, and gaseous emissions. Recent research has concentrated on using food waste as an energy source (e.g. for the manufacture of bioethanol and biodiesel) rather than disposing of and decomposing it. Utilizing a biorefinery or biotechnology, organic waste can also be used to produce valuable organic chemicals, such as succinic acid and/or bioplastics. The chapter further provides a detailed explanation of mapping secondary data from different departments and contrasts it with previously conducted research studies. This chapter is intended to summarise the facts on the efficient use of biochar for pollution control and environmental management.

Keywords Biochar · Pyrolysis · Rural · Solid wastes · Thermo-chemical · Valorization

---

M. Anil Kumar (✉) · K. Manikanda Bharath · P. Jakulin Divya Mary · K. S. Giridharan
Centre for Rural and Entrepreneurship Development, National Institute of Technical Teachers Training and Research, Chennai, Tamil Nadu 600113, India
e-mail: anilkumar@nitttrc.edu.in

P. G. Moradeeya
Department of Environmental Science and Engineering, Marwadi University, Rajkot, Gujarat 360003, India

G. Baskar et al. (eds.), *Circular Bioeconomy Perspectives in Sustainable Bioenergy Production*, Energy, Environment, and Sustainability,
https://doi.org/10.1007/978-981-97-2523-6_11

11.1 Introduction

The effective and sustainable management of biomass waste has become more significant within the framework of environmental preservation and the transition towards an environmentally friendly economy [51]. Rural areas often generate significant volumes of different natural waste biomass originating from a range of sources, including agriculture, forestry, and other organic materials. Inadequate handling of biomass waste can lead to the contamination of the environment, the release of greenhouse gases, and the potential for health risks [43]. However, it is important to note that these waste materials possess the potential to be converted into useful goods through the utilization of pyrolytic processes. Various agriculture and forestry methods in rural communities worldwide result in significant amounts of organic waste biomass being generated [27]. Rural regions often exhibit a diverse range of biomass resources, including agricultural residues, forest debris, and organic wastes. According to Schipfer et al. [60], inadequate management of resources can lead to both environmental contamination and economic losses. Pyrolysis is a chemical process that facilitates the conversion of various biomass feedstocks into three distinct products, namely biochar, bio-oil, and syngas. The implementation of this conversion yields numerous benefits for both the environment and the local economy, as evidenced by studies conducted by Chormare et al. [20] and Seah et al. [61]. The accumulation of biomass poses environmental concerns and represents a missed opportunity for the valorization of resources [14]. Pyrolytic conversion is a thermochemical process characterized by the application of heat to biomass in the absence of oxygen, as described by Sobamowo and Ojolo [62]. The aforementioned study conducted by Kassim et al. [33] demonstrates the promising capability of this procedure to transform diverse forms of organic waste biomass into valuable commodities, while also efficiently mitigating environmental issues.

The study conducted by Bandgar et al. [10] examined the thermochemical process of various natural waste biomass derived from rural areas, as well as the following valorization of the resulting products. The term "heterogenic" is used to emphasize the diverse and mixed nature of biomass, encompassing agricultural wastes, forest trash, and other organic components [13]. The implementation of pyrolysis provides these communities with an effective means of managing their trash in an environmentally sustainable manner, while also taking advantage of economic opportunities through the extraction of valuable products from the process [5]. This article examines the most recent advancements in pyrolysis technology, the optimization of the process, and the utilization of the resulting products. The paper also investigates the challenges and opportunities associated with the implementation of pyrolytic conversion systems in rural regions, taking into account the unique situations, socioeconomic considerations, and environmental variables [53].

The global yearly production of agricultural biomass is estimated to be over 140 billion metric tonnes. A significant quantity of biomass has the potential to produce enormous quantities of valuable energy and minerals. A considerable proportion of the global populace residing in economically disadvantaged nations persists in confronting a dearth of energy accessibility, impacting approximately 1.6 billion individuals. One possible approach involves the conversion of agricultural biomass waste into energy, as proposed by Mishra et al. [46]. The aforementioned methodology possesses the capacity to significantly diminish dependence on non-renewable energy sources, alleviate the emission of greenhouse gases, and offer a viable and enduring energy solution. The utilization of biomass waste has considerable promise as a feasible source of raw materials for both extensive industrial manufacturing and localized small-scale businesses [12]. Biomass include residual materials from agricultural and animal husbandry practices, such as discarded stalks, straws, leaves, roots, husks, nut and seed shells, scrap wood, and animal excrement. The utilization of waste biomass is highly regarded due to its abundant availability, renewable nature, and cost-efficiency [32]. Nations are currently engaged in the active exploration of alternative energy sources as a means to reduce the release of carbon dioxide and other greenhouse gases, in response to the worldwide endeavour to address and mitigate the impacts of climate change. The utilization of biomass as a source of energy generation has minimal implications for the global carbon cycle, thereby aiding in the reduction of our reliance on fossil fuels. The simultaneous advantages of this duality are beneficial for both the economy and the environment.

Pyrolytic conversion refers to a thermochemical process wherein biomass undergoes heating under conditions of limited oxygen, leading to the decomposition of the biomass into various by-products such as biochar, bio-oil, and syngas [7]. The heterogeneous natural waste biomass found in rural areas encompasses a range of organic waste components, including agricultural residues, forest remnants, animal excrement, and similar items [11, 33]. The pyrolysis process entails the concurrent generation of many useful chemicals. For instance, the utilization of biochar as a soil amendment, the augmentation of bio-oil for the production of biofuels, and the harnessing of syngas for the generation of heat and power have been identified as potential applications [14].

The pyrolytic conversion process entails the conversion of diverse natural waste biomass derived from rural regions, with the concomitant extraction of value from this biomass [71]. This study examines the latest advancements in pyrolysis technology, the process of selecting appropriate feedstock, the optimization of the pyrolysis process, and the effective utilization of pyrolysis products. Furthermore, this study investigates the social and environmental implications associated with the utilization of pyrolytic conversion systems in rural regions [13]. There is a rising concern regarding the degradation of the environment and a pressing need for sustainable energy solutions. As a result, there has been a notable increase in the exploration of renewable and eco-friendly alternatives for the utilization of biomass. Biomass, as an abundant and diversified natural resource, holds considerable promise in its capacity to contribute to a sustainable energy framework. The utilization of waste biomass derived from agricultural, forestry, and organic origins is widespread among rural communities.

However, the efficient management and utilization of this diverse biomass are often overlooked, leading to environmental pollution, and missed opportunities for resource recovery [11, 33, 63]. The utilization of pyrolytic conversion technology offers a feasible approach to efficiently address the simultaneous issues of waste management and energy demands in rural areas [22]. Pyrolysis is a thermally driven chemical process whereby biomass undergoes transformation into several beneficial outputs, such as charcoal, bio-oil, and syngas, through exposure to elevated temperatures inside an oxygen-limited environment. This methodology has several advantages, including the production of biochar, which can be utilized as a soil supplement to enhance soil fertility and sequester carbon, the conversion of bio-oil into biofuels, and the utilization of syngas for heat and power generation. The utilization of various types of natural waste biomass derived from rural communities through the process of pyrolytic conversion presents a unique opportunity for the establishment of a circular economy [19]. The aforementioned methodology entails the transformation of waste materials into valuable resources, hence promoting the principles of environmental sustainability and facilitating the growth of local economies. The utilization of underutilized waste biomass is an opportunity for rural communities to reduce their reliance on fossil fuels, mitigate the emission of greenhouse gases, and enhance their socioeconomic resilience.

The main aim of this chapter is to provide important insights into the pyrolytic conversion of diverse natural waste biomass derived from rural communities [27]. Furthermore, it aims to emphasize the importance of simultaneous valorization within the framework of sustainable waste management and resource utilization. Furthermore, the primary objective of this study is to examine the recent progress and scholarly investigations pertaining to the pyrolytic conversion of various types of organic waste biomass derived from rural areas. Additionally, the study seeks to assess the potential of this conversion process in terms of generating economic value, as highlighted by my researcher [32, 48, 56, 63].

The objective of this study is to provide a comprehensive analysis of the potential advantages and drawbacks associated with this distinctive technique through an examination of prominent research findings, technical advancements, and challenges. Furthermore, the main aim of this study is to identify regions with a dearth of information and possible directions for future research. These endeavours possess the capacity to augment the sustainable utilization of biomass resources, stimulate rural development, and promote the advancement of environmental sustainability. The main aim of this study is to provide a significant contribution to the advancement of renewable energy technologies, the improvement of rural livelihoods, and the conservation of natural resources for the benefit of future generations.

11.2 Circularity of Biomass Waste and Plastic Waste via Co-pyrolysis

The increasing global concern about environmental degradation and the depletion of finite resources has spurred the exploration of sustainable strategies for waste management and energy production [48, 60]. The handling of biomass waste and plastic rubbish presents significant challenges in the field of waste management. Nevertheless, the utilization of innovative technologies like co-pyrolysis [43] presents the possibility of circularity. Circularity refers to a deliberate approach aimed at creating self-sustaining systems that effectively reuse or convert waste materials into useful resources. This technique serves to reduce environmental damage and promote the adoption of sustainable practices. According to Bhatt et al. [12], pyrolysis, a thermochemical process, offers a promising solution for effectively tackling the challenges associated with both biomass and plastic waste. By means of co-processing the waste streams, the method of pyrolysis has the capacity to diminish the quantities of garbage and concurrently produce useful commodities, including biofuels, charcoal, and valuable chemicals. Co-pyrolysis refers to a thermochemical process in which biomass and plastic waste are subjected to pyrolysis either concurrently or consecutively [28, 49]. The aforementioned procedure yields significant outputs, such as biofuels, charcoal, and various chemical compounds. This process enables the efficient utilization of biomass and plastic waste, thereby limiting environmental impacts and producing a self-sustaining system in which waste materials are transformed into valuable resources.

11.2.1 Importance of Circular Bioeconomy

The contemporary global community is presently faced with complex challenges, including climate change, resource depletion, and environmental degradation. The circular bioeconomy has emerged as a transformative solution within this context, effectively addressing the pressing concerns at hand [19]. The concept of the circular bioeconomy combines the principles of circular economy with the responsible utilization of biological resources, presenting a holistic and innovative strategy for promoting a resilient and regenerative society [9]. The circular bioeconomy places significant focus on the efficient utilization of renewable biological resources, such as biomass, agricultural produce, and organic waste, with the aim of generating value-added products, energy, and services [12]. The main goal is to reduce reliance on finite resources, mitigate the impacts of climate change, promote biodiversity conservation, and foster the creation of environmentally sustainable innovations. The circular bioeconomy seeks to transition away from a linear framework typified by the "take-make-dispose" approach, towards a regenerative system that prioritizes the reuse, repurposing, and regeneration of resources [19]. The circular bioeconomy's importance in addressing pressing global issues has been discussed by Lopes et al. [39]. This research investigates the various aspects of circularity and bio-based practices, highlighting their potential benefits for the environment, economy, and society. The assessment will additionally assess the alignment between the circular bioeconomy and sustainable development goals, as well as its role in bolstering resilience in the context of worldwide crises. Through a comprehensive examination of existing literature and the integration of findings from multiple scholarly investigations, the principal aim of this research endeavour is to provide a comprehensive understanding of the significance of the circular bioeconomy and its potential to foster transformative and innovative progress towards a sustainable future.

11.2.1.1 Significance of Globular Bioeconomy

Three key concepts can be employed to mitigate the ecological consequences of human activities: the circular economy, the bioeconomy, and the bioeconomy of circularity [65]. These concepts can be employed in the pursuit of establishing a sustainable human civilization and safeguarding the environment. The circular economy is characterized by the utilization of non-hazardous materials and energy sources that do not exhaust natural resources. It aims to facilitate the reduction, minimization, and eventual closure of all material resource loops. Within the societal, economic, and environmental framework, the concept of "bioeconomy" pertains to the transformation of renewable biological resources into biofuels, bioproducts, and biopower in a manner that ensures long-term sustainability [58, 65]. The circular bioeconomy, alternatively referred to as the bio-based circular carbon economy, represents a comprehensive framework that effectively integrates the benefits associated with both of these ideas. This paradigm is alternatively referred to as the circular carbon economy. The circular bioeconomy is strongly dependent on carbon as a fundamental constituent. According to Seah et al. [61] and Tan and Lamers [67], in order to achieve a sustainable future, it is imperative to undertake measures such as the implementation of strategies to "close the carbon cycle" and the development of carbon sinks inside the technosphere.

11.2.2 Biomass Waste

Biomass waste refers to organic resources that are discarded or classified as waste materials due to various human activities, such as agriculture, forestry, industrial processes, and municipal waste [46]. The organic constituents being examined are derived from organisms that are presently extant or have recently become extinct, encompassing flora, fauna, and microorganisms. In contrast to the protracted production process spanning millions of years from decomposed organic material that fossil fuels experience, biomass waste is a renewable and sustainable resource that may be continuously supplied by inherent natural mechanisms. Biomass waste, a significant by-product arising from our activities and natural phenomena, exhibits enormous promise as a valuable asset for the promotion of sustainable development and the preservation of the environment. Biomass waste encompasses a diverse range of organic materials that are derived from living creatures or have been recently produced by them. This category encompasses a range of materials, such as agricultural residues, forestry by-products, organic substances, and other biodegradable entities. In contrast to finite fossil fuels, biomass waste presents itself as a viable and ecologically sustainable option due to its renewability and lack of contribution to climate change when subjected to combustion [43].

Only forty per cent of the trash that is produced by the agriculture industry each year is recycled [35, 69]. This results in an annual production of around 140 billion tonnes of biomass leftovers. Only 4.5% of the lignocellulosic biomass that is produced each year from forestry and agricultural leftovers is exploited [61]. This waste contributes approximately 181.5 billion tonnes of lignocellulosic biomass. In addition to the biomass that is obtained from the agricultural and forestry industries, an additional significant source of biomass is the organic waste that is produced as a result of human activities. As the human population continues to rise, the amount of garbage produced across the globe has also increased, becoming a problem on a global scale. According to Maalouf and Mavropoulos [40], billion tonnes of municipal solid trash are generated every year all over the world. However, a sizeable amount of this rubbish is still not collected by local governments. According to Seah et al. [61], the collected municipal solid trash, only around 19% is recycled, 11% is transformed to electricity, and the other 70% is placed in landfills and dumpsites. Organic trash, which is also known as biodegradable waste, is a word that is used to describe the undesired natural products that are produced by animals and plants. Human waste, sewage, food waste, green trash, abattoir waste, wet market waste, and other types of waste are included in this category of undesirable items. These wastes pose a risk to human health if they are not effectively managed because they can cause unpleasant odours to be released, they can encourage the growth of pathogenic coliform bacteria, and they can make it easier for illnesses to be passed from one person to another [8, 52, 68]. An example of this would be the dangerous bacteria *Escherichia coli* (EC), which produces the toxins and has been linked to a significant number of cases of food-borne illness in human beings. The majority of EC originates from ruminants, with cattle being the most important contributor. Raw meat items, along with freshly harvested fruits and vegetables, have been discovered to be contaminated with EC. Diseases brought on by EC range from relatively harmless bloody diarrhoea (haemorrhagic colitis) to the potentially lethal haemolytic uremic syndrome [30, 55]. The landfills receive a significant amount of organic waste, most of which is comprised of discarded food. According to research conducted by the Food and Agriculture Organisation of the United Nations [31, 59], each year, the food supply chain results in the loss of 1.3 billion tonnes of food, which is equivalent to one-third of the world's total food output. According to Seah et al. [61], the spoiled food is either thrown out at the consumer level or lost at the harvesting and processing stages. According to research by Melikoglu et al. [45] and Schanes et al. [59] up to 95% of the food that is thrown away ends up in landfills, where it serves as a source of methane and other greenhouse gases. According to Schanes et al. [59], this not only helps to ensure the availability of food but also contributes to the acceleration of climate change. Therefore, it is of the utmost importance that global biomass wastes (including forestry residues, agricultural residues, and organic wastes) be treated in an appropriate and secure manner in order to minimise the risks to people's health, decrease their impact on the ecosystem, prevent the depletion of resources, and recycle plant material waste products [23, 37, 70].

The environmental challenges linked to biomass waste encompass the build-up and improper disposal of this waste, leading to the release of greenhouse gases, contamination of soil and water, and degradation of habitats. However, the diligent management and utilization of biomass waste can offer various benefits, such as the production of sustainable energy, improvement of soil fertility, reduction of waste, and mitigation of climate change. When effectively managed, biomass waste can be utilized as a valuable resource, offering various environmental, economic, and social benefits [39]. Communities have the potential to adopt more sustainable and circular resource use and waste management strategies by using biomass waste. The present study investigates diverse approaches for the utilization of biomass waste, encompassing biomass-to-energy technology, composting, and biochar production. Moreover, this review aims to examine the diverse challenges faced in the management of biomass waste, spanning technological, environmental, and policy-related considerations.

11.2.3 Plastic Waste

The problem of plastic trash has emerged as a significant environmental issue in modern times, with far-reaching impacts on ecosystems, human welfare, and the global environment as a whole. Plastics, widely recognized for their durability and versatility, have become essential components of our daily lives, serving various purposes in packaging, consumer goods, electronics, and various industrial domains [12, 28]. However, the widespread utilization and improper disposal of these substances have resulted in the accumulation of plastic waste in many settings, including landfills, oceans, and natural ecosystems, hence leading to substantial ecological consequences.

Plastic waste has been piling up at an alarming rate in the contemporary world as a consequence of the meteoric rise in the manufacture of plastic and the pitiful rate at which plastic gets recycled. According to Maquart et al. [44], around 55% of the world's plastic trash is disposed of in landfills or the natural environment. Every year, one million tonnes of waste are discharged into the ocean, with the majority of that waste coming from Asian countries [50]. According to estimates provided by Lebreton and Andrady [36], the global production of MPW in 2015 was anywhere from 60 and 99 million metric tonnes (Mt). This quantity might triple to 155–265 Mt y^{-1} by the year 2060 if everything continues as it has been up until that point [36]. Even in the next years, the load of future MPW will still be disproportionately large in the continents of Africa and Asia. According to Kibria et al. [34], the large quantity of waste plastics has a negative impact on the ecosystem due to the contamination of the soil caused by landfilling, the contamination of the marine environment caused by ocean dumping, and the pollution of the air caused by open dumping. Plastics have been responsible for a rise in the amount of pollution caused by plastics and have contributed to a wide variety of significant environmental issues.

Plastic wastes can seep from landfills and bioaccumulate; these wastes contain flame retardants, bisphenol A (BPA), phthalates, and heavy metals such as lead and cadmium. Because of this, the consumption of marine species by humans has been linked to an increased risk of developing cardiovascular disease, reproductive problems, and obesity [16]. Therefore, it is absolutely necessary to implement plastic waste management solutions such as recycling or conversion to energy so that each plastic product can be used to its maximum potential and circularity can be achieved. The nation produces 160,038.9 TPD of solid waste, of which 152,749.5 TPD is collected with a collection efficiency of 95.4%, 79,956.3 TPD (50%) is processed, and 29,427.2 TPD (18.4%) is landfilled. The collection efficiency for solid waste is 95.4% and unaccounted for is 50655.4 TPD, equal to 31.7% of the total quantity of solid waste produced. Its state-wise details are provided in Table 11.1.

An analysis has been conducted on the data pertaining to solid waste management (SWM) for the period spanning from 2015 to 2021 is given in Table 11.2.

The extended durability of plastics in the natural environment, with several centuries of degradation processes, has prompted apprehension regarding their detrimental impacts on fauna, aquatic organisms, and the ecological food web. Microplastics, which are little plastic particles that emerge due to the degradation of larger plastic objects, have permeated even the most isolated regions of the planet, presenting potential hazards to both human well-being and biodiversity. This study investigates the diverse environmental consequences of plastic trash, encompassing marine pollution, habitat degradation, and the potential implications for climate change. Furthermore, this analysis examines the socioeconomic dimensions of managing plastic trash, considering the financial implications of waste management systems and the possibilities for implementing circular economy strategies [19]. The text also delves into examining the role played by technology and innovation in managing plastic waste. This includes an exploration of the progress made in recycling technologies and processes that convert plastic into energy. Gaining comprehension of the obstacles and investigating prospective remedies is vital to formulate efficacious policies, advancing public consciousness, and cultivating cooperative endeavours to achieve a sustainable future devoid of plastic.

11.2.4 Co-pyrolysis of Biomass and Plastic

The co-pyrolysis of biomass and plastic is a recently developed and promising thermochemical process that involves the simultaneous or sequential pyrolysis of these different types of feedstocks [1, 49]. Pyrolysis is a thermochemical conversion process that takes place in the absence of oxygen, resulting in the production of useful by-products such biochar, bio-oil, and syngas. Number of benefits can be obtained from the simultaneous thermal breakdown of plastic waste and biomass, including the co-production of energy and high-value products, as well as possible uses that effectively solve waste management issues and advance the development of a more sustainable and circular economic model [46]. Table 11.3 gives the different reports

Table 11.1 Overall solid waste management status in India

No.	State	TPD of wastes			
		Solid waste generated	Collected	Treated	Landfilled
1.	Andhra Pradesh	6898	6829	1133	205
2.	Arunachal Pradesh	236.51	202.11	Nil	27.5
3.	Assam	1199	1091	41.4	0
4.	Bihar	4281.27	4013.55	Not provided	No
5.	Chhattisgarh	1650	1650	1650	0
6.	Goa	226.87	218.87	197.47	22.05
7.	Gujarat	10,373.79	10,332	6946	3385.82
8.	Haryana	5352.12	5291.41	3123.9	2167.51
9.	Himachal Pradesh	346	332	221	111
10.	Jammu & Kashmir	1463.23	1437.28	547.5	376
11.	Jharkhand	2226.39	1851.65	758.26	1086.33
12.	Karnataka	11,085	10,198	6817	1250
13.	Kerala	3543	964.76	2550	Not provided
14.	Madhya Pradesh	8022.5	7235.5	6472	763.5
15.	Maharashtra	22,632.71	22,584.4	15,056.1	1355.36 (unscientifically disposed = 6221.5)
16.	Manipur	282.3	190.3	108.6	81.7
17.	Meghalaya	107.01	93.02	9.64	83.4
18.	Mizoram	345.47	275.92	269.71	0
19.	Nagaland	330.49	285.49	122	7.5
20.	Odisha	2132.95	2097.14	1038.31	1034.33
21.	Punjab	4338.37	4278.86	1894.04	2384.82
22.	Rajasthan	6897.16	6720.476	1210.46	5082.16
23.	Sikkim	71.9	71.9	20.35	51.55
24	Tamil Nadu	13,422	12,844	9430.35	2301.04
25	Telangana	9965	9965	7530	991
26	Tripura	333.9	317.69	214.06	12.9
27	Uttarakhand	1458.46	1378.99	779.85	–
28	Uttar Pradesh	14,710	14,292	5520	0
29	West Bengal	13,709	13,356	667.6	202.23
30	Andaman and Nicobar Islands	89	82	75	7
31	Chandigarh	513	513	69	444

(continued)

Table 11.1 (continued)

No.	State	TPD of wastes			
		Solid waste generated	Collected	Treated	Landfilled
32	DDDNH	267	267	237	14.5
33	Delhi	10,990	10,990	5193.57	5533
34	Lakshadweep	35	17.13	17.13	Nil
35	Puducherry	504.5	482	36	446
Total		**160,038.9**	**152,749.5**	**79,956.3**	**29,427.2**

Source Annual Report on Solid Waste Management (2020–21), CPCB, Delhi

Table 11.2 Trends in SWM of last six years (2015–21) in India

Year	SW generation per capita (g/d)	Gap in SWM (%)
2015–16	118.68	40.99
2016–17	132.78	79.78
2017–18	98.79	9.38
2018–19	121.54	30.35
2019–20	119.26	25.82
2020–21	119.07	31.70

on the co-pyrolysis of waste biomass and biomass. The escalating environmental concerns at a worldwide level, such as the widespread occurrence of plastic pollution and the increase in garbage generation, have inspired a quest for innovative and environmentally friendly strategies for waste management.

The co-pyrolysis of biomass and plastic waste has evolved as an innovative technology that presents a potential solution to tackle these pressing concerns concurrently [60]. The thermochemical process described herein entails the concurrent or

Table 11.3 Co-pyrolysis of natural biomass and plastic waste

Feedstock	Heating rate (°C/min)	Liquid yield (wt%)	References
PP and saw dust	30	24–48	Ye et al. [75]
Palm shell and PS		61.6	Abnisa et al. [2]
Rice husk and PP	–	74.0	Costa et al. [21]
Red oak and HDPE		57.6	Xue et al. [73]
Lignin and PP	–	47.4	Duan et al. [24]
Waste cooking oil and waste plastic	50	62	Mahari et al. [41]
Grape seeds and PS	100	56	Sanahuja-Parejo et al. [57]
Rice straw and PP		35.7	Suriapparao et al. [66]

sequential pyrolysis of biomass and plastic waste, yielding valuable commodities such as biochar, bio-oil, and syngas. The co-pyrolysis of biomass and plastic waste presents several advantages compared to separate pyrolysis methods [18, 49]. This approach not only redirects plastic waste away from landfills and incineration facilities but also utilises biomass waste's energy and high carbon content. Implementing this dual waste valorisation strategy effectively contributes to establishing a circular economy by converting waste materials into valuable resources [19]. This technique not only mitigates environmental pollution but also enhances resource efficiency. When two components combine to modify hydrogen and oxygen levels, synergistic effects result. Plastic wastes' high hydrogen/carbon and low oxygen/carbon rates function synergistically with lignocellulosic biomass's high and low.

One potential strategy for addressing the growing issue of plastic waste management involves the utilization of co-pyrolysis with biomass to convert plastic trash into bio-oil and biochar. Pyrolysis is a thermochemical phenomenon characterized by the thermal decomposition of substances at elevated temperatures, occurring in an oxygen-deprived environment, resulting in the generation of valuable end products. The aforementioned products encompass gas, liquid bio-oil, and solid charcoal [25]. The pyrolysis process of waste plastic is influenced by several process factors, including high temperature and high pressure, leading to different oil yields [4, 6]. The process of plastic pyrolysis yields an oil with favourable attributes, such as a significant heating value and a substantial proportion of hydrocarbons, which are akin to those found in normal gasoline. In contrast, the process of plastic pyrolysis yields significant amounts of ash, and the resultant oil exhibits elevated levels of oxygen and moisture, as well as increased corrosiveness and viscosity [4, 6, 64]. Nevertheless, the adverse characteristics mentioned above can be mitigated with the incorporation of organic matter obtained from the biomass source, as demonstrated by Esso et al. [26] in their study on co-pyrolysis. The findings of recent studies conducted by Esso et al. [26] and Wang et al. [72] indicate that the co-pyrolysis of biomass with plastic has been shown to significantly improve the quality and yield of biofuels, without any observed adverse consequences, in comparison with the pyrolysis of biomass alone.

Co-pyrolysis is widely considered to be a straightforward and safe process for the generation of biofuels of superior quality, eliminating the necessity for bio-oil treatment through high-pressure hydrogenation [15, 3, 73, 80]. Yaman et al. [74] suggest that an additional advantage of co-pyrolysis lies in its capacity to facilitate the retrieval of significant molecules, hence concurrently diminishing dependence on fossil fuels. The customization of co-pyrolysis products according to specific application requirements has been explored in several studies [19, 29, 42, 47, 54]. Several examples of uses include gas turbines, diesel engines, and generators. In contrast to the fractional distillation process employed for fossil fuels, co-pyrolysis technologies yield reduced amounts of carbon emissions. The reason for this is that co-pyrolysis does not necessitate the utilization of solvents, has the ability to operate with or without catalysts, and reduces the quantity of hydrogen needed for the hydrotreatment of crude bio-oil [17, 77, 78]. The co-pyrolysis technique yields biodiesel with equivalent characteristics to petroleum diesel. Moreover, the biodiesel's low viscosity enhances its potential for many applications and augments its combustion efficiency. This finding provides evidence that the utilization of biofuels generated via co-pyrolysis has promise in serving as feasible alternatives to conventional fossil fuels.

Nonetheless, certain obstacles arise in the context of co-pyrolysis and these issues encompass optimizing process conditions, ensuring compatibility among various feedstocks as well as establishing suitable technology and policy frameworks. Continued research and development endeavours are currently being undertaken to address these obstacles and harness the complete capabilities of co-pyrolysis as an environmentally sound waste management and energy generation technology. The simultaneous thermal decomposition of biomass and plastic presents a potentially advantageous method for transforming a wide range of waste materials into valuable commodities, thereby tackling the complexities associated with waste management and promoting the principles of sustainability and circularity within the economy. Through ongoing research and substantial investment in this technology, co-pyrolysis can assume a substantial role in valorizing waste materials and generating renewable energy.

11.2.5 Influential Parameters in Co-pyrolysis

The term "co-pyrolysis" refers to the concurrent thermal decomposition of various materials, resulting in the production of a wide range of products. The process of co-pyrolysis has a significant influence on the quantity, content, and properties of the resulting products. There exist several notable aspects in co-pyrolysis that have the potential to influence the overall process and its outcomes [14]. Pyrolysis is a thermal decomposition process that occurs in the absence of oxygen, resulting in the production of various by-products such as biochar, bio-oil, and syngas. The utilization of co-pyrolysis has generated significant attention owing to its numerous advantages in comparison with pyrolysis employing a singular feedstock. The aforementioned benefits entail enhanced production output and quality, heightened energy efficiency, and the utilization of waste resources for added value. The identification and analysis of relevant parameters in co-pyrolysis provide valuable insights for the optimization of thermochemical conversion processes [38].

The composition of the feedstock mixture has an impact on the distribution of the end product. The attainment of an optimal feedstock ratio holds promise for generating synergistic effects, hence enhancing the overall efficiency of the process. The aforementioned contributions facilitate a deeper understanding of the complex relationship between feedstocks, reaction conditions, and resulting products, hence promoting the development of co-pyrolysis technologies that are more efficient and sustainable. However, it is crucial to engage in ongoing research and experimentation in order to improve our comprehension of the key elements associated with co-pyrolysis and make progress in this field [79]. The optimization of crucial parameters in co-pyrolysis has the potential to promote the development of effective and economically viable approaches for producing value biofuels, boosting soil quality through biochar, and deriving other valuable products from renewable resources. The temperature at which a reaction is conducted has a critical role in determining the rate at which the reaction proceeds as well as the amount of desired products that are formed. Higher temperatures have the ability to promote secondary reactions and cause alterations in the distribution of products, while lower temperatures may potentially result in enhanced char formation. The choice of reactor type, such as fixed-bed, fluidized-bed, or rotary kiln, can have a significant impact on the heat and mass transport characteristics, ultimately affecting the yields and compositions of the final products. The reaction pathways and product yields can be influenced by the pressure conditions experienced during co-pyrolysis. Various external forces can lead to the production of a variety of gaseous and liquid substances.

The duration of the feedstock's exposure to the pyrolysis temperature directly influences the magnitude of the pyrolysis reactions. Increased residence periods can result in more significant reaction extents and alterations in the distribution of products. The energy consumption during pyrolysis and the resulting product distribution can be significantly influenced by the moisture content present in the feedstock, particularly in the case of feedstocks derived from biomass [14]. In order to achieve sustainability and promote eco-friendliness, co-pyrolysis processes must consider environmental factors such as greenhouse gas emissions, waste minimisation, and energy utilization. Pre-treatment techniques, such as drying or grinding, can potentially alter the feedstock's characteristics, hence impacting the behaviour of co-pyrolysis. The utilisation of catalysts or additives during co-pyrolysis can alter the pathways of the reaction and the selectivity of the resulting products, thereby enhancing both the quality and quantity of the products obtained. The rate of temperature increase during co-pyrolysis affects the kinetics and outcomes of various reactions and the composition of the finished product. The concurrent thermal decomposition of biomass and plastic has the potential to decrease the ash content of the resulting biochar. The inhibitory effect of plastic on ash formation during pyrolysis produces a carbonaceous substance with higher purity.

11.3 Choice of Feedstocks

The careful choice of feedstocks is a crucial aspect of the co-pyrolysis process as it has a direct influence on the product distribution and its characteristics. The selection of feedstocks should take into consideration many factors such as the availability and cost of the feedstocks, the desired end products, and the overall sustainability of the process [39]. Proximate analysis is a method employed to ascertain the quantities of moisture, volatile matter, fixed carbon, and ash present in the feedstock. The ultimate analysis provides data on the elemental composition of the feedstock, encompassing carbon, hydrogen, oxygen, nitrogen, and sulphur. The aforementioned evaluations are crucial for acquiring a comprehensive comprehension of the energy content of the feedstock and its subsequent behaviour throughout the pyrolysis process. When making choices for feedstocks for co-pyrolysis, it is crucial to conduct a comprehensive examination of the feedstock's chemical composition, accessibility, cost-effectiveness, suitability for sustainable utilization, and compatibility with the desired final products. The customization of co-pyrolysis processes has been shown to effectively produce valuable products while simultaneously solving environmental and economic challenges [49]. This objective is achieved by a meticulous process of feedstock selection and the optimization of product ratios.

11.3.1 Biomass Feedstock

Biomass feedstocks encompass a wide array of organic resources, which encompass, but are not restricted to, wood, agricultural residues (e.g. straw and husks), energy crops (e.g. switchgrass), and organic waste (e.g. food waste and yard trimmings) [12]. Biomass possesses the inherent advantage of being a renewable resource and exhibits the potential to attain carbon neutrality, rendering it a favourable choice for co-pyrolysis from an environmental perspective. The potential of biomass feedstock in influencing the global transition towards sustainable and renewable energy systems holds significant promise. The importance of biomass feedstock is anticipated to persist as a vital and sustainable resource for bioenergy, encompassing biofuels such as bioethanol and biodiesel, as well as biogas. The advancement of technology and research is anticipated to bolster the efficiency and viability of biomass-derived energy production, hence augmenting its competitiveness relative to conventional fossil fuels. The incorporation of biomass feedstock holds considerable significance within the context of the circular economy paradigm, wherein organic waste materials undergo a conversion process to yield valuable commodities including biochar, bioplastics, and biochemicals. The implementation of this integration will enable the enhancement of resource utilization and the mitigation of waste generation. According to Lopes et al. [39], the implementation of sustainable strategies in the management of biomass feedstock has promise for making a substantial contribution to the process of carbon sequestration. The reason for this phenomenon is that the increase in biomass promotes the uptake of carbon dioxide from the atmosphere.

Integrating bioenergy production with carbon capture and storage (BECCS) has the potential to result in a net removal of carbon dioxide from the atmosphere, contributing to the mitigation of climate change. This initiative provides decentralized energy solutions explicitly focusing on rural and isolated regions. The use of locally accessible biomass resources has the potential to enhance energy accessibility and decrease dependence on centralized power networks. The primary objective of the research and development activities will be to investigate and analyse different feedstock sources for biomass conversion. This encompasses many sources of biomass, such as non-food biomass, algae, agricultural wastes, and waste materials, mitigating the risk of rivalry with food production. Biomass feedstock is paramount in facilitating the shift towards a sustainable energy paradigm. Using this resource as a renewable and carbon–neutral source shows potential in mitigating greenhouse gas emissions, enhancing energy security, and facilitating rural development. In order to fully harness the potential of biomass feedstock, it is imperative to effectively tackle the difficulties associated with land usage, feedstock supply, and environmental impact. The implementation of sustainable practices, such as the adoption of responsible land management techniques, the practice of crop rotation, and the conversion of waste into energy, will play a crucial role in guaranteeing the enduring sustainability of biomass feedstock.

11.3.2 Plastic Feedstock

Plastic waste, including polyethylene, polypropylene, and polystyrene, is commonly utilized as a feedstock for co-pyrolysis. Plastic garbage not only plays a significant role in promoting efficient waste management strategies but also presents the potential for the production of valuable hydrocarbon-rich goods. There exists a necessity to develop scientifically engineered landfills in order to facilitate appropriate waste management, as the current practice involves the disposal of garbage in dumpsites. Based on the data provided by State Pollution Control Boards (SPCBs) and Pollution Control Committees (PCCs), it has been determined that the country possesses a total of 3184 landfill sites. Out of the total, a total of 234 dumpsites have undergone reclamation attempts, and the remaining 8 have been converted into landfills. Uttar Pradesh exhibits the highest prevalence of dumpsites, with a cumulative count of 609, while Madhya Pradesh and Maharashtra trail behind with 326 and 237 dumpsites, respectively. In relation to the reclamation of dumpsites, the state of Maharashtra has witnessed the most number of sites being reclaimed, with a total of 141. This is followed by Madhya Pradesh with 50 sites, Tamil Nadu with 23 sites, and Telangana with 6 sites, in descending order. In the state of Andhra Pradesh, three dumpsites have undergone a transformation into landfills. Furthermore, it should be noted that a landfill has been established in each of the states of Meghalaya, Rajasthan, Sikkim, Telangana, and Chandigarh. Moreover, Fig. 11.1 illustrates the overall solid waste deposition and market share of the waste problem in India.

The inclusion of plastic waste in the pyrolysis process has been observed to positively impact the production of biochar by facilitating the formation of additional char. The outcome of this process yields a greater overall biochar production compared to the exclusive utilization of biomass pyrolysis. The presence of a significant amount of hydrocarbons in plastic waste has the potential to augment the carbon content of biochar, hence enhancing its stability and ability to sequester carbon over an extended period of time. The future prospects for plastic feedstock involve the analysis of environmental challenges linked to plastic waste, as well as the investigation of innovative approaches for converting plastics into valuable commodities. The examination of recycling technologies, including chemical recycling and pyrolysis, as means to convert plastic waste into primary resources for the production of new plastics, fuels, and other important commodities, has garnered significant attention

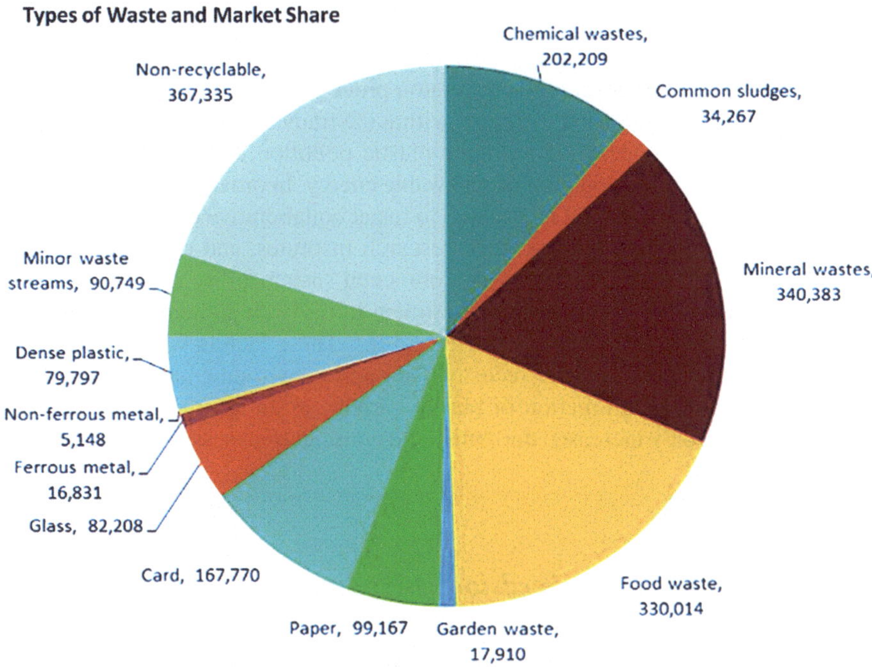

Fig. 11.1 Economic prospectives of the total waste and market share

and generated a sense of optimism. According to Schipfer et al. [60], these technologies have the ability to alleviate the environmental impacts associated with plastic waste and promote the adoption of a circular economy framework. The progress in the development of bio-based and biodegradable plastics holds promise for reducing reliance on polymers derived from fossil fuels and addressing their environmental impacts. The anticipated rise in acceptance and popularity is attributed to the use of sustainable plastic manufacturing methods that employ renewable feedstocks, namely those from plant-based sources. The utilization of plastic feedstock in waste-to-energy conversion methods, such as co-pyrolysis with biomass, has the potential to augment the valorization of trash and facilitate the retrieval of energy resources. The aforementioned technology possesses the capacity to aid the amelioration of the accumulation of plastic trash in landfills and promote the progression of renewable energy generation. The transmission of knowledge regarding the ecological consequences linked to plastic waste and the unexplored possibilities of plastic feedstock in waste management and energy production should be given utmost priority [76].

The dissemination of knowledge to the general population about the responsible utilisation of plastic, recycling practices, and the reduction of waste would facilitate modifications in behaviour and contribute to the advancement of sustainable solutions. The pressing environmental issues associated with plastic waste need creative strategies to transform plastic trash into valuable resources. By adopting sophisticated recycling technologies, using sustainable plastic manufacturing techniques, and implementing waste-to-energy conversion processes, the plastic feedstock may be effectively used as a valuable resource within the framework of circular economy practises [19]. These endeavours mitigate plastic pollution, enhance resource efficiency, and foster the development of renewable energy. In order to fully harness the potential of plastic feedstock, it is essential to foster collaborative endeavours among governmental bodies, industrial sectors, research institutes, and consumers. Minimizing plastic waste and promoting environmental sustainability may be achieved by implementing policies, increasing investment in research and development, and widespread adoption of sustainable practices [76]. These factors will be crucial in designing a future where plastic feedstock supports environmental sustainability. By reconceptualizing the function of plastic garbage as an asset, we may foster a future characterized by increased durability and conscientiousness, benefiting future generations.

11.3.3 Biomass-Plastic Feedstock Blending

The utilization of biomass and plastic waste as co-feedstocks in co-pyrolysis represents a novel approach, wherein these distinct materials are combined and subjected to thermal degradation in the absence of oxygen. The concurrent thermal decomposition of these particular sources of biomass offers a promising prospect for the utilization of discarded resources and the generation of environmentally friendly power. The composition of feedstocks plays a significant role in determining the outcomes of the pyrolysis process and the resulting yields of products. Biomass feedstocks encompass a plentiful supply of cellulose, hemicellulose, and lignin, while plastic waste predominantly consists of extended hydrocarbon chains. Different types of feedstocks can lead to a diverse array of products. In recent years, there has been a significant increase in the adoption of biomass conversion technologies, including the combustion of rice husk and sugarcane bagasse, as well as the gasification of diverse agricultural leftovers. Nevertheless, a considerable proportion of this invaluable resource is still being wasted as a result of its utter indifference or unregulated combustion in agricultural fields. The aforementioned issue is notably widespread in poor nations, when the lack of a suitable institutional structure impedes endeavours aimed at mitigating pollution.

The typical practice of immediately incinerating agricultural leftovers is a commonly utilized technique, it results in the emission of air pollutants, which can have adverse impacts on both human health and the environment. The inadequate utilization of biomass, a resource that is capable of renewal, presents a number of issues. The issue under consideration relates to the efficient utilization of biomass as a feasible energy source and other necessary commodities. Biomass feedstocks, such as agricultural residues, energy crops, and organic waste, are widely recognized as renewable and abundant sources of carbon-based materials. On the other hand, plastic waste, which poses a notable ecological concern, is comprised of elongated hydrocarbon chains derived from polymers sourced from petroleum. The combination of biomass and plastic feedstocks produces synergistic effects, resulting in enhanced process efficiency and unique product outcomes.

The interplay between biomass and plastic during the process of pyrolysis can impact the rate at which reactions occur and the distribution of products, ultimately leading to enhanced yields and modified characteristics of the resultant products compared to pyrolysis conducted using each feedstock separately. Combining these feedstocks and putting them into pyrolysis, various valuable products may be derived, including bio-oil, biochar, and syngas. The combination of biomass and plastic feedstocks for co-pyrolysis has substantial promise in waste valorisation, enhanced product yields, and the generation of bio-oil and biochar of superior quality. Gin et al. [28] examined the impact of the initial plastic-biomass mixture to catalyst ratio, while others examined catalytic pyrolysis of plastic alone. To lower operating costs, pyrolysis of high-density polyethylene utilizing zeolite as the catalyst can be done with low catalyst content. The study found that reducing catalyst content reduces system activity, but raising process temperature can compensate. Research indicates that the catalyst amount used in co-pyrolysis of plastic and biomass impacts product yield and distribution by causing secondary cracking of pyrolyzates, resulting in the production of polyaromatic hydrocarbons and hydrogenated bio-oil fractions used in conventional fuels.

The utilization of biomass presents numerous benefits. Biomass, particularly the biomass resources obtained from agricultural activities, is a viable and plentiful kind of renewable energy. The utilization of this chemical does not result in more carbon dioxide being released into the Earth's atmosphere, rendering it a promising substitute for fossil fuels and a valuable instrument in the mitigation of greenhouse gas emissions. The utilization of biomass has promise in enhancing farmers' economic resources while ensuring the preservation of their ability to cultivate vital food and non-food crops. This issue is currently under serious deliberation over prioritization. However, additional research is necessary to improve the selection of feedstock combinations, develop appropriate catalysts and additives, and address the economic and environmental considerations related to the extensive adoption of this methodology. The combination of biomass and plastic feedstock exhibits significant potential in the realm of sustainable waste management and the production of renewable energy. However, the combination of biomass and plastic feedstocks in co-pyrolysis presents certain challenges that need to be resolved. The aforementioned concerns encompass determining appropriate ratios for the feedstock, addressing potential by-products, such as halogens derived from plastics, and assessing the economic feasibility and environmental implications of this methodology.

11.4 Synergistic Effects of Co-pyrolysis

The phenomenon of synergistic effects in co-pyrolysis refers to the enhanced performance and benefits that arise from the simultaneous pyrolysis of multiple diverse feedstocks. The co-pyrolysis process entails the thermal decomposition of feedstocks, which can lead to improved reactions, increased product yields, and enhanced properties of the resulting products when compared to the pyrolysis of individual feedstocks. The phenomenon of synergy in co-pyrolysis arises from the intrinsic complementarity observed among various feedstocks, which possess unique chemical compositions and features. The simultaneous pyrolysis of different feedstocks can have a reciprocal influence on their individual characteristics, leading to a thermal conversion procedure characterized by improved efficiency and efficacy. Overall, the synergistic impacts of co-pyrolysis offer favourable potential for augmenting the efficacy of transforming diverse feedstocks into valuable commodities. However, the realization of these benefits relies on a thorough comprehension of feedstock interactions and the careful selection of feedstock combinations.

The progress made in co-pyrolysis research holds great significance in the advancement of sustainable and efficient pyrolysis methods for a range of applications, such as bioenergy production, waste disposal, and soil improvement. The incorporation of synergistic effects will play a pivotal role in attaining these aims. The utilization of biomass and plastic feedstocks in a mutually beneficial manner holds promise for the production of bio-oils with enhanced energy density. The observed rise in calorific value can be attributed to the integration of hydrocarbons derived from the plastic feedstocks into the bio-oil. The integration of biomass and plastic feedstock blending holds considerable promise in addressing the growing complexities associated with plastic waste management and the demand for environmentally viable energy options. The utilization of biomass and plastic feedstocks in co-pyrolysis represents an innovative and environmentally aware approach to converting waste materials into valuable resources. The co-pyrolysis process, namely the combination of biomass and plastic feedstocks, has demonstrated the manifestation of synergistic effects. The aforementioned impacts have the potential to yield increased bio-oil output, improved characteristics, and tailored biochar properties. As previously said, the advantages demonstrate the ability of this approach to make a substantial contribution to the production of sustainable and efficient biofuels, the valorization of waste, and the promotion of environmental stewardship.

11.4.1 Synergistic Mechanism

The phenomenon in which the combined effect of multiple components or factors exceeds the sum of their individual effects is commonly known as the synergistic mechanism. The concept of synergistic mechanism holds significant importance and can be observed across several scientific and technical contexts. The statement underscores the significance of collaboration and the mutually beneficial improvement of separate elements. The concept of synergy is widely acknowledged across diverse disciplines, encompassing chemistry, biology, ecology, and the social sciences. In the fields of chemistry and biology, the interplay of various chemicals or molecules can give rise to effects that exhibit greater potency or novelty compared to the effects produced by individual substances in isolation.

In the field of ecology, the coexistence of different species can give rise to a harmonious ecosystem in which each species plays a role in supporting the health and functioning of others, so enhancing the overall stability and resilience of the system. Gaining comprehension of synergistic mechanisms inside industrial applications and devising strategies to harness their potential are pivotal measures in enhancing process efficiency, facilitating product development, and refining system architecture. In the context of co-pyrolysis, as discussed earlier, the amalgamation of biomass and plastic feedstocks has the potential to generate synergistic outcomes, hence enhancing the overall efficiency and product yields of the pyrolysis procedure as shown in Fig. 11.2.

The synergistic process encompasses not only positive interactions but also negative interactions, when the combined effect is more detrimental than the total of the separate effects. The utilization of the term "antagonistic mechanism" may be deemed suitable for application in such circumstances. The concept of synergistic processes underscores the importance of holistic consideration of a system, rather than focusing solely on its individual components. In order to enhance their ability to develop more efficient and effective solutions, scientists and engineers might strategically leverage synergistic interactions. This phenomenon has the potential to drive advancements across multiple industries and facilitate the progress of sustainable and innovative technologies.

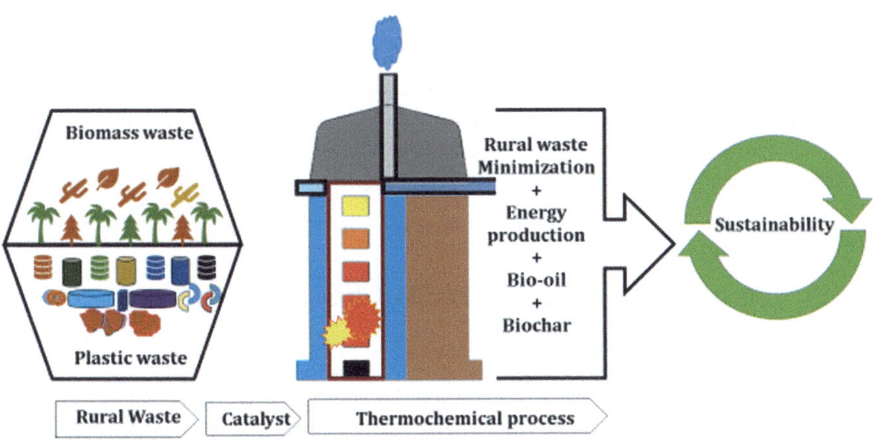

Fig. 11.2 Waste utilization using synergistic thermochemical co-processing

11.4.2 Impact on Quality and Yield of Biofuels

The potential influence of synergistic effects during co-pyrolysis, particularly in the blending of biomass and plastic feedstocks, has the capacity to significantly impact the quality and yield of the resulting biofuels. The inclusion of plastic feedstocks in the co-pyrolysis process has the potential to enhance the yield of bio-oil. During the pyrolysis process, plastics, which contain a significant amount of hydrocarbons, introduce additional volatile compounds into the mixture. Consequently, the overall bio-oil production is greater in comparison with biomass pyrolysis. The presence of biomass and plastic feedstocks may result in a reduction of oxygen-containing molecules in the bio-oil compared to previous conditions.

As a result, the process yields bio-oil with higher energy content, increased stability, reduced acidity, and improved properties, rendering it highly suitable for utilization as a renewable fuel source. The inclusion of plastic in the feedstock mix during the co-pyrolysis process has the potential to mitigate the formation of excessive char. As a consequence of this phenomenon, there is a potential decrease in the quantity of solid char generated, leading to an increased production of valuable liquid goods, notably bio-oil. By adjusting the feedstock ratio and modifying the processing conditions, it is possible to tailor the composition of the resulting bio-oil to meet specific application requirements. This allows for the production of biofuels that possess desired characteristics suitable for transportation or industrial purposes. This characteristic enables the utilization of bio-oil in a diverse range of applications. The co-pyrolysis process entails the interaction between biomass and plastic, which has the potential to alter the properties of the resulting biochar. This might potentially lead to the production of biochar characterized by an increased carbon content, reduced ash content, and improved surface area. Consequently, it would be more suitable for enhancing soil quality and facilitating carbon sequestration. It is imperative to emphasize that the influence of synergistic effects on the quality and quantity of biofuels generated through co-pyrolysis is governed by a multitude of criteria. The elements encompassed in this study comprise the composition of the feedstock, the ratio of feedstock, the temperature at which pyrolysis occurs, and the potential influence of catalysts or additives. Therefore, it is imperative to optimize these parameters in order to fully exploit the potential of the biomass-plastic feedstock blending process and achieve the desired outcomes in biofuel production.

11.4.3 Impact on Quality and Yield of Biochar

The potential impact of synergistic effects during co-pyrolysis, namely the combination of biomass and plastic feedstocks, can also have implications for the characteristics and quantity of the resulting biochar. Biochar is generated by the process of biomass pyrolysis, resulting in the formation of a solid carbonaceous material. Biochar has the potential to serve as a soil amendment, a carbon sequestration agent, or a precursor for several other applications. Enhancing the surface area of biochar has the potential to enhance its ability to absorb and retain water and nutrients, hence augmenting its efficacy as a constituent of soil amendments. The combination of biomass and plastic feedstock can result in modified attributes of biochar, perhaps enhancing its capacity for nutrient retention. Biochar is found to be a more effective soil conditioner, hence promoting plant growth and enhancing soil fertility. The composition of biochar can be influenced by the ratio of feedstocks employed in the co-pyrolysis process as well as the processing parameters. The interplay between the two components, which adjusts the amount of hydrogen and oxygen in the system, is what creates the synergistic effects and makes them possible. For instance, the high hydrogen/carbon (H/C) ratios of plastic wastes and the low oxygen/carbon (O/C) ratios of lignocellulosic biomass operate synergistically together. Similarly, the high oxygen/carbon (O/C) ratios of lignocellulosic biomass and the low hydrogen/carbon (H/C) ratios of lignocellulosic biomass also work well together.

The characteristics of biochar can be adjusted to meet the specific needs of a given soil type or agricultural pursuit through the manipulation of these variables. It is imperative to bear in mind that the impact of synergistic effects on the quality and productivity of biochar is contingent upon various factors. The factors encompassed in this category consist of the various types of biomass and plastic feedstocks utilized, the ratio of feedstocks employed, the temperature at which the pyrolysis process is conducted, and the potential inclusion of catalysts or additives. It is imperative to optimize these parameters in a suitable manner in order to fully exploit the potential benefits that may be derived from the combination of biomass and plastic feedstock in the process of biochar synthesis. The aforementioned advantages exemplify the potential of this approach in producing biochar that may be tailored to improve soil health and trap carbon, thereby promoting environmentally sustainable agriculture and conserving natural resources.

11.5 Conclusions and Future Perspectives

Pyrolytic conversion and valorization of rural waste biomass provides a sustainable, decentralized waste-to-energy option. Pyrolysis technology in rural areas can improve resource efficiency and ecological sensitivity. Management of diverse waste biomass in rural areas is difficult. Pyrolysis converts organic waste into valuable products, reducing landfill trash and pollution. The thermal breakdown of waste biomass in pyrolysis can provide sustainable energy to rural areas. This could boost energy security, rural development, and economic prosperity. Pyrolysis biochar can improve soil fertility and water retention. The benefits include increased agricultural yield and less chemical fertilizer use. Pyrolysis turns waste biomass into biochar, bio-oil, and syngas. Circular economy strives to improve resource efficiency and reduce fossil fuel use. Rural pyrolysis technology can help spread knowledge and skills. Innovative waste management solutions could empower local populations and boost rural development.

Pyrolytic conversion and valorization of heterogeneous rural waste biomass provides a gateway to a cleaner, self-sufficient future. Pyrolyzing waste biomass can provide renewable energy, soil health, and environmental protection to rural areas. Technical accessibility, infrastructure development, and community participation are unique challenges to implementation. Government, research, and business must collaborate for innovation, customization, and sustainability. Pyrolysis technology can help rural communities promote environmental stewardship, socioeconomic development, and resilient, self-sustaining communities that contribute to the low carbon, circular economy, and global sustainability. The review covers feedstock selection, pyrolysis process optimization, product characterization, environmental impacts, and socioeconomic impacts. This literature analysis shows how pyrolytic conversion can turn agricultural waste into a valuable resource while resolving environmental concerns and fostering a sustainable future. Pyrolytic conversion and valorization of waste biomass could improve rural communities' environmental sustainability and inclusion.

References

1. Abnisa, F., & Daud, W. M. A. W. (2014). A review on co-pyrolysis of biomass: An optional technique to obtain a high-grade pyrolysis oil. *Energy Conversion and Management, 87*, 71–85. https://doi.org/10.1016/j.enconman.2014.07.007
2. Abnisa, F., Daud, W. W., Ramalingam, S., Azemi, M. N. B. M., & Sahu, J. N. (2013). Co-pyrolysis of palm shell and polystyrene waste mixtures to synthesis liquid fuel. *Fuel, 108*, 311–318. https://doi.org/10.1016/j.fuel.2013.02.013
3. Alabdrabalnabi, A., Gautam, R., & Sarathy, S. M. (2022). Machine learning to predict biochar and bio-oil yields from co-pyrolysis of biomass and plastics. *Fuel, 328*, 125303. https://doi.org/10.1016/j.fuel.2022.125303
4. Alam, S. S., Husain Khan, A., & Khan, N. A. (2022). Plastic waste management via thermochemical conversion of plastics into fuel: A review. *Energy Sources, Part A: Recovery, Utilization, and Environmental Effects, 44*(3), 1–20. https://doi.org/10.1080/15567036.2022.2097750
5. Alazaiza, M. Y., Albahnasawi, A., Ali, G. A., Bashir, M. J., Nassani, D. E., Al Maskari, T., & Abujazar, M. S. S. (2022). Application of natural coagulants for pharmaceutical removal from water and wastewater: A review. *Water, 14*, 140. https://doi.org/10.3390/w14020140
6. Al-Rumaihi, A., Shahbaz, M., Mckay, G., Mackey, H., & Al-Ansari, T. (2022). A review of pyrolysis technologies and feedstock: A blending approach for plastic and biomass towards optimum biochar yield. *Renewable and Sustainable Energy Reviews, 167*, 112715. https://doi.org/10.1016/j.rser.2022.112715
7. Anand, A., Kumar, V., & Kaushal, P. (2022). Biochar and its twin benefits: Crop residue management and climate change mitigation in India. *Renewable and Sustainable Energy Reviews, 156*, 111959. https://doi.org/10.1016/j.rser.2021.111959
8. Anand, U., Reddy, B., Singh, V. K., Singh, A. K., Kesari, K. K., Tripathi, P., Kumar, P., Tripathi, V., & Simal-Gandara, J. (2021). Potential environmental and human health risks caused by antibiotic-resistant bacteria (ARB), antibiotic resistance genes (ARGs) and emerging contaminants (ECs) from municipal solid waste (MSW) landfill. *Antibiotics, 10*(4), 374. https://doi.org/10.3390/antibiotics10040374
9. Armenise, S., SyieLuing, W., Ramírez-Velásquez, J. M., Launay, F., Wuebben, D., Ngadi, N., & Muñoz, M. (2021). Plastic waste recycling via pyrolysis: A bibliometric survey and literature

review. *Journal of Analytical and Applied Pyrolysis, 158*, 105265. https://doi.org/10.1016/j.jaap.2021.105265
10. Bandgar, P. S., Jain, S., & Panwar, N. L. (2022). A comprehensive review on optimization of anaerobic digestion technologies for lignocellulosic biomass available in India. *Biomass and Bioenergy, 161*, 106479. https://doi.org/10.1016/j.biombioe.2022.106479
11. Ben-Iwo, J., Manovic, V., & Longhurst, P. (2016). Biomass resources and biofuels potential for the production of transportation fuels in Nigeria. *Renewable and Sustainable Energy Reviews, 63*, 172–192. https://doi.org/10.1016/j.rser.2016.05.050
12. Bhatt, M., Chakinala, A. G., Joshi, J. B., Sharma, A., Pant, K. K., Shah, K., & Sharma, A. (2021). Valorization of solid waste using advanced thermo-chemical process: A review. *Journal of Environmental Chemical Engineering, 9*, 105434. https://doi.org/10.1016/j.jece.2021.105434
13. Butler, E., Devlin, G., Meier, D., & McDonnell, K. (2011). A review of recent laboratory research and commercial developments in fast pyrolysis and upgrading. *Renewable and Sustainable Energy Reviews, 15*, 4171–4186. https://doi.org/10.1016/j.rser.2011.07.035
14. Caputo, C., & Mašek, O. (2021). Spear (solar pyrolysis energy access reactor): Theoretical design and evaluation of a small-scale low-cost pyrolysis unit for implementation in rural communities. *Energies, 14*(8), 2189. https://doi.org/10.3390/en14082189
15. Çepelioğullar, Ö., & Pütün, A. E. (2013). Thermal and kinetic behaviors of biomass and plastic wastes in co-pyrolysis. *Energy Conversion and Management, 75*, 263–270. https://doi.org/10.1016/j.enconman.2013.06.036
16. Chen, H. L., Nath, T. K., Chong, S., Foo, V., Gibbins, C., & Lechner, A. M. (2021). The plastic waste problem in Malaysia: Management, recycling and disposal of local and global plastic waste. *SN Applied Sciences, 3*, 1–15. https://doi.org/10.1007/s42452-021-04234-y
17. Chen, M., Zhang, S., Su, Y., Niu, X., Zhu, S., & Liu, X. (2022). Catalytic co-pyrolysis of food waste digestate and corn husk with CaO catalyst for upgrading bio-oil. *Renewable Energy, 186*, 105–114. https://doi.org/10.1016/j.renene.2021.12.139
18. Chen, W., Shi, S., Zhang, J., Chen, M., & Zhou, X. (2016). Co-pyrolysis of waste newspaper with high-density polyethylene: Synergistic effect and oil characterization. *Energy Conversion and Management, 112*, 41–48. https://doi.org/10.1016/j.enconman.2016.01.005
19. Chew, K. W., Chia, S. R., Chia, W. Y., Cheah, W. Y., Munawaroh, H. S. H., & Ong, W. J. (2021). Abatement of hazardous materials and biomass waste via pyrolysis and co-pyrolysis for environmental sustainability and circular economy. *Environmental Pollution, 278*, 116836. https://doi.org/10.1016/j.envpol.2021.116836
20. Chormare, R., Sahoo, T. P., Chanchpara, A., Saravaia, H. T., & Madhava, A. K. (2023). Thermochemical conversion and kinetic evaluation of *Casuarina equisetifolia* pines to biochar and their utilization in sequestering toxic metal ions. *Biomass Conversion and Biorefinery, 13*, 9423–9434. https://doi.org/10.1007/s13399-022-03526-6
21. Costa, P., Pinto, F., Miranda, M., André, R., & Rodrigues, M. (2014). Study of the experimental conditions of the co-pyrolysis of rice husk and plastic wastes. *Chemical Engineering Transactions, 39*, 1639–1644. https://doi.org/10.3303/CET1439274
22. Demirbas, A., Pehlivan, E., & Altun, T. (2006). Potential evolution of Turkish agricultural residues as bio-gas, bio-char and bio-oil sources. *International Journal of Hydrogen Energy, 31*(5), 613–620. https://doi.org/10.1016/j.ijhydene.2005.06.003
23. Dhanya, B. S., Mishra, A., Chandel, A. K., & Verma, M. L. (2020). Development of sustainable approaches for converting the organic waste to bioenergy. *Science of the Total Environment, 25*, 138109. https://doi.org/10.1016/j.scitotenv.2020.138109
24. Duan, D., Wang, Y., Dai, L., Ruan, R., Zhao, Y., Fan, L., Tayier, M., & Liu, Y. (2017). Ex-situ catalytic copyrolysis of lignin and polypropylene to upgrade bio-oil quality by microwave heating. *Bioresource Technology, 241*, 207–213. https://doi.org/10.1016/j.biortech.2017.04.104
25. Durak, H. (2023). Comprehensive assessment of thermochemical processes for sustainable waste management and resource recovery. *Processes, 11*(7), 2092. https://doi.org/10.3390/pr11072092

26. Esso, S. B. E., Xiong, Z., Chaiwat, W., Kamara, M. F., Longfei, X., Xu, J., Ebako, J., Jiang, L., Su, S., Hu, S., Wang, Y., & Xiang, J. (2022). Review on synergistic effects during co-pyrolysis of biomass and plastic waste: Significance of operating conditions and interaction mechanism. *Biomass and Bioenergy, 159*, 106415. https://doi.org/10.1016/j.biombioe.2022.106415
27. Foong, S. Y., Liew, R. K., Yang, Y., Cheng, Y. W., Yek, P. N. Y., Mahari, W. A. W., Lee, W. Y., Han, C. S., Vo, D.-V. N., Le, Q. V., Aghbashlo, M., Tabatabaei, M., Sonne, C., Peng, W., & Lam, S. S. (2020). Valorization of biomass waste to engineered activated biochar by microwave pyrolysis: Progress, challenges, and future directions. *Chemical Engineering Journal, 389*, 124401. https://doi.org/10.1016/j.cej.2020.124401
28. Gin, A. W., Hassan, H., Ahmad, M. A., Hameed, B. H., & Din, A. M. (2021). Recent progress on catalytic co-pyrolysis of plastic waste and lignocellulosic biomass to liquid fuel: The influence of technical and reaction kinetic parameters. *Arabian Journal of Chemistry, 14*(4), 103035. https://doi.org/10.1016/j.arabjc.2021.103035
29. Guo, M., & Bi, J. C. (2015). Characteristics and application of co-pyrolysis of coal/biomass blends with solid heat carrier. *Fuel Processing Technology, 138*, 743–749. https://doi.org/10.1016/j.fuproc.2015.07.018
30. Hunt, J. M. (2010). Shiga toxin–producing *Escherichia coli* (STEC). *Clinics in Laboratory Medicine, 30*, 21–45. https://doi.org/10.1016/j.cll.2009.11.001
31. Ishangulyyev, R., Kim, S., & Lee, S. H. (2019). Understanding food loss and waste—Why are we losing and wasting food? *Foods, 8*(8), 297. https://doi.org/10.3390/foods8080297
32. Kartik, S., Balsora, H. K., Sharma, M., Saptoro, A., Jain, R. K., Joshi, J. B., & Sharma, A. (2022). Valorization of plastic wastes for production of fuels and value-added chemicals through pyrolysis—A review. *Thermal Science and Engineering Progress, 32*, 101316. https://doi.org/10.1016/j.tsep.2022.101316
33. Kassim, F. O., Thomas, C. P., & Afolabi, O. O. (2022). Integrated conversion technologies for sustainable agri-food waste valorization: A critical review. *Biomass and Bioenergy, 156*, 106314. https://doi.org/10.1016/j.biombioe.2021.106314
34. Kibria, M. G., Masuk, N. I., Safayet, R., Nguyen, H. Q., & Mourshed, M. (2023). Plastic waste: Challenges and opportunities to mitigate pollution and effective management. *International Journal of Environmental Research, 17*(1), 20. https://doi.org/10.1007/s41742-023-00507-z
35. Kumar, S., Lohan, S. K., & Parihar, D. S. (2023). Biomass energy from agriculture. In A. Rakshit, A. Biswas, D. Sarkar, V. S. Meena, & R. Datta (Eds.), *Handbook of energy management in agriculture*. Springer. https://doi.org/10.1007/978-981-19-7736-7_10-1
36. Lebreton, L., & Andrady, A. (2019). Future scenarios of global plastic waste generation and disposal. *Palgrave Communications, 5*(1), 1–11. https://doi.org/10.1057/s41599-018-0212-7
37. Lee, S. Y., Sankaran, R., Chew, K. W., Tan, C. H., Krishnamoorthy, R., Chu, D. T., & Show, P. L. (2019). Waste to bioenergy: A review on the recent conversion technologies. *BMC Energy, 1*(1), 1–22. https://doi.org/10.1186/s42500-019-0004-7
38. Li, X., Li, J., Zhou, G., Feng, Y., Wang, Y., Yu, G., Deng, S., Huang, J., & Wang, B. (2014). Enhancing the production of renewable petrochemicals by co-feeding of biomass with plastics in catalytic fast pyrolysis with ZSM-5 zeolites. *Applied Catalysis A: General, 481*, 173–182. https://doi.org/10.1016/j.apcata.2014.05.015
39. Lopes, T. F., Carvalheiro, F., Duarte, L. C., Gírio, F., Quintero, J. A., & Aroca, G. (2019). Techno-economic and life-cycle assessments of small-scale biorefineries for isobutene and xylo-oligosaccharides production: A comparative study in Portugal and Chile. *Biofuels, Bioproducts and Biorefining, 13*(5), 1321–1332. https://doi.org/10.1002/bbb.2036
40. Maalouf, A., & Mavropoulos, A. (2023). Re-assessing global municipal solid waste generation. *Waste Management & Research, 41*(4), 936–947. https://doi.org/10.1177/0734242X221074116
41. Mahari, W. A. W., Tung, C., & Haur, W. (2018). Microwave copyrolysis of waste polyolefins and waste cooking oil: Influence of N_2 atmosphere versus vacuum environment. *Energy Conversion and Management, 171*, 1292–1301. https://doi.org/10.1016/j.enconman.2018.06.073
42. Mahari, W. A. W., Azwar, E., Foong, S. Y., Ahmed, A., Peng, W., Tabatabaei, M., Aghbashlo, M., Park, Y.-K., Sonne, C., & Lam, S. S. (2021). Valorization of municipal wastes using

co-pyrolysis for green energy production, energy security, and environmental sustainability: A review. *Chemical Engineering Journal, 421*, 129749. https://doi.org/10.1016/j.cej.2021.129749

43. Manikandan, S., Subbaiya, R., Biruntha, M., Krishnan, R. Y., Muthusamy, G., & Karmegam, N. (2022). Recent development patterns, utilization and prospective of biofuel production: Emerging nanotechnological intervention for environmental sustainability—A review. *Fuel, 314*, 122757. https://doi.org/10.1016/j.fuel.2021.122757
44. Maquart, P. O., Froehlich, Y., & Boyer, S. (2022). Plastic pollution and infectious diseases. *The Lancet Planetary Health, 6*(10), e842–e845. https://doi.org/10.1016/S2542-5196(22)00198-X
45. Melikoglu, M., Lin, C. S. K., & Webb, C. (2013). Analysing global food waste problem: Pinpointing the facts and estimating the energy content. *Central European Journal of Engineering, 3*, 157–164. https://doi.org/10.2478/s13531-012-0058-5
46. Mishra, R., Ong, H. C., & Lin, C. W. (2023). Progress on co-processing of biomass and plastic waste for hydrogen production. *Energy Conversion and Management, 284*, 116983. https://doi.org/10.1016/j.enconman.2023.116983
47. Mohamed, B. A., O'Boyle, M., & Li, L. Y. (2023). Co-pyrolysis of sewage sludge with lignocellulosic and algal biomass for sustainable liquid and gaseous fuel production: A life cycle assessment and techno-economic analysis. *Applied Energy, 346*, 121318. https://doi.org/10.1016/j.apenergy.2023.121318
48. Mondal, P. (2021). Techno-economic and environmental performance assessment of a MSW to energy plant for Indian urban sectors. *Thermal Science and Engineering Progress, 21*, 100777. https://doi.org/10.1016/j.tsep.2020.100777
49. Neha, S., Ramesh, K. P. K., & Remya, N. (2022). Techno-economic analysis and life cycle assessment of microwave co-pyrolysis of food waste and low-density polyethylene. *Sustainable Energy Technologies and Assessments, 52*, 102356. https://doi.org/10.1016/j.seta.2022.102356
50. Ng, C. H., Mistoh, M. A., Teo, S. H., Galassi, A., Ibrahim, A., Sipaut, C. S., Foo, J., Seay, J., Taufiq-Yap, Y. H., & Janaun, J. (2023). Plastic waste and microplastic issues in Southeast Asia. *Frontiers in Environmental Science, 11*, 427. https://doi.org/10.3389/fenvs.2023.1142071
51. Okolie, J. A., Epelle, E. I., Tabat, M. E., Orivri, U., Amenaghawon, A. N., Okoye, P. U., & Gunes, B. (2022). Waste biomass valorization for the production of biofuels and value-added products: A comprehensive review of thermochemical, biological and integrated processes. *Process Safety and Environmental Protection, 159*, 323–344. https://doi.org/10.1016/j.psep.2021.12.049
52. Omang, D. I., John, G. E., Inah, S. A., & Bisong, J. O. (2021). Public health implication of solid waste generated by households in Bekwarra Local Government area. *African Health Sciences, 21*(3), 1467–1473. https://doi.org/10.4314/ahs.v21i3.58
53. Ong, H. C., Chen, W. H., Farooq, A., Gan, Y. Y., Lee, K. T., & Ashokkumar, V. (2019). Catalytic thermochemical conversion of biomass for biofuel production: A comprehensive review. *Renewable and Sustainable Energy Reviews, 113*, 109266. https://doi.org/10.1016/j.rser.2019.109266
54. Rajkumar, P., & Murugavelh, S. (2022). Co-pyrolysis of wheat husk and residual tyre: Techno-economic analysis, performance and emission characteristics of pyro oil in a diesel engine. *Bioresource Technology Reports, 19*, 101164. https://doi.org/10.1016/j.biteb.2022.101164
55. Ray, R., & Singh, P. (2022). Prevalence and implications of shiga toxin-producing *E. coli* in farm and wild ruminants. *Pathogens, 11*(11), 1332. https://doi.org/10.3390/pathogens11111332
56. Rezaei, M., Ghobadian, B., Samadi, S. H., & Karimi, S. (2018). Electric power generation from municipal solid waste: A techno-economical assessment under different scenarios in Iran. *Energy, 152*, 46–56. https://doi.org/10.1016/j.energy.2017.10.109
57. Sanahuja-Parejo, O., Veses, A., Navarro, M. V., López, J. M., Murillo, R., Callén, M. S., & García, T. (2019). Drop-in biofuels from the co-pyrolysis of grape seeds and polystyrene. *Chemical Engineering Journal, 377*, 120246. https://doi.org/10.1016/j.cej.2018.10.183

58. Scarlat, N., Dallemand, J. F., Monforti-Ferrario, F., & Nita, V. (2015). The role of biomass and bioenergy in a future bioeconomy: Policies and facts. *Environmental Development, 15*, 3–34. https://doi.org/10.1016/j.envdev.2015.03.006
59. Schanes, K., Dobernig, K., & Gözet, B. (2018). Food waste matters—A systematic review of household food waste practices and their policy implications. *Journal of Cleaner Production, 182*, 978–991. https://doi.org/10.1016/j.jclepro.2018.02.030
60. Schipfer, F., Pfeiffer, A., & Hoefnagels, R. (2022). Strategies for the mobilization and deployment of local low-value, heterogeneous biomass resources for a circular bioeconomy. *Energies, 15*(2), 433. https://doi.org/10.3390/en15020433
61. Seah, C. C., Tan, C. H., Arifin, N. A., Hafriz, R. S. R. M., Salmiaton, A., Nomanbhay, S., & Shamsuddin, A. H. (2023). Co-pyrolysis of biomass and plastic: Circularity of wastes and comprehensive review of synergistic mechanism. *Results in Engineering*, 100989. https://doi.org/10.1016/j.rineng.2023.100989
62. Sobamowo, G. M., & Ojolo, S. J. (2018). Techno-economic analysis of biomass energy utilization through gasification technology for sustainable energy production and economic development in Nigeria. *Journal of Energy*. https://doi.org/10.1155/2018/4860252
63. Somorin, T. O., Adesola, S., & Kolawole, A. (2017). State-level assessment of the waste-to-energy potential (via incineration) of municipal solid wastes in Nigeria. *Journal of Cleaner Production, 164*, 804–815. https://doi.org/10.1016/j.jclepro.2017.06.228
64. Soni, V. K., Singh, G., Vijayan, B. K., Chopra, A., Kapur, G. S., & Ramakumar, S. S. V. (2021). Thermochemical recycling of waste plastics by pyrolysis: A review. *Energy & Fuels, 35*(16), 12763–12808. https://doi.org/10.1021/acs.energyfuels.1c01292
65. Stephenson, P. J., & Damerell, A. (2022). Bioeconomy and circular economy approaches need to enhance the focus on biodiversity to achieve sustainability. *Sustainability, 14*(17), 10643. https://doi.org/10.3390/su141710643
66. Suriapparao, D. V., Vinu, R., Shukla, A., & Haldar, S. (2020). Effective deoxygenation for the production of liquid biofuels via microwave assisted co-pyrolysis of agro residues and waste plastics combined with catalytic upgradation. *Bioresource Technology, 302*, 122775. https://doi.org/10.1016/j.biortech.2020.122775
67. Tan, E. C., & Lamers, P. (2021). Circular bioeconomy concepts—A perspective. *Frontiers in Sustainability, 2*, 701509. https://doi.org/10.3389/frsus.2021.701509
68. Tomita, A., Cuadros, D. F., Burns, J. K., Tanser, F., & Slotow, R. (2020). Exposure to waste sites and their impact on health: A panel and geospatial analysis of nationally representative data from South Africa, 2008–2015. *The Lancet Planetary Health, 4*(6), e223–e234. https://doi.org/10.1016/S2542-5196(20)30101-7
69. Tripathi, N., Hills, C. D., Singh, R. S., & Atkinson, C. J. (2019). Biomass waste utilisation in low-carbon products: Harnessing a major potential resource. *NPJ Climate and Atmospheric Science, 2*(1), 35. https://doi.org/10.1038/s41612-019-0093-5
70. Uddin, M. N., Siddiki, S. Y. A., Mofijur, M., Djavanroodi, F., Hazrat, M. A., Show, P. L., Ahmed, S.F., & Chu, Y. M. (2021). Prospects of bioenergy production from organic waste using anaerobic digestion technology: A mini review. *Frontiers in Energy Research, 9*, 627093. https://doi.org/10.3389/fenrg.2021.627093
71. van Schalkwyk, D. L., Mandegari, M., Farzad, S., & Görgens, J. F. (2020). Techno-economic and environmental analysis of bio-oil production from forest residues via non-catalytic and catalytic pyrolysis processes. *Energy Conversion and Management, 213*, 112815. https://doi.org/10.1016/j.enconman.2020.112815
72. Wang, Z., Burra, K. G., Lei, T., & Gupta, A. K. (2021). Co-pyrolysis of waste plastic and solid biomass for synergistic production of biofuels and chemicals—A review. *Progress in Energy and Combustion Science, 84*, 100899. https://doi.org/10.1016/j.pecs.2020.100899
73. Xue, Y., Zhou, S., Brown, R. C., Kelkar, A., & Bai, X. (2015). Fast pyrolysis of biomass and waste plastic in a fluidized bed reactor. *Fuel, 156*, 40–46. https://doi.org/10.1016/j.fuel.2015.04.033
74. Yaman, E., Ulusal, A., & Uzun, B. B. (2021). Co-pyrolysis of lignite and rapeseed cake: A comparative study on the thermal decomposition behavior and pyrolysis kinetics. *SN Applied Sciences, 3*, 1–15. https://doi.org/10.1007/s42452-020-04040-y

75. Ye, J. L., Cao, Q., & Zhao, Y. S. (2008). Co-pyrolysis of polypropylene and biomass. *Energy Sources, Part A, 30*(18), 1689–1697. https://doi.org/10.1080/15567030701268419
76. Zhang, Y., Yu, Z., Zhang, X., & Ma, X. (2023). Comparative study on the synergistic co-pyrolysis of *Thlaspi arvense* L. seed with different plastics: Thermal behaviors, product distributions, and kinetics analysis. *Biomass Conversion and Biorefinery, 13*, 6197–6211. https://doi.org/10.1007/s13399-021-01712-6
77. Zhao, Y., Wang, Y., Duan, D., Ruan, R., Fan, L., Zhou, Y., Dai, L., Lv, J., & Liu, Y. (2018). Fast microwave-assisted ex-catalytic co-pyrolysis of bamboo and polypropylene for bio-oil production. *Bioresource Technology, 249*, 69–75. https://doi.org/10.1016/j.biortech.2017.09.184
78. Zhong, S., Zhang, B., Liu, C., Mwenya, S., & Zhang, H. (2022). A minireview on catalytic fast co-pyrolysis of lignocellulosic biomass for bio-oil upgrading via enhancing monocyclic aromatics. *Journal of Analytical and Applied Pyrolysis, 164*, 105544. https://doi.org/10.1016/j.jaap.2022.105544
79. Zhou, G., Li, J., Yu, Y., Li, X., Wang, Y., Wang, W., & Komarneni, S. (2014). Optimizing the distribution of aromatic products from catalytic fast pyrolysis of cellulose by ZSM-5 modification with boron and co-feeding of low-density polyethylene. *Applied Catalysis A: General, 487*, 45–53. https://doi.org/10.1016/j.apcata.2014.09.009
80. Zhou, L., Wang, Y., Huang, Q., & Cai, J. (2006). Thermogravimetric characteristics and kinetic of plastic and biomass blends co-pyrolysis. *Fuel Processing Technology, 87*(11), 963–969. https://doi.org/10.1016/j.fuproc.2006.07.002

Chapter 12
Improving the Biogas Generation Potential from Organic Wastes Using Hydrochar as an Additive Lab-Scale Case Study from Central Sweden: Part 1

Maria Kristoffersson, Maria Sandberg, and G. Venkatesh

Abstract At Biogasbolaget AB in Karlskoga in south-central Sweden, organic wastes like food waste, manure, and silage are digested anaerobically to yield biogas, which subsequently can be upgraded to biomethane, and used as a replacement for fossil-diesel in public transport. The digesters at the firm are currently operating below their maximum capacity. This chapter deals with the evaluation of the potential of hydrochar to augment biogas production in a batch process. Hydrochar produced from two sources—forestry sector and municipal organic wastes—were compared, and using the Automatic Methane Potential Testing System (AMPTS II) in the lab at Karlstad University, the optimal dosage was determined. Experiments were also conducted with hydrochar alone, to verify if the hydrochar was being anaerobically digested to yield biogas. The hydrochar sourced from municipal waste, when dosed at 8 g/l, produced 841 Nml of biogas/gram of VS (volatile solids) in the substrate, 93% greater than the reference case of no addition of hydrochar. The forestry-sector-sourced hydrochar on the other hand, at the same dosage, registered an increase of just 16.6%. A streamlined environmental life cycle analysis showed that significant climate-benefits can be availed of, implying environmental sustainability, when the additional biogas is refined and used to replace fossil-diesel in public bus transport. Hydrochar-assisted anaerobic digestion of organic wastes may be posited as a technology which may entrench itself in the circular bio-economies of tomorrow, and by doing so, contribute to a set of sustainable development goals. While these were batch-digestion experiments, this part of the two-part series recommends more-realistic continuous-digestion experiments which incidentally form the focus of Part 2.

M. Kristoffersson · M. Sandberg · G. Venkatesh (✉)
Energy and Environmental Systems Group, Department of Engineering and Chemical Sciences, Karlstad University, Karlstad, Sweden
e-mail: venkatesh.govindarajan@kau.se

M. Sandberg
e-mail: maria.sanberg@kau.se

G. Baskar et al. (eds.), *Circular Bioeconomy Perspectives in Sustainable Bioenergy Production*, Energy, Environment, and Sustainability,
https://doi.org/10.1007/978-981-97-2523-6_12

Keywords Anaerobic digestion · Automatic methane potential testing system · Batch process · Biogas · Circular bioeconomy · Hydrochar · Hydrothermal carbonisation

12.1 Introduction

In the decade 2010–2019, the total anthropogenic greenhouse gas (GHG) emissions amounted to approximately 54.4 billion ton-CO_2-equivalents (CO_2-e). It is by now a clichéd fact that the annual GHG emissions hereafter must be halved (in 2030 with respect to 2019), if we intend and hope to keep the global average temperature rise below 1.5 °C. If 'net-zero' is the target set, the GHG emissions which will inevitably occur, will have to be compensated by 'negative GHG emissions' or in other words, carbon-negative solutions and alternatives. The United Nations Environmental Programme (UNEP) estimates that by the end of this century, the temperature would have risen by between 2.4 and 2.6 °C [1].

While that is an overview of the world, in Sweden (where this case study is based), over the period 1990–2021, GHG emissions have decreased by 33% [1], from 71.4 to 47.8 million tons CO_2-eq., and Sweden has set for itself an ambitious target of 'net-zero GHG emissions' in 2045 [2]. To reach there, among other things, the transport sector which currently accounts for one-third of the country's GHG emissions, ought to be decarbonised. A 70% decrease in GHG emissions from the transport sector has been targeted in 2030, with respect to 2019 [3]. In 2022, non-fossil fuels powered about 93% of public transport in Sweden—39% hydrogenated vegetable oil, 31% biogas, 14% rapeseed methyl ester, and 8% electricity (hydropower + nuclear energy) [4]. Fossil-diesel accounted for approximately 7.2% of the total vehicle-kilometres travelled by buses in 2022—equivalent to an energy use of 342.8 GWh.

Biogas (or biomethane when refined) is an output of anaerobic digestion of organic materials (wastes, more often; and thereby a permanent fixture in the circular bio-economies of the future). Compressed biogas finds use as fuel for cars, light trucks and buses, while liquid biogas is more suitable for heavy vehicles, ships and industrial applications, and as a substitute for fossil-based natural gas (which also contains methane), it is poised to play a key role in combating climate change and truncating the anthropogenic greenhouse gas footprint [5], and thereby contributing to nearing the targets set under sustainable development goal (SDG) 13. Anaerobic digestion of organic wastes lies at the confluence of sustainable waste management and renewable energy generation (SDG 7), both of which are sine qua non is a circular bioeconomy.

Biochar and hydrochar, which will entrench themselves as useful and valuable biorefinery outputs in future, can be produced from renewable organic wastes. They will find use as fuels, adsorbents, soil conditioners, additives (catalysts if we may say so) for composting and anaerobic digestion [6]. Hydrochar is produced through the hydrothermal carbonisation (HTC) route. As opposed to biochar which is a product of pyrolysis of organic wastes which need to be dried before being fed into the furnace, moisture content is not a hindrance to the HTC process. Additives like activated carbon, graphene, nanotubes, biochar, and hydrochar have merited attention of late; the last two being posited as substitutes for other carbonaceous materials. Biochar and hydrochar confer multiple benefits—low cost, diversity of sources from which they can be produced, environmental sustainability, and the possibility of valorising digestates.

The physico-chemical properties of hydrochar can be modified to suit different end-use applications [7]. HTC is a suitable method for handling wastewater sludge, but in that case, it has a tendency of being enriched in heavy metals (HM) like silver, cadmium, chromium, copper, mercury, nickel, lead, and zinc [8]. They [8] have also pointed out that if the starting material is sludge from the forestry sector, or digestate from food waste anaerobic digester, the HM concentrations therein are much lower vis-à-vis municipal wastewater sludge. The hydrochar may resist inhibitory agents in a digester, enrich the microbial diversity in it, augment the activity of the acidogenic bacteria, catalyse electron transfers among reactants, and promote the hydrolysis of polysaccharides, proteins and lipids, and thus sustain biogas production [6].

Hydrochar addition to anaerobic digesters is a relatively nascent field of research, with most of the peer-reviewed publications being less than five years old. All of them are lab-scale studies conducted in a batch reactor, with the objective of estimating the change in the biogas generation potential of the substrate being digested. Hydrochar sourced from the digestate of an anaerobic digester [9], bamboo [10], rice straws [11], and wastewater sludge [12], augmented biogas production by 49%, 131%, 127%, and 18.4%, respectively, with respect to reference cases without hydrochar addition.

One of the partners in this research project—the Biogasbolaget (Biogas Company) in Karlskoga in central Sweden—generates biogas by digesting a blend of silage, slurry, and food waste gathered from the food processing sector and slaughterhouses [13]. It supplies 48 GWh worth of transport fuel every year (equivalent to, and thereby supplanting 5 million litres of fossil-diesel). In addition to producing fuel for the transport sector, Biogasbolaget also supplies fertiliser in the form of the digestate from the anaerobic digester. By augmenting the output of biogas, and thereby the fuel it is refined to, more buses in the region can be powered by biomethane, bringing down in the process, the consumption of fossil-diesel.

As mentioned earlier, the relative newness of this field of research warrants the conduct of more lab-scale trials which would eventually lead to pilot-scale studies to establish hydrochar in the circular bio-economies of the future, as a product which would enable the valorisation of a range of organic wastes into biomethane. As there are no studies conducted till date, with hydrochar produced from sludge sourced from the forestry sector (which incidentally is a significant contributor to the Swedish economy), this study—batch anaerobic digestion of organic substrate with hydrochar from two provenances—HTC of forestry sludge and HTC of municipal sludge—can be considered as a valuable addition to the knowledge base.

The specific goals set by the authors for this study are as under:

- To investigate if hydrochar augments biomethane production. If yes, to determine the dosage at which the yield is the maximum (4, 8, or 10 g/l)

- To compare the effect of hydrochar originating from two different provenances—forestry sector sludge and municipal sludge—on the augmentation of the yield of biomethane
- To carry out a simple environmental life cycle analysis (E-LCA) to study the effects of hydrochar-assisted anaerobic digestion on the global warming potential, and thereby study its potential to contribute to the sustainable development goals 7 and 13.

A brief background focusing on food waste (the substrate in the digester) and the anaerobic digestion process itself, follows the introduction, before the methodology is briefly described. The results are presented and discussed succinctly before concluding with some take-home messages in support of a circular bioeconomy, and the interventions needed to advance towards one sooner than later.

12.2 Background

12.2.1 Food Waste as Substrate to the Digester

The global generation of food waste in 2025 would amount to 2.2 billion tons [14]. The quantities of food waste can be decreased, but cannot be entirely avoided. A category called labelled as 'inedible and unavoidable food waste' which includes items like fruit peels and skins, bone-bits (meat products), and spent coffee grounds has been defined in literature [15]. There is also a huge fraction of avoidable food waste, which can be decreased. While challenges posed by pollutants and pathogens will loom larger, there will also be greater opportunities to extract value out of the biodegradable fractions. While landfilling organic biodegradable wastes is a sheer destruction of value, incineration for energy recovery and composting is better approaches to adopt. However, the most environment-friendly method of handling food waste is anaerobic digestion to yield biogas [16].

About 100,000 tons of food waste are collected annually in Sweden, for valorisation—corresponding to about 10 kg per capita. However, there is still a fraction of food waste which, owing to the absence of proper source segregation, does not find its way to the biological valorisation facilities—biogas generation and composting [17]. When food waste is subjected to source segregation, pre-treatment in the form of pulverisation, dilution, and separation from plastic and metal is mandatory. The biogas yield in the digester depends to a great extent on the segregation and pre-treatment methods adopted.

12.2.2 Anaerobic Digestion and Biogas Production

From Hydrolysis to Methanogenesis: Biogas production in an anaerobic digester is a series of steps which begins with hydrolysis of the substrates, proceeds to the formation of volatile fatty acids, therefrom to acetic acid and hydrogen, before yielding methane and carbon dioxide (the mixture we refer to as biogas).

- *Hydrolysis*: Complex organic molecules like carbohydrates, fats, and proteins which cannot serve as usable substrates are disintegrated enzymatically by hydrolysing microorganisms to smaller ones like simpler sugars, fatty acids, and alcohols. In general, when the substrate is particulate in nature, this turns out to be the rate-limiting step in the anaerobic digestion process.
- *Acidogenesis (Fermentation)*: The products of hydrolysis are acted upon by acidogenic bacteria to yield volatile fatty acids (VFA) like acetic acid, propionic acid, butyric acid, etc., and ammonia, carbon dioxide, and hydrogen.
- *Acetogenesis*: This is a reductive process in which carbon dioxide and the lower-molecular weight organic VFAs are converted to acetate by anaerobic acetogenic bacteria. Autotrophic acetogens as opposed to the heterotrophic ones we have been referring to heretofore can produce acetate from carbon dioxide and hydrogen (both products of acidogenesis). It is noteworthy that acetate production and methanogenesis take place simultaneously.
- *Methanogenesis*: Biogas can be formed either by the disintegration of acetate into carbon dioxide and methane by acetoclastic ('acetate breaking') methanogens; or by the reduction of carbon dioxide by hydrogen by the hydrogenotrophic ('hydrogen consuming') methanogens. The former account for close to 70% of the total biogas yield. When the substrate is food (and vegetable) waste, this becomes the rate-limiting step, owing to the high activity of the acidogens which results in rapid accumulation of VFAs, and a conspicuous drop in the pH which damps down methanogenic activity. This is the rate-limiting step and is sensitive to changes in operating conditions like temperature, pH, substrate composition, feed rate, and leakage of oxygen into the chamber.

Role of Heavy Metals in the Biogas Generation Process: Heavy metals like cobalt (Co), iron (Fe), and nickel (Ni)—micronutrients—are considered to play an enabling role in the anaerobic digestion process by virtue of their reductive-oxidative properties. While a deficiency of Ni and Co depresses methane generation by 10% and 25%, respectively, heavy metals in very high concentrations—as is the case in industrial wastewater and municipal sludge—are toxic to microorganisms. Copper (Cu), Ni, and zinc (Zn) in the right amounts catalyse hydrolysis and sustain microbial activity in the digester, in general. Acidogenesis is favoured by lower concentration of cadmium (Cd) and higher concentration of chromium (Cr) [18], while additions of Co, Zn, magnesium (Mg), niobium (Nb), calcium (Ca), Ni, Cd, and Fe in the form of salts and metal oxides are common, to aid the four steps of anaerobic digestion—described in the previous sub-section—by catalysing electron transfer processes during enzymatic degradation [6].

The Residual Digestate: The macronutrients—nitrogen and phosphorus—which entered the anaerobic digester are left behind in the residual digestate, which thereby has fertiliser value. As more and more of the organics are hydrolysed, and eventually converted to methane and carbon dioxide, the nitrogen and phosphorus become more easily accessible and available to the plants, when the digestate is transferred to the soil on cropland. However, the presence of undesirables like heavy metals, pesticides, pharmaceuticals, and disease-causing pathogens may make the use of this digestate as fertiliser, questionable. If the digestate traces its origins to animal dung, source-segregated food wastes, wastes from farms, slaughterhouses and the food processing sector, and other types of easily biodegradable organic wastes, it is labelled as a 'bio-fertiliser', and has to be certified by Avfall Sverige (Waste Management Sweden, www.avfallsverige.se) as being fit for use as a fertiliser.

12.3 Methodology

12.3.1 Sourcing of Substrate, Graft, and Hydrochar

The study was conducted in the laboratory on the premises of Karlstad University in early 2023. The substrate—a blend of food waste, slurry, and slaughterhouse waste—was obtained from Biogasbolaget AB (one of the firms associated with this study) and mixed with glycerol and refrigerated to ensure that it did not biodegrade. The graft (active bacteria + organics) to be used along with hydrochar (sometimes referred to as HC hereafter in the chapter) produced from forest-sector sludge was placed in a warm compartment at 37 °C for 13 days, so that the organics could be digested and converted to biogas which could be released out to the atmosphere, and the residual active bacteria could be subsequently added on to the substrate. The second graft to be used along with the HC sourced from municipal sludge, on the other hand, was kept warm for just a day before being added on to the substrate. The pulp and paper mill contributing sludge for hydrothermal carbonisation (HTC) was Stora Enso's Heinola-based (Finland) plant, and the HC-producer was C-Green.

12.3.2 Estimation of the Total Solids and Volatile Solids

Figure 12.1 is a self-explanatory simple sketch, which shows how the total solids (TS) and volatile solids (VS) content of the different substrates and the two grafts were calculated. While TS, as shown in Fig. 12.1, is the dry solids content of the sample, the difference between TS and the Ashes (shown in the split bar on the far right of Fig. 12.1) gives one the VS content (which can be expressed as a percentage

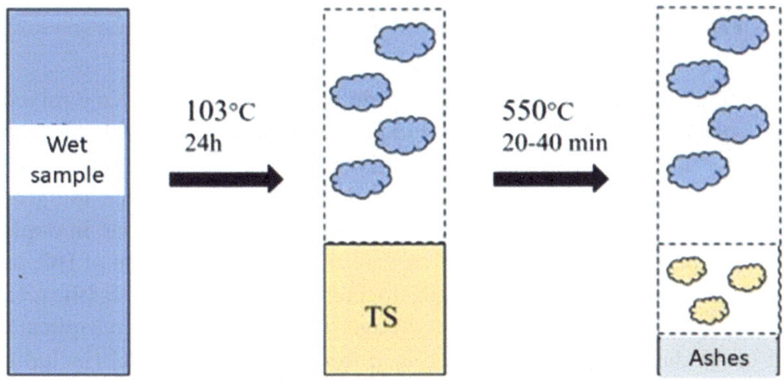

Fig. 12.1 Illustration of the TS and VS estimation tests

Table 12.1 Results of the TS and VS tests (values as % of total wet mass of sample)

Material	Total solids (TS)	Volatile solids (VS)
Substrate (average of all the types)	8.8	7.8
Graft (average)—used with HC from forest-sector sludge	4.9	3.6
Graft (average)—used with HC from municipal sludge	5.5	4.1
Silage (average)	26.5	5.2
Glycerol (average)	88	88

of TS, as well as a percentage of the total mass of the wet sample). The results—averages calculated from multiple trials—are given in Table 12.1, in the Results and Discussion section.

12.3.3 Setup of the Automatic Methane Potential Testing System II

The laboratory-scale experiments in batch anaerobic digestion were conducted using the Automatic Methane Potential Testing System (AMPTS II). This system utilises fourteen 500-ml flasks (80% working volume), which are placed in a water bath warmed and held at a mesophilic temperature of 37 °C. Mixing during the digestion within each flask is accomplished by stirrers which operate continuously on a stir-pause cycle (stirring for a minute and pausing for the next). The resultant biogas is then 'refined' by dissolving the carbon dioxide in it, in sodium hydroxide. The quantity of methane generated is measured continuously, by the AMPTS II software programme, and the cumulative generation curve is plotted.

12.3.4 Testing with Forest-Sector Hydrochar

Hydrochar having its provenance in paper mill sludge is pulverised and mixed with substrate from Biogasbolaget. Studies done in the past have shown that the optimal dosage of HC lies between 4 and 10 g/l. This motivated the selection of dosages 4, 8, and 10 g/l, for this set of experiments. As also mentioned earlier, the graft was digested separately to determine its methane gas yield (this was done in duplicate). The substrate mixed with the graft was digested without the addition of HC, and the results thereof served as the reference case for comparison. All the tests (the reference and those with the additions of different dosages of HC) were done in triplicate, over a period of 18 days. For each of the four different dosages of HC (including the reference case of 0 g/l), 54.8 g of graft and 11.86 g of substrate were employed. About 57 g of graft were added to the two flasks designated for the determination of the methane-yielding potential of the graft alone. De-ionised water was added to all the flasks to bring the working volume in each of them to 400 ml, before measuring the pH.

An additional test was conducted with graft + HC, without any substrate, to see if the hydrochar itself yielded biogas. The leachability of VFAs from the HC was also tested by mixing the HC (8 g/l) with de-ionised water, and letting it stay in the warm water bath. This was necessitated to establish if the HC contributed VFAs to the acetogens in the digester, to be converted eventually to methane and carbon dioxide.

12.3.5 Testing with Hydrochar Produced from Municipal Sludge

All the tests described in Sect. 3.4 were repeated by simply changing the provenance of the hydrochar from forest-sector sludge to municipal sludge. The digestion happened over a period of 15 days, as opposed to 18 days for the previous set. C-Green delivered the HC needed for this second set of tests in pulverised form, thus obviating the need for an extra pulverisation step in the lab. The mass of graft needed in each flask for the second set of tests was less by 5.53 g, vis-à-vis the first set (49.27 g instead of 54.8 g).

12.3.6 Streamlined Environmental-LCA (E-LCA)

The hydrochar production through the HTC-route (at C-Green) uses electricity (500 kWh per ton of HC total solids). At the rate of 8 g/l HC dosage to anaerobic digestion at Biogasbolaget, about 628 tons of HC would need to be produced and supplied by C-Green, annually. It was deemed necessary and interesting to analyse the HC addition process, and the far-reaching consequences thereof in the wider economy, from an environmental perspective. The functional unit chosen was the excess refined biogas resulting from the addition of 628 tons of HC per year, and the corresponding avoided impacts from the production and use of an equivalent amount of diesel–fuel to power buses was calculated. The HC production process using the HTC-route, anaerobic digestion to produce biogas, refining of the biogas to biomethane, distribution of the biomethane, and its combustion in the use phase was included within the system boundary. The life cycle of diesel likewise—from rigs-to-road, so to say—was also accounted for. Figure 12.2 shows the processes in the life cycles of diesel and HC-assisted biogas production, refining, and use, and Table 12.2 summarises the relevant data.

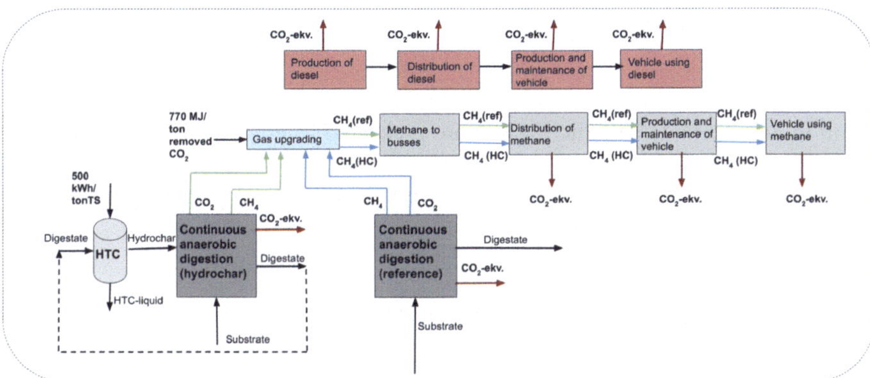

Fig. 12.2 Processes in the life cycles of biomethane and diesel, in the estimation of the net greenhouse gas footprint

Table 12.2 Biogas-related data [22, 23]

Data	Value	Units
Low heating value of biogas (97% methane)	48.5	MJ/kg
Density of biogas (97% methane)	0.75	Kg/m^3
Specific fuel usage (biogas bus)	19.4	MJ/km
Reference value of annual biogas production	5 million	Nm^3/year
Biogas production when hydrochar produced from forest-sector sludge is added	5.83 million	Nm^3/year
Biogas production when hydrochar produced from municipal sludge is added	9.67 million	Nm^3/year
Life cycle GHG emissions of refined biogas	0.28	kg CO_2-eq/km
Life cycle GHG emissions of fossil-diesel	1.1	kg CO_2-eq/km
Additional energy usage for refining the extra biogas produced (HPWS process)	845	GJ

High Pressure Water Scrubbing (HPWS) is used at the Biogasbolaget to refine biogas to biomethane, and electricity use amounts to 770 MJ/ton-CO_2-removed) [19]. The value for the GHG footprint of electricity (Swedish mix) used for the HTC and HPWS processes was obtained from the Ecoinvent database [20] in SimaPro [21]. It was assumed that the marginal increase in the GHG footprint was equal to that added on by the HPWS refining of the additional biogas generated by adding HC. The net avoided GHG emissions in a year, when the extra biomethane produced by adding hydrochar is used to substitute fossil-diesel as transport fuel for buses was calculated.

12.4 Results and Discussion

The results for biomethane production, pH pre- and post-digestion, the VFA concentration in the digestate, and the E-LCA are presented as compared between the first and second sets of experiments—performed with HC from forest-sector sludge and municipal sludge, respectively. Microscopy results pertain only to the second set.

12.4.1 Biomethane Production (BMP)

Figures 12.3 and 12.4 present the cumulative methane production from 12 and 14 flasks of the AMPTS II system, for municipal-sludge-sourced and forest-sludge-sourced HC, respectively. The red line depicts the production of methane in the flask with only the graft in it; the other values are adjusted by subtracting this value from them. In Fig. 12.3, among the two 8 g/l yellow lines, the upper one depicts methane production equal to 881 ml/g VS. The uppermost of the three blue lines (10 g/l addition) corresponds to a production of 904 ml/g VS. These have to be analysed vis-à-vis the orange reference line (no HC addition) which reaches 485 ml/g VS. Also observed in Fig. 12.3 is the fact that the production tapers off after 5 days for the reference case, while the others where HC is added to the substrate-graft blends in the flasks, register a late rise to much higher values than the reference case at the end of the 15-day period.

As seen in Fig. 12.4, the 8 g/l addition of HC from forest-sector sludge (yellow line) yielded the greatest amount of methane (560 ml/gVS) over the 18-day period—about 100 ml/gVS greater than the reference case (orange line). It is also clear that 4 mg/l HC addition yielded less methane than the reference case, the reason thereof not being very clear.

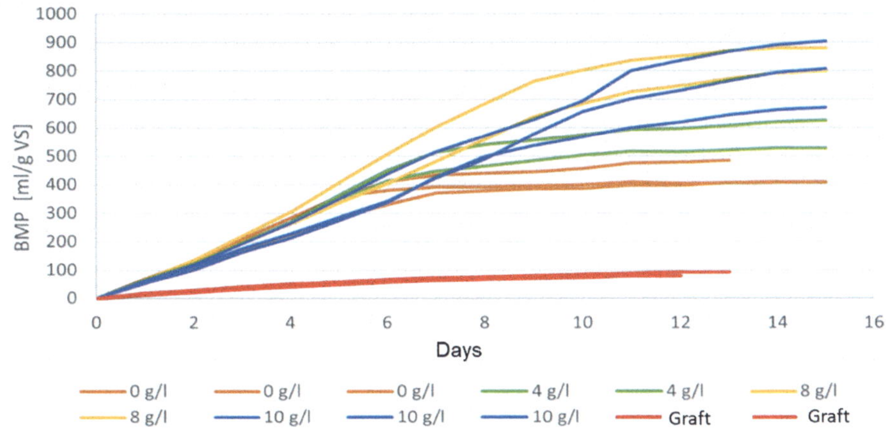

Fig. 12.3 Biomethane potential (BMP) for reference case, graft, and different concentrations of hydrochar from municipal sludge, over 15 days

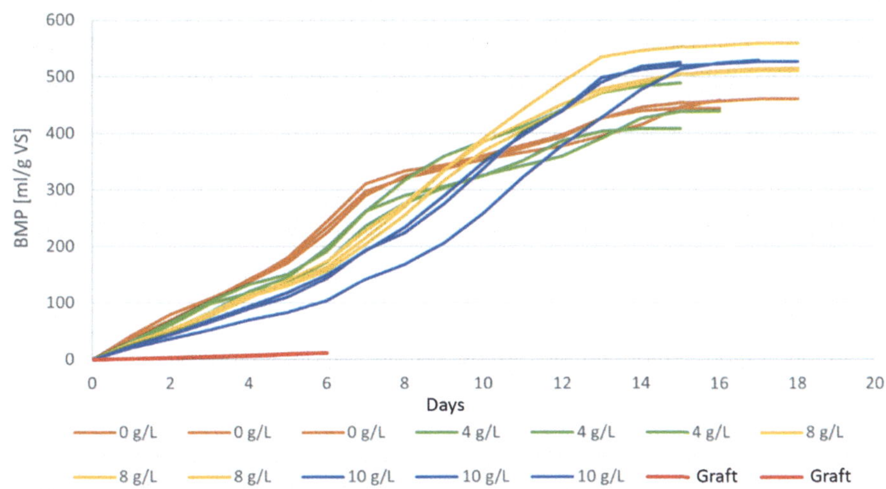

Fig. 12.4 Biomethane potential (BMP) for reference case, graft, and different concentrations of hydrochar from forest-sector sludge, over 18 days

12.4.2 PH Before and After Digestion

The average pH values recorded before and after the digestion are given in Table 12.3. For the set of experiments done with HC from municipal sludge, the pH decreased slightly for the reference and the HC addition cases, oscillating around neutrality for the reference. As pointed out by authors earlier [24], the optimal pH should be in the range of 6.5–7.2. The pH in the flask with only the graft decreased, but stayed slightly alkaline all throughout. For the other set, with HC sourced from forest-sector

Table 12.3 pH values pre- and post-digestion

Case	Hydrochar (HC) from municipal sludge			Hydrochar (HC) from forest-sector sludge		
	pH pre-digestion	pH post-digestion	% change	pH pre-digestion	pH post-digestion	% change
0 g/l HC	7.2	6.92	– 3.9	7.36	7.5	1.9
4 g/l HC	7.31	6.88	– 5.9	7.44	7.46	0.3
8 g/l HC	7.29	6.83	– 6.3	7.5	7.54	0.5
10 g/l HC	7.25	6.82	– 5.9	7.5	7.49	– 0.1
Only graft	7.95	7.22	– 9.2	8.25	8.59	4.1

sludge, the pH showed a tendency to increase slightly, while staying always in the slightly alkaline range. Markedly, the decreases in the pH for the experiments done with HC from municipal sludge were higher in general than the increases in the same for those done with HC from forest-sector sludge. Nutrient addition (for nitrogen) along with the HC from the forest-sector sludge to the anaerobic digester is common [8]. This results in a higher ammonium concentration in the digester and thereby an increase in pH (as seen in Table 3) that there was a difference in the duration over which the digestion proceeded—15 days where HC from municipal sludge was used, and 18 days for the other—may also be a contributing factor to the increase in the pH for the latter.

12.4.3 Volatile Fatty Acids (VFAs)

The VFA concentration in the digestate, as seen in Fig. 12.5, generally increased as the dosage of the HC rose. However, it is unclear why there was a drop in the VFA concentration with respect to the reference case, from 360 to 327 mg/l, when 4 mg/l HC from forest-sector sludge was added. The span between the reference case and 10 mg/l addition is 100 mg/l and 77 mg/l, respectively, for the bars shown in blue (HC from forest-sector sludge) and red (HC from municipal sludge), respectively. The leaching test referred to earlier and showed that HC did indeed release VFAs to the de-ionised water—an average of 77.5 mg/l and 84 mg/l, respectively, for 8 mg/l HC from municipal sludge and forest-sector sludge.

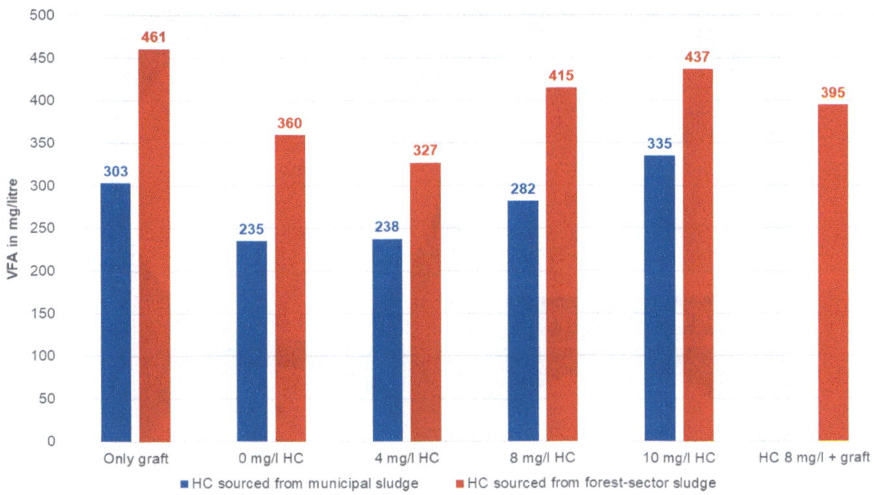

Fig. 12.5 VFA concentrations in the digestate, for both sets of experiments

12.4.4 Observations from Microscopy

The digestate from the experiments conducted with HC from municipal sludge was observed under the microscope. The dark blobs in Fig. 12.6 shown by red arrows are the residual fragments of HC in the digestate. Around the blobs, and on their surfaces, one can notice what presumably are bacterial colonies in biofilms. Quite obviously, it was easier to locate HC blobs when HC was initially added at higher concentrations (8 and 10 mg/l), as is also clear from Fig. 12.6.

12.4.5 E-LCA Results

The E-LCA results are presented in Fig. 12.7. The increase in biogas generation by adding HC (produced from forest-sector sludge or municipal sludge) entails additional resource usage and thereby GHG emissions during the biogas to biomethane refining step (refer Fig. 12.2). As seen from Table 12.2, the additional biogas generated by adding HC sourced from municipal sludge (9.67 million Nm^3) is much greater than that generated by adding HC from the other provenance (5.83 million Nm^3). This is reflected in the difference between the upper and central black bars in Fig. 12.7 (3091 tons and 5164 tons of CO_2-eq.). Additional biomethane production, as per the system expansion assumption made in the E-LCA, implies substitution of a greater quantity of fossil-diesel as transport fuel for buses. This is reflected by the two darker green bars (upper and central), and the difference between them. The net avoided GHG emissions, as shown in Fig. 12.7, are 14,783 tons (for HC from municipal sludge) and 8937 tons CO_2-eq (for HC from forest-sector sludge).

Fig. 12.6 Microscopy of the digestate, clearly showing residual fragments of hydrochar and biofilms in their vicinity (HC sourced from municipal sludge)

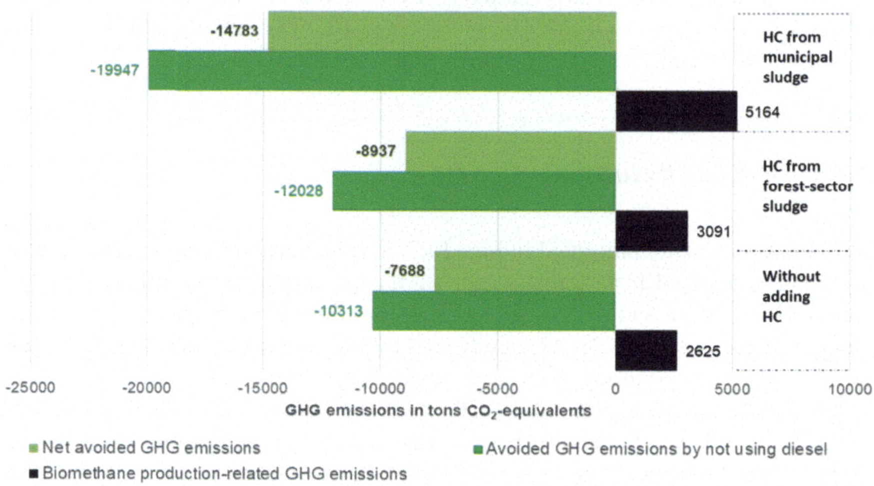

Fig. 12.7 Comparative E-LCA results for the GHG emissions and net avoided emissions

12.4.6 Overview of the Findings

It has been clearly established that addition of HC to the anaerobic digester yielded more biogas, and that is an interesting finding for a circular bioeconomy which is gradually taking shape. The effect of the provenance of the HC has also been clearly demonstrated through the trials conducted in the laboratory. HC addition to the substrate provides more VFA to the methanogenic bacteria, as noted in earlier studies [6, 11], and thereby the possibility of producing more biogas. The fact that VFAs (which could have been converted to biogas by the methanogens) were found in the digestate after the 15-day and 18-day trials (see Fig. 12.5) can be attributed to the fact that the trials could have been prolonged by a few more days to further enhance the output of biogas.

The microscopy test revealed that bacterial biofilms find a habitat for themselves on the surfaces of the HC chunks in the digestate. The E-LCA showed clearly that there are huge environmental benefits to be availed of, by investing in the HTC process for HC production, and subsequently using it to augment the anaerobic digestion of substrates of different types. A clear contribution to the Sustainable Development Goals 7 and 13 has been demonstrated.

12.5 Conclusions and Recommendations

The batch-digestion experiments presented in this chapter (the first of a two-part series) augur well as a justification for the production of HC and its utilisation in anaerobic digestion. Municipal sludge emerges as a better option vis-a-vis forest-sector sludge, as it yields much more biogas.

The utilisation of the digestate from Karlskoga's Biogasbolaget (thereby introducing circularity within the system) as the in-feed to the HTC equipment to produce hydrochar is also something which may be of interest to the firm. One of the recommendations of the first part—testing by mixing HC from municipal sludge with substrate in a more-realistic continuous-digestion process—leads seamlessly to Part 2, which also presents results of a life cycle costing analysis to establish the economic sustainability of this intervention which has now been proven to be environmentally sustainable.

Acknowledgements The authors would like to thank everyone at C-Green, Biogasbolaget and Karlstad University for their help and support with this study, at different points of time in the first half of 2023.

References

1. Naturvårdsverket [Nature Conservation Agency], n.d. (2023). *Globala utsläpp av växthusgaser [Global greenhouse gas emissions]*. Accessed on March 30, 2023 at https://www.naturvardsverket.se/amnesomraden/klimatomstallningen/det-globala-klimatarbetet/globala-utslapp-av-vaxthusgaser/
2. Sverigesmiljomål [Sweden's environmental targets]. (2023). *Begränsad klimatpåverkan - Sveriges miljömål [Limiting climate effects—Sweden's environmental targets]*. Accessed on March 31, 2023 at https://sverigesmiljomal.se/miljomalen/begransad-klimatpaverkan/
3. Naturvårdsverket [Nature Conservation Agency], n.d. (2023). *B. Sveriges utsläpp och upptag av växthusgaser [Sweden's GHG emissions and capture]*. Accessed on the March 31, 2023, at https://www.naturvardsverket.se/data-och-statistik/klimat/sveriges-utslapp-och-upptag-av-vaxthusgaser/
4. Bussar-Miljöbarometern [Buses-The Eco-barometer]. (2023). Accessed at https://2030.miljobarometern.se/nationella-indikatorer/bilen/andel-fossiloberoende-fordon-i-trafik-b1c/bussar/, on April 27, 2023.
5. Norrman, J., Arnell, J., Belhaj, M., & Flodström, E. (2005). Biogas som drivmedel för bussar i kollektivtrafik [Biogas as fuel in the public transportation sector]. IVL Svenska Miljöinstitutet. Accessible at https://www.ivl.se/publikationer/publikationer/Biogassomdrivmedelförbussarikollektivtrafik
6. Cavali, M., Libardi Junior, N., Mohedano, R. A., Belli Filho, P., da Costa, R. H. R., & de Castilhos Junior, A. B. (2022). Biochar and hydrochar in the context of anaerobic digestion for a circular approach: An overview. *Science of The Total Environment, 822*, 153614. https://doi.org/10.1016/j.scitotenv.2022.153614
7. Quintana-Najera, J., Blacker, A. J., Fletcher, L. A., & Ross, A. B. (2021). The effect of augmentation of biochar and hydrochar in anaerobic digestion of a model substrate. *Bioresource Technology, 321*, 124494. https://doi.org/10.1016/j.biortech.2020.124494; Starr, K., Gabarrell, X., Villalba, G., Talens, L., & Lombardi, L. (2012). Life cycle assessment of biogas upgrading technologies. *Waste Management, 32*, 991–999.
8. Baresel, C., Axegård, P., Lazic, A., Bornold, N., Yang, J.-J., & Malovanyy, A. (2023). *Framtidens slamhantering vid Roslagsvatten (The future of sludge handling at Roslagsvatten)*. Accessible at https://ivl.se/publikationer/publikationer/framtidens-slamhantering-vid-roslagsvatten-behandling-av-kommunalt-orotat-slam-med-htc-teknik-oxypower-htctm-och-rening-av-htc-vatten-med-sbr-och-mbbr.html. IVL svenska miljöinstitutet, Stockholm.
9. Xu, S., Wang, C., Duan, Y., & Wong, J.W.-C. (2020). Impact of pyrochar and hydrochar derived from digestate on the co-digestion of sewage sludge and swine manure. *Bioresource Technology, 314*, 123730. https://doi.org/10.1016/j.biortech.2020.123730
10. Choe, U., Mustafa, A. M., Lin, H., Xu, J., & Sheng, K. (2019). Effect of bamboo hydrochar on anaerobic digestion of fish processing waste for biogas production. *Bioresource Technology, 283*, 340–349. https://doi.org/10.1016/j.biortech.2019.03.084
11. Xu, J., Mustafa, A. M., Lin, H., Choe, U. Y., & Sheng, K. (2018). Effect of hydrochar on anaerobic digestion of dead pig carcass after hydrothermal pretreatment. *Waste Management, 78*, 849–856. https://doi.org/10.1016/j.wasman.2018.07.003
12. Wu, B., Yang, Q., Yao, F., Chen, S., He, L., Hou, K., Pi, Z., Yin, H., Fu, J., Wang, D., & Li, X. (2019). Evaluating the effect of biochar on mesophilic anaerobic digestion of waste activated sludge and microbial diversity. *Bioresource Technology, 294*, 122235. https://doi.org/10.1016/j.biortech.2019.122235
13. Karlskoga Energi. (2018). *Så gör vi biogas [This is how we produce biogas]*. Accessed on the January 2, 2023 at http://www.karlskogaenergi.se/om-oss/vara-verksamheter/biogas/
14. Xu, Q., Luo, L., Li, D., Johnravindar, D., Varjani, S., Wong, J. W. C., & Zhao, J. (2022). Hydrochar prepared from digestate improves anaerobic co-digestion of food waste and sewage sludge: Performance, mechanisms, and implication. *Bioresource Technology, 362*, 127765. https://doi.org/10.1016/j.biortech.2022.127765

15. Teigiserova, D. A., Hamelin, L., & Thomsen, M. (2019). Review of high-value food waste and food residues biorefineries with focus on unavoidable wastes from processing. *Resources, Conservation and Recycling, 149*, 413–426. https://doi.org/10.1016/j.resconrec.2019.05.003
16. Zhou, Y., Engler, N., Li, Y., & Nelles, M. (2020). The influence of hydrothermal operation on the surface properties of kitchen waste-derived hydrochar: Biogas upgrading. *Journal of Cleaner Production, 259*, 121020. https://doi.org/10.1016/j.jclepro.2020.121020
17. Avfall Sverige. (2023). *I mål med matavfallet [On track with food waste]*. Accessed at https://www.avfallsverige.se/fakta-statistik/insamling/matavfall/i-mal-med-matavfallet/, on April 3, 2023.
18. Kadam, R., Khanthong, K., Jang, H., Lee, J., & Park, J. (2022). Occurrence, fate, and implications of heavy metals during anaerobic digestion: A review. *Energies, 15*, 8618. https://doi.org/10.3390/en15228618
19. Starr, K., Gabarrell, X., Villalba, G., Talens, L., & Lombardi, L. (2012). Life cycle assessment of biogas upgrading technologies. *Waste Management, 32*, 991–999. https://doi.org/10.1016/j.wasman.2011.12.016
20. Swiss Center for Life-Cycle Inventories. Ecoinvent database V3.4 (2019). Technical report. Accessible at www.ecoinvent.org
21. PRé Consultants. (2023). *SimaPro 9.4*. https://doi.org/10.1016/j.wasman.2011.12.016
22. Biogasbolaget i Mellansverige AB. (2018). KarlskogaEnergi. Accessed on February 1, 2023 at http://www.karlskogaenergi.se/Vara-tjanster/biogas/biogasbolaget/
23. Lyng, K.-A., & Brekke, A. (2019). Environmental life cycle assessment of biogas as a fuel for transport compared with alternative fuels. *Energies, 12*, 532. https://doi.org/10.3390/en12030532
24. Ayodele, O. O., Adekunle, A. E., Adesina, A. O., Pourianejad, S., Zentner, A., & Dornack, C. (2021). Stabilization of anaerobic co-digestion of biowaste using activated carbon of coffee ground biomass. *Bioresource Technology, 319*, 124247. https://doi.org/10.1016/j.biortech.2020.124247

Chapter 13
Improving the Biogas Generation Potential from Organic Wastes Using Hydrochar as an Additive Lab-Scale Case Study from Central Sweden: Part 2

Annette Liisa Kariis, Maria Sandberg, and G. Venkatesh

Abstract In Part 2 of the two-part series, single-stage anaerobic co-digestion in two continuously fed reactors replaced the batch process of Part 1. A life-cycle costing analysis was carried out to determine if, investing in a hydrothermal carbonization (HTC) system to produce hydrochar in-plant to augment biogas production will be economically feasible. Hydrochar addition resulted in a 59% rise in biogas yield (and 53.5% in methane yield). The study confirmed the techno-economic feasibility for coupling an HTC plant with a digester supplying 25% of the digestate it produces, to the former, as the raw material for hydrochar production. The rest of the digestate (rich in carbon, nitrogen, and phosphorus) can be used as fertiliser. Investing in an HTC plant contributing to a rise in methane production of 17% (or 53%) will result in a net profit of 363 million SEK (or 1237 million SEK) over a 20-year period. If the Karlskoga biogas plant decides to rely on purchasing hydrochar from the external market instead, the corresponding net profit will be 177 million SEK (or 1052 million SEK) over the same 20-year period, implying that a decision to integrate and interconnect is likely to be economically more feasible, in a circular bioeconomy in the future.

Keywords Anaerobic digestion · Continuous reactor · Biogas · Circular bioeconomy · Life-cycle costing · Hydrochar · Hydrothermal carbonisation · Net present value

---

A. L. Kariis · M. Sandberg · G. Venkatesh (✉)
Energy and Environmental Systems Group, Department of Engineering and Chemical Sciences, Karlstad University, Karlstad, Sweden
e-mail: venkatesh.govindarajan@kau.se

M. Sandberg
e-mail: maria.sanberg@kau.se

G. Baskar et al. (eds.), *Circular Bioeconomy Perspectives in Sustainable Bioenergy Production*, Energy, Environment, and Sustainability,
https://doi.org/10.1007/978-981-97-2523-6_13

13.1 Introduction

In the second decade of this century, the average annual greenhouse gas (GHG) emissions rose to a record high in human history [1]. As indicated in Part I of this series, the realisation that a decoupling between socio-economic development and dependence on fossil fuels is indispensable in the years to come has been a shot in the arm for a circular bioeconomy, which is gradually entrenching itself.

Food waste is among the handful of substitutes for fossil fuels, one can avail of. In 2019, 931 million tons of food waste were generated globally, of which 61% emanated from households [2]. The Swedish Food Agency has repeatedly emphasised on the indispensability of both decreasing avoidable food waste, and valorising the unavoidable fraction to the greatest extent possible, in biorefineries of a future circular bioeconomy [3].

The most common approach to the valorisation of food wastes globally has been anaerobically digesting (or co-digesting with wastewater sludge, and agricultural wastes) them to yield biogas (or biomethane, after refining). Biogas is an energy source for the generation of heat and electricity, as well as the production of transport fuel. The resulting digestate which is rich in nutrients (phosphorus especially) finds use as organic fertiliser. Substituting conventional fossil fuels hereby, and decreasing the need for the production of synthetic fertilisers, uncovers significant environmental benefits, while generating additional income to more people along the value chain—in other words, being socially desirable, economically attractive, and environment-friendly. Over the 10-year period between 2009 and 2018, the number of biogas plants in Europe (micro-, small-, and industrial scales of production) increased rapidly, tripling from 6200 to 18,202 [4]. As they have proliferated, new challenges have emerged. These have spurred more research in a multitude of aspects related to anaerobic digestion of organic wastes—optimisation of the operational variables, configuration of the bioreactors, opportunities for and the effects of co-digesting, control of nutrients, pretreatment of the substrates to improve biodegradability (digestability in other words), and the effect of additives like hydrochar and biochar on the yield of biogas.

13.1.1 Hydrochar as an Additive

Hydrochar—one of the additives referred to in the previous paragraph—is a carbonaceous material produced using biomass as the starting material, through a process called hydrothermal carbonisation (HTC). The carbonisation occurs in an aqueous medium ('hydro'), which serves both as solvent and catalyst, at around 160–280 °C ('thermal'), at above the saturation pressure at the operating temperature/s. At a supercritical state, water contributes to an increase in the solubility and reactivity of the biomass. The carbonisation mechanisms include hydrolysis, dehydration, decarboxylation, condensation, polymerisation, and aromatisation [5, 6]. The products comprise a gaseous fraction (mostly carbon dioxide), a liquid fraction which is a mixture of organic compounds, and a solid fraction which is the hydrochar in question. While the need for drying the biomass prior to the process is obviated (this being a process which happens anyway in an aqueous medium), the resulting slurry needs to be dewatered. However, the energy requirement for the dewatering of the slurry is much less than what would have been required for pre-process-drying if the latter had been necessary [7] (This has a favourable effect on the environmental footprint of the HTC process, which has been dwelt upon in Part I).

An addition of 10 g/l of hydrochar (from rice straw) augmented the biogas yield, in a study [8], by 90% when the substrate was pre-treated pork wastes. Two years thereafter, a 19% rise in biogas production was observed when hydrochar produced by hydrothermally carbonising food wastes was added to an anerobic digester which was using the same type of food wastes, as substrates [9]. A majority of the studies have resorted to the batch process (adopted in Part I of the current study). Among the published studies which have adopted a 25-day continuous digestion process [10, 11], the former reported a 5% rise in biogas yield and a 0% rise in methane yield [with municipal organic waste as substrate and hydrochar sourced from spent coffee grounds] while the latter reported a 12% increase in methane yield [air-dried chicken manure as substrate and pyrochar from fruit wastes].

Hydrochar is reported to have a strong buffering capacity and an ability to tone down the effects of inhibitory agents in the digester [4, 12], a larger surface area, greater pore sizes, and pore volume vis-à-vis pyrochar [13]. While being a carrier for the substrate, it tends to promote the colony-formation (and sequestration or sheltering) of bacteria on the aforesaid large surface area, facilitating quicker and easier direct interspecies electron transfer (abbreviated as DIET by Cavali et al., Ayodele et al., [1, 14]), leading consequently to improved hydrolysis, acidogensis, acetogenesis and subsequently, a rise in biogas production [12]. Colony formation concentrates active microorganisms—all the different types described in Part I; an intercellular (bacterial) distance of less than 1 μm is deemed to be necessary to ensure oxidation of the volatile fatty acids and hydrogen production.

DIET is further enhanced by the presence of oxygenated functional groups (C–O, C=O, and C–OH among others) which abound in hydrochar [1]. The fact that the HTC process is conducted at a lower temperature than pyrolysis is the reason behind the presence of many more oxygenated functional groups in hydrochar as compared to pyrochar. These oxygenated functional groups have both reductive and oxidative abilities and thereby are able to both accept electrons from bacteria and donate electrons to methanogens [15]. The alkaline nature of hydrochar (pH remaining close to 7.4) also contributes to optimising the pH in the digester, and thus favours methanogenesis (methanogens are sensitive to lower pH values) [8]. Slightly alkaline pH values promote the formation of more VFAs, thanks to greater activity on the part of acidogens. If the hydrochar is produced from wastewater sludge, humus-like constituents in it improve the oxidative ability of the hydrochar and augment the production of short-chain fatty acids like acetic acid and propionic acid. Hydrochar which remains in the digestate improves the fertiliser value of the same, by ensuring that nutrients—nitrogen and phosphorus—are sequestered within the matrix. It also immobilises heavy metals, contributing in its own way to reduced leaching of the same to the water bodies [4].

13.1.2 Objectives

In that backdrop generated by referring to past studies conducted with hydrochar, and as a continuation from Part I of this two-part series, the objectives of this study are as under:

- To investigate how the addition of hydrochar influences biogas (and methane) production, and the stability of a more-realistic continuous anaerobic digestion process, as opposed to the batch process adopted in Part I.
- To determine if it would be possible to interconnect the anaerobic digester with an HTC-reactor and study the associated material flows.
- To assess the economic feasibility of producing hydrochar in-plant vis-à-vis purchasing it from an external source, with biogas being refined to biomethane (as detailed in Part I) and sold to the transport sector.

13.2 Methodology

13.2.1 Characterisation of the Hydrochar Used in the Study

The hydrochar used in the experiments, similar to Part I, was sourced from C-Green. Data about its characteristics [16] have been listed in Table 13.1.

Table 13.1 Characteristics of the hydrochar used in the study

Specific area (BET)	35 m^2/g
Conductivity	947 μS/cm
Moisture content	0.9%
pH	5.3
Ash content	22.8%
Metal	**Concentration (mg/kg TS)**
Silver	0.1
Cadmium	1.42
Chromium	19
Copper	27.5
Nickel	35.6
Zinc	392
Lead	4.2

13.2.2 Setup in the Laboratory

Two lab-scale single-stage continuous anaerobic digestion reactors (one of them, the reference; the other for tests with hydrochar addition) were set up at Karlstad University (refer to Fig. 13.1). Tests were performed between 24 February and 2 May 2023. The tanks were fed and emptied regularly, and the volume of biogas collected was measured on a prescheduled daily basis.

The reactors shown in Fig. 13.1 (volume of 15 L each) had an inlet for the substrate (which was injected in) and outlets for the digestate and the biogas. The digestate outlet was coupled to a collection tank by a tube. The biogas which was generated was collected at the backside of the reactors in a plastic bag as shown. The reactors had their dedicated stirrers driven by motors placed atop them, and immersion heaters maintained the temperature at 35 ± 1 °C required for mesophilic digestion.

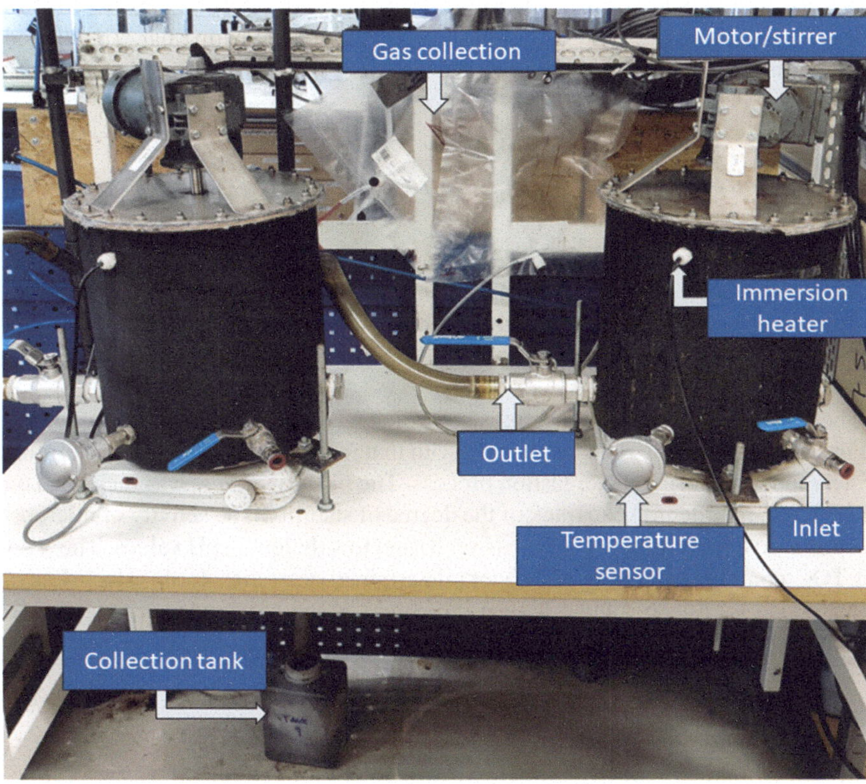

Fig. 13.1 The laboratory setup for the continuous anaerobic digestion trials

13.2.3 Substrate and Graft Acquisition from Biogasbolaget

As with the experiments in Part I, Biogasbolaget provided the graft (digestate from its anaerobic digester) and substrate (a mixture of food waste, glycerol, manure, and silage). Glycerol, as is commonly known, aids biogas production. The Total Solids (TS) content and Volatile Solids (VS) content were estimated (refer to Part I). As gathered from a handbook online [17], a reasonable organic loading rate (OLR) would be 3–5 g VS per litre graft per day. However, to adapt to the conditions in the laboratory reactors, 10 g VS per litre graft was chosen as the OLR. The composition of the mixture fed to the reactor was 90.9% food waste + manure, 8.5% silage, and 0.6% glycerol; equivalent to 25 L of substrate, 1.5 dl of glycerol, and 351 g of silage.

For the reference case (without addition of hydrochar), 118 ml of substrate + silage + glycerol in the proportion stated in the previous paragraph were mixed with 481 ml deionised water. For the test case, 4.8 g of finely ground hydrochar (corresponding to 8 g/l) were added to the substrate-mixture first, before dilution with deionised water, to ensure complete dissolution of the hydrochar. The diluted substrate-mix was introduced into the reactors using a 100 ml plastic syringe.

13.2.4 Measurements in the Laboratory

The volume of biogas was measured every day using a 100 ml air-tight glass syringe. In the last week of the 68-day long experiment, the methane content in the biogas was also measured using a gas chromatograph with a flame ionising detector (Clarus 480; non-polar capillary column; J&W Scientific, Elite-5, 30 m × 0.25 mm × 0.25 μm; using helium at 7 bar pressure).

The TS in the filtrate from the digestate which was collected in the tank depicted in Fig. 13.1 was measured and compared with that in the substrate-mix, to determine the TS reduction during the digestion process. The pH of the same was measured on a daily basis, in order to keep track of the degree of stability. The activity of microorganisms is adversely affected at high (over 8) and low (below 7) pH values. The VFA and TOC contents in the digestate were determined according to the Hach Lange LCK365 and LCK 386 standards respectively. LCK365 is intended for concentrations in the range of 75–3600 mg/l propionic acid, while LCK 386 is meant for 30–300 mg/l.

13.2.5 Life-Cycle Costing

Life-cycle costing (LCC) of a project or a business undertaking accounts for the present value (PV, determined on the basis of a chosen discount rate) of all the expenditure and revenue streams over its operating life-cycle and calculates the net present value (NPV) as the difference between the PV of the revenue streams and that of the expenditure streams [18]. An LCC was carried out for two different scenarios, to determine which of the two would be more feasible/profitable economically.

- **Scenario 1**: Biogasbolaget will have its own captive hydrochar-producing HTC-unit which will supply the hydrochar to be added to the anaerobic digester. A greater portion of the hydrochar will be sold on the market, while biogas will be sold to the public transport sector.
- **Scenario 2**: Biogasbolaget will purchase the amount of hydrochar (HC used as acronym at some junctures in the text) needed to be added to the digester, from an external supplier. The biogas produced, will be sold to the transport sector.

Data pertaining to C-Green [19] and Biogasbolaget [20] relevant for the LCC calculations are summarised in Table 13.2.

Table 13.2 Data used in the LCC analysis

Data item	Value	Source/Reference
Capital investment for an HTC-reactor	80 million SEK	[19]
Functional lifetime of the HTC-reactor	20 years	
Operation and maintenance expenses	1.6 million SEK per year (estimated in the year prior to the start of the functional lifetime, so-called year-zero in LCC)	
Energy demand during the operation phase	500 kWh electricity/ton HC	
Total annual output	5000 ton HC total solids	
Electricity tariff	1.37 SEK/kWh	[21]
Biomethane produced as transport fuel	40 GWh	[20]
Conversion factor—Nm^3biomethane fuel to energy	1 Nm^3biomethane fuel = 9.67 kWh	[22]
Price of biomethane (as fuel)	17.5 SEK/Nm^3	[20]
Estimated price of purchased hydrochar	10,000 SEK/ton HC solids (estimated in the year prior to the start of the functional lifetime, so-called year-zero in LCC)	[19]
Total annual HC demand at Biogasbolaget	623 ton HC solids/year	Calculated value
Discount rate	$r = 5\%$	[23]
Inflation rate	$i = 10\%$	Swedish Statistics Agency (www.scb.se)

Corresponding to a given demand for hydrochar at Biogasbolaget (623 tons HC solids per year), the rise in biomethane yield in Part I (batch process) was 17%. This study (as stated later on in the article) showed that a rise of 53% could be achieved by availing of continuous anaerobic digestion.

The present value (PV) of all the costs associated with the building, operation, and maintenance of an HTC-facility at Biogasbolaget, when apportioned to the 623 tons of HC supplied to the digester can be calculated according to Eq. 13.1. Note that 623 tons account for 12.4% of the total estimated annual output of 5000 tons.

$$PV_{HTC} = \left(\left(I + (O+M)\left(\frac{(1+i)^1}{(1+r)^1} + \frac{(1+i)^2}{(1+r)^2} + \ldots + \frac{(1+i)^{20}}{(1+r)^{20}}\right)\right)/5000\right)*623 \tag{13.1}$$

In Eq. 13.1 (pertaining to Scenario 1), 'I' stands for the initial capital investment, '$O+M$' stands for the annual operation and maintenance expenses estimated in the year prior to the start of operation (the so-called year-zero in LCC). This increases from year to year, at the inflation rate. The values 'i' and 'r' represent the inflation rate and the discount rate respectively. The former is assumed to be 10% (even though the average in Sweden till before 2020 has been around 2%; inflation has been very high in the third decade of this century following the pandemic and the Ukraine crisis), and the latter 5%.

The PV of the costs associated with the purchase of hydrochar from an external supplier over the 20-year period (the same as the functional lifetime of the HTC-facility in Scenario 1) can be calculated using Eq. 13.2.

$$PV_{\text{purchase, HC}} = \text{Annual HC cost}\left(\frac{(1+i)^1}{(1+r)^1} + \frac{(1+i)^2}{(1+r)^2} + \cdots + \frac{(1+i)^{20}}{(1+r)^{20}}\right) \tag{13.2}$$

'Annual HC cost' in Eq. 13.2, as listed in Table 13.2, is the cost of purchase as determined in year-zero, similar to '$O+M$' in Eq. 13.1. It increases from year to year at the rate of inflation. The PV of the marginal increase in annual benefit owing to the rise in the production of biomethane fuel (17% in Part I and 53% in this study) can be calculated using Eq. 13.3.

$$PV_{\text{additional benefit}} = \text{Annual add. benefit}\left(\frac{(1+i)^1}{(1+r)^1} + \frac{(1+i)^2}{(1+r)^2} + \cdots + \frac{(1+i)^{20}}{(1+r)^{20}}\right) \tag{13.3}$$

'Annual additional benefit' is determined on the basis of known price of biomethane fuel in the year prior to the start of operation. This, like '$O+M$' in Eq. 13.1 and 'Annual HC cost' in Eq. 13.2, will increase from year to year at the rate of inflation.

The Net Present Value (NPV) was subsequently determined for both scenarios. For the first scenario, 'NPV_{HTC}' is the difference between $PV_{additional\ benefit}$ (Eq. 13.2) and PV_{HTC} (Eq. 13.1). The other one 'NPV_{HC}' is the difference between $PV_{additional\ benefit}$ (Eq. 13.2) and $PV_{purchase,\ HC}$ (Eq. 13.3).

$$NPV_{HTC} = PV_{additional\ benefit} - PV_{HTC} \tag{13.4}$$

$$NPV_{HC} = PV_{additional\ benefit} - PV_{purchase,\ HC} \tag{13.5}$$

The capital investment listed in Table 13.2 corresponds to a total annual production of 5000 tons of HC solids. At a concentration of 8 g HC/l, one would need 623 tons of HC solids per year for the digester. If Biogasbolaget would design an HTC-facility exactly similar to the one at C-green (including the capacity and the output), it would have to deal with surplus hydrochar. It can very well scout for potential buyers in Sweden. One can very well expect a rise in demand for hydrochar in the years to come, as more biogas plants decide to augment biogas (and thereby biomethane) production by adding hydrochar to their digesters. The LCC was extended to determine the effect of the unit selling price of the surplus hydrochar [(5000–623) tons] on the NPV of the 'production and sale of 4377 tons of hydrochar'.

$$PV_{sale,\ HC} = Benefit_{sale,\ HC}\left(\frac{(1+i)^1}{(1+r)^1} + \frac{(1+i)^2}{(1+r)^2} + \cdots + \frac{(1+i)^{20}}{(1+r)^{20}}\right) \tag{13.6}$$

'$Benefit_{sale,\ HC}$' is the total benefit from selling 4377 tons of HC annually. This is estimated in year-zero and increases from year to year at the rate of inflation.

$$NPV_{sale,\ HC} = PV_{sale,\ HC} - PV_{total\ production\ cost,\ surplus\ HC} \tag{13.7}$$

$PV_{total\ production\ cost,\ surplus\ HC}$ is the present value of the life-cycle cost of producing 4377 tons of HC solids per year, for 20 years. This is 7 times the value of PV_{HTC} calculated in Eq. 13.1, corresponding to 623 tons of HC solids per year. The unit selling price of HC was set at 10,000 SEK/ton (the same as what Biogasbolaget would have to pay if they decided to purchase the 623 tons they would need annually).

Sensitivity Analysis: As indicated earlier, inflation in Sweden has fluctuated over time and has had an average value of 2%, as per the Swedish Statistics Central Bureau, and at the time of writing, stands at 10%. The discount rate impacts the NPV of business projects, and since that can also vary (quite like the inflation rate does), a sensitivity analysis was deemed to be necessary, to bolster the LCC. In addition to the values of 'i' and 'r', the unit purchase price of hydrochar (Scenario 2) which was assumed to be 10,000 SEK/ton HC solids was varied, and the sensitivity of the NPV to the variation was tested. Likewise, a sensitivity analysis of the unit selling price of the surplus hydrochar (Scenario 1) was also carried out.

13.3 Results and Discussion

Some of the key results from this study have been presented hereunder.

13.3.1 Biogas and Methane Production

Figure 13.2 shows the results for daily biogas production, for a total period of 42 days. On average, the daily production of biogas over these 6 weeks was 533 ml/g VS. This was over 7.6 times more than the corresponding average value for the reference case (without the addition of hydrochar in other words).

The reason why Fig. 13.2 starts off from day-27 was the detection of leakage in the equipment, which took some time to be remedied. Once it was remedied, the daily biogas production rose and remained at an average value of 533 ml/g VS. While the problem was resolved for the reactor to which hydrochar was added, no marked change in the biogas generation was recorded for the reference reactor. It is seen from Fig. 13.2 that the yield dropped on day-33 before picking up again and stabilising on day-41. The reasons for this behaviour are not very clear, as the pH remained fairly stable throughout the experiment. Clearly, a 663% increase in biogas production resulted thanks to the addition of hydrochar into the reactor.

Biogas can have varying compositions of carbon dioxide and methane, and from an energetic point of view, it is the methane content which is of importance. The reference case registered an average of 24.6% methane in the biogas, while addition of hydrochar increased this share to an average of 60.9% in this study which was equivalent to 367 ml/g VS. It is thus clear that the methane generation potential is more than doubled by the addition of hydrochar to the substrate in the digester.

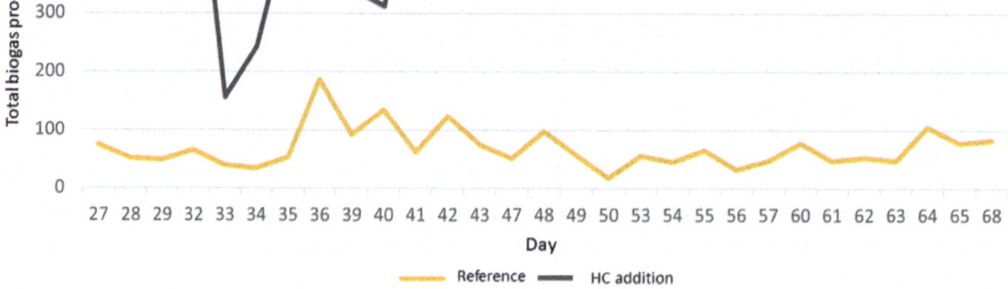

Fig. 13.2 Daily biogas production depicted in Fig. 13.1, between day-27 and day-68

There were some practical difficulties with availing of the gas chromatograph which resulted in limiting the readings to 7 (within the said 9-day period towards the fag end of the experiment).

13.3.2 *Material Flow Analysis*

Figure 13.3 presents a material flow analysis (MFA) diagram for Biogasbolaget with an in-house HTC-facility. The substrate-mix (23.27 tons per day, with 11% total solids content) flows into the digester along with deionised water (192.05 tons per day). On the outlet side, there are two streams—190 tons per day of water and 10 tons per day of dry solids. To maintain an HC concentration of 8 g/l in the digester, if the HTC-reactor has a 70% efficiency of conversion of biomass to solid hydrochar, 2.46 tons (per day) of digestate solids will have to be recirculated to the HTC-reactor. Biogas, digestate as fertiliser, and the 190 tons per day of water are the outflows from the system (the sale of surplus hydrochar which is not connected to the digestion process within the firm has not been indicated as an outflow).

The fertiliser stream which leaves the system on the right has more carbon and nutrients (nitrogen and phosphorus; as well as some essential micronutrients), when compared to a similar stream from a digester which does not avail of hydrochar addition. Evidently, this is also a saleable product for Biogasbolaget, in addition to biogas (or refined biomethane) and surplus hydrochar. The fact that the digestate with hydrochar is richer in carbon and nutrients also motivates the recirculation shown in Fig. 13.3.

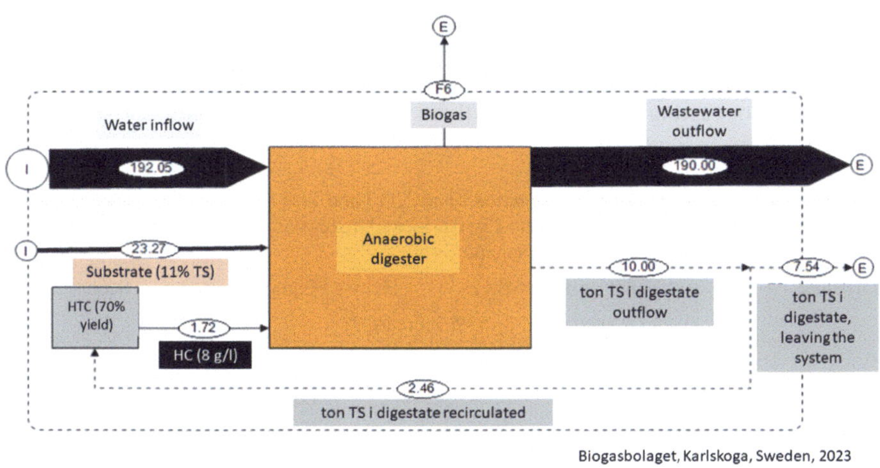

Fig. 13.3 Material flow diagram for the HTC-digester combination at Biogasbolaget

13.3.3 Life-Cycle Costing

Table 13.3 presents the PV and NPV values for Scenarios 1 and 2. Values corresponding to rise in biomethane production of 17 and 53% are both represented adjacent to each other.

If Biogasbolaget decides to invest in the HTC-facility yielding 5000 tons of HC solids per year, and the captive consumption remains constant at 623 tons per year, business sense would dictate finding a market for the surplus 4377 tons (which has a total life-cycle cost associated with it). While the PV of 31 million SEK (Table 13.3) corresponds to the 623 tons used in the digester. The PV of life-cycle costs associated with the surplus would be 219 million SEK. Considering a unit sale price for the surplus hydrochar (determined in and for year-zero), of 10,000 SEK per ton HC solids, the PV of the total benefits over the 20-year period would equal 1479 million SEK, giving a NPV for the 'produce-and-sell-surplus-hydrochar' part of the project equal to 1260 million SEK. This NPV then would be added on to the NPV_{HTC} reported in Table 13.3, to obtain the total NPV of the investment in the HTC-facility, intended to produce two saleable outputs.

In Fig. 13.4, the effect of variations in the discount rate (between 2 and 6%), and the inflation rate (between 2 and 10%) has been depicted graphically, for Scenario 1. The NPV_{HTC} in this case corresponds to the apportioned present value of the life-cycle costs to the production of 623 tons of HC solids per year, and a rise in biomethane production by 17%. In other words, the NPV of the 'produce-and-sell-surplus-hydrochar' part of the project has not been included.

The NPV for Scenario 2 will depend on the purchase price of hydrochar and there is an uncertainty surrounding that value as there is no market as such for hydrochar. However, as mentioned earlier, one expects this to become a reality in the years to come, as a circular bioeconomy entrenches itself around the world. With an increase in this purchase price, the NPV will tend to decrease. As seen in Fig. 13.5, as long as the purchase price is below 18,000 SEK/ton HC solids, the NPV will be positive (suggesting that it would still not be unprofitable to purchase hydrochar

Table 13.3 PV and NPV values for Scenarios 1 and 2, (17 and 53% rise in the biomethane production) [sale of surplus hydrochar in Scenario 1 is overlooked in this comparison, but becomes realistic when there is a burgeoning market and demand for it]

Variables	Value (million SEK)		Equations used	Scenarios
	17% increase	53% increase		
PV_{HTC}	31	31	Equation 13.1	1
$PV_{purchase, HC}$	217	217	Equation 13.2	2
$PV_{additional\ benefit}$	394	1268	Equation 13.3	1 and 2
NPV_{HTC}	363	1237	Equation 13.4	1
NPV_{HC}	177	1051	Equation 13.5	2
$NPV_{HTC} - NPV_{HC}$	**186**		Scenario 1 economically more attractive	

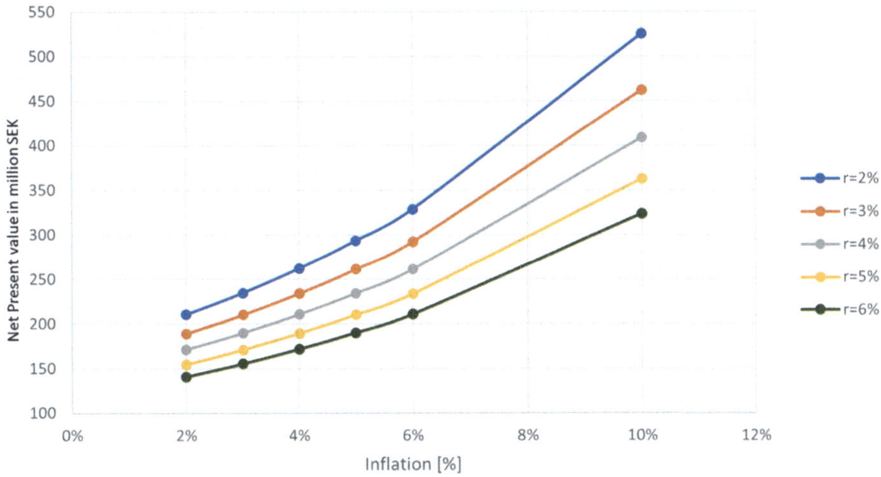

Fig. 13.4 Variations in NPV with discount and inflation rates, for a 17% rise in biomethane production for Scenario 1 (not considering any sale of surplus hydrochar)

from external suppliers—as, if and when they begin to exist—and use it to augment biogas production).

One can now turn back to Scenario 1 and consider both the sale of surplus hydrochar and the additional biomethane fuel. To exactly break even, the surplus hydrochar would have to be sold at 1479 SEK/ton HC solids. This is the point at which the value of $NPV_{sale,\,HC}$ (calculated using Eq. 13.7) drops to zero. At this point, the total NPV, as seen in Fig. 13.6, is 363 million SEK (the same as NPV_{HTC} from Table 13.3). If there is no or less demand for hydrochar, and Biogasbolaget is compelled to sell it at a lower price than the aforesaid breakeven price, $NPV_{sale,\,HC}$

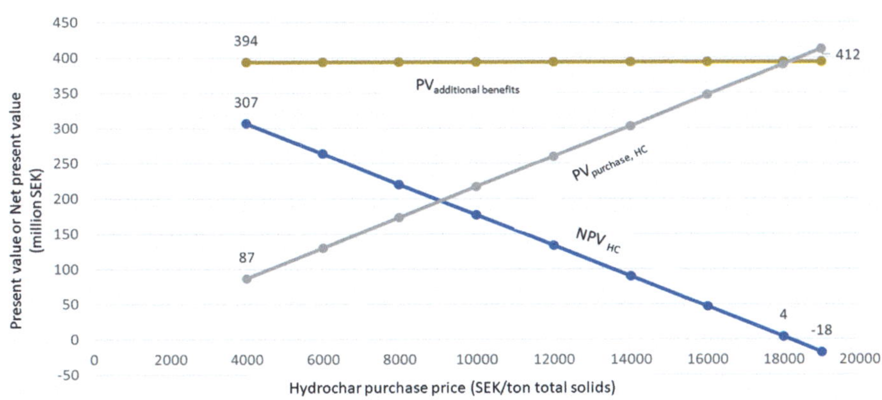

Fig. 13.5 The sensitivity of the NPV in Scenario 2 to the purchase price of hydrochar, for a 17% rise in biomethane production

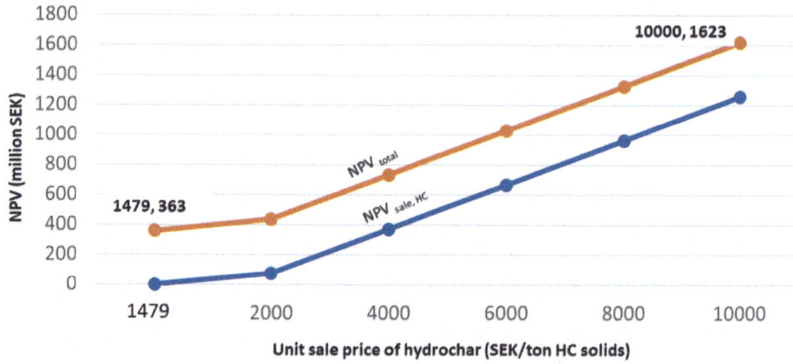

Fig. 13.6 The sensitivity of the NPV to the unit sale price of hydrochar

will be negative, and this will pull down the total NPV of the in-house HTC-facility installation project. At a selling price of 10,000 SEK/ton, the total NPV stands at 1623 million SEK.

It must be mentioned, however, that most of the data obtained [19, 20] are studied approximations made by experts in the business. There are very few HTC-facilities in operation at the time of writing, and access to realistic data is difficult. This study uncovers something which has the potential to attract interest among biogas plants in the future. But, on date, no biogas plant purchases hydrochar from C-green. While it would have been reasonable to envisage an HTC-facility at Biogasbolaget producing precisely what the digester would need (623 tons of HC solids per year), the authors have designed a scenario which is ideal, desirable and realistic, when viewed from the perspective of a circular bioeconomy in which hydrochar would be a saleable product, sought-after by biogas plants in Europe and beyond. The electricity tariff may not increase at the general inflation rate considered in the analysis (10%); it is possible that it may even drop slightly. The selling price of biomethane fuel (supplied to the transport sector), determined at year-zero, is 17.5 SEK/Nm^3. This is also assumed to increase at the general inflation rate, which may not be the case in reality.

Uncertainty surrounds the selling price of the surplus hydrochar in Scenario 1. As gathered from Peter Axegård of C-Green [19], there is a demand for hydrochar at combined heat and power plants, at the time of writing. If new demand at biogas plants is uncovered, courtesy studies like this one, the value and thereby the selling price of hydrochar will increase. But to meet the increasing demand, if more HTC-facilities are set up, and supply tends to rise faster than demand (when all the facilities are operating at 100% capacity), the price will drop down. The possibilities though are difficult to predict. The sensitivity analysis conducted contributes a little to dealing with this uncertainty.

13.4 Conclusions and Recommendations

The findings of this study, as well as those in Part I, augur well for a circular bioeconomy, like the one Sweden is keen on establishing in the years to come. By extension, it will motivate trials and experiments in other parts of the world, to verify, confirm, and subsequently commercialise the utilisation of hydrochar in biogas plants. Biogas plants like Biogasbolaget for instance, can diversify into the production, use, and sale of hydrochar, and also deliver richer organic fertilisers to farmers as one of the three saleable outputs, thereby contributing to phosphorus recovery. The integration between the HTC-reactor and the anaerobic digester, supplying to each other (as depicted in Fig. 13.4), uncovers environmental and techno-economic advantages.

The LCC and the accompanying sensitivity analysis 'speak out' in support of captive production of hydrochar by the HTC process, to be partly used in-house to produce more biogas and thus sell more refined biomethane fuel, and largely sold to potential users in the external marketplace (CHP plants, biogas plants, etc.)

Valorising what would otherwise be dealt with in non-profitable and undesirable ways, to replace and substitute what is not desired in a circular bioeconomy in the future, is indisputably a win–win situation, en route to the attainment of a clutch of complementary sustainable development goals.

Acknowledgements The authors would like to thank everyone at C-Green, Biogasbolaget and Karlstad University for their help and support with this study, at different points of time in the first half of 2023.

References

1. Cavali, M., Junior, N. L., de Almeida Mohedano, R., Belli Filho, P., da Costa, R. H. R., & de Castilhos Junior, A. B. (2022). Biochar and hydrochar in the context of anaerobic digestion for a circular approach: An overview. *Science of the Total Environment, 822*, 153614. https://doi.org/10.1016/j.scitotenv.2022.153614
2. European Commission. (n.d.). *Food waste*. Available at: https://food.ec.europa.eu/safety/food-waste_en. Accessed on February 7, 2023.
3. Livsmedelsverket. (2022). Därför ska vi minska matsvinnet [Therefore, shall we reduce food waste] Available at: https://www.livsmedelsverket.se/matvanor-halsa--miljo/maltider-i-vard-skola-och-omsorg/matsvinn-i-storkok/handbok-for-minskat-matsvinn/darfor-ska-vi-minska-matsvinnet. Accessed on February 7, 2023.
4. Chiappero, M., Norouzi, O., Hu, M., Demichelis, F., Berruti, F., Di Maria, F., Mašek, O., & Fiore, S. (2020). Review of biochar role as additive in anaerobic digestion processes. *Renewable and Sustainable Energy Reviews, 131*, 110037. https://doi.org/10.1016/j.rser.2020.110037
5. Merzari, F., Goldfarb, J., Andreottola, G., Mimmo, T., Volpe, M., & Fiori, L. (2020). Hydrothermal carbonization as a strategy for sewage sludge management: Influence of process withdrawal point on hydrochar properties. *Energies, 13*, 2890. https://doi.org/10.3390/en13112890
6. Assis, E. I. N. C., Gidudu, B., & Chirwa, E. M. N. (2022). Hydrothermal carbonisation of paper sludge: Effect of process conditions on hydrochar fuel characteristics and energy recycling

efficiency. *Journal of Cleaner Production, 373*, 133775. https://doi.org/10.1016/j.jclepro.2022.133775

7. Pecchi, M., & Baratieri, M. (2019). Coupling anaerobic digestion with gasification, pyrolysis or hydrothermal carbonization: A review. *Renewable and Sustainable Energy Reviews, 105*, 462–475. https://doi.org/10.1016/j.rser.2019.02.003
8. Xu, J., Mustafa, A. M., Lin, H., Choe, U. Y., & Sheng, K. (2018). Effect of hydrochar on anaerobic digestion of dead pig carcass after hydrothermal pretreatment. *Waste Management, 78*, 849–856. https://doi.org/10.1016/j.wasman.2018.07.003
9. Zhou, Y., Engler, N., Li, Y., & Nelles, M. (2020). The influence of hydrothermal operation on the surface properties of kitchen waste-derived hydrochar: Biogas upgrading. *Journal of Cleaner Production, 259*, 121020. https://doi.org/10.1016/j.jclepro.2020.121020
10. Ayodele, O. O., Adekunle, A. E., Adesina, A. O., Pourianejad, S., Zentner, A., & Dornack, C. (2021). Stabilization of anaerobic co-digestion of biowaste using activated carbon of coffee ground biomass. *Bioresource Technology, 319*, 124247. https://doi.org/10.1016/j.biortech.2020.124247
11. Ma, J., Pan, J., Qiu, L., Wang, Q., & Zhang, Z. (2019). Biochar triggering multipath methanogenesis and subdued propionic acid accumulation during semi-continuous anaerobic digestion. *Bioresource Technology, 293*, 122026. https://doi.org/10.1016/j.biortech.2019.122026
12. Wang, R., Peng, P., Song, G., Zhao, Z., & Yin, Q. (2022). Effect of corn Stover hydrochar on anaerobic digestion performance of its associated wastewater. *Environmental Pollution, 315*, 120430. https://doi.org/10.1016/j.envpol.2022.120430
13. Wu, B., Yang, Q., Yao, F., Chen, S., He, L., Hou, K., Pi, Z., Yin, H., Fu, J., Wang, D., & Li, X. (2019). Evaluating the effect of biochar on mesophilic anaerobic digestion of waste activated sludge and microbial diversity. *Bioresource Technology, 294*, 122235. https://doi.org/10.1016/j.biortech.2019.122235
14. Ayodele, O. O., Adekunle, A. E., Alagbe, O. A., Anguruwa, G. T., Ademola, A. A., Odega, C. A., & Dornack, C. (2022). Application of biomass-derived hydrochar in process stability of anaerobic digestion. *Bioresour. Technol. Rep., 17*, 100903. https://doi.org/10.1016/j.biteb.2021.100903
15. Xu, Q., Yang, G., Liu, X., Wong, J. W. C., & Zhao, J. (2023). Hydrochar mediated anaerobic digestion of bio-wastes: Advances, mechanisms and perspectives. *Science of the Total Environment, 884*, 163829. https://doi.org/10.1016/j.scitotenv.2023.163829
16. Axegård, P., Baresel, C., Lazic, A., Bornold, N., Yang, J. -J., & Malovanyy, A. (2023). Framtidens slamhantering vid Roslagsvatten (Sludge handling in the future at Roslagsvattten). Ivl Sven. Miljöinstitutet. Accessible at www.ivl.se
17. Schnürer, A., & Jarvis, Å. (2009). *Microbiology handbook for biogas plants*. Accessible at: https://www.build-a-biogas-plant.com/PDF/Microbiology_handbook_for-Biogas_plants. Published by Avfall Sverige and Svenskt Gasteknisk Center AB.
18. Venkatesh, G. (2019). *Life-cycle costing: A primer* (1st ed.). Bookboon. Available at: https://bookboon.com/en/life-cycle-costing-ebook?mediaType=ebook
19. Axegård, P. (2023). *C-Green*. Personal Communication in March-May 2023.
20. Lagerblad, J. (2023). *Biogasbolaget*. Personal Communication in April 2023
21. Swedish Energy Agency (Energimyndigheten). (2023). *Energiläget*. Available at: https://www.energimyndigheten.se/statistik/energilaget/?currentTab=1. Accessed on May 9, 2023.
22. Svenskt Gastekniskt Center AB. (2012). *Basic data on biogas*. Accessible at www.sgc.se/ckfinder/userfiles/files/BasicDataonBiogas2012.pdf
23. Nordlöf, P. (2014). Summary of the Swedish debate on the discount rate. In *Powerpoint presentation on the 6th of November 2014 for Trafikverket* (Swedish Transport Administration). Accessed in May 2023 on www.trafikverket.se

Chapter 14
Heterogeneous Hydrochar-Based Catalysts for Biodiesel Production

Muhammad Aliyu, Umer Rashid, Wan Azlina Wan Ab Karim Ghani, Muhamad Amran bin Mohd Salleh, Balkis Hazmi, Ibrahim Garba Shitu, and Ali Salisu

Abstract The potential of heterogeneous hydrochar-based catalysts as a desirable technology for biodiesel production is highlighted in this book chapter. These hydrochar-based catalysts have renewable features, are cost-effective, and have efficient catalytic properties, making them an excellent choice in the field due to their high carbon content and unique physicochemical features. This chapter aims to provide a comprehensive review of the feedstock of hydrochar from different biomass sources including agricultural waste, food waste, and forestry waste. The hydrochar production technologies and factors affecting the properties of hydrochar during production were also discussed. The catalytic performance to produce biodiesel can be considerably improved by the development of types of heterogeneous hydrochar-based catalysts via introducing active species such as metal nanoparticles or acid- or base-functional groups onto the hydrochar surface, and the parameters that influence

M. Aliyu · U. Rashid (✉) · B. Hazmi
Institute of Nanoscience and Nanotechnology, Universiti Putra Malaysia (UPM), 43400 Serdang, Selangor, Malaysia
e-mail: umer.rashid@upm.edu.my; dr.umer.rashid@gmail.com

M. Aliyu
Department of Material Engineering, Graduate School of Engineering, Kyushu Institute of Technology, 1-1 Sensuicho, Tobata-ku, Kitakyushu, Fukuoka 804-8550, Japan

U. Rashid
Center of Excellence in Catalysis for Bioenergy and Renewable Chemical (CBRC), Faculty of Science, Chulalongkorn University, Pathumwan, Bangkok 10330, Thailand

W. A. W. A. K. Ghani · M. A. M. Salleh
Department of Chemical and Environmental Engineering, Faculty of Engineering, Universiti Putra Malaysia, 43400 Serdang, Selangor, Malaysia

I. G. Shitu
Department of Physics, Faculty of Natural and Applied Science, Sule Lamido University, Kafin Hausa, Jigawa State, Nigeria

A. Salisu
Department of Chemistry, Faculty of Natural and Applied Science, Sule Lamido University, Kafin Hausa, Jigawa State, Nigeria

G. Baskar et al. (eds.), *Circular Bioeconomy Perspectives in Sustainable Bioenergy Production*, Energy, Environment, and Sustainability,
https://doi.org/10.1007/978-981-97-2523-6_14

biodiesel yield and selectivity during biodiesel production utilizing heterogeneous hydrochar-based catalysts were discussed. The multifaceted nature of hydrochar, combined with its renewable origin and distinctive physicochemical properties, positions it as a valuable resource for advancing sustainable and efficient catalytic processes. Finally, this chapter advances knowledge on heterogeneous hydrochar-based catalysts and encourages further research in this promising area while outlining current research and future directions.

Keywords Hydrochar · Biomass · Hydrothermal carbonization · Heterogeneous catalysts · Biodiesel production

14.1 Introduction

Energy has always been recognized as a fundamental driver for sustaining economic growth in any nation, with coal, natural gas, and crude oil serving as major sources to meet this energy demand [1]. However, the rapid expansion of industrialization and urbanization is projected to lead to a 33% increase in global energy consumption by 2035 [2]. The depletion of non-renewable resources and their contribution to environmental issues like global warming have fueled the desire for alternative energy conservation and utilization approaches [3]. The world's focus has shifted toward sustainable and eco-friendly strategies for producing renewable energy sources. The objective is to develop non-toxic and biodegradable alternatives that offer environmentally friendly and economically viable solutions [4]. One highly preferred renewable energy source in the present era is biodiesel, with biodiesel standing out as a non-toxic, biodegradable, and renewable energy option [5].

Biodiesel plays a crucial role in addressing the pressing need for sustainable and renewable energy sources worldwide. Biodiesel is produced from renewable sources including soybean oil, canola oil, palm oil, or waste cooking oil [6].

Biodiesel can be used in diesel engines in replacement of petroleum-based diesel fuel without any modifications. It has various advantages over conventional diesel fuel, including decreased carbon dioxide, particulate matter, and sulfur dioxide emissions [7]. Biodiesel is also less hazardous and combustible than diesel fuel, which makes it easier to handle and transport. The biodiesel production process of biodiesel involves a chemical reaction called transesterification or esterification, which converts vegetable oil or animal fat into biodiesel [8]. During this process, the vegetable oil or animal fat is mixed with alcohol and a catalyst, which breaks down the oil molecules and produces biodiesel glycerin and water [9]. With the increasing demand for biodiesel, researchers are actively pursuing innovative low-cost catalysts that can optimize the efficiency and sustainability of the biodiesel production process.

In recent years, heterogeneous hydrochar-based catalysts have emerged as a promising solution within this field of research [10]. These catalysts have garnered significant attention due to their inherent advantages, which make them highly attractive for biodiesel production. One of the key advantages of heterogeneous hydrochar-based catalysts is their renewable nature [11]. The use of hydrochar as a precursor for catalyst synthesis capitalizes on the abundant carbon content present in biomass, effectively utilizing renewable resources [12]. This aligns perfectly with the goal of sustainable energy production and contributes to reducing the reliance on non-renewable fossil fuels [13]. Additionally, the cost-effectiveness of heterogeneous hydrochar-based catalysts enhances their appeal for biodiesel production. The availability of biomass feedstock and the relatively simple synthesis process of hydrochar-based catalysts result in lower production costs compared to conventional catalysts [14].

The term "hydrochar" encompasses a carbon-rich substance that is derived through the process of hydrothermal carbonization (HTC), often known as "wet carbonization," from various biomass sources ranging from agricultural residues and food waste to forestry by-products, contributing to the utilization of diverse organic materials [15]. HTC includes heating biomass at high pressures and temperatures in the presence of water, forming hydrochar, a carbon-rich solid material [16]. Hydrochar usually contains a high percentage of carbon range between 50 and 90%, as well as small amounts of hydrogen, oxygen, and nitrogen. It might also have ash, which contains minerals and trace elements from the original feedstock [16]. Hydrochar has several possible applications, including its use as a fuel source for power generation, as a soil supplement in agriculture and as a catalyst. Hydrochar serves as an exceptional precursor for catalyst synthesis, primarily due to its abundant carbon content and distinctive physicochemical properties [2].

Hydrochar has a variety of properties and a porous structure with a large surface area, making it an attractive choice for catalyst production [17]. The chemical composition of hydrochar can be modified through the introduction of active species onto its surface [18]. This procedure involves the addition of materials such as metal nanoparticles or acid–base functional groups to the hydrochar matrix, which improves its catalytic capabilities for biodiesel production [3]. Therefore, the catalyst becomes capable of enabling and accelerating the conversion of triglycerides or fatty free acids (FFA) into biodiesel, leading to a significant improvement in the overall efficiency of the biodiesel production process [19]. Furthermore, the incorporation of active species onto the hydrochar matrix also contributes to the enhanced stability of the catalyst, enabling it to maintain its catalytic activity over prolonged reaction periods. Another significant advantage of this approach is its potential to mitigate the environmental impact associated with biodiesel production [5].

This book chapter delves into the improvements of heterogeneous hydrochar-based catalysts for biodiesel production. This study investigates multiple components, such as their activation, modification, catalytic performance on biodiesel production, and ongoing research efforts aimed at enhancing their performance and usability in real-world biodiesel production scenarios. We hope to provide significant insights and contribute to the improvement of biodiesel production methods using hydrochar-based catalysts by investigating the most recent results and breakthroughs in this field.

14.2 Hydrochar

14.2.1 Feedstock, Composition, and Properties of Hydrochar

Biomass serves as the primary feedstock for hydrochar production, encompassing various organic materials derived from agricultural waste, food waste, and forestry waste as shown in Fig. 14.1 [20]. These diverse biomass feedstocks contribute to the composition of hydrochar, which is a carbon-rich material obtained through hydrothermal carbonization [21]. The composition of hydrochar depends on the specific biomass source used and the process parameters employed during its production. It typically contains carbon, hydrogen, oxygen, and small amounts of other elements, such as nitrogen and sulfur [22]. The elemental composition and chemical properties of hydrochar can vary, influencing its reactivity, surface area, and porosity. Therefore, the selection of biomass feedstock and control of process parameters play a crucial role in determining the composition and properties of hydrochar [23].

14.2.1.1 Agricultural Waste

Agricultural waste is the organic by-products that remain after the cultivation and harvesting of crops [24]. These residues encompass a wide range of plant materials such as stems, leaves, husks, shells, and straws, which are typically left behind in the fields after the primary crop has been collected [25]. With the large-scale agricultural practices prevalent in many regions, agricultural residues have become a significant and easily accessible source of biomass [26]. Their abundance and availability make them attractive feedstock options for various applications, including hydrochar production. Rice straws, corn stalks, sugarcane bagasse, and wheat straw are among the most utilized agricultural residues in hydrochar production processes [27].

The potential of these agricultural wastes is maximized by using them as feedstock hydrochar production, converting what would otherwise be considered waste into a lucrative resource [28]. The use of agricultural residues for hydrochar production has various advantages. For instance, it provides a sustainable and renewable alternative to existing fossil fuel-based resources, helping to establish a more ecologically friendly energy sector [29]. Using agricultural wastes reduces the demand for non-renewable resources, promotes resource efficiency, and reduces the environmental effect of fossil fuel consumption [30]. Furthermore, using agricultural waste for producing hydrochar helps the efficient use of agricultural waste, and instead of being dumped or left in the fields, these wastes are turned into valuable goods with potential applications in a variety of industries. This minimizes waste and provides potential for economic growth and resource efficiency in the agricultural sector.

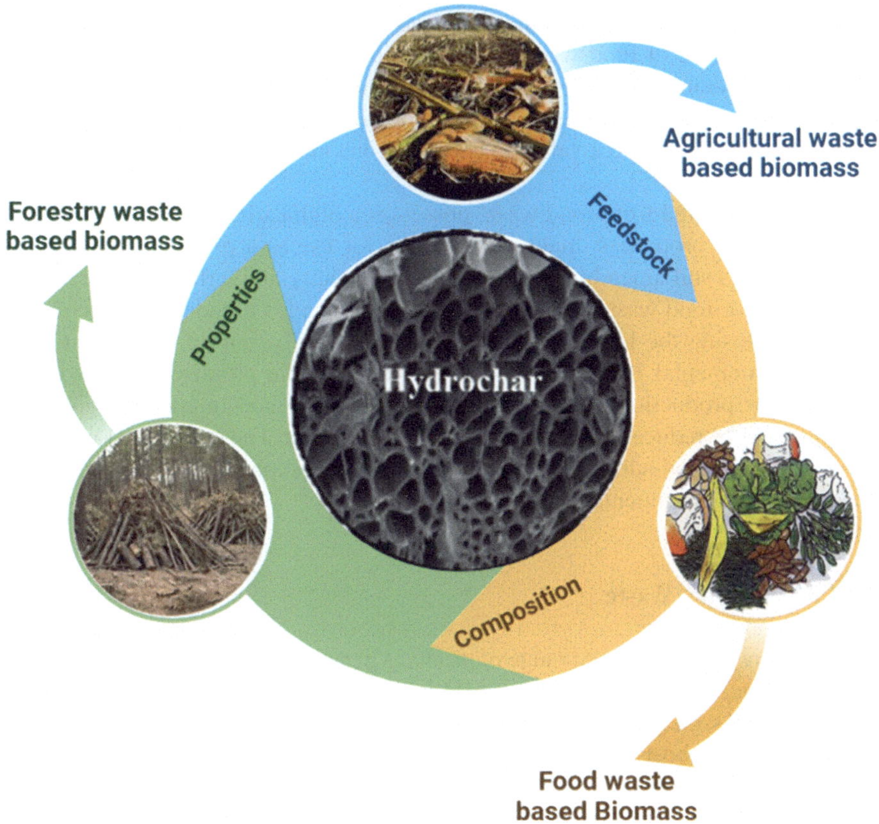

Fig. 14.1 The distribution of feedstock for hydrochar production

14.2.1.2 Food Waste

Food waste encompasses various organic materials that are no longer consumed or used, originating from households, restaurants, food processing industries, and other sources [31]. It poses a significant environmental challenge due to the substantial quantities of food waste that end up in landfills, resulting in the generation of greenhouse gas emissions, particularly methane (CH_4). To address this issue, utilizing food waste as a feedstock for hydrochar production offers an eco-friendly solution to convert this waste into a valuable resource [32]. Food waste consists of a wide range of materials, including fruit and vegetable peels, expired or spoiled food, and discarded leftovers. These materials, if not properly managed, would contribute to the increasing volume of waste and its associated environmental impacts [33]. On the other hand, redirecting food waste to hydrochar production can successfully turn its organic content into a carbon-rich material via the process of HTC. Using food waste as a feedstock for hydrochar production has various advantages [34]. Firstly, it helps to address the environmental concerns associated with food waste disposal in landfills. The use of food waste for hydrochar production reduces the emission of methane, a potent greenhouse gas, lowering the carbon footprint and mitigating climate change. Moreover, the conversion of food waste into hydrochar presents an opportunity to generate a valuable carbonaceous material that can be utilized in various applications [35].

Hydrochar derived from food waste possesses a high carbon content and unique physicochemical properties, making it suitable for use as a precursor for catalyst synthesis or as a soil amendment to improve soil quality and fertility [36]. Furthermore, the use of food waste for hydrochar production adds to the circular economy concept by closing the loop in the food system [37]. To sum up, food waste is a serious environmental issue, but it also has considerable promise as a feedstock for hydrochar production. Not only may the environmental implications of food waste disposal be reduced by converting it into hydrochar. The use of food waste for hydrochar production adheres to circular economy principles and adds to resource efficiency and environmental sustainability.

14.2.1.3 Forestry Waste

Forestry waste involves a wide range of products generated during forestry activities, including branches, bark, sawdust, wood chips, and other woody waste. In traditional forestry practices, these materials are often considered waste or by-products, and their disposal may cause cost and ecological challenges [38]. However, through the utilization of forestry residue as feedstocks for hydrochar production, these materials can be converted into valuable resources with a wide range of applications [39]. Due to activities such as tree harvesting, logging, and wood processing, the forestry industry generates a substantial amount of wood waste. Previously, these wastes were either left to disintegrate in the forest or disposed of by burning or landfilling, which can add to air pollution, soil degradation, and misused resources. These materials can be effectively exploited by channeling forestry waste toward hydrochar production [15]. The use of forestry wastes as feedstock for hydrochar production has various advantages. For instance, it contributes to addressing environmental challenges linked with forestry residue disposal [40]. The release of harmful pollutants from the burning or decomposition of forestry waste is decreased by turning these wastes into hydrochar, contributing to improved air quality and lower greenhouse gas emissions. Furthermore, hydrochar made from forestry residue can be used in a variety of applications [41].

14.2.1.4 Composition and Properties of Hydrochar

The composition of hydrochar obtained from different biomass sources can vary depending on the feedstock used and the specific hydrothermal carbonization process employed [16]. Generally, the hydrochar is predominantly composed of carbon, with content of volatile matter, moisture, and other organic compounds compared to the original biomass [2]. Hydrochar exhibits unique physicochemical properties, including high carbon content, low ash content, and stability. Its characteristics make it an excellent precursor for catalyst synthesis and various other applications, such as soil amendment, carbon sequestration, and energy storage [17]. The composition and properties of hydrochar are influenced by several factors, including the type of biomass feedstock, HTC process parameters, and post-treatment techniques [42]. The composition of hydrochar varies depending on the biomass feedstock used. Different biomass sources, such as agricultural waste, food waste, and forestry waste, have distinct chemical compositions, including varying levels of carbon, hydrogen, oxygen, nitrogen, and other trace elements [43]. These elements contribute to the overall composition and can impact the properties and potential applications of hydrochar. The carbon content of hydrochar is a crucial property that determines its energy content and stability. Hydrochar typically has a high carbon content, ranging from 50 to 90%, which makes it a potential source of renewable energy [44]. The carbon concentration of hydrochar also determines its calorific value, making it a desirable option for energy production [3].

The surface area and pore structure of hydrochar are also essential properties. Hydrochar possesses a porous structure with a high surface area, which provides numerous sites for adsorption and chemical reactions. The pore size distribution, including micropores, mesopores, and macropores, affects the reactivity of the hydrochar [19]. These properties make hydrochar suitable for applications such as adsorbents, soil amendments, and catalyst support. The presence of functional groups on the surface of hydrochar is another significant aspect of its composition. These functional groups, including hydroxyl, carboxyl, and phenolic groups, contribute to the reactivity and surface chemistry of hydrochar. They can serve as active sites for various applications or be used to attach different functional groups, leading to increased interactions with various reactants in various environmental applications [5]. The thermal stability of hydrochar is also an important property. Hydrochar exhibits good thermal stability, which enables its use in high-temperature processes without significant degradation [45]. This property is advantageous for applications such as energy conversion, where hydrochar can be utilized as a solid fuel or feedstock for thermochemical processes. Table 14.1 shows the feedstocks, elemental compositions, and properties of hydrochar.

Other properties of hydrochar include its pH, ash content, and density. The pH of hydrochar can influence its behavior in soil applications and its interaction with other substances [46]. The ash content represents the inorganic components present in hydrochar, which can impact its combustion characteristics and the release of nutrients when used as a soil amendment. The density of hydrochar influences its handling and transportation, with lower densities being preferable for easier handling [11]. Finally, the composition and qualities of hydrochar are determined by the biomass feedstock, HTC process parameters, and post-treatment procedures [47]. The current study focuses on modifying the composition and features of hydrochar to optimize its potential as a sustainable and adaptable material in the pursuit of a greener and more sustainable future.

Table 14.1 The feedstocks, elemental compositions, and properties of hydrochar

Biomass	Feedstocks	C (%)	O (%)	H (%)	N (%)	S (%)	Ash content (%)	BET surface area (m^2/g)	Pore volume (cm^2/g)	Average pore diameter (nm)	Yield (%)	Refs.
Agricultural waste	Walnut shell	58.32	11.85	3.17	2.53	1.22	1.49	18.89	0.09	5.37	54.34	[5]
	Millet stalk	53.23	12.15	6.73	4.03	1.74	1.34	10.34	0.08	7.24	51.67	[48]
	Wheat straw	57.76	11.45	6.21	1.43	0.87	1.67	20.43	0.05	5.43	66.34	[49]
	Rice straw	62.32	13.45	4.87	2.03	1.02	1.09	11.89	0.07	4.67	59.43	[50]
Food waste	Banana peels	66.13	11.35	5.23	4.03	1.94	1.84	11.34	0.07	5.24	52.43	[51]
	Watermelon peels	56.43	14.35	7.43	3.23	1.34	2.34	13.34	0.03	8.34	51.23	[52]
	Grapefruit peels	67.1	16.5	5.2	2.8	1.8	2.4	18.23	0.08	7.98	53.67	[13]
	Cassava rhizome	59.83	17.15	9.43	5.33	1.74	3.24	16.54	0.06	9.14	49.67	[53]
Forestry waste	Bamboo sawdust	70.43	11.55	4.13	2.23	0.84	1.14	33.34	0.05	7.04	57.34	[54]
	Wood sawdust	68.22	15.14	5.53	2.73	2.04	2.04	26.34	0.02	4.14	55.76	[55]
	Pinus Caribbean	62.31	13.12	7.36	3.83	1.71	1.94	17.14	0.07	5.30	58.54	[56]
	Spent liquor	65.71	12.24	6.41	2.83	2.11	1.88	25.61	0.09	9.04	52.43	[57]

14.2.2 Hydrochar Production

Hydrochar production encompasses various techniques, including conventional hydrothermal carbonization (HTC) and microwave-assisted hydrothermal carbonization (MA-HTC), which offer efficient and sustainable pathways for converting biomass into a valuable carbon-rich material.

14.2.2.1 Conventional Hydrothermal Carbonization (HTC)

Hydrothermal carbonization (HTC) has recently gained recognition as a practical and established conventional thermochemical technique for producing char [58]. In this method, biomass materials are transformed into a carbon-rich solid substance referred to as hydrochar. This transformation occurs within an aqueous environment, typically at temperatures ranging from 175 to 250 °C, with specific timeframes and pressures that vary based on the type of feedstock biomass used [46]. One of the potential HTC's significant benefits is its compatibility with wet biomass, which eliminates the requirement to dry the biomass prior to carbonization. HTC achieves this by processing moist biomass in hot, pressurized water, leading to significant energy and cost savings that would otherwise be required for pre-drying the biomass before undergoing thermochemical conversion [58]. The reaction conditions in conventional HTC, including temperature, pressure, reaction time, and biomass-to-water ratio, are crucial in influencing the properties and yield of hydrochar [45]. Higher temperatures and longer reaction times generally lead to increased carbonization and the formation of a more carbon-rich hydrochar. Adjusting the pressure and biomass-to-water ratio can also impact the chemical reactions and the final characteristics of the hydrochar produced [59]. As a process of producing hydrochar, conventional HTC has several advantages. A significant advantage is its adaptability to a wide variety of biomass feedstocks. This adaptability enables the use of a wide range of organic materials, boosting the efficient use of biomass resources and reducing waste [60].

14.2.2.2 Microwave-Assisted Hydrothermal Carbonization (MA-HTC)

Microwave-assisted hydrothermal carbonization (MA-HTC) is an emerging technique that combines the principles of conventional HTC with the efficient heating capabilities of microwaves [61]. In MA-HTC, biomass is mixed with water and subjected to microwave irradiation, which rapidly heats the reaction mixture and accelerates the carbonization process [62]. The localized and rapid heating provided by microwaves promotes efficient biomass conversion and reduces reaction times compared to conventional HTC. MA-HTC offers advantages such as energy efficiency, shorter processing times, and improved control over reaction conditions [33]. Both HTC and MA-HTC present efficient and sustainable methods for hydrochar production. They offer opportunities to convert various types of biomasses, including agricultural residues, food waste, and forestry by-products, into valuable carbon-rich material. These techniques contribute to the utilization of biomass resources, reducing waste and promoting a circular economy. Furthermore, they have the potential for integration with existing waste management systems and may be adapted for large-scale production, making them viable solutions for sustainable biomass. However, to get the appropriate hydrochar qualities, temperature, biomass-to-water ratio, and reaction time must be properly optimized. Table 14.2 illustrates a comparison of microwave-assisted HTC against conventional HTC.

Table 14.2 A comparison between micro-assisted HTC and conventional HTC

	MA-HTC	Conventional HTC
Heating method	In MA-HTC, microwave radiation is used to rapidly heat the reaction mixture. It allows for precise and efficient heating of the reactants	HTC relies on external heating sources like electric heaters, ovens, or autoclaves to raise the temperature of the reaction vessel. It is a slower and less precise heating method
Reaction speed	Microwave heating is generally faster, and reactions can be completed more quickly compared to conventional HTC. It is ideal for achieving rapid temperature ramps	Conventional HTC usually requires longer reaction times due to the slower heating process
Temperature control	Microwave heating offers better temperature control and uniformity within the reaction vessel, resulting in consistent reaction conditions	Temperature control in conventional HTC can be less precise, potentially leading to temperature gradients within the vessel
Energy efficiency	Microwave heating is energy efficient as it directly heats the reactants, reducing energy wastage	Conventional heating methods may be less energy-efficient because they heat the surroundings as well as the reactants
Scalability	Scaling up MA-HTC can be challenging and expensive due to the need for specialized equipment and potential issues with uniform heating at larger scales	Conventional HTC is more easily scalable and adaptable to larger reactors and industrial processes
Reactor design	Reactor design for MA-HTC is more complex and requires materials compatible with microwave radiation	Conventional HTC can use standard reactor vessels, making it simpler in terms of equipment
Safety	Safety precautions are necessary when working with microwaves, including the risk of thermal runaway reactions	Conventional HTC is considered safer but still requires standard safety measures

14.2.2.3 The Reaction Mechanism Occurs During Hydrothermal Carbonization

The mechanism of carbonization and the synthesis route of hydrochar via the hydrothermal carbonization (HTC) process of biomass involves a series of complex thermochemical reactions [15]. These processes are as follows: **Depolymerization** is the stage of HTC, and complex organic polymers within the biomass, such as hemicellulose, cellulose, and lignin, undergo depolymerization. This means that large and intricate molecules break down into smaller, simpler compounds [41]. **Hydrolysis** is a fundamental reaction in HTC. It involves the cleavage of chemical bonds within the biomass due to the presence of water under high-temperature and high-pressure conditions. This process produces simpler organic compounds, including sugars and organic acids [18]. **Dehydration** reactions occur as the temperature within the HTC reactor rises. These reactions lead to the removal of water molecules from the biomass. The removal of water makes an environment conducive to carbonization. **Decarboxylation** reactions are particularly important in the HTC process. During this step, certain organic compounds within the biomass, such as carboxylic acids, release carbon dioxide (CO_2) when exposed to high temperatures. This process reduces the oxygen content in the biomass and contributes to the formation of carbon-rich structures [19]. **Direct carbonization** is a key transformation step in HTC. It involves the conversion of the remaining organic components or lignin into carbon-rich solids. Complex organic molecules break down, and the resulting carbonaceous materials resemble charcoal. **Isomerization** reactions involve the rearrangement of atoms within molecules, resulting in the formation of molecules with the same molecular formula but different structural arrangements. **Rearrangement** reactions involve the reshuffling of atoms and chemical groups within the biomass components. These rearrangements lead to the formation of more stable carbon structures [5]. **Polymerization** is the process through which smaller organic molecules formed during the previous reactions reassemble into larger, more complex carbon structures. This step contributes to the development of the hydrochar's solid structure. **Surface diffusion** involves the movement of molecules on the surface of the developing hydrochar. This can lead to the rearrangement of atoms and functional groups, further enhancing the hydrochar's properties [63]. Figure 14.2 depicts the mechanism of carbonization and the synthesis route of hydrochar via the HTC process of biomass.

14.2.2.4 Parametric Factors Affecting Hydrochar Production

Hydrochar production is a complex process influenced by various parametric factors that significantly impact the properties and yield of hydrochar. Proper understanding and control of these factors are essential for optimizing the production process and obtaining hydrochar with desired characteristics [64]. The following are some of the key parametric factors that play a crucial role in hydrochar production. Figure 14.3 depicts the parametric factors influencing hydrochar production.

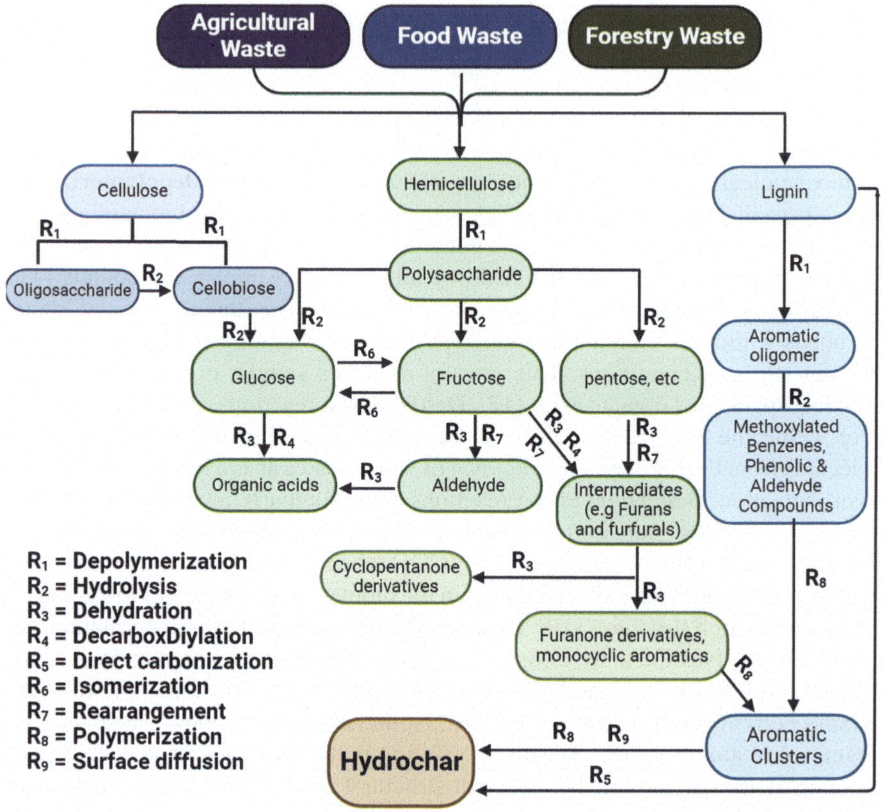

Fig. 14.2 The mechanism of carbonization and the synthesis route of hydrochar via HTC process of biomass

Temperatures

The reaction temperature during hydrochar production is a crucial parameter that significantly influences the quality and properties of the resulting hydrochar [62]. Higher temperatures generally accelerate the reaction kinetics and promote the carbonization process, leading to increased carbon content and enhanced thermal stability of the hydrochar [33]. The higher temperatures promote the formation of carbon-rich structures and accelerate the breakdown of complex organic molecules. However, excessively high temperatures can have negative consequences, such as the loss of vital organic components through volatilization and the beginning of undesired side reactions that can compromise the quality of the hydrochar [65]. The optimal temperature range for hydrochar production varies depending on the specific feedstock being utilized and the desired properties of the hydrochar [2]. Different biomass feedstocks have different thermal degradation characteristics and require

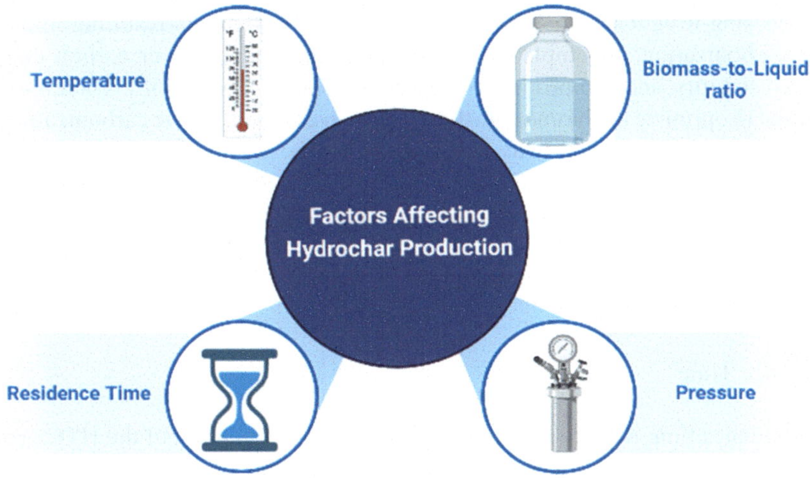

Fig. 14.3 Illustration of the parametric factors affecting hydrochar production

specific temperature conditions to maximize carbonization while minimizing undesirable reactions. Furthermore, the desired properties of the hydrochar, such as its carbon content, porosity, and stability, also influence the optimal temperature range [18]. To determine the optimal temperature for hydrochar production, a thorough understanding of the feedstock characteristics and the desired hydrochar properties is necessary. Experimental studies and optimization techniques are often employed to identify the temperature range that yields hydrochar with the desired properties [3].

Biomass-to-Liquid Ratio

The biomass-to-liquid ratio is a critical parameter that influences the HTC process and ultimately affects the properties of the resulting hydrochar [19]. This ratio refers to the proportion of biomass material relative to the liquid medium, typically water, used in the process. The biomass-to-liquid ratio plays a crucial role in determining the extent of carbonization and the characteristics of hydrochar [5]. A higher biomass-to-liquid ratio generally leads to increased carbonization of the biomass and higher carbon content in the hydrochar. This is because a higher proportion of biomass material facilitates a more extensive reaction with the liquid medium, promoting the conversion of organic compounds into carbon-rich structures [63]. Therefore, the hydrochar produced at higher biomass-to-liquid ratios tends to have higher carbon content, increased stability, and enhanced energy content.

However, it is important to strike a balance when selecting the biomass-to-liquid ratio. The presence of a large amount of biomass can hinder the efficient transfer of heat, causing irregular temperature distribution and potentially resulting in incomplete carbonization. Incomplete carbonization can lead to lower carbon content, reduced stability, and compromised properties of the hydrochar [58]. Therefore, it is essential to optimize the biomass-to-liquid ratio to ensure effective carbonization and the production of high-quality hydrochar [45]. The best ratio may change depending on the biomass feedstock, the desired hydrochar features, and the processing conditions. Experimental studies and optimization approaches are widely utilized to determine a suitable biomass-to-liquid ratio to produce hydrochar with the required characteristics and maximize the overall efficiency of the HTC process.

Residence Time

The residence time is a key parameter that controls the duration of the HTC process and has a considerable impact on the degree of carbonization as well as the qualities of the resulting hydrochar, which is an essential variable in the production of hydrochar [4]. A longer residence time provides more time for the biomass to undergo the complex chemical reactions involved in carbonization. This extended duration allows for a more thorough conversion of organic compounds into carbon-rich structures, resulting in higher carbon content and increased stability of the hydrochar [66]. It enables the formation of a well-developed and interconnected carbon matrix, which enhances the overall quality and desired properties of hydrochar.

However, it is crucial to establish a balance when determining the time of residence. Excessive residence times can result in over-carbonization, whereby the hydrochar turns overly carbonized and loses some of its desirable properties such as hydrochar reactivity, increased brittleness, reduced porosity, and limiting its application [43]. The optimal residence time is determined by several variables, including the biomass feedstock, the desired properties of the hydrochar, and the process conditions [4]. The suitable residence time for hydrochar production is determined through experimental research and process optimization approaches. It is feasible to achieve the necessary level of carbonization, stability, and other critical characteristics of the hydrochar by carefully controlling the residence time.

Pressure

The pressure used during the hydrochar production process is an important parameter that has a large impact on the reaction kinetics and the properties of the produced hydrochar [63]. They improve the interaction of biomass and water by increasing the solubility of biomass components in the liquid phase. This increased solubility enhances reactant diffusion and accelerates reaction rates, resulting in faster carbonization. Therefore, hydrochar generated at greater pressures has a higher degree of carbonization and a denser structure [45]. However, the pressure levels employed in hydrochar production must be carefully considered. Excessive pressure can result in energy-intensive activities and higher expenses for the equipment needed to handle high-pressure conditions [60].

Finding the perfect pressure range will be critical for balancing the desired carbonization level with the practical efficiency of the production process. The pressure conditions required for optimal hydrochar production are determined by several factors, including the kind of biomass feedstock, desired hydrochar qualities, and overall process design [47]. It is feasible to achieve the appropriate level of carbonization and modify the properties of the hydrochar to match specific requirements by carefully managing the pressure applied during hydrochar production. This enables the production of hydrochar with desirable properties such as better stability, higher energy content, improved porosity, and higher carbon yield [42]. These parametric factors interact with each other, and their optimal values depend on the specific feedstock, desired hydrochar properties, and process conditions.

14.2.3 Activation and Modification Hydrochar as a Promising Catalysts

Hydrochar activations and modifications have sparked substantial interest in the field of biodiesel production due to their ability to transform hydrochar into effective catalysts. Hydrochar produced by the HTC of biomass is unique by its high carbon content and porous structure, making it a desirable option for catalyst production. Figure 14.4 depicts the activation and modification of hydrochar.

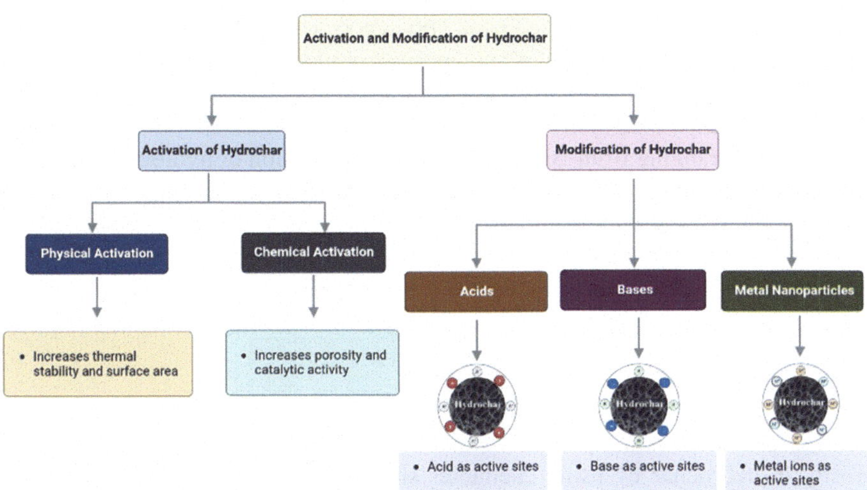

Fig. 14.4 Illustration of the activation and modification of hydrochar

14.2.3.1 Activation of Hydrochar

Activation is a crucial post-treatment process employed to enhance the properties of hydrochar obtained through HTC. The objective of activation is to improve the surface area, porosity, and catalytic activity of hydrochar [67]. Activation methods include physical activation (e.g., using carbon dioxide or steam) and chemical activation (e.g., using phosphoric acid or potassium hydroxide). These processes produce a highly porous structure with a huge surface area known as activated carbon or activated hydrochar, which can serve as a perfect surface for catalytic reactions [63].

Physical Activation

Physical activation processes involve the use of carbon dioxide or steam to produce pores within the hydrochar structure [15]. Carbon dioxide activation, also known as gas activation, involves subjecting hydrochar to a stream of carbon dioxide at high temperatures, resulting in the formation of micropores and mesopores within the hydrochar [40]. On the other hand, the physical activation process involves heating the hydrochar for several hours in an inert atmosphere, such as nitrogen or argon, or in the presence of steam or carbon dioxide [17]. This process reduces the volatile components and generates micropores and mesopores on the surface of the hydrochar, leading to a highly porous structure with a large surface area and high catalytic activity. The product of this process is known as activated carbon or activated hydrochar [44].

Activating hydrochar with steam is a physical process that is both more secure and beneficial to the environment than chemical activation. After the HTC of the biomass, an endothermic process of steam activation is performed [68]. This process takes place at temperatures ranging from 200 to 800 °C, with a period of 30–60 min and a heating rate of 5–10 °C/min. CO_2 activation is an endothermic process in which carbon dioxide is used as an activation gas rather than steam due to its less reactive nature and increased oxidation reaction [3]. To make activated HC, CO_2 is activated at temperatures ranging from 200 to 900 °C. CO_2 activation of hydrochar improves porosity and overall surface area with oxygen-containing functional groups, which boosts the catalytic activity of the hydrochar. The principle of gasification and the removal of volatile compounds govern the physical activation process, leading to the formation of a larger pore structure [19].

Chemical Activation

Chemical activation is the process of introducing functional groups that generate porosity in hydrochar by using chemicals such as phosphoric acid or potassium hydroxide. Phosphoric acid activation is a popular process that includes activating hydrochar with phosphoric acid and then carbonizing it at high temperatures [69]. The phosphoric acid reacts with the hydrochar, causing pores to form and a porous structure to develop. Similarly, the activation of hydrochar with potassium hydroxide, which includes the carbonaceous material, forms pores and increases surface area [66]. Chemical activation methods allow for control over the pore size and distribution of functional groups in the hydrochar, allowing for customization of its catalytic capabilities [5].

Hydrochar can also be chemically activated by impregnating it with different chemicals, and lower activation temperatures, increased carbon yield, greater surface area, formation of diverse functional groups, and well-developed micro- and meso-porous structure are the multiple benefits of the chemical activation of hydrochar [70]. Activated hydrochar has various applications for catalysts and other applications, including water treatment, air pollution management, and energy storage [71]. The activation processes, whether physical or chemical, result in the formation of activated carbon, which possesses a highly porous structure with a large surface area. The increased surface area provides abundant active sites for reactant adsorption and interaction, leading to enhanced catalytic performance [72]. The presence of pores and functional groups within the activated carbon facilitates the diffusion of reactants and promotes desirable catalytic reactions [68]. In general, activation methods are critical in converting hydrochar into activated carbon with increased surface area and reactivity. The resulting activated hydrochar serves as an appropriate framework for catalytic processes, providing improved catalytic performance and efficiency [73].

14.2.3.2 Modification of Hydrochar

Modification of hydrochar involves the introduction of active species onto the surface of hydrochar, which can significantly enhance its catalytic performance for various applications, including biodiesel production [15]. Three common types of active species used in hydrochar modification are acid- or base-functional groups and metal nanoparticles.

Modification of Hydrochar with Acid

Modification of hydrochar with acid involves introducing acid-functional groups onto the hydrochar surface. This modification process enhances the catalytic properties of hydrochar and expands its potential applications in various fields, including biodiesel production [43]. Acid modification of hydrochar can be achieved by treating it with acid solutions, such as sulfuric acid (H_2SO_4) or hydrochloric acid (HCl) [2]. The acid reacts with the hydrochar surface, leading to the formation of acid-functional groups, such as carboxylic (–COOH) or phenolic (–OH) groups. These acid groups act as catalysts in heterogeneous acid-catalyzed reactions, promoting the necessary chemical transformations in biodiesel production [17].

The modification process involves several steps, including the preparation of the acid solution, the immersion or impregnation of hydrochar in the acid solution, and subsequent washing and drying to remove excess acid [43]. The concentration of the acid solution and the duration of the modification process can be optimized to achieve the desired level of acid-functional groups on the hydrochar surface. These acid groups provide active sites for the esterification and transesterification reactions involved in the conversion of feedstock into biodiesel. They facilitate the necessary chemical reactions, improve reaction rates, and enhance the selectivity of biodiesel production [3]. The modified hydrochar with acid-functional groups offers several advantages. It can improve the efficiency of biodiesel production by promoting the conversion of triglycerides or fatty free acids (FFA) into biodiesel and reducing the formation of unwanted by-products [19]. Additionally, the acid modification of hydrochar can enhance the stability of the catalyst, ensuring its long-term performance during repeated use. The acidic modification of hydrochar enables the development of more sustainable and effective catalysts for biodiesel production [69].

Modification of Hydrochar with Base

Modification of hydrochar with base involves the introduction of base-functional groups onto the hydrochar surface. This modification process enhances the catalytic properties of hydrochar and broadens its applications in various fields, including biodiesel production [45]. Base modification of hydrochar can be achieved by treating it with base solutions, such as sodium hydroxide (NaOH) or potassium hydroxide (KOH). The base reacts with the hydrochar surface, leading to the formation of base-functional groups, such as hydroxyl (–OH) or amine ($-NH_2$) groups [74]. These base groups act as catalysts in heterogeneous base-catalyzed reactions, facilitating the necessary chemical transformations in biodiesel production. The modification process typically involves the preparation of the base solution, immersion, or impregnation of hydrochar in the base solution, and subsequent washing and drying to remove the excess base [46]. The concentration of the base solution and the duration of the modification process can be optimized to achieve the desired level of base-functional groups on the hydrochar surface [11].

The incorporation of base-functional groups onto the hydrochar surface enhances its catalytic activity in biodiesel production. These base groups provide active sites for the transesterification reactions involved in converting feedstock into biodiesel [50]. They promote the conversion of triglycerides or FFA into biodiesel and contribute to increased reaction rates and improved selectivity. Modified hydrochar with base-functional groups offers several advantages. It enhances the efficiency of biodiesel production by facilitating the desired chemical reactions and minimizing unwanted side reactions [67]. Additionally, the base modification of hydrochar can improve the stability and reusability of the catalyst, ensuring consistent performance over multiple reaction cycles, to further improve the catalytic performance of base-modified hydrochar in biodiesel production [20].

Modification of Hydrochar with Metal Nanoparticles

The introduction of metal species into the hydrochar surface is essential for the modification of hydrochar with metal nanoparticles. This modification procedure improves hydrochar's catalytic features and improves its potential applications, including biodiesel production [23]. The modification process begins with the preparation of metal precursor solutions, which contain metal ions of interest, such as nickel (Ni), palladium (Pd), or iron (Fe). These metal precursors are typically in the form of salts, such as nickel nitrate or palladium chloride. The hydrochar is then immersed or impregnated in the metal precursor solution, allowing the metal ions to adsorb onto its surface [27]. After impregnation, the hydrochar is subjected to a reduction step, typically using a reducing agent such as sodium borohydride ($NaBH_4$), to convert the metal ions into metal nanoparticles or subjected to calcination for certain temperatures. The reduction reaction results in the deposition of well-dispersed metal nanoparticles on the hydrochar surface [30].

The incorporation of metal nanoparticles onto the hydrochar surface enhances its catalytic activity in biodiesel production. These metal nanoparticles act as catalysts, promoting the necessary chemical reactions involved in the conversion of feedstock into biodiesel [75]. They provide active sites for the transesterification or esterification reactions, facilitating the breakdown of triglycerides or FFA and improving the efficiency and selectivity of the biodiesel production process. The metal nanoparticles provide increased surface area and improve the accessibility of reactants to catalytic sites, enhancing the overall catalytic performance [76]. Additionally, metal nanoparticles can exhibit specific catalytic properties, such as high activity and selectivity, enabling more efficient and effective biodiesel production [77], exploring various metal precursors, optimization strategies for nanoparticle size and distribution, and the influence of different metals on the catalytic performance of heterogeneous hydrochar-based catalysts.

14.2.4 Catalytic Performance of Heterogeneous Hydrochar-Based Catalysts in Biodiesel Production

The catalytic performance of heterogeneous hydrochar-based catalysts in biodiesel production has gained significant attention due to their potential to improve the efficiency and sustainability of the process [78]. Here are some key aspects of their catalytic performance in biodiesel production:

14.2.4.1 Transesterification/Esterification Activity

Heterogeneous hydrochar-based catalysts have shown remarkable catalytic activity in the transesterification and esterification reactions, which are crucial steps in biodiesel production [2]. Transesterification involves the conversion of triglycerides or FFA present in the feedstock, such as vegetable oils or animal fats with alcohol, typically methanol or ethanol, to produce biodiesel or fatty acid methyl esters (FAME) or fatty acid ethyl esters (FAEE) [79]. On the other hand, esterification refers to the reaction between free fatty acids (FFA) and alcohol to form FAME. The surface functional groups of hydrochar materials, such as hydroxyl (–OH) or carboxyl (–COOH) and sulfonate (–SO_3H) groups, play a crucial role in facilitating the transesterification and esterification reaction [80]. These functional groups can act as acid or base sites, effectively catalyzing the conversion of triglycerides into FAMEs. The hydroxyl groups can serve as acid sites by protonating the alcohol, while the carboxyl groups can act as base sites by deprotonating the carbonyl group of the triglycerides or FFA [81].

The presence of these functional groups on the hydrochar's surface enhances the catalytic activity by promoting the activation of reactants and facilitating the desired reaction pathways. The acid–base catalysis promotes the breaking of the ester bonds in the triglycerides and the subsequent esterification of the fatty acids with the alcohol to form FAME [82]. Furthermore, the hydrochar's surface functional groups also interact with the reactants through adsorption and surface reactions. This interaction enhances the contact between the reactants and the catalyst, improving the reaction kinetics and increasing the conversion efficiency [83]. The catalytic activity of hydrochar-based catalysts in transesterification or esterification reactions is closely related to the concentration and accessibility of the surface functional groups. Factors such as the hydrochar preparation conditions, precursor material, and post-treatment methods can influence the density and distribution of these functional groups on the hydrochar surface [84].

In general, hydroxyl and carboxyl groups on the surface of heterogeneous hydrochar-based catalysts play an important role in accelerating the transesterification reaction in biodiesel production [2]. Its acid–base properties promote the activation of reactants and provide the essential catalytic sites for the conversion of triglycerides and FFA into FAME. The catalytic activity of hydrochar-based catalysts can be increased by modifying the hydrochar composition and surface properties, leading to increased efficiency and yield in biodiesel production processes [85]. Figure 14.5 depicts an illustration of biodiesel production through esterification/transesterification of triglycerides or free fatty acids.

14.2.4.2 Biodiesel Yield and Selectivity

The catalytic activity of heterogeneous hydrochar-based catalysts in biodiesel production is influenced by several key factors, including the catalyst composition, surface area, porosity, and surface functional groups [86]. Each of these factors contributes to the overall performance and effectiveness of the catalysts in promoting the transesterification reaction and influencing the biodiesel yield and selectivity [6]. To begin, the catalyst composition is critical in determining catalytic

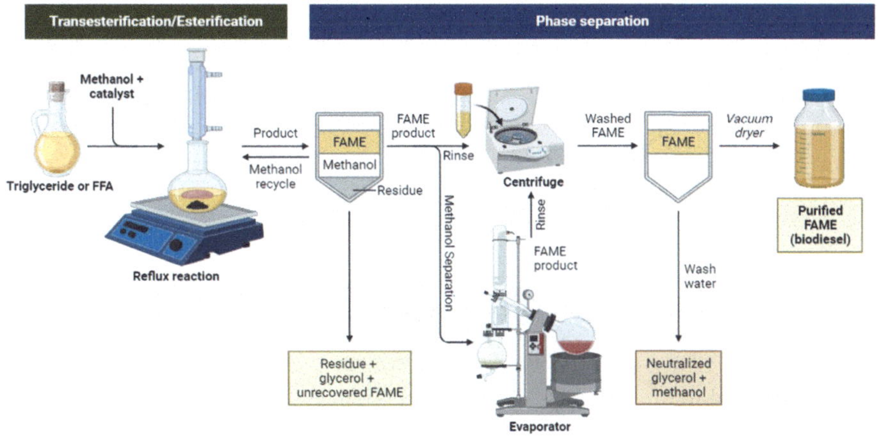

Fig. 14.5 Illustration of production of biodiesel via esterification/transesterification of triglyceride or free fatty acids

activity. Hydrochar-based catalysts can be made from a variety of precursor materials, including biomass which is the main source [9]. The catalyst composition is influenced by the precursor material used and the HTC treatment conditions. The carbon concentration, elemental composition, and surface functional groups of the hydrochar can all have an impact on its catalytic efficiency [87]. Catalysts with larger areas of surface and porosity have more active sites for transesterification and esterification reactions to take place. The increased availability of active sites improves the interaction between the catalyst and the reactants, leading to higher biodiesel yield [9].

Lastly, the presence of certain surface functional groups on hydrochar-based catalysts, such as hydroxyl (–OH) or carboxyl (–COOH), influences reaction kinetics and selectivity toward FAME production. These functional groups may serve as acid or basic sites, allowing the transesterification of triglycerides and alcohol [8]. The acidic or basic character of the functional groups facilitates reactant activation, allowing for efficient esterification and enhancing selectivity toward FAME. To improve the catalytic activity of hydrochar-based catalysts in biodiesel production, the composition, surface area, porosity, and functional groups of the catalyst need to be optimized [88]. By adjusting these factors, the amount of biodiesel that can be made can be increased by increasing the number of active sites, making the mass transfer as efficient as possible, and promoting the desired reaction pathways [89].

Optimization of reaction conditions is crucial for achieving high biodiesel yields and desired selectivity in biodiesel production using hydrochar-based catalysts. Several key parameters, including catalyst loading, reaction temperature, alcohol-to-oil molar ratio, and reaction time, play significant roles in determining the efficiency of the process. Figure 14.6 depicts an illustration of the variables influencing biodiesel yield.

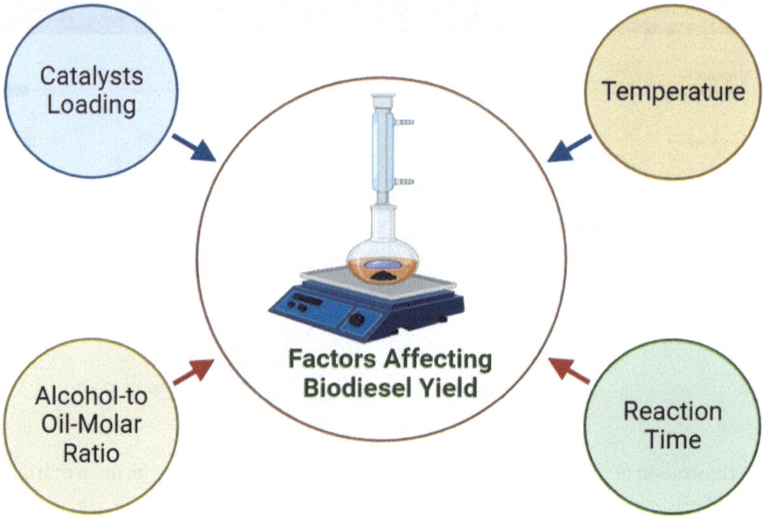

Fig. 14.6 Schematic of the factors affecting biodiesel yield

Catalyst Loading

The dosage of hydrochar-based catalyst used in the transesterification and esterification reaction is referred to as catalyst loading. It is an important parameter to consider achieving a balance between attaining the desired catalytic activity and reducing the costs associated with excessive catalyst usage [90]. Catalyst loading plays a significant role in determining the number of active sites available for the transesterification or esterification reaction to occur. Increasing the catalyst loading generally leads to a higher density of active sites on the catalyst's surface, facilitating the conversion of triglycerides into biodiesel and glycerol. This can result in improved reaction kinetics and higher biodiesel yields [89].

Optimizing the catalyst loading is critical for maximizing biodiesel production and ensuring optimal usage of the hydrochar-based catalyst. It requires finding an appropriate balance between providing enough active sites for the reaction and reducing catalyst costs [91]. Biodiesel production can improve the efficiency of the transesterification and esterification processes by carefully determining the optimum catalyst loading and optimizing the conversion of triglycerides or FFA into biodiesel while keeping catalyst consumption at an economically sustainable amount. The suitable catalyst loading depends on a variety of factors, including the type of hydrochar-based catalyst utilized, the properties of the feedstock, and the required reaction conditions [92]. Typically, experimental investigations and optimization approaches are used to find the best catalyst loading for a certain biodiesel production system. These studies entail evaluating catalytic performance at various catalyst loadings, analyzing biodiesel yields, considering financial implications, and comparing entire process efficiency.

Reaction Temperature

The reaction temperature is a key factor that significantly impacts the efficiency and selectivity of the transesterification and esterification reactions in biodiesel production. It plays a crucial role in controlling the reaction kinetics and influencing the distribution of products formed [93]. Typically, these reactions are carried out at elevated temperatures within a range of 50–90 °C. Higher reaction temperatures result in higher reaction rates due to higher molecular activity and collision frequency of the reactant molecules. This faster reaction kinetics can be favorable in terms of reducing reaction time and increasing the production of biodiesel. However, it is vital to note that extremely high temperatures can also enhance the occurrence of side reactions, resulting in the formation of unwanted by-products [94].

Optimizing the reaction temperature is essential for achieving the desired reaction kinetics and product selectivity. Biodiesel production may optimize the conversion of triglycerides or FFA into biodiesel while reducing the formation of undesirable by-products by carefully managing the temperature [95]. Finding the optimum temperature for the reaction requires considering several factors, including the nature of the catalyst, the type of feedstock utilized, and the desired qualities of the biodiesel. To determine the appropriate reaction temperature for biodiesel production via evaluating the reaction performance and product distribution at various temperatures, consider both the biodiesel yield and the quality of the generated product [80]. It also helps to ensure the economic viability and environmental sustainability of biodiesel production by ensuring that the process operates under the best possible conditions in terms of reaction kinetics and product selectivity.

Alcohol-to-Oil Molar Ratio

The alcohol-to-oil molar ratio is a crucial parameter that significantly influences the transesterification and esterification reaction in biodiesel production. It refers to the ratio of alcohol, typically methanol or ethanol, to triglyceride or FFA feedstock, which is commonly derived from vegetable oils or animal fats [96]. This ratio plays a vital role in determining the extent of transesterification or esterification and directly affects the conversion efficiency and biodiesel yield. A higher alcohol-to-oil molar ratio commonly improves the completion of the transesterification and esterification reactions. The excess alcohol ensures that there are enough alcohol molecules available to react with the triglycerides or FFA, resulting in a higher conversion of triglycerides or FFA into biodiesel. This increased conversion efficiency is advantageous since it increases biodiesel yield, resulting in more efficient feedstock utilization [97].

However, it is essential to realize that excessive alcohol use might have negative effects. Initially, increasing the alcohol-to-oil molar ratio can result in increased costs due to the reaction requiring more alcohol. Alcohol is frequently more expensive than the feedstock itself, and utilizing too much can have a substantial impact on the overall economics of biodiesel production [81]. Furthermore, a high alcohol amount in the reaction mixture can complicate the separation and purification processes during the biodiesel product's downstream processing. This can result in higher energy usage and additional purification processes, increasing the entire process's complexity and cost. Finding the optimum alcohol-to-oil molar ratio is important for balancing conversion efficiency, and yield [86]. Considerations include the specific feedstock characteristics, the catalyst used, and the intended quality criteria for the biodiesel yield. This optimization contributes to the overall economic feasibility and sustainability of biodiesel production by providing effective resource utilization and limiting the environmental impact associated with excessive alcohol consumption and purification challenges.

Reaction Time

The reaction time in the production of biodiesel refers to the period that the transesterification or esterification reaction is allowed to proceed. It is an essential variable that has a direct impact on the level of conversion and the overall yield of biodiesel. The response time is essential in obtaining the target biodiesel production efficiency while limiting unwanted side effects [98]. Prolonged reaction times regularly improve triglyceride or FFA conversion into biodiesel by allowing more time for the transesterification or esterification reaction to occur. This expanded duration allows for better exploitation of the available reactants, perhaps leading to higher biodiesel yield. On the other hand, the longer reaction times could raise the possibility of side reactions such as hydrolysis and saponification. These side reactions might result in the formation of undesired compounds, lowering the quality of the biodiesel [99].

Optimizing the reaction time is vital for generating high biodiesel yields while reducing unwanted reactions. The optimum reaction time is determined by taking numerous aspects into account, including the specific hydrochar-based catalyst employed, the properties of the feedstock, and the desired quality criteria for the biodiesel product. Experimentation and process optimization approaches are frequently used to determine the best reaction time for a certain biodiesel production process. Efforts to optimize the reaction time not only led to increased biodiesel yields but also added to the economic viability and environmental sustainability of biodiesel production.

Optimizing the reaction time, as well as other reaction variables like catalyst loading, reaction temperature, and alcohol-to-oil molar ratio, aims to improve the efficiency and selectivity of biodiesel production [100]. Biodiesel production can enhance mass transfer efficiency, maximize the usage of active sites on the catalyst, and achieve an appropriate acid–base balance by carefully adjusting these parameters. This enhancement increases biodiesel production, selectivity for high-quality FAME, and overall process economics and sustainability. Table 14.3 summarizes the activation, modification, and parametric parameters influencing biodiesel yield.

Table 14.3 The activation, modification, and parametric factors affecting biodiesel yield

Hydrochar feedstock	Activation/ Modification	Biodiesel feedstock	Type of reaction	Catalyst loading (wt.%)	Temperature/ time (°C)	Reaction time (min)	Alcohol-oil-molar ratio	Biodiesel yield (%)	Refs.
Musa Champa peduncle waste	K_2CO_3/CaO	Jatropha oil	Esterification	7	65	60	12:1	98.38	[101]
Coconut endocarp	KOH	Safflower oil	Transesterification	10	70	120	19:1	99.00	[102]
Orange peel	H_2SO_4	Oleic acid	Esterification	3	80	180	20:1	92.80	[103]
Ziziphus Mauritania L	H_2SO_4	Oleic acid	Esterification	10	80	240	20:1	91.00	[104]
Corncob residue	H_2SO_4	Palm fatty acid distillate	Esterification	5	70	300	20:1	92.00	[105]
Mesocarp fibers	$K_2CO_3/$ $Cu(NO_3)_2$	Waste cooking oil	Transesterification	4	70	120	12:1	95.36	[98]
Water hyacinth	$CH_3C_6H_4SO_2$	Oleic acid	Esterification	3	75	90	15:1	95.56	[106]
Empty fruit bunch	$K_2CO_3/$ $Cu(NO_3)_2$	Waste cooking oil	Esterification–transesterification	5	70	120	12:1	97.1	[96]

14.2.4.3 Catalyst Stability and Regeneration

Catalyst stability and regeneration are essential considerations in the practical application of hydrochar-based catalysts in biodiesel production. The carbonaceous nature of hydrochar provides inherent stability, allowing it to withstand the harsh reaction conditions involved in the transesterification and esterification process [82]. The robust carbon matrix of hydrochar ensures its structural integrity and maintains catalytic activity even at high temperatures and in the presence of active sites. The carbon-based structure of hydrochar exhibits excellent thermal stability, preventing catalyst degradation and minimizing catalyst deactivation mechanisms such as coke formation or sintering of active sites. This stability allows hydrochar-based catalysts to maintain their catalytic activity over prolonged reaction times, ensuring consistent performance and reliable biodiesel production [107]. Table 14.4 summarizes the comparison of hydrochar-based catalysts with conventional catalysts in terms of biodiesel yields, productivity, and recyclability.

Furthermore, the reusability of heterogeneous hydrochar-based catalysts is advantageous for their practical application. Hydrochar-based catalysts can be easily regenerated and reused, reducing the need for frequent catalyst replacement. Regeneration methods typically involve cleaning or rejuvenating the catalyst to restore its initial activity [108]. Techniques such as washing with solvents, thermal treatment, or chemical reactivation can effectively remove contaminants or reactant residues and restore the catalyst's surface properties. The reusability of hydrochar-based catalysts not only reduces overall process costs but contributes to the economic viability and sustainability of biodiesel production. The negative environmental effect of the process is decreased waste formation and increased catalyst usage [87].

It is important to note that the stability and reusability of hydrochar-based catalysts can depend on various factors, including the catalyst composition, preparation method, reaction conditions, and feedstock characteristics [7]. Optimization of these parameters is crucial to maximize catalyst stability and ensure efficient regeneration,

Table 14.4 The comparison of hydrochar-based catalysts with conventional catalysts in terms of biodiesel yields, productivity, and recyclability

	Heterogeneous hydrochar-based catalysts	Conventional (homogeneous) catalysts:
Biodiesel yield	Hydrochar-based catalysts often offer high yields in biodiesel production, contributing to efficient processes	Yields can vary among different homogeneous catalysts; some may offer high yields, while others may not be as effective
Productivity	They can enhance productivity due to their porous structure and active sites, promoting faster reactions	Productivity depends on the specific catalyst used; it can range from highly efficient to less effective
Recyclability	Hydrochar-based catalysts are generally recyclable, reducing costs and waste in the long term	The recyclability of homogeneous catalysts varies widely; some may degrade over time, potentially increasing expenses and waste

enabling extended catalyst lifetimes and reducing the overall operating costs. In a nutshell, the carbonaceous characteristics of hydrochar-based catalysts permit them to withstand strong reaction conditions in biodiesel production [9]. The reusability of hydrochar-based catalysts reduces the need for frequent replacement and helps the process's economic feasibility. Hydrochar-based catalysts can provide long-lasting catalytic activity by optimizing catalyst stability and regeneration techniques, ensuring constant and sustainable biodiesel production.

14.2.4.4 Reaction Mechanisms

The production of biodiesel involves the conversion of triglycerides and FFA, which are commonly found in vegetable oils or animal fats, into biodiesel and glycerol or water through transesterification and esterification reactions. This process is typically carried out in the presence of a catalyst to enhance the reaction rate and efficiency [8]. Figure 14.7 depicts a schematic of the reaction mechanism of the active site on the surface hydrochar-based catalysts.

In the adsorption step, the triglycerides or FFA and alcohol molecules are adsorbed onto the surface of the heterogeneous hydrochar-based catalyst. Hydrochar possesses a porous structure that provides a large surface area with abundant adsorption sites. This porous nature facilitates the interaction between the reactants and the catalyst, promoting their adsorption and subsequent reaction [88]. Once the reactants are adsorbed, the transesterification and esterification reactions take place. The alcohol molecules, such as methanol or ethanol, react with the triglycerides bound to the catalyst's surface. This results in the formation of fatty acid alkyl esters (biodiesel) and glycerol or water as a by-product. The hydrochar-based catalyst plays a crucial role by providing active sites or functional groups on its surface, such as hydroxyl (–OH) or carboxyl (–COOH) groups [90]. These active sites act as acid-based catalysts or base-based catalysts, facilitating the breaking ester bonds in the triglyceride molecules and forming ester bonds in the biodiesel molecules.

After the transesterification and esterification reactions occur, the biodiesel and glycerol molecules need to be desorbed from the catalyst's surface. The porous structure of the hydrochar-based catalyst aids in the desorption process, allowing the products to be released from the catalyst [85]. After the reaction, the biodiesel and glycerol must be separated. Separation can be performed in a variety of ways due to their various physical and chemical properties, such as boiling points and densities. Distillation and centrifugation are two methods widely used to separate biodiesel from glycerol. It should be noted that the composition and properties of the heterogeneous hydrochar-based catalyst utilized can have an impact on the reaction mechanism. Different catalysts may have different active sites or functions, leading to differences in the reaction route and kinetics [109].

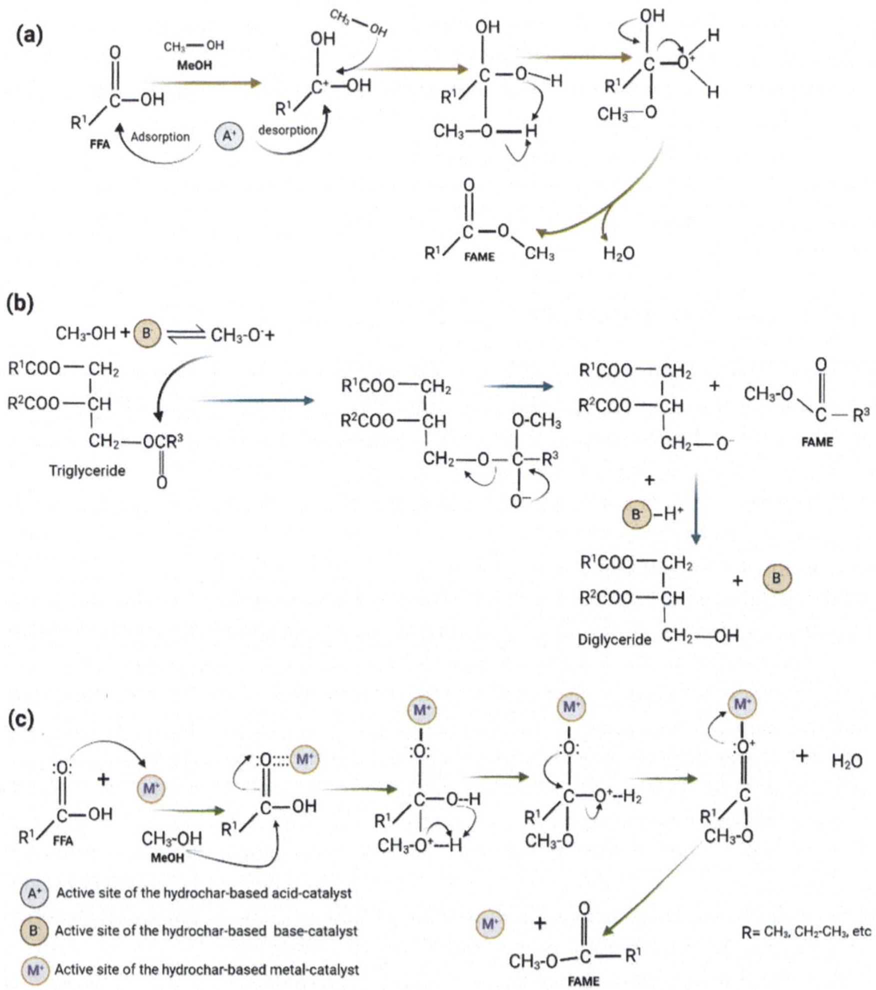

Fig. 14.7 Schematic of the reaction mechanism of the active site on the surface hydrochar-based catalysts

14.3 Present Research Status and Future Research Targets

Research in hydrochar production is currently focused on utilizing various biomass feedstocks, such as agricultural waste, food waste, and forestry waste, to produce carbon-rich hydrochar through hydrothermal carbonization (HTC) and microwave-assisted hydrothermal carbonization (MA-HTC) processes. The composition and properties of hydrochar are influenced by the specific feedstock used and the process parameters employed. Key factors affecting the quality and yield of hydrochar include temperature, biomass-to-liquid ratio, residence time, and pressure. Optimization of these parameters is being studied to obtain hydrochar with desirable characteristics, including high carbon content, stability, and porosity. The research aims to optimize the hydrochar production process and explore its potential application as a catalyst for biodiesel production. The current study focuses on enhancing the catalytic activity of hydrochar through activation and modification methods. Physical and chemical activation processes are explored to increase the surface area and porosity of hydrochar, while modification involves introducing active species or metal nanoparticles onto its surface.

In future, research in the field should concentrate on many different areas. Initially, there is a need to investigate and evaluate various biomass feedstocks for hydrochar production in terms of availability, sustainability, and required hydrochar qualities. This includes investigating new and unconventional biomass sources. Process optimization is another important target, involving the fine-tuning of parameters such as temperature, biomass-to-liquid ratio, residence time, and pressure in the HTC process. The goal is to enhance carbonization efficiency, increase carbon content, and optimize the properties of hydrochar. Advanced characterization techniques, including surface analysis, should be further developed to gain a deeper understanding of hydrochar's composition, structure, and properties, as well as its interactions with other substances. Additionally, there is a need to explore and expand the potential applications of hydrochar in various fields as a part of catalyst support such as soil amendment, carbon sequestration, and wastewater treatment. This involves investigating its performance in different scenarios and optimizing its properties for specific applications. Comprehensive life cycle assessments and environmental impact studies are essential to evaluate the environmental footprint and sustainability of hydrochar production processes, including assessing its potential benefits in terms of reducing greenhouse gas emissions, managing waste, and improving resource efficiency. To bridge the gap between laboratory-scale research and large-scale production, efforts should be directed toward developing scalable and cost-effective hydrochar production technologies and overcoming technical, economic, and logistical challenges. Furthermore, the optimization of activation and modification processes for hydrochar to enhance catalytic activity is important, including fine-tuning physical and chemical activation methods and exploring novel strategies for more efficient and sustainable catalyst production. Understanding the structure, morphology, and surface chemistry of activated and modified hydrochar-based catalysts through advanced analytical techniques will contribute to optimizing catalytic performance and designing more effective catalysts. The stability and regenerability of hydrochar-based catalysts should also be studied to develop long-lasting catalysts with minimal performance decline over multiple reaction cycles, ensuring economic viability and sustainability. Process optimization and scale-up studies are needed to optimize reaction conditions and assess the feasibility of hydrochar-based catalysts in large-scale biodiesel production, aiming for high yields, improved selectivity, and energy-efficient processes.

14.4 Conclusions

In conclusion, the choice of biomass feedstock and process parameters significantly influence the composition and properties of hydrochar produced through hydrothermal carbonization, while hydrochar activations and modifications hold promise for transforming hydrochar into effective catalysts for biodiesel production. The utilization of agricultural waste, food waste, and forestry waste as feedstock offers a sustainable solution for waste management and resource utilization. Optimizing temperature, biomass-to-liquid ratio, residence time, and pressure is crucial for obtaining hydrochar with desired characteristics. Hydrothermal carbonization and microwave-assisted hydrothermal carbonization provide efficient and environmentally friendly methods for hydrochar production, contributing to biomass conversion and waste reduction. On the other hand, physical and chemical activation methods, as well as modifications with acid/base-functional groups or metal nanoparticles, enhance the catalytic performance of hydrochar. Heterogeneous hydrochar-based catalysts exhibit remarkable activity in biodiesel production reactions, with catalyst composition, surface area, porosity, and functional groups influencing the overall performance. The stability and reusability of hydrochar-based catalysts further contribute to their long-term effectiveness and cost-effectiveness. Ultimately, continued research and optimization of hydrochar activation, modification, and catalytic performance are necessary for advancing the field and facilitating the commercialization of hydrochar-based catalysts in the biodiesel industry.

References

1. Rattanaya, T., Kongjan, P., Cheewasedtham, C., Bunyakan, C., Yuso, P., Cheirsilp, B., & Jariyaboon, R. (2022). Application of palm oil mill waste to enhance biogas upgrading and hornwort cultivation. *Journal of Environmental Management, 309*, 114678. https://doi.org/10.1016/j.jenvman.2022.114678
2. Alfredo Quevedo-Amador, R., Elizabeth Reynel-Avila, H., Ileana Mendoza-Castillo, D., Badawi, M., & Bonilla-Petriciolet, A. (2022). Functionalized hydrochar-based catalysts for biodiesel production via oil transesterification: optimum preparation conditions and performance assessment. *Fuel, 312*, 122731. https://doi.org/10.1016/j.fuel.2021.122731
3. Munir, M. T., Ul Saqib, N., Li, B., & Naqvi, M. (2023). Food waste hydrochar: an alternate clean fuel for steel industry. *Fuel, 346*, 128395. https://doi.org/10.1016/j.fuel.2023.128395
4. Liu, F., Yu, R., Ji, X., & Guo, M. (2018). Hydrothermal carbonization of holocellulose into hydrochar: Structural, chemical characteristics, and combustion behavior. *Bioresource Technology, 263*, 508–516. https://doi.org/10.1016/j.biortech.2018.05.019
5. Başakçılardan Kabakcı, S., & Baran, S. S. (2019). Hydrothermal carbonization of various lignocellulosics: Fuel characteristics of hydrochars and surface characteristics of activated hydrochars. *Waste Management, 100*, 259–268. https://doi.org/10.1016/j.wasman.2019.09.021
6. Yaashikaa, P. R., Kumar, P. S., & Karishma, S. (2022). Bio-derived catalysts for production of biodiesel: a review on feedstock, oil extraction methodologies, reactors and lifecycle assessment of biodiesel. *Fuel, 316*, 123379. https://doi.org/10.1016/j.fuel.2022.123379

7. Fonseca, J. M., Spessato, L., Cazetta, A. L., da Silva, C., & Almeida, V. D. C. (2022). Sulfonated carbon: synthesis, properties and production of biodiesel. *Chemical Engineering and Processing-Process Intensification, 170*, 108668.
8. Ravikumar, R., Kiran, K., Hebbar, G. S., Naresh, H., & Panigrahi, S. P. (2022). An overview of nano-catalysts in biodiesel production. *Journal of Mines, Metals and Fuels*. https://doi.org/10.18311/jmmf/2022/31998
9. Adhikesavan, C., Ganesh, D., & Charles Augustin, V. (2022). Effect of quality of waste cooking oil on the properties of biodiesel, engine performance and emissions. *Cleaner Chemical Engineering, 4*, 100070. https://doi.org/10.1016/j.clce.2022.100070
10. Sztancs, G., Juhasz, L., Nagy, B. J., Nemeth, A., Selim, A., Andre, A., Toth, A. J., Mizsey, P., & Fozer, D. (2020). Co-hydrothermal gasification of *Chlorella vulgaris* and hydrochar: The effects of waste-to-solid biofuel production and blending concentration on biogas generation. *Bioresource Technology, 302*, 122793. https://doi.org/10.1016/j.biortech.2020.122793
11. Nzediegwu, C., Naeth, M. A., & Chang, S. X. (2021). Carbonization temperature and feedstock type interactively affect chemical, fuel, and surface properties of hydrochars. *Bioresource Technology, 330*, 124976. https://doi.org/10.1016/j.biortech.2021.124976
12. Safari, F., Javani, N., & Yumurtaci, Z. (2018). Hydrogen production via supercritical water gasification of almond shell over algal and agricultural hydrochars as catalysts. *International Journal of Hydrogen Energy, 43*, 1071–1080. https://doi.org/10.1016/j.ijhydene.2017.05.102
13. Inkoua, S., Li, C., Zhang, L., Li, X., Zhang, S., Xiang, J., Hu, S., Wang, Y., & Hu, X. (2023). Hydrothermal carbonization of grapefruit peels in swine manure-derived liquid: Property of resulting hydrochars and activated carbons. *Fuel, 351*, 128889. https://doi.org/10.1016/j.fuel.2023.128889
14. Marrakchi, F., Sohail Toor, S., Haaning Nielsen, A., Helmer Pedersen, T., & Aistrup Rosendahl, L. (2023). Bio-crude oils production from wheat stem under subcritical water conditions and batch adsorption of post-hydrothermal liquefaction aqueous phase onto activated hydrochars. *Chemical Engineering Journal, 452*, 139293. https://doi.org/10.1016/j.cej.2022.139293
15. Luo, X., Huang, Z., Lin, J., Li, X., Qiu, J., Liu, J., & Mao, X. (2020). Hydrothermal carbonization of sewage sludge and in-situ preparation of hydrochar/MgAl-layered double hydroxides composites for adsorption of Pb(II). *Journal of Cleaner Production, 258*, 120991. https://doi.org/10.1016/j.jclepro.2020.120991
16. Pereira, G. R., Lopes, R. P., Wang, W., Guimarães, T., Teixeira, R. R., & Astruc, D. (2022). Triazole-functionalized hydrochar-stabilized Pd nanocatalyst for ullmann coupling. *Chemosphere, 308*, 136250. https://doi.org/10.1016/j.chemosphere.2022.136250
17. Zhuang, X., Liu, J., & Ma, L. (2022). Facile synthesis of hydrochar-supported catalysts from glucose and its catalytic activity towards the production of functional amines. *Green Energy and Environment*. https://doi.org/10.1016/j.gee.2022.01.012
18. Vega, M. F., Florentino-Madiedo, L., Díaz-Faes, E., & Barriocanal, C. (2021). Influence of feedwater pH on the CO_2 reactivity of hydrochars: Co-carbonisation with a bituminous coal. *Renewable Energy, 170*, 824–831. https://doi.org/10.1016/j.renene.2021.01.100
19. Zhuang, X., Song, Y., Zhan, H., Bi, X. T., Yin, X., & Wu, C. (2020). Pyrolytic conversion of biowaste-derived hydrochar: Decomposition mechanism of specific components. *Fuel, 266*, 117106. https://doi.org/10.1016/j.fuel.2020.117106
20. Bona, D., Scrinzi, D., Tonon, G., Ventura, M., Nardin, T., Zottele, F., Andreis, D., Andreottola, G., Fiori, L., & Silvestri, S. (2022). Hydrochar and hydrochar co-compost from OFMSW digestate for soil application: 2. Agro-environmental properties. *Journal of Environmental Management, 312*, 114894. https://doi.org/10.1016/j.jenvman.2022.114894
21. Scrinzi, D., Bona, D., Denaro, A., Silvestri, S., Andreottola, G., & Fiori, L. (2022). Hydrochar and hydrochar co-compost from OFMSW digestate for soil application: 1. Production and chemical characterization. *Journal of Environmental Management, 309*, 114688. https://doi.org/10.1016/j.jenvman.2022.114688
22. Wortmann, M., Keil, W., Brockhagen, B., Biedinger, J., Westphal, M., Weinberger, C., Diestelhorst, E., Hachmann, W., Zhao, Y., Tiemann, M., Reiss, G., Hüsgen, B., Schmidt, C., Sattler,

K., & Frese, N. (2022). Pyrolysis of sucrose-derived hydrochar. *Journal of Analytical and Applied Pyrolysis, 161*, 105404. https://doi.org/10.1016/j.jaap.2021.105404
23. Wang, Z., Zhai, Y., Wang, T., Wang, B., Peng, C., & Li, C. (2020). Pelletizing of hydrochar biofuels with organic binders. *Fuel, 280*, 118659. https://doi.org/10.1016/j.fuel.2020.118659
24. Magdziarz, A., Wilk, M., & Wądrzyk, M. (2020). Pyrolysis of hydrochar derived from biomass—Experimental investigation. *Fuel, 267*, 117246. https://doi.org/10.1016/j.fuel.2020.117246
25. Shao, Y., Long, Y., Wang, H., Liu, D., Shen, D., & Chen, T. (2019). Hydrochar derived from green waste by microwave hydrothermal carbonization. *Renewable Energy, 135*, 1327–1334. https://doi.org/10.1016/j.renene.2018.09.041
26. Zhang, Y. N., Guo, J. Z., Wu, C., Huan, W. W., Chen, L., & Li, B. (2022). Enhanced removal of Cr(VI) by cation functionalized bamboo hydrochar. *Bioresource Technology, 347*, 126703. https://doi.org/10.1016/j.biortech.2022.126703
27. Huezo, L., Vasco-Correa, J., & Shah, A. (2021). Hydrothermal carbonization of anaerobically digested sewage sludge for hydrochar production. *Bioresource Technology Reports, 15*, 100795. https://doi.org/10.1016/j.biteb.2021.100795
28. Al Afif, R., Tondl, G., & Pfeifer, C. (2023). Experimental and simulation study of hydrochar production from cotton stalks. *Energy, 276*, 127573. https://doi.org/10.1016/j.energy.2023.127573
29. Luo, Y., Lan, Y., Liu, X., Xue, M., Zhang, L., Yin, Z., He, X., Li, X., Yang, J., Hong, Z., Naushad, M., & Gao, B. (2023). Hydrochar effectively removes aqueous Cr(VI) through synergistic adsorption and photoreduction. *Separation and Purification Technology, 317*, 123926. https://doi.org/10.1016/j.seppur.2023.123926
30. He, J., Ren, S., Zhang, S., & Luo, G. (2021). Modification of hydrochar increased the capacity to promote anaerobic digestion. *Bioresource Technology, 341*, 125856. https://doi.org/10.1016/j.biortech.2021.125856
31. Sridhar, A., Kapoor, A., Senthil Kumar, P., Ponnuchamy, M., Balasubramanian, S., & Prabhakar, S. (2021). Conversion of food waste to energy: A focus on sustainability and life cycle assessment. *Fuel, 302*, 121069. https://doi.org/10.1016/j.fuel.2021.121069
32. Sharma, P., Gaur, V. K., Sirohi, R., Varjani, S., Hyoun Kim, S., & Wong, J. W. C. (2021). Sustainable processing of food waste for production of bio-based products for circular bioeconomy. *Bioresource Technology, 325*, 124684.
33. Xie, X., Peng, C., Song, X., Peng, N., & Gai, C. (2022). Pyrolysis kinetics of the hydrothermal carbons derived from microwave-assisted hydrothermal carbonization of food waste digestate. *Energy, 245*, 123269. https://doi.org/10.1016/j.energy.2022.123269
34. Kumar, M., Dutta, S., You, S., Luo, G., Zhang, S., Show, P. L., Sawarkar, A. D., Singh, L., & Tsang, D. C. W. (2021). A critical review on biochar for enhancing biogas production from anaerobic digestion of food waste and sludge. *Journal of Cleaner Production, 305*, 127143.
35. Ul Saqib, N., Sarmah, A. K., & Baroutian, S. (2019). Effect of temperature on the fuel properties of food waste and coal blend treated under co-hydrothermal carbonization. *Waste Management, 89*, 236–246. https://doi.org/10.1016/j.wasman.2019.04.005
36. Usmani, Z., Sharma, M., Awasthi, A. K., Sharma, G. D., Cysneiros, D., Nayak, S. C., Thakur, V. K., Naidu, R., Pandey, A., & Gupta, V. K. (2021). Minimizing hazardous impact of food waste in a circular economy—Advances in resource recovery through green strategies. *Journal of Hazardous Materials, 416*, 126154. https://doi.org/10.1016/j.jhazmat.2021.126154
37. Wang, L., Chi, Y., Du, K., Zhou, Z., Wang, F., & Huang, Q. (2022). Hydrothermal treatment of food waste for bio-fertilizer production: Formation and regulation of humus substances in hydrochar. *Science of the Total Environment, 838*, 155900. https://doi.org/10.1016/j.scitotenv.2022.155900
38. Cao, L., Zhang, C., Chen, H., Tsang, D. C. W., Luo, G., Zhang, S., & Chen, J. (2017). Hydrothermal liquefaction of agricultural and forestry wastes: State-of-the-art review and future prospects. *Bioresource Technology, 245*, 1184–1193.
39. Deng, F., Jiang, H., Xie, Z., Chen, Y., Zhou, P., Liu, X., & Li, D. (2022). Nutrient recovery from biogas slurry via hydrothermal carbonization with different agricultural and forestry

residue. *Industrial Crops and Products, 189*, 115891. https://doi.org/10.1016/j.indcrop.2022.115891
40. Liu, Z., Quek, A., Kent Hoekman, S., Srinivasan, M. P., & Balasubramanian, R. (2012). Thermogravimetric investigation of hydrochar-lignite co-combustion. *Bioresource Technology, 123*, 646–652. https://doi.org/10.1016/j.biortech.2012.06.063
41. Li, S., Ma, Q., Chen, L., Yang, Z., Aqeel Kamran, M., & Chen, B. (2022). Hydrochar-mediated photocatalyst Fe_3O_4/BiOBr@HC for highly efficient carbamazepine degradation under visible LED light irradiation. *Chemical Engineering Journal, 433*, 134492. https://doi.org/10.1016/j.cej.2021.134492
42. Donar, Y. O., Çağlar, E., & Sinağ, A. (2016). Preparation and characterization of agricultural waste biomass based hydrochars. *Fuel, 183*, 366–372. https://doi.org/10.1016/j.fuel.2016.06.108
43. Huangfu, X., Xu, N., Yang, J., Yang, H., Zhang, M., Ye, Z., Wang, S., & Chen, J. (2020). Transport and retention of hydrochar-diatomite nanoaggregates in water-saturated porous sand: Effect of montmorillonite and phosphate at different ionic strengths and solution pH. *Science of the Total Environment, 703*, 134487. https://doi.org/10.1016/j.scitotenv.2019.134487
44. Fakudze, S., Zhang, Y., Wei, Y., Li, Y. H., Chen, J., Wang, J., & Han, J. (2023). Taguchi-optimized oxy-combustion of hydrochar/coal blends for CO_2 capture and maximized combustion performance. *Energy., 267*, 126602. https://doi.org/10.1016/j.energy.2022.126602
45. Santos, M. M., Marques Sierra, A. L., Amado-Fierro, Á., Suárez, M., Blanco, F., La Fuente, J. M. G., Diez, M. A., & Centeno, T. A. (2023). Reducing cement consumption in mortars by waste-derived hydrochars. *Journal of Building Engineering, 75*, 106987. https://doi.org/10.1016/j.jobe.2023.106987
46. Celletti, S., Bergamo, A., Benedetti, V., Pecchi, M., Patuzzi, F., Basso, D., Baratieri, M., Cesco, S., & Mimmo, T. (2021). Phytotoxicity of hydrochars obtained by hydrothermal carbonization of manure-based digestate. *Journal of Environmental Management, 280*, 111635. https://doi.org/10.1016/j.jenvman.2020.111635
47. Roman, F. F., Diaz De Tuesta, J. L., Praça, P., Silva, A. M. T., Faria, J. L., & Gomes, H. T. (2021). Hydrochars from compost derived from municipal solid waste: Production process optimization and catalytic applications. *Journal of Environmental Chemical Engineering, 9*, 104888. https://doi.org/10.1016/j.jece.2020.104888
48. Dong, X., Guo, S., Ma, M., Zheng, H., Gao, X., Wang, S., & Zhang, H. (2020). Hydrothermal carbonization of millet stalk and dilute-acid-impregnated millet stalk: combustion behaviors of hydrochars by thermogravimetric analysis and a novel mixed-function fitting method. *Fuel, 273*, 117734. https://doi.org/10.1016/j.fuel.2020.117734
49. Reza, M. T., Rottler, E., Herklotz, L., & Wirth, B. (2015). Hydrothermal carbonization (HTC) of wheat straw: Influence of feedwater pH prepared by acetic acid and potassium hydroxide. *Bioresource Technology, 182*, 336–344. https://doi.org/10.1016/j.biortech.2015.02.024
50. Li, Y., Tsend, N., Li, T. K., Liu, H., Yang, R., Gai, X., Wang, H., & Shan, S. (2019). Microwave assisted hydrothermal preparation of rice straw hydrochars for adsorption of organics and heavy metals. *Bioresource Technology, 273*, 136–143. https://doi.org/10.1016/j.biortech.2018.10.056
51. Yusuf, I., Flagiello, F., Ward, N. I., Arellano-García, H., Avignone-Rossa, C., & Felipe-Sotelo, M. (2020). Valorisation of banana peels by hydrothermal carbonisation: Potential use of the hydrochar and liquid by-product for water purification and energy conversion. *Bioresource Technology Reports, 12*, 100582. https://doi.org/10.1016/j.biteb.2020.100582
52. Azaare, L., Commeh, M. K., Smith, A. M., & Kemausuor, F. (2021). Co-hydrothermal carbonization of pineapple and watermelon peels: Effects of process parameters on hydrochar yield and energy content. *Bioresource Technology Reports, 15*, 100720. https://doi.org/10.1016/j.biteb.2021.100720
53. Nakason, K., Panyapinyopol, B., Kanokkantapong, V., Viriya-empikul, N., Kraithong, W., & Pavasant, P. (2018). Characteristics of hydrochar and liquid fraction from hydrothermal carbonization of cassava rhizome. *Journal of the Energy Institute, 91*, 184–193. https://doi.org/10.1016/j.joei.2017.01.002

54. Li, Y., Meas, A., Shan, S., Yang, R., & Gai, X. (2016). Production and optimization of bamboo hydrochars for adsorption of Congo red and 2-naphthol. *Bioresource Technology, 207*, 379–386. https://doi.org/10.1016/j.biortech.2016.02.012
55. Wang, C., Zhang, S., Wu, S., Sun, M., & Lyu, J. (2020). Multi-purpose production with valorization of wood vinegar and briquette fuels from wood sawdust by hydrothermal process. *Fuel, 282*, 118775. https://doi.org/10.1016/j.fuel.2020.118775
56. Andrade, J. G. S., Porto, C. E., Moreira, W. M., Batistela, V. R., & Scaliante, M. H. N. O. (2023). Production of hydrochars from *Pinus caribaea* for biosorption of methylene blue and tartrazine yellow dyes. *Cleaner Chemical Engineering, 5*, 100092. https://doi.org/10.1016/j.clce.2022.100092
57. Zhao, K., Li, Y., Zhou, Y., Guo, W., Jiang, H., & Xu, Q. (2018). Characterization of hydrothermal carbonization products (hydrochars and spent liquor) and their biomethane production performance. *Bioresource Technology, 267*, 9–16. https://doi.org/10.1016/j.biortech.2018.07.006
58. Khosravi, A., Zheng, H., Liu, Q., Hashemi, M., Tang, Y., & Xing, B. (2022). Production and characterization of hydrochars and their application in soil improvement and environmental remediation. *Chemical Engineering Journal, 430*, 133142. https://doi.org/10.1016/j.cej.2021.133142
59. Cheng, H., Ji, R., Yao, S., Song, Y., Sun, Q., Bian, Y., Wang, Z., Zhang, L., Jiang, X., & Han, J. (2021). Potential release of dissolved organic matter from agricultural residue-derived hydrochar: Insight from excitation emission matrix and parallel factor analysis. *Science of the Total Environment, 781*, 146712. https://doi.org/10.1016/j.scitotenv.2021.146712
60. Karatas, O., Khataee, A., & Kalderis, D. (2022). Recent progress on the phytotoxic effects of hydrochars and toxicity reduction approaches. *Chemosphere, 298*, 134357.
61. Ma, X. Q., Liao, J. J., Chen, D. B., & Xu, Z. X. (2021). Hydrothermal carbonization of sewage sludge: Catalytic effect of Cl^- on hydrochars physicochemical properties. *Molecular Catalysis, 513*, 111789. https://doi.org/10.1016/j.mcat.2021.111789
62. Li, C., Feng, Y., Zhong, F., Deng, J., Yu, T., Cao, H., & Niu, W. (2022). Optimization of microwave-assisted hydrothermal carbonization and potassium bicarbonate activation on the structure and electrochemical characteristics of crop straw-derived biochar. *Journal of Energy Storage, 55*, 105838. https://doi.org/10.1016/j.est.2022.105838
63. Zulkornain, M. F., Shamsuddin, A. H., Normanbhay, S., Md Saad, J., & Ahmad Zamri, M. F. M. (2022). Optimization of rice husk hydrochar via microwave-assisted hydrothermal carbonization: Fuel properties and combustion kinetics. *Bioresource Technology Reports, 17*, 100888. https://doi.org/10.1016/j.biteb.2021.100888
64. Gajera, Z. R., Mungray, A. A., Rene, E. R., & Mungray, A. K. (2023). Hydrothermal carbonization of cow dung with human urine as a solvent for hydrochar: An experimental and kinetic study. *Journal of Environmental Management, 327*, 116854. https://doi.org/10.1016/j.jenvman.2022.116854
65. Cruz, O. F., Silvestre-Albero, J., Casco, M. E., Hotza, D., & Rambo, C. R. (2018). Activated nanocarbons produced by microwave-assisted hydrothermal carbonization of Amazonian fruit waste for methane storage. *Materials Chemistry and Physics, 216*, 42–46. https://doi.org/10.1016/j.matchemphys.2018.05.079
66. Xu, J., Zhang, J., Huang, J., He, W., & Li, G. (2020). Conversion of phoenix tree leaves into hydro-char by microwave-assisted hydrothermal carbonization. *Bioresource Technology Reports, 9*, 100353. https://doi.org/10.1016/j.biteb.2019.100353
67. Khan, M. A., Hameed, B. H., Siddiqui, M. R., Alothman, Z. A., & Alsohaimi, I. H. (2022). Physicochemical properties and combustion kinetics of food waste derived hydrochars. *Journal of King Saud University-Science, 34*(4), 101941. https://doi.org/10.1016/j.jksus.2022.101941
68. Wang, F., Guo, C., Liu, X., Sun, H., Zhang, C., Sun, Y., & Zhu, H. (2022). Revealing carbon-iron interaction characteristics in sludge-derived hydrochars under different hydrothermal conditions. *Chemosphere, 300*, 134572. https://doi.org/10.1016/j.chemosphere.2022.134572

69. Peng, H., Li, Y., Wen, J., & Zheng, X. (2021). Synthesis of $ZnFe_2O_4$/B, N-codoped biochar via microwave-assisted pyrolysis for enhancing adsorption-photocatalytic elimination of tetracycline hydrochloride. *Industrial Crops and Products, 172*, 114066. https://doi.org/10.1016/j.indcrop.2021.114066
70. Yu, J., Xie, S., & Zhang, T. (2022). Influences of hydrothermal carbonization on phosphorus availability of swine manure-derived hydrochar: Insights into reaction time and temperature. *Materials Science for Energy Technologies, 5*, 416–423. https://doi.org/10.1016/j.mset.2022.10.001
71. Roman, F. F., Diaz, J. L., Tuesta, D., Praça, P., Faria, J. L., & Gomes, H. T. (2021). Hydrochars from compost derived from municipal solid waste: Production process optimization and catalytic applications. *Journal of Environmental Chemical Engineering, 9*, 1–9. https://doi.org/10.1016/j.jece.2020.104888
72. Xiong, J., Pan, Z., Xiao, X., Huang, H., Lai, F., Wang, J., & Chen, S. (2019). Study on the hydrothermal carbonization of swine manure: The effect of process parameters on the yield/properties of hydrochar and process water. *Journal of Analytical and Applied Pyrolysis, 144*, 104692. https://doi.org/10.1016/j.jaap.2019.104692
73. Aliyu, M., Abdullah, A. H., & Tahir, M. I. B. M. (2022). A potential approach for converting rubber waste into a low-cost polymeric adsorbent for removing methylene blue from aqueous solutions. *Indonesian Journal of Chemistry, 22*, 653–665. https://doi.org/10.22146/ijc.69674
74. Aliyu, M., Abdullah, A. H., & Tahir, M. I. B. M. (2022). Adsorption tetracycline from aqueous solution using a novel polymeric adsorbent derived from the rubber waste. *Journal of the Taiwan Institute of Chemical Engineers, 136*, 104333. https://doi.org/10.1016/j.jtice.2022.104333
75. McIntosh, S., Padilla, R. V., Rose, T., Rose, A. L., Boukaka, E., & Erler, D. (2022). Crop fertilisation potential of phosphorus in hydrochars produced from sewage sludge. *Science of the Total Environment., 817*, 153023. https://doi.org/10.1016/j.scitotenv.2022.153023
76. Zhang, S., Sheng, K., Yan, W., Liu, J., Shuang, E., Yang, M., & Zhang, X. (2021). Bamboo derived hydrochar microspheres fabricated by acid-assisted hydrothermal carbonization. *Chemosphere, 263*, 128093. https://doi.org/10.1016/j.chemosphere.2020.128093
77. Hurst, G., Ruiz-Lopez, S., Rivett, D., & Tedesco, S. (2022). Effect of hydrochar from acid hydrolysis on anaerobic digestion of chicken manure. *Journal of Environmental Chemical Engineering, 10*, 108343. https://doi.org/10.1016/j.jece.2022.108343
78. Yu, J., Zhu, Z., Zhang, H., Chen, T., Qiu, Y., Xu, Z., & Yin, D. (2019). Efficient removal of several estrogens in water by Fe-hydrochar composite and related interactive effect mechanism of H_2O_2 and iron with persistent free radicals from hydrochar of pinewood. *Science of the Total Environment, 658*, 1013–1022. https://doi.org/10.1016/j.scitotenv.2018.12.183
79. Li, H., Cao, M., Watson, J., Zhang, Y., & Liu, Z. (2021). In Situ hydrochar regulates Cu fate and speciation: Insights into transformation mechanism. *Journal of Hazard Materials, 410*, 124616. https://doi.org/10.1016/j.jhazmat.2020.124616
80. Coral, N., Brasil, H., Rodrigues, E., da Costa, C. E. F., & Rumjanek, V. (2019). Microwave-modified hydrotalcites for the transesterification of soybean oil. *Sustainable Chemistry and Pharmacy, 11*, 49–53. https://doi.org/10.1016/j.scp.2019.01.002
81. Woranuch, W., Ngaosuwan, K., Kiatkittipong, W., Wongsawaeng, D., Appamana, W., Powell, J., Lalthazuala Rokhum, S., & Assabumrungrat, S. (2022). Fine-tuned fabrication parameters of CaO catalyst pellets for transesterification of palm oil to biodiesel. *Fuel, 323*, 124356. https://doi.org/10.1016/j.fuel.2022.124356
82. Cao, Y., Dhahad, H. A., Esmaeili, H., & Razavi, M. (2022). MgO@CNT@K_2CO_3 as a superior catalyst for biodiesel production from waste edible oil using two-step transesterification process. *Process Safety and Environmental Protection, 161*, 136–146. https://doi.org/10.1016/j.psep.2022.03.026
83. Xia, S., Hu, Y., Chen, C., Tao, J., Yan, B., Li, W., Zhu, G., Cheng, Z., & Chen, G. (2022). Electrolytic transesterification of waste cooking oil using magnetic Co/Fe–Ca based catalyst derived from waste shells: A promising approach towards sustainable biodiesel production. *Renewable Energy, 200*, 1286–1299. https://doi.org/10.1016/j.renene.2022.10.071

84. Takase, M., Bryant, I. M., Essandoh, P. K., & Amankwa, A. E. K. (2023). A comparative study on performance of KOH and 32%KOH/ZrO 2–7 catalysts for biodiesel via transesterification of waste *Adansonia digitata* oil. *Green Technologies and Sustainability, 1*, 100004. https://doi.org/10.1016/j.grets.2022.100004
85. Zhang, W., Wang, C., Luo, B., He, P., Zhang, L., & Wu, G. (2022). Efficient and economic transesterification of waste cooking soybean oil to biodiesel catalyzed by outer surface of ZSM-22 supported different Mo catalyst. *Biomass and Bioenergy, 167*, 106646. https://doi.org/10.1016/j.biombioe.2022.106646
86. Sukpancharoen, S., Katongtung, T., Rattanachoung, N., & Tippayawong, N. (2023). Unlocking the potential of transesterification catalysts for biodiesel production through machine learning approach. *Bioresource Technology, 378*, 128961. https://doi.org/10.1016/j.biortech.2023.128961
87. Khanian-Najaf-Abadi, M., Ghobadian, B., Dehghani-Soufi, M., & Heydari, A. (2023). Experimental evaluation of simultaneous variations in biodiesel yield and color using choline hydroxide catalyst in an ultrasonic reactor. *Journal of Cleaner Production, 382*, 134767. https://doi.org/10.1016/j.jclepro.2022.134767
88. Ningaraju, C., Yatish, K. V., & Sakar, M. (2022). Simultaneous refining of biodiesel-derived crude glycerol and synthesis of value-added powdered catalysts for biodiesel production: A green chemistry approach for sustainable biodiesel industries. *Journal of Cleaner Production, 363*, 132448. https://doi.org/10.1016/j.jclepro.2022.132448
89. Parida, S., Singh, M., & Pradhan, S. (2022). Biomass wastes: A potential catalyst source for biodiesel production. *Bioresource Technology Reports, 18*, 101081.
90. Ashine, F., Kiflie, Z., Prabhu, S. V., Tizazu, B. Z., Varadharajan, V., Rajasimman, M., Joo, S. W., Vasseghian, Y., & Jayakumar, M. (2023). Biodiesel production from *Argemone mexicana* oil using chicken eggshell derived CaO catalyst. *Fuel, 332*, 126166. https://doi.org/10.1016/j.fuel.2022.126166
91. Kedir, W. M., & Asere, T. G. (2022). Biodiesel production from waste frying oil using catalysts derived from waste materials. *Journal of the Turkish Chemical Society Section A: Chemistry, 9*, 939–952. https://doi.org/10.18596/jotcsa.997456
92. Sami, H. Q. A., Hanif, M. A., Rashid, U., Nisar, S., Bhatti, I. A., Rokhum, S. L., Tsubota, T., & Alsalme, A. (2022). Environmentally safe magnetic nanocatalyst for the production of biodiesel from *Pongamia pinnata* oil. *Catalysts, 12*, 1266. https://doi.org/10.3390/catal12101266
93. Jaruwat, D., Udomsap, P., Chollacoop, N., & Eiad-ua, A. (2019). Preparation of carbon supported catalyst from cattail leaves for biodiesel fuel upgrading application. *Materials Today: Proceedings, 17*, 1319–1325.
94. Amenaghawon, A. N., Obahiagbon, K., Isesele, V., & Usman, F. (2022). Optimized biodiesel production from waste cooking oil using a functionalized bio-based heterogeneous catalyst. *Cleaner Engineering and Technology, 8*, 100501. https://doi.org/10.1016/j.clet.2022.100501
95. Tosun, Z., & Aydin, H. (2022). Combustion, performance and emission analysis of propanol addition on safflower oil biodiesel in a diesel engine. *Cleaner Chemical Engineering., 3*, 100041. https://doi.org/10.1016/j.clce.2022.100041
96. Yusuff, A. S., & Owolabi, J. O. (2019). Synthesis and characterization of alumina supported coconut chaff catalyst for biodiesel production from waste frying oil. *South African Journal of Chemical Engineering, 30*, 42–49. https://doi.org/10.1016/j.sajce.2019.09.001
97. Abdullah, R. F., Rashid, U., Hazmi, B., Ibrahim, M. L., Tsubota, T., & Alharthi, F. A. (2022). Potential heterogeneous nano-catalyst via integrating hydrothermal carbonization for biodiesel production using waste cooking oil. *Chemosphere, 286*, 131913. https://doi.org/10.1016/j.chemosphere.2021.131913
98. Parandi, E., Safaripour, M., Mosleh, N., Saidi, M., Rashidi Nodeh, H., Oryani, B., & Rezania, S. (2023). Lipase enzyme immobilized over magnetic titanium graphene oxide as catalyst for biodiesel synthesis from waste cooking oil. *Biomass and Bioenergy, 173*, 106794. https://doi.org/10.1016/j.biombioe.2023.106794

99. Abdullah, R. F., Rashid, U., Ibrahim, M. L., NolHakim, M. A. H. L., Moser, B. R., & Alharthi, F. A. (2021). Bifunctional biomass-based catalyst for biodiesel production via hydrothermal carbonization (HTC) pretreatment—Synthesis, characterization and optimization. *Process Safety and Environmental Protection, 156*, 219–230. https://doi.org/10.1016/j.psep.2021.10.007
100. Rajendran, N., Kang, D., Han, J., & Gurunathan, B. (2022). Process optimization, economic and environmental analysis of biodiesel production from food waste using a citrus fruit peel biochar catalyst. *Journal of Cleaner Production, 365*, 132712. https://doi.org/10.1016/j.jclepro.2022.132712
101. Naidu, B. R., & Venkateswarlu, K. (2022). Dried water extract of pomegranate peel ash (DWEPA) as novel and biorenewable heterogeneous catalyst for biodiesel production and biopotent quinoxalines synthesis. *Bioresource Technology Reports, 18*, 101107. https://doi.org/10.1016/j.biteb.2022.101107
102. Nath, B., Basumatary, B., Brahma, S., Das, B., Kalita, P., Rokhum, S. L., & Basumatary, S. (2023). Musa champa peduncle waste-derived efficient catalyst: Studies of biodiesel synthesis, reaction kinetics and thermodynamics. *Energy, 270*, 126976. https://doi.org/10.1016/j.energy.2023.126976
103. Tu, R., Sun, Y., Wu, Y., Fan, X., Wang, J., Shen, X., He, Z., Jiang, E., & Xu, X. (2019). Effect of surfactant on hydrothermal carbonization of coconut shell. *Bioresource Technology, 284*, 214–221. https://doi.org/10.1016/j.biortech.2019.03.120
104. Espro, C., Satira, A., Mauriello, F., Anajafi, Z., Moulaee, K., Iannazzo, D., & Neri, G. (2021). Orange peels-derived hydrochar for chemical sensing applications. *Sensors and Actuators B: Chemical, 341*, 130016. https://doi.org/10.1016/j.snb.2021.130016
105. Zhang, C., Ma, X., Huang, T., Zhou, Y., & Tian, Y. (2020). Co-hydrothermal carbonization of water hyacinth and polyvinyl chloride: Optimization of process parameters and characterization of hydrochar. *Bioresource Technology, 314*, 123676. https://doi.org/10.1016/j.biortech.2020.123676
106. Siregar, A. G. A., Manurung, R., & Taslim, T. (2021). Synthesis and characterization of sodium silicate produced from corncobs as a heterogeneous catalyst in biodiesel production. *Indonesian Journal of Chemistry, 21*, 88–96. https://doi.org/10.22146/ijc.53057
107. Yani, F. T., Ulhaqi, R., Pratiwi, W. P., Pontas, K., & Husin, H. (2019). Utilization of water hyacinth-based biomass as a potential heterogeneous catalyst for biodiesel production. *Journal of Physics: Conference Series, 1402*(5), 055005.
108. Nahas, L., Dahdah, E., Aouad, S., El Khoury, B., Gennequin, C., Abi Aad, E., & Estephane, J. (2023). Highly efficient scallop seashell-derived catalyst for biodiesel production from sunflower and waste cooking oils: Reaction kinetics and effect of calcination temperature studies. *Renewable Energy, 202*, 1086–1095. https://doi.org/10.1016/j.renene.2022.12.020
109. Adepoju, T. F., Victor, E., Ekop, E. I., Emberru, R. E., Balogun, T. A., & Adeniyi, A. D. (2022). Residual wood ash powder: A predecessor for the synthesis of $CaO–K_2O–SiO_2$ base catalyst employed for the production of biodiesel from *Asimina triloba* oil seed. *Case Studies in Chemical and Environmental Engineering, 6*, 100252. https://doi.org/10.1016/j.cscee.2022.100252

Chapter 15
Circular Bioeconomy Approaches for Valorizing Waste Streams into Bio-jet Fuel

Louella Concepta Goveas, S. M. Vidya, Ramesh Vinayagam, and Raja Selvaraj

Abstract As the aviation sector is experiencing phenomenal growth, the substantial demand for aviation fuel is also increasing. Synthesis of sustainable aviation fuel would help curb this enhanced demand while reducing greenhouse gas (GHGs) emissions. In this regard, bio-jet fuels have shown promising potential as they are renewable, release less CO_2 on combustion, highly compatible with aviation engines, and can be used without any modifications. Several researchers have exploited sustainable waste sources for bio-jet fuel synthesis by various conversion technologies, which utilize the principles of circular bioeconomy. This book chapter particulars on the aspects of circular bioeconomy along with the conversion technologies, advantages, and disadvantages of bio-jet fuel. Additionally, circular economy-based production of bio-jet fuel from various waste streams, as demonstrated by researchers are thoroughly reviewed. The various challenges to bio-jet fuel production and their possible solutions are assessed to facilitate the suitable implementation of a circular bioeconomy.

Keywords Circular bioeconomy · Bio-jet fuel · Bioenergy · Waste stream · Valorization

L. C. Goveas (✉) · S. M. Vidya
Department of Biotechnology Engineering, NMAM Institute of Technology, Nitte (Deemed to be University), Mangalore, Karnataka 574110, India
e-mail: goveas.louella@nitte.edu.in

R. Vinayagam · R. Selvaraj (✉)
Department of Chemical Engineering, Manipal Institute of Technology, Manipal Academy of Higher Education, Manipal, Karnataka 576104, India
e-mail: raja.s@manipal.edu

G. Baskar et al. (eds.), *Circular Bioeconomy Perspectives in Sustainable Bioenergy Production*, Energy, Environment, and Sustainability,
https://doi.org/10.1007/978-981-97-2523-6_15

15.1 Introduction

The aviation sector is budding rapidly and is anticipated to continue to grow. The demand for air transport is rising at an average of 4.3% per year and by 2030 at least 200,000 flights are predicted to fly around the world daily [1]. Subsequently, a 5% increase per annum in the fossil fuel requirement for the aviation sector by 2030 is projected [2]. Depletion of fossil fuels is presently the cause for alarm and additionally, using non-renewable energy sources results in the emission of greenhouse gases (GHGs), mostly CO_2, which contributes to global warming. The CO_2 emissions from the worldwide aviation sector accounted for 2.3% of global energy-related emissions in 2020 and are expected to reach 4.3–4.8% by 2030 [3]. Therefore, the development of renewable aviation fuel which can aid in the facilitation of continued energy supply with reduced environmental load is vital.

Bio-jet fuel is the most promising renewable substitute and has been considered to replace up to 30% of the fuel for the aviation industry by 2030. The utilization of this fuel offers many advantages such as reduced CO_2 emissions, non-requirement of engine modifications, and lesser aromatic components [4]. Raw materials such as corn grains, oil seeds, and non-edible crops were primarily used for bio-jet fuel production by various conversion technologies, however, recent times have seen utilization of waste resources such as agricultural and forest residues, municipal solid waste streams, oil industry wastewaters as well as wet wastes [5]. Not only does this contribute to the generation of a sustainable energy source, but also helps implement the concept of circular bioeconomy [6]. Circular bioeconomy refers to the utilization of waste into useful products including energy, thereby creating a zero-waste environment. This consequentially reduces natural energy and resource depletion and thereby enforces the concept of recycling and reuse [7].

Although the need for bio-jet fuel is vast and its utilization is highly advantageous, many challenges affect its large-scale manufacture and subsequent marketing. A constant supply of feedstock, price difference, well-researched manufacturing process, and availability of cheap hydrogen are some of the numerous challenges that must be worked upon to facilitate the full-fledged use of bio-jet fuel as an alternative [8].

The present review discusses the significance of bio-jet fuels as an alternate energy source for the aviation industry. It also substantiates the principles of circular bioeconomy and its emphasis on recycling wastes and reuses into useful products. The main focus of this review is on the circular economy-based generation of bio-jet fuels from various waste streams, while addressing the challenges faced for substantial bio-jet fuel production.

15.2 Circular Bioeconomy—The Concept

Circular bioeconomy is the merger of bioeconomy and circular economy, two prominent approaches with several intersecting topics (Fig. 15.1) [9]. A thorough understanding of circular bioeconomy requires the reader to comprehend the basic concepts of these two approaches.

The idea of bioeconomy was introduced in the 2000s by European Union to propagate the implementation of biotechnology for valuable product generation. Bioeconomy considers the utilization and subsequent conversion of renewable bio-based resources and their waste streams into valuable products. Bioeconomy also aims at the development of economy by generating employment opportunities, and by responding to social and environmental challenges [10]. It majorly depends on sustainable techniques for the utilization of biological sources, scientific technologies, and advancements in all economic sectors leading to a sustainable economy [7]. It provides solutions to worldwide problems by delivering safe and healthy food, bio-energy as a substitute to fossil fuels, novel processing techniques and chemicals, new product characteristics, sustainable bio-farming practices, and green chemistry, thereby resulting in a reduction of GHG emissions leading to minimization of global warming [11]. The sustainable development goals (SDGs) are a set of 17 specific objectives implemented by the United Nations to achieve prosperity while promoting a safe and sustainable environment [12]. Bioeconomy is considered highly important as it can help establish several SDGs by aiding in achieving the goal of zero hunger by providing safe and healthy food, enabling clean water for consumption by wastewater treatment, helping to generate bioenergy resulting in the reduction of climate change, and also promoting sustainability in all the aspects of living [10].

The circular economy is the concept of manufacture and utilization, wherein the prevailing constituents and goods are reused, restored, revamped, and recycled for prolonged periods. In this way, the life cycle of products is extended [13]. It implies minimizing waste, by recycling the materials from a product at the end of its life cycle. These materials are further reused to generate new value. Circular economy hence

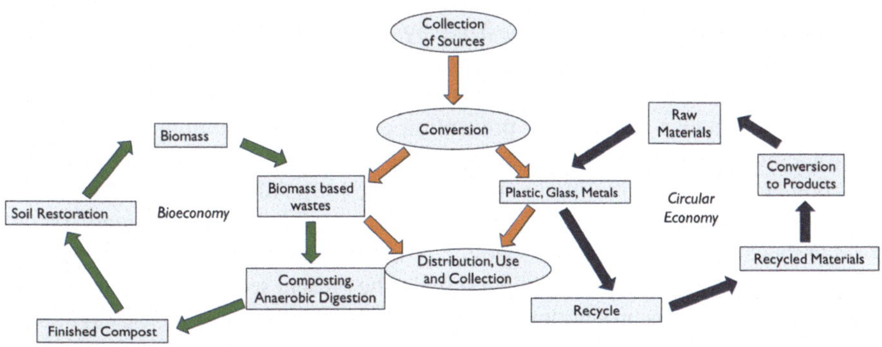

Fig. 15.1 Concepts of bioeconomy and circular economy

lessens the utilization of raw materials, energy, and water by conversion of generated debris into useful products [14]. The basic principles of the circular economy include reuse, recycling, and repair thereby minimizing the total amount of debris that is generated, circulation of goods and resources at their highest value, and regenerating the environment. The major focus of circular economy is the continuous utilization of wastes thereby limiting the input of energy and raw materials [15]. This concept aims to offer solutions to two issues, i.e., conservation of resources, and substantially minimizes the adverse influence on environment and climate. The circular economy is distinct from the concept of linear economy, wherein the goods are discarded at the end of their life cycle, which results in a lot of expenditure of energy and resources [9]. The conversion from a linear to a circular economy would result in environmental protection by reduction of waste eliminating pollution, reducing the requirement of resources in this era of population crisis and boosting pecuniary growth.

The merger of bioeconomy and circular economy substantiates the sustainable use of biological sources, to generate goods, improving their yield, with a simultaneous decrease in detrimental effects on the environment. This incorporation, thereby assists in the development of the present economic strategy [13]. If the implementation of a linear economy is persistent, it results in the depletion of existing natural sources, unraveling into an ecological crisis. Therefore, a bioeconomy-dependent circular economy will enable us to restrict the expenditure of natural resources, by recycling and reuse of generated biological-based goods and their wastes [16]. Hence, circular bioeconomy emphasizes that the utilization of biological resources does not adversely affect the global climate, biodiversity, and the environment. Circularity and sustainable development are both promoted by circular bioeconomy, which ensures that the biological resources are utilized to their maximum strength [17]. This results in the complete minimization of waste products and restricts the utilization of fossil fuels.

15.3 Bio-jet Fuel—Synthesis from Waste Streams

Commercial aviation fuel is synthesized by distilling crude oil at 205 °C and comprises n-paraffins, branched iso-alkanes, naphthenes (cyclic alkanes), and aromatics. Naphthene content reduces the freezing point of the fuel, while the presence of aromatics helps in the prevention of leakages. The number of alkanes contributes to the energy density. The major problem with the utilization of this fuel is its carbon emission contributing to climate change [18]. Additionally, the depleting sources of non-renewable reserves make it extremely challenging to fend for the world in these times of booming economy. A suitable substitute is the conversion of waste biomasses into sustainable aviation fuel. The present section deliberates on the basics of bio-jet fuel, its production routes, and circular bioeconomy-based synthesis through various waste streams.

15.3.1 Bio-jet Fuel—Composition and Advantages

Bio-jet fuel is also called bio-kerosene or synthetic paraffinic kerosene and consists of n-paraffins, cyclo-alkanes, and branched iso-alkanes. Aromatics may or may not be present and is dependent on the synthesis route and raw materials. Hence, bio-jet fuel has to be blended up to a maximum of 50% with conventional jet fuel [4]. The utilization of bio-jet fuel substantially reduces CO_2 release and helps achieve zero-carbon emissions. As this fuel is manufactured from biomass, it reduces CO_2 emissions between 20 and 90% when compared with jet fuels. Air-craft engines do not require any modifications to sustain this fuel and hence could be blended with fossil fuels, without any adaptation. Additionally, these fuels possess fewer sulfur amounts, exhibit lesser emissions from the tailpipe, have exceptional cold flow characteristics, and have greater heat and oxidation stability [19]. One major drawback of bio-jet fuels is their absence of aromatics which can result in fuel leaking, due to shrinkage and hardening of seals. The addition of aromatics could help solve this problem, but combustion causes soot formation [20].

15.3.2 Production Routes

Bio-based raw materials are converted into bio-jet fuel via several conversion technologies which predominantly depend on the composition of raw materials. Four classes of conversion technologies include oil-to-jet, gas-to-jet, sugar-to-jet, and alcohol-to-jet which multiple techniques available under each class [21] (Fig. 15.2).

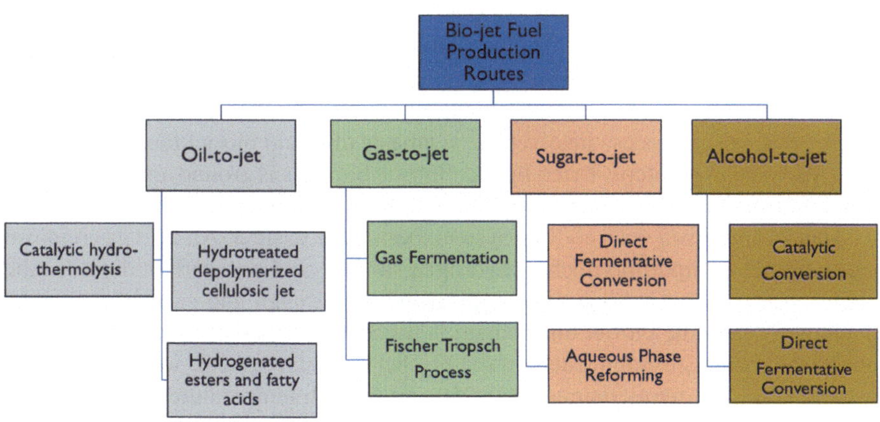

Fig. 15.2 Classification of bio-jet fuel production routes

Oil-to-jet: Hydrogenated esters and fatty acids (HEFA), catalytic hydro-thermolysis (CHT), and hydro-treated depolymerized cellulosic jet (HDCJ) are the three techniques involved under oil-to-jet technologies [21].

Triglycerides, unsaturated and saturated fats usually present in vegetable and animal-based oils are treated via HEFA to produce bio-jet fuels. HEFA is a two-step process, the saturated fats, obtained by the conversion of unsaturated fats and triglycerides via catalytic hydrogenation, are further converted into straight-chain alkanes and propane along with water and CO_2. The straight alkanes are cracked and isomerized into mixed alkane liquid fuel, later fractionated to generate bio-jet fuel along with naphtha, paraffinic diesel, and light gases. The first step is catalyzed by Ni, Co, Mo, or zeolites of noble metals while, metals such as platinum, and nickel associated with zeolite and activated carbon catalyze the second step [8].

CHT or hydrothermal liquefaction is utilized for the conversion of plant-based or algal oils into bio-jet fuel with aromatics. Processing steps, such as conjugation, cyclization, and cross-linking are carried out on triglycerides, which are then subjected to hydrolysis and cracking in the presence of catalysts and water. A mixture of hydrocarbons is obtained by decarboxylation and dehydration, later fractionated into naphtha, diesel, and bio-jet fuel [21, 22].

HDCJ is a two-step technique where the bio-oil obtained from pyrolysis of lignocellulosic biomass along with biochar and biogas is converted into bio-jet fuel. Catalytically hydro-treated bio-oil is subjected to hydrogenation at high temperatures to achieve hydrocarbon fuel. This mixture is then fractionated to separate the bio-jet fuel [23].

Although the techniques of production of hydrocarbon mixtures are different in these three techniques, the downstream processing step of fractionating into bio-jet fuel is the same.

Gas-to-jet: Several types of gas such as syngas, natural gas, or bio-gas can be converted into bio-jet fuel via Fischer Tropsch and gas fermentation techniques. Syngas and biogas are produced by gasification and pyrolysis of lignocellulosic biomass.

Fischer Tropsch process involves the synthesis of liquid bio-jet fuel from syngas and is achieved in six steps. Dried lignocellulosic biomass is ground and subjected to gasification under steam and pure oxygen at high temperatures. Ash and tar present in the syngas is removed via a cooling system, beside the gasifier. An acid gas removal system removes impurities such as hydrogen sulfide, acid gas, carbon dioxide, and sulfides. The purified syngas then is mixed with hydrogen and carbon monoxide, which then enters the bio-jet oil production process. A series of catalytic processes convert the syngas to liquid fuels in the Fischer Tropsch system, which can either be operated at low temperatures (200–250 °C) or high temperatures (300–350 °C) with iron-based and cobalt-based oxide nanoparticle-based catalysts, respectively. The liquid fuel is then subjected to hydrocracking, and distilled to achieve bio-jet fuel. Excess gas is used for the generation of electricity and is utilized to operate the system [24].

The syngas produced from lignocellulosic biomass could be microbially fermented to produce bio-jet fuel instead of catalytic conversion. This process, known as gas fermentation, usually involves the fermentation of cold syngas into butanol/acetone by *Clostridium* species, an acetogenic bacteria. These alcohols could be then converted into bio-jet fuel via an alcohol-to-jet-based process involving several steps including oligomerization, distillation, and hydrogenation. Raw materials such as organic waste arising from households and industries could also be refurbished into bio-jet fuel via gas fermentation [21].

Sugar-to-jet: The conversion of lignocellulosic biomass containing multiple components is processed to sugars which are then converted to bio-jet oil by two prominent processes, i.e., aqueous phase reforming of sugars to hydrocarbons and direct fermentative conversion of sugars to bio-jet fuel.

The aqueous phase reforming process includes pre-treatment of lignocellulosic biomass via enzymes to obtain cellulose and hemicellulose, which are further acid or enzymatically hydrolyzed to obtain short-length sugars (5–6 carbon). Hydrogenation or hydrogenolysis is then performed to convert these sugars to polyhydric alcohols and short-chain oxygenated compounds, respectively [21]. These intermediates are then converted to oxygenates via aqueous phase reforming which includes several steps such as dehydrogenation, deoxygenation, hydrogenation, hydrogenolysis, and cyclization. The oxygenates formed are then processed into alkanes via any of three reactions, *i.e.*, direct catalytic condensation, acid condensation, and hydrogenation/dehydrogenation. The generated alkanes are then converted to bio-jet fuel via a series of steps [25].

Genetically engineered microbial strains ferment sugars directly to alkane kind fuels, bypassing the alcohol step. Raw materials such as lignocellulosic biomass or crops are primarily pre-treated to produce small-chain sugars enzymatically or mechanically [26]. These sugars are then converted to acetyl CoA via glycolysis which are then converted to isopentyl pyrophosphate, a liquid fuel precursor, via either of two pathways, i.e., mevalonic acid or deoxyxylulose-5-phosphate pathway. This precursor is then converted to bio-jet fuel post-hydro-processing [21].

Alcohol-to-jet: Several alcohols such as methanol, ethanol, and higher alcohols are processed into bio-jet fuels by catalytic conversion or fermentation. The lignocellulosic biomass is pre-treated by acid or enzymes to obtain sugars which are then subsequently fermented to alcohols by microbes or yeasts. During catalytic conversion, the olefins produced from the catalytic dehydration of alcohols are oligomerized to produce an intermediate distillate, which is then converted to hydrocarbons via hydrogenation. In the final step, these hydrocarbons are then distilled to obtain the bio-jet fuel [8].

15.3.3 Conversion of Waste Streams to Bio-jet Fuels

The raw materials used for the production of bio-jet fuels are starch, sugars, oils, and triglycerides. The first-generation jet fuel was produced from food crops, but due to overwhelming demand, the present focus is on agricultural wastes, forest, and woody by-products, organic matter from municipal and industrial wastes, animal fats, and used vegetable oils (Table 15.1). In the present times, the focus is on the use of algal oils, a promising raw material for biofuel production, and also aids in carbon sequestration.

Lignocellulosic Wastes: The cellulose and hemicellulose present in lignocellulosic wastes can be essentially converted into bio-jet fuel by techniques such as HDCJ, Fischer Tropsch process, direct fermentation, and catalytic conversion. Agricultural by-products, forest residues, and several waste biomasses have been exploited by several researchers to produce bio-jet fuel with detailed economic analysis. Oxygenates were obtained from hemicellulose and cellulose present in corncobs, which were subsequently converted into bio-jet fuel by hydrogenation and hydrodeoxygenation at a yield of 0.125 kg/kg of corncob, with an operating cost of 1540 dollars per ton [37]. Cellulose and hemicellulose from corn stalks were utilized as a source for bio-jet fuel production by aqueous conversion, wherein the lignin which usually goes as waste was also used for H_2 production. This hydrogen was subsequently involved in hydrogenation during the fuel production process [38]. Rice husk, a major agricultural waste byproduct in Taiwan was pyrolyzed and converted into liquid bio-jet fuel by the Fischer Tropsch process at a final yield of 52%. Analysis of the fuel by gas chromatography–mass spectrometry revealed the presence of iso- and cyclo-alkanes which made it appropriate for usage as a bio-jet fuel [39]. In a recent study, poplar biomass was treated by enzymes and dilute acids, the released sugars were fermented by *Clostridium thermoaceticum,* to acetic acid. Acetic acid

Table 15.1 Bio-jet fuel from various lignocellulosic and oil-based biomass

Raw material	Production technique	Yield	References
Chicken fat	Hydro-treating	46.47%	[27]
Microalgae, *Pongamia* seeds, and sugarcane	HEFA, fermentation	–	[28]
Waste cooking oil	Hydro-processing	–	[29]
Grindelia squarrosa	Biphasic tandem catalytic process	90% carbon	[30]
Jatropha curcas fruit	Fermentation, alcohol to jet	40.37%	[31]
Rice husk	Fischer Tropsch process	42.50%	[32]
Douglas fir sawdust pellets	Tandem catalytic conversion	–	[33]
Vegetable oil, sugarcane juice	HEFA, alcohol to jet	–	[34]
Sawdust	Fischer Tropsch process	–	[35]
Poplar biomass	Fermentation, alcohol to jet	330 L/ton	[36]

thus produced was converted into hydrocarbon fractions via reactive distillation, hydrotreatment, dehydration, and oligomerization. The unused lignin content may be burned for electricity generation or hydrogen production leading to the reusing of by-products generated during this process [36]. Waste woody biomass obtained post-forest harvesting was pre-treated to release sugars, which were then consequently converted to iso-butanol by fermentation. The produced iso-butanol was oligomerized and hydrogenized to generate iso-paraffinic kerosene. The residue obtained post this process was further utilized for biogas production, whereas the wastewater was treated and recycled back for the pre-treatment step [40]. Similarly, another study in British Columbia utilized wood pellets and forest residues that were pyrolyzed to produce bio-oil which was upgraded and converted into bio-jet fuel. Additionally, bio-gas and bio-char were produced which were utilized for other purposes. Lifecycle analysis revealed that about 300 million liter per year of bio-jet fuel could be produced per year from forest residue which could reduce the carbon intensity by 110% [41]. When potato by-products and sugar beets from processing industries were used as raw materials for bio-jet fuel production, the GHG emissions were reduced by 52% and 44% correspondingly. The production proceeded by alcohol-to-jet conversion, where the pre-treated sugars were fermented to acetone, butanol, and ethanol by *Clostridium* strains. These alcohols were then catalytically converted to bio-jet fuel by condensation and hydro-treatment [42]. Eucalyptus wood chips were converted to bio-jet oil by pre-treatment, fermentation into alcohols, catalytic hydrogenation, and oligomerization. Fermentation was performed by *Saccharomyces cerevisiae* and *Clostridium* strains. It was observed that the eucalyptus-based fuel showed 93% less GHG emissions than the fossil fuel-based jet fuel [43].

Oil and Triglyceride-Based Wastes: Processes such as HEFA and CHT convert the triglycerides and esters present in oil-based wastes into sustainable bio-jet fuel. The most popular oil-based waste streams utilized by researchers are non-edible oil seeds, waste cooking oils, and animal fats.

Waste cooking oils contain free fatty acids which are a valuable reserve for conversion into bio-based fuels. As it is generated worldwide on a large-scale basis, its repeated consumption and further, its disposal results in severe health and environmental hazards [44]. Waste cooking oil obtained from Egyptian restaurants was subjected to hydrothermal catalytic cracking in the presence of a nano zinc aluminate catalyst. Bio-jet fuel was then obtained by fractional distillation and blended at 5% with Jet A1 fuel [45]. Waste cooking oil was pre-treated and converted into bio-jet fuel by a one-step hydrogenation reaction. The carbon intensity of this fuel was 63% lesser than that of conventional fuel, while the GHG emissions were a mere 18% [46].

Non-edible oil seeds, oily biomass, and residues depict maximum potential to be converted into biofuels as their production is cost-effective without competition to food security. Jatropha and castor oil were converted into bio-jet fuel by HEFA and resulting in reduced environmental impacts in China [47]. Oil, extracted from *Jatropha curcas* dehulled fruits, was subjected to hydrogenation and converted to propane. This propane was then converted into bio-jet fuel via a series of reactions through the HEFA process [47]. Bio-jet fuel was produced from non-edible date palm seed oil by a one-step process catalyzed by tantalum phosphate. The hydrocarbon mixture obtained as a product contained 53.6% bio-jet fuel, while the rest was green diesel fraction [48]. Sunflower oil was extracted from sunflower residual wastes by the cold press and treated by catalytic cracking under a ZSM-5 catalyst. The upgraded oil was then distilled to obtain hydrocarbons, which on hydrogenation yielded bio-jet fuel. The physical and chemical properties were comparable with conventional aviation fuel [49]. Similarly, waste soybean oil and palm oil residues were converted via a single step into good quality bio-jet fuel-based hydrocarbons through a Pd-based zeolite catalyst without hydrogen addition [50].

Animal carcasses are generated in large amounts owing to increased consumption of meat and poultry, due to the global population. About 25% of the carcass contains fat, which could be easily utilized for the production of biofuels. The fuels obtained from these triglycerides release lower emissions when compared with oil seed-based crops. In a recent study, chicken fat was hydro-processed by a single step into bio-jet fuel with 94% conversion and reduced carbon emissions [27]. Bio-oil produced by pyrolysis of lignocellulosic waste was blended with poultry fat, to give better quality. Subsequent conversion and hydro-treatment by biochar-based catalyst resulted in the generation of bio-jet fuel [51].

15.4 Challenges—Emphasis on Economy and Environment

Although the generation of bio-jet fuel is valuable for circular bioeconomy-based sustainable development directing to reduced carbon emissions, several problems affect its large-scale production and utilization. The primary challenge is the uninterrupted supply and availability of feedstock. If non-edible crops are utilized as raw materials, their growth would lead to food and water scarcity [52]. Additionally, the application of fertilizers and pesticides causes groundwater and soil contamination. If agricultural by-products, forest residues, industrial processing wastes, or waste oils are to be employed as a source, it must be ensured that a continuous supply with consistent quality must always be accessible which is rather challenging [53]. Feedstocks such as weeds, algae, and halophytes having the potential to grow in wastes and waste regions could be used as sources for bio-jet fuel production. This will solve the difficulty of competition for food crops and water usage, also, they are not harmful to the environment. On the contrary, they sequester atmospheric CO_2 and help in the prevention of climate change [54].

Production of bio-jet fuel requires a large water footprint, which is dependent on the raw material. The water footprint was considerably low when microalgae were utilized and drastically increased when crops, such as sugarcane, rapeseed oil, jatropha, and soybean oils were used as raw materials, as irrigation contributed to the water usage. On the other hand, the energy requirement for the growth of microalgae is high contributing to the increased costs of bio-jet fuel [55]. Another challenge is the increased production costs which make bio-jet fuel expensive when compared with conventional aviation fuels. The increased cost is mostly due to the present availability of feedstocks and the continued development of production processes. Additionally, factors, such as feedstock pre-treatment, hydrogen and catalyst consumption, and operational costs also contribute to the high costs. A ton of synthetic jet fuel costs 317 dollars per ton in 2017, while the cost of raw materials for bio-jet fuel such as animal waste, and algal and vegetable oils is higher (550–1200 dollars per ton) [56]. Lignocellulosic waste requires processing by pre-treatment, fermentation, or pyrolysis to be made suitable for bio-jet fuel production. The residual waste generated consists of soil contaminants and micro-organisms that must be treated before disposal which may contribute to economic costs. Hence the production costs of bio-jet fuel exceed 7–8 times more than those of conventional jet fuels [23].

Renewable jet fuels have to be blended with synthetic fuels, as utilization of pure bio-jet fuel may freeze at flight operational conditions. Bio-jet fuels also have less thermal stability properties at high temperatures and may be non-stable at continued storage conditions. These fuels may contain small fractions of metals, impurities, and micronutrients from the raw materials, requiring rigorous processing [56]. Subsequent quality check is performed to ensure its usage in aircraft. The produced fuel should be resistant to a wide range of operational conditions and offer good performance. Presently, bio-jet fuels blend with Jet A-1 at 1:1 without any issues and most aircrafts have been certified to fly on 50% alternate fuels [23].

15.5 Conclusions

Production of bio-jet fuels from lignocellulosic and triglyceride-based waste streams serves to solve multiple issues of non-renewable energy scarcity, meeting the demands of the aviation industry, environmental pollution, and global climate change. The utilization of wastes to produce a useful product includes aspects of reuse, recycling, refurbishment, and reduction leading to the implementation of a circular bioeconomy. This also contributes to the United Nation's Sustainable Development Goals and helps build a secure future for the upcoming generations. The large-scale production and utilization of bio-jet fuel is yet to be achieved owing to several challenges such as the non-availability of stable feedstocks and environmental and economic issues. Efficient technology and feedstock could be achieved through research with support from the government, leading to sustainable production of bio-jet fuels.

References

1. Fageda, X., & Teixidó, J. J. (2022). Pricing carbon in the aviation sector: Evidence from the European emissions trading system. *Journal of Environmental Economics and Management, 111*, 102591. https://doi.org/10.1016/J.JEEM.2021.102591

2. Timmis, A. J., et al. (2015). Environmental impact assessment of aviation emission reduction through the implementation of composite materials. *International Journal of Life Cycle Assessment, 20*(2), 233–243. https://doi.org/10.1007/S11367-014-0824-0/METRICS
3. Börjesson, P., Björnsson, L., Ericsson, K., & Lantz, M. (2023). Systems perspectives on combined production of advanced biojet fuel and biofuels in existing industrial infrastructure in Sweden. *Energy Conversion and Management X, 19*, 100404. https://doi.org/10.1016/J.ECMX.2023.100404
4. Lim, J. H. K., et al. (2021). Utilization of microalgae for bio-jet fuel production in the aviation sector: Challenges and perspective. *Renewable and Sustainable Energy Reviews, 149*, 111396. https://doi.org/10.1016/J.RSER.2021.111396
5. Moreno-Gómez, A. L., Gutiérrez-Antonio, C., Gómez-Castro, F. I., & Hernández, S. (2020). Production of biojet fuel from waste raw materials. *Process System Engineering for Biofuels Development*. https://doi.org/10.1002/9781119582694.CH6
6. Koytsoumpa, E. I., Magiri-Skouloudi, D., Karellas, S., & Kakaras, E. (2021). Bioenergy with carbon capture and utilization: A review on the potential deployment towards a European circular bioeconomy. *Renewable and Sustainable Energy Reviews, 152*, 111641. https://doi.org/10.1016/J.RSER.2021.111641
7. Goveas, L. C., Nayak, S., Vinayagam, R., Loke Show, P., & Selvaraj, R. (2022). Microalgal remediation and valorisation of polluted wastewaters for zero-carbon circular bioeconomy. *Bioresource Technology, 365*, 128169. https://doi.org/10.1016/J.BIORTECH.2022.128169
8. Tiwari, R., et al. (2023). Environmental and economic issues for renewable production of bio-jet fuel: A global prospective. *Fuel, 332*, 125978. https://doi.org/10.1016/J.FUEL.2022.125978
9. Tan, E. C. D., & Lamers, P. (2021). Circular bioeconomy concepts—A perspective. *Frontiers in Sustainability*. https://doi.org/10.3389/FRSUS.2021.701509
10. Eversberg, D., & Fritz, M. (2022). Bioeconomy as a societal transformation: Mentalities, conflicts and social practices. *Sustainable Production and Consumption, 30*, 973–987. https://doi.org/10.1016/J.SPC.2022.01.021
11. Wei, X., et al. (2022). From biotechnology to bioeconomy: A review of development dynamics and pathways. *Sustainability, 14*(16), 10413. https://doi.org/10.3390/SU141610413
12. Wang, X., Khurshid, A., Qayyum, S., & Calin, A. C. (2022). The role of green innovations, environmental policies and carbon taxes in achieving the sustainable development goals of carbon neutrality. *Environmental Science and Pollution Research, 29*(6), 8393–8407. https://doi.org/10.1007/S11356-021-16208-Z/TABLES/9
13. D'Amato, D., & Korhonen, J. (2021). Integrating the green economy, circular economy and bioeconomy in a strategic sustainability framework. *Ecological Economics, 188*, 107143. https://doi.org/10.1016/J.ECOLECON.2021.107143
14. Škrinjarí, T. (2020). Empirical assessment of the circular economy of selected European countries. *Journal of Cleaner Production, 255*, 120246. https://doi.org/10.1016/J.JCLEPRO.2020.120246
15. D'Urzo, M., & Campagnaro, C. (2023). Design-led repair and reuse: An approach for an equitable, bottom-up, innovation-driven circular economy. *Journal of Cleaner Production, 387*, 135724. https://doi.org/10.1016/J.JCLEPRO.2022.135724
16. Carus, M., & Dammer, L. (2018). The circular bioeconomy—Concepts, opportunities, and limitations. *Industrial Biotechnology, 14*(2), 83–91. https://doi.org/10.1089/IND.2018.29121.MCA
17. Sharma, R., & Malaviya, P. (2023). Ecosystem services and climate action from a circular bioeconomy perspective. *Renewable and Sustainable Energy Reviews, 175*, 113164. https://doi.org/10.1016/J.RSER.2023.113164
18. Goh, B. H. H., et al. (2022). Recent advancements in catalytic conversion pathways for synthetic jet fuel produced from bioresources. *Energy Conversion and Management, 251*, 114974. https://doi.org/10.1016/J.ENCONMAN.2021.114974
19. Lahijani, P., Mohammadi, M., Mohamed, A. R., Ismail, F., Lee, K. T., & Amini, G. (2022). Upgrading biomass-derived pyrolysis bio-oil to bio-jet fuel through catalytic cracking and hydrodeoxygenation: A review of recent progress. *Energy Conversion and Management, 268*, 115956. https://doi.org/10.1016/J.ENCONMAN.2022.115956

20. Lin, C. H., & Wang, W. C. (2020). Direct conversion of glyceride-based oil into renewable jet fuels. *Renewable and Sustainable Energy Reviews, 132*, 110109. https://doi.org/10.1016/J.RSER.2020.110109
21. Wang, W. C., & Tao, L. (2016). Bio-jet fuel conversion technologies. *Renewable and Sustainable Energy Reviews, 53*, 801–822. https://doi.org/10.1016/J.RSER.2015.09.016
22. Wicker, R. J., Kwon, E., Khan, E., Kumar, V., & Bhatnagar, A. (2023). The potential of mixed-species biofilms to address remaining challenges for economically-feasible microalgal biorefineries: A review. *Chemical Engineering Journal, 451*, 138481. https://doi.org/10.1016/J.CEJ.2022.138481
23. Hussain, I., Ganiyu, S. A., Alasiri, H., & Alhooshani, K. (2022). A state-of-the-art review on waste plastics-derived aviation fuel: Unveiling the heterogeneous catalytic systems and techno-economy feasibility of catalytic pyrolysis. *Energy Conversion and Management, 274*, 116433. https://doi.org/10.1016/J.ENCONMAN.2022.116433
24. Collis, J., Duch, K., & Schomäcker, R. (2022). Techno-economic assessment of jet fuel production using the Fischer-Tropsch process from steel mill gas. *Frontiers in Energy Research, 10*, 1049229. https://doi.org/10.3389/FENRG.2022.1049229/BIBTEX
25. Zhu, X., Xiao, J., Wang, C., Zhu, L., & Wang, S. (2022). Global warming potential analysis of bio-jet fuel based on life cycle assessment. *Carbon Neutrality, 1*(1), 1–13. https://doi.org/10.1007/S43979-022-00026-4/TABLES/8
26. Majidian, P., Tabatabaei, M., Zeinolabedini, M., Naghshbandi, M. P., & Chisti, Y. (2018). Metabolic engineering of microorganisms for biofuel production. *Renewable and Sustainable Energy Reviews, 82*, 3863–3885. https://doi.org/10.1016/J.RSER.2017.10.085
27. Moreno-Gómez, A. L., Gutiérrez-Antonio, C., Gómez-Castro, F. I., & Hernández, S. (2021). Modelling, simulation and intensification of the hydroprocessing of chicken fat to produce renewable aviation fuel. *Chemical Engineering and Processing—Process Intensification, 159*, 108250. https://doi.org/10.1016/J.CEP.2020.108250
28. Klein-Marcuschamer, D., et al. (2013). Technoeconomic analysis of renewable aviation fuel from microalgae, *Pongamia pinnata*, and sugarcane. *Biofuels, Bioproducts and Biorefining, 7*(4), 416–428. https://doi.org/10.1002/BBB.1404
29. Chen, R. X., & Wang, W. C. (2019). The production of renewable aviation fuel from waste cooking oil. Part I: Bio-alkane conversion through hydro-processing of oil. *Renewable Energy, 135*, 819–835. https://doi.org/10.1016/J.RENENE.2018.12.048
30. Yang, X., et al. (2018). Production of high-density renewable aviation fuel from arid land crop. *ACS Sustainable Chemistry and Engineering, 6*(8), 10108–10119. https://doi.org/10.1021/ACSSUSCHEMENG.8B01433/SUPPL_FILE/SC8B01433_SI_001.PDF
31. Romero-Izquierdo, A. G., Gutiérrez-Antonio, C., Gómez-Castro, F. I., & Hernández, S. (2022). Synthesis and intensification of a biorefinery to produce renewable aviation fuel, biofuels, bioenergy and chemical products from *Jatropha curcas* fruit. *IET Renewable Power Generation, 16*(14), 2988–3008. https://doi.org/10.1049/RPG2.12388
32. Wang, W. C., Liu, Y. C., & Nugroho, R. A. A. (2022). Techno-economic analysis of renewable jet fuel production: The comparison between Fischer-Tropsch synthesis and pyrolysis. *Energy, 239*, 121970. https://doi.org/10.1016/J.ENERGY.2021.121970
33. Zhang, X., et al. (2016). Enhancement of jet fuel range alkanes from co-feeding of lignocellulosic biomass with plastics via tandem catalytic conversions. *Applied Energy, 173*, 418–430. https://doi.org/10.1016/J.APENERGY.2016.04.071
34. Diederichs, G. W., Ali Mandegari, M., Farzad, S., & Görgens, J. F. (2016). Techno-economic comparison of biojet fuel production from lignocellulose, vegetable oil and sugar cane juice. *Bioresource Technology, 216*, 331–339. https://doi.org/10.1016/J.BIORTECH.2016.05.090
35. Zhang, Y., et al. (2015). Production of jet and diesel biofuels from renewable lignocellulosic biomass. *Applied Energy, 150*, 128–137. https://doi.org/10.1016/J.APENERGY.2015.04.023
36. Crawford, J. T., Shan, C. W., Budsberg, E., Morgan, H., Bura, R., & Gustafson, R. (2016). Hydrocarbon bio-jet fuel from bioconversion of poplar biomass: techno-economic assessment. *Biotechnology and Biofuels*. https://doi.org/10.1186/s13068-016-0545-7

37. Li, Y., et al. (2017). Process and techno-economic analysis of bio-jet fuel-range hydrocarbon production from lignocellulosic biomass via aqueous phase deconstruction and catalytic conversion. *Energy Procedia, 105*, 675–680. https://doi.org/10.1016/J.EGYPRO.2017.03.374
38. Hao, J., Xiao, J., Song, G., & Zhang, Q. (2021). Energy and exergy analysis of bio-jet fuel production from lignocellulosic biomass via aqueous conversion. *Case Studies in Thermal Engineering, 26*, 101006. https://doi.org/10.1016/J.CSITE.2021.101006
39. Chen, Y. K., Lin, C. H., & Wang, W. C. (2020). The conversion of biomass into renewable jet fuel. *Energy, 201*, 117655. https://doi.org/10.1016/J.ENERGY.2020.117655
40. Ganguly, I., Pierobon, F., Bowers, T. C., Huisenga, M., Johnston, G., & Eastin, I. L. (2018). 'Woods-to-Wake' Life Cycle Assessment of residual woody biomass based jet-fuel using mild bisulfite pretreatment. *Biomass and Bioenergy, 108*, 207–216. https://doi.org/10.1016/J.BIOMBIOE.2017.10.041
41. Ringsred, A., van Dyk, S., & Saddler, J. (2021). Life-cycle analysis of drop-in biojet fuel produced from British Columbia forest residues and wood pellets via fast-pyrolysis. *Applied Energy, 287*, 116587. https://doi.org/10.1016/J.APENERGY.2021.116587
42. Moretti, C., López-Contreras, A., de Vrije, T., Kraft, A., Junginger, M., & Shen, L. (2021). From agricultural (by-)products to jet fuels: Carbon footprint and economic performance. *Science of the Total Environment, 775*, 145848. https://doi.org/10.1016/J.SCITOTENV.2021.145848
43. Silva Braz, D., & Pinto Mariano, A. (2018). Jet fuel production in eucalyptus pulp mills: Economics and carbon footprint of ethanol vs. butanol pathway. *Bioresource Technology, 268*, 9–19. https://doi.org/10.1016/J.BIORTECH.2018.07.102
44. Goh, B. H. H., et al. (2020). Progress in utilisation of waste cooking oil for sustainable biodiesel and biojet fuel production. *Energy Conversion and Management, 223*, 113296. https://doi.org/10.1016/J.ENCONMAN.2020.113296
45. El-Araby, R., Abdelkader, E., El Diwani, G., & Hawash, S. I. (2020). Bio-aviation fuel via catalytic hydrocracking of waste cooking oils. *Bulletin of the National Research Centre, 44*, 1–9. https://doi.org/10.1186/S42269-020-00425-6
46. Zhang, Z., Wei, K., Li, J., & Wang, Z. (2022). Life-cycle assessment of bio-jet fuel production from waste cooking oil via hydroconversion. *Energies, 15*(18), 6612. https://doi.org/10.3390/EN15186612/S1
47. Liu, H., Zhang, C., Tian, H., Li, L., Wang, X., & Qiu, T. (2021). Environmental and techno-economic analyses of bio-jet fuel produced from Jatropha and castor oilseeds in China. *International Journal of Life Cycle Assessment, 26*(6), 1071–1084. https://doi.org/10.1007/S11367-021-01914-0/METRICS
48. Rambabu, K., et al. (2023). Sustainable production of bio-jet fuel and green gasoline from date palm seed oil via hydroprocessing over tantalum phosphate. *Fuel, 331*, 125688. https://doi.org/10.1016/J.FUEL.2022.125688
49. Zhao, X., Wei, L., Julson, J., Qiao, Q., Dubey, A., & Anderson, G. (2015). Catalytic cracking of non-edible sunflower oil over ZSM-5 for hydrocarbon bio-jet fuel. *New Biotechnology, 32*(2), 300–312. https://doi.org/10.1016/J.NBT.2015.01.004
50. Choi, I. H., Hwang, K. R., Han, J. S., Lee, K. H., Yun, J. S., & Lee, J. S. (2015). The direct production of jet-fuel from non-edible oil in a single-step process. *Fuel, 158*, 98–104. https://doi.org/10.1016/J.FUEL.2015.05.020
51. Roy, P., et al. (2023). Hydrotreatment of pyrolysis bio-oil with non-edible carinata oil and poultry fat for producing transportation fuels. *Fuel Processing Technology, 245*, 107753. https://doi.org/10.1016/J.FUPROC.2023.107753
52. Zhang, L., Butler, T. L., & Yang, B. (2020). Recent trends, opportunities and challenges of sustainable aviation fuel. In *Green energy to sustainability: Strategies for global industries* (pp. 85–110). https://doi.org/10.1002/9781119152057.CH5
53. Why, E. S. K., Ong, H. C., Lee, H. V., Gan, Y. Y., Chen, W. H., & Chong, C. T. (2019). Renewable aviation fuel by advanced hydroprocessing of biomass: Challenges and perspective. *Energy Convers. Manag., 199*, 112015. https://doi.org/10.1016/J.ENCONMAN.2019.112015
54. Dolah, R., Zafar, S., & Hassan, M. Z. (2022) Alternative jet fuels: biojet fuels' challenges and opportunities. In *Value-Chain Biofuels Fundamentals, Technology and Standardization* (pp. 181–194). https://doi.org/10.1016/B978-0-12-824388-6.00022-1

55. Deane, J. P., & Pye, S. (2018). Europe's ambition for biofuels in aviation—A strategic review of challenges and opportunities. *Energy Strategy Reviews, 20*, 1–5. https://doi.org/10.1016/J.ESR.2017.12.008
56. Zhang, C., Hui, X., Lin, Y., & Sung, C. J. (2016). Recent development in studies of alternative jet fuel combustion: Progress, challenges, and opportunities. *Renewable and Sustainable Energy Reviews, 54*, 120–138. https://doi.org/10.1016/J.RSER.2015.09.056

Chapter 16
Sustainable Bioethanol Production from the Pretreated Waste Lignocellulosic Feedstocks

Belete Tessema Asfaw, Meroda Tesfaye Gari, Mani Jayakumar, and Gurunathan Baskar

Abstract Due to its production from renewable feedstocks, usage as an alternative fuel, and favorable effects on the environment, bioethanol is one of the most intriguing biofuels currently on the market. It is currently mostly made using basic ingredients that contain sugar and starch. However, a variety of readily available lignocellulosic biomass feedstocks are used to produce bioethanol, including waste from forestry, agriculture, and the processing of cereals. The complex mixture of carbohydrates known as lignocellulose needs to undergo an efficient pretreatment practice in order to generate the pathways necessary for the synthesis of fermentable sugars, which are subsequently hydrolyzed and fermented into bioethanol. Renewable lignocellulosic raw materials are cheap feedstocks that do not take away market share from the food and feed sector, promoting sustainability despite financial and technological barriers. The procedures for bioethanol extraction and purification, as well as the creation of bioethanol from renewable raw materials, have also been covered. This paper examines the developments in the manufacturing of bioethanol as a fuel from

B. T. Asfaw · M. Jayakumar (✉)
Department of Chemical Engineering, Haramaya University, Haramaya Institute of Technology, Haramaya, Ethiopia
e-mail: drjayakumarmani@haramaya.edu.et

B. T. Asfaw · M. T. Gari
College of Biological and Chemical Engineering, Addis Ababa Science and Technology University, Addis Ababa, Ethiopia

M. Jayakumar
Department of Biotechnology, Faculty of Engineering, Karpagam Academy of Higher Education, Coimbatore 641 021, Tamilnadu, India

Centre for Natural Products and Functional Foods, Karpagam Academy of Higher Education, Coimbatore 641 021, Tamilnadu, India

G. Baskar
Department of Biotechnology, St. Joseph's College of Engineering, Chennai 600119, India

School of Engineering, Lebanese American University, Byblos 1102 2801, Lebanon

G. Baskar et al. (eds.), *Circular Bioeconomy Perspectives in Sustainable Bioenergy Production*, Energy, Environment, and Sustainability,
https://doi.org/10.1007/978-981-97-2523-6_16

a variety of renewable feedstocks, as well as the methodologies for purification and separation, prospects, and challenges.

Keywords Bioethanol · Biofuels · Pretreatment · Sustainable feedstocks · Waste biomass

16.1 Introduction

Only a small portion of fossil fuels is used to develop new chemicals and materials, while the majority—roughly 75%—are still used to generate heat and energy. Fossil fuels are naturally replenished by the carbon cycle at a rate that is significantly slower than the rate at which they are now used [29]. Fossil fuel production is also unsustainable because the majority of its reserves are concentrated in a small number of countries. Additionally, burning fossil fuels and altering land use brought on by human activities result in higher greenhouse gas emissions, which worsen the crisis caused by global warming [6, 28, 34]. Much work has gone into creating other energy sources, such as biofuels like biodiesel and bioethanol, in anticipation of their unavoidable deaths. The usage of both as liquid fuels, particularly for transportation, makes them both enticing, however, bioethanol is used more frequently than the other [54]. Other potential uses for bioethanol include industrial processes and heating.

Finding alternative fuels that are practical from a commercial and environmental standpoint is a hot topic right now [3]. Biogas, bioethanol, biodiesel, and biobutanol are the most ideal biofuels to replace fossil fuels [66]. Bioethanol is one of the potential fuel alternatives and is produced from a range of sources [57]. By substituting bioethanol for gasoline, more than 80% will reduce carbon emissions and overall gasoline consumption will go down by 30%. One technique for studying the generation of bioethanol is continuous fermentation [57]. The output of bioethanol can increase by up to four times when it is produced continuously rather than using centrifuges and settling tanks [71].

Because they are better sources of the readily available starch and sugar needed for fermentation, food crops like wheat, corn, potatoes, beets, and sugarcane have historically been utilized as feedstocks for alcoholic fermentation [11]. But as the population of the world increases and the amount of arable land decreases, worries about the production of fuel from food crops have risen [14]. Researchers are looking into non-edible biomass sources such as wheat straw, grass, crop leftovers, lignocellulosic materials, and algae in order to produce eco-friendly bioethanol [26, 49]. As a result, a greater variety of feedstock materials can now be used to produce bioethanol. Technological breakthroughs in the field have made possible the ability to create ethanol derived from a wider variety of lignocellulosic biomass feedstocks. The use of fermentation practice now allows for the exploitation of a hitherto untapped biomass resource to produce bioethanol [69]. This chapter focused to summarize the current state of knowledge on biomass raw materials feedstocks, bioethanol synthesis, various by-products and future prospects and challenges were addressed.

16.2 Potential Biomass and Its Sources

The preparation of bioethanol is a possibility for various biomass types. Some of the chosen biomasses can be categorized based on raw materials that comprise sugar, starch, and lignocellulosic because of the chemical makeup of the carbohydrate sources [58]—materials with sugar, including sugar beet, sugarcane, molasses, cheese processing waste (whey), and sweet sorghum, materials with starch, including corn, wheat, and cassava; and materials with lignocellulosic biomass, including straw and leftover crop and agricultural waste [6].

16.2.1 Sugar

The two major sugar-producing plants within the world are sugar beet and sugar cane. Sugarcane and sugar beet are used to manufacture around 40% of the bioethanol consumed worldwide [6]. The vast majority of *Saccharomyces* species produce Invertase enzyme, which is readily capable to hydrolyze them. The chemical Invertase cleaves the glycosidic bond between the glucose and fructose monomers amid the primary arrange of alcoholic maturation. In a method known as glycolysis, each glucose particle is broken into two pyruvate particles [60]. This bioprocess is more feasible than when producing bioethanol from feedstocks containing starch since pretreatment is not required when producing bioethanol from feedstocks containing sugar (sucrose). To extract sugars from sugar crops and turn them into a fermentation medium, all that is required is a milling process. In this situation, juice or molasses can be used to directly generate ethanol [73].

Sugarcane is collected in the fields used in the initial stage of ethanol manufacture—the roots, stalk, crowns, and leaves of the sugarcane plant [37]. Since the stalks contain the majority of the sugars, they are valuable for commercial processing. The majority of the leaves and tops are cut off the stalks and left on the fields during mechanical harvest, safeguarding the soil and inhibiting the growth of weeds and other plant species. A stem plus a straw make up sugarcane (or trash). Before the cane is ground into juice, the sugarcane stem is removed to produce sugar (sucrose) or alcohol (ethanol). The waste material produced when the juice from sugarcane stalks is extracted is known as bagasse. Straw is made up of accessible tops, dry leaves, and fresh leaves that have not yet been harvested (or garbage) [43]. The tops of the cane plant have desiccated, often brownish-colored leaves that are located between the top end and the last stalk node. Fresh leaves are colored green and golden. The leaves may be utilized for the following purposes: (1) as a fuel, (2) transformation into oil, gas, and char; and (3) gasification-based conversion to methanol. The tops could be fed to ruminants dry or fresh, utilized as a raw material to manufacture methane through anaerobic fermentation, or utilized for energy after having the water content reduced. Normally, when the crop is harvested, SB and SS are burned in the open countryside or used as an untapped source of simple sugars that might be converted into alcohol [6].

Sugar beet is the primary raw material used to make sugar in Europe. Both the by-products of sugar beet processing, high-purity crystal sugar and thin and thick juice, can be transformed into bioethanol or other bio-based goods [36]. Raw sugar beet cassettes are also effective building blocks for the synthesis of bioethanol. The configuration of the bioprocess, microbiological stability, and transport properties are all impacted by the employment of intermediates in the sugar processing process. Granulated sugar and sugar syrup can both be used all year as the primary ingredients in the synthesis of bioethanol. Additionally, they can act as a base for a number of chemical end products or intermediates [6].

When fresh juice is turned into sugar, molasses is produced. It is a thick liquid that can be mechanically divided. It contains larger concentrations of the microorganisms needed to make alcohol, ethanol, or ethanol fuel. Sugarcane typically yields 23 L of molasses. The classification process, cane variety, and soil type all affect the molasses composition [10].

Using the juice created by the extraction process, sugar and ethanol are created. Molasses, a leftover from the manufacturing of sugar, is combined with the ethanol juice when preparing the mash for ethanol production [73]. The cogeneration system uses bagasse from the extraction system to produce steam and power [15]. Significant amounts of fermentable sugars can be found in molasses that has been diluted (thrice) with high-quality water and permitted to age while being within the nearness of a yeast culture (*Saccharomyces cerevisiae*), either in batches or continuously [52]. Lower orders of sugar (such as pentoses, tetroses, trioses, and dioses, etc.), water solvent amino acids, lignin, and other natural divisions, etc., remain unfermented and are left in spent wash after the fermentable sugars (FS) have been synthesized by the action of yeast as an alcohol rectified spirit (RS) or ethanol. Greater amounts of organic components undergo a reduction process that results in an unpleasant odor [10].

16.2.2 Starch

Starch was a high-yield feedstock for ethanol manufacturing that was a polysaccharide made up only of glucose molecules and used to store energy in plants ($(C_6H_{10}O_5)n$, $n = 300$–360) [75]. Currently, starch from cereal crops and juice or molasses from sugarcane plants are the two main feedstock types used in industrial bioethanol production facilities. Sugarcane accounts for about 60% of the world's ethanol production, with starchy cereals accounting for the other 40% [4].

Making bioethanol from starch required the following three steps: fermentation of glucose to produce ethanol and carbon dioxide; hydrolysis of higher sugars to monosaccharides (like glucose); and product purification. Around 60% of the bioethanol generated worldwide is made from starch-containing feedstocks [27].

16.2.3 Lignocellulose

Due to the renewable nature of lignocellulosic biomass and its lack of competition with food crops, producing bioethanol from these raw materials is desirable and sustainable [40]. Additionally, using bioethanol made from lignocellulosic biomass can significantly reduce greenhouse gas emissions. Agricultural wastes like corn stover, crop straws, sugar cane bagasse, agricultural and forestry residues, herbal plants (alfalfa, switchgrass), short-rotation woody crops instead of grass, pulp and paper mill waste, forestry wastes, cornstalks, waste paper, and paper products, as well as food industry trash and municipal solid waste incorporate waste [21]. The production of bioethanol from these feedstocks may be a preferable and useful alternative to the disposal of these by-products. Cellulose, hemicellulose, lignin, protein, ash, and trace amounts of extractives make up lignocellulosic biomass. Due to its availability, accessibility, and sustainable supply, lignocellulosic biomass is being taken into consideration as a source of feedstock for the manufacture of bioethanol [13]. Typically, lignin, hemicellulose, and cellulose make up around 10–25%, 20–30%, and 40–50%, respectively, of the agricultural lignocellulosic biomass [2]. Investigation of several sugar juices used as feedstocks in ethanol synthesis from sugar crops are illustrated in Table 16.1.

Lignocellulose is a complicated mixture of carbohydrates that requires effective pretreatment in order to allow enzymes to produce fermentable sugars, which are then hydrolyzed and fermented into ethanol [6]. Lignocellulosic resources like wood and agricultural waste must be converted into sugars by mineral acids [59]. The four steps that followed one another to create bioethanol from lignocellulose were pretreatment, hydrolysis, fermentation, and separation/purification [1].

16.3 Pretreatment

The complicated holocellulose polymers have been broken down into straightforward fermentable sugars using a variety of pretreatment techniques [19]. Pretreatment practice should obtain lignin for the creation of valuable final products, utilize as little heat and power as feasible, reduce inhibitor formation, prevent the breakdown of pentose sugars, and make the process more economical [45]. To get rid of lignin, an assortment of pretreatment strategies is being utilized, counting physical, chemical, physicochemical, and organic (biological) ones [22]. The size and crystallinity of the biomass are reduced by physical pretreatment techniques, including milling and grinding [61].

Table 16.1 Investigation of several sugar juices used as feedstocks in ethanol synthesis from sugar crops

Name of the crops	Significant investigation	Major accomplishments	References
Sweet sorghum	Estimates of yield, yield components, and quality were taken into account, and they varied significantly between the two seasons	Extraction of 38.89 gallons of ethanol per ton of stalk	[25]
Sugarcane	A thermotolerant yeast was isolated using an enrichment approach from cane juice for the synthesis of ethanol at a high temperature	Isolation and selection of a thermotolerant yeast strain from the juice that produced large amounts of ethanol at both 40 and 45°	[32]
	Cells were acclimated to galactose medium, and juices were added	With adapted cells, there was observed to be more ethanol (30%) than with unadapted cells	[9]
Sugar beet	Four sugar beet substrates' potential is being investigated	Ethanol production (> 40 g/l)	[48]
Dates	Date genetic variations' juices were extracted, and fermentation was tested at 300 °C and pH 7	All cultivars generated ethanol at a maximum volumetric concentration of 25.0%	[72]
Watermelon	For maximum sugar recovery, the trials were run at a variety of transmembrane pressures (100–300 kPa)	After the concentration procedure, the final concentrate's sugar level dramatically rose	[5]

16.3.1 Physical Pretreatment

By chipping, grinding, shearing, or milling, lignocellulosic biomass can be reduced to smaller fragments [46]. This increases surface area and reduces particle size, increasing cellulose conversion by making it easier for cellulases to access the surface of the biomass [76]. To produce particles with a three- to five-mm-diameter sieve, primary size reduction is done using hammer mills or Wiley mills. Sonication, microwave, infrared, gamma radiation, pyrolysis, and radiation with gamma radiation are additional efficient physical therapeutic methods [55]. In order to liberate the starch that is bound to the corn fiber, achieve homogeneity in the material's dispersion, and achieve a dramatic rise in the rate of enzymatic hydrolysis, mechanical instruments and processes are commonly employed to cut or grind the corn fiber. Physical preparation, such as grinding, however, consumes a lot of energy and is not economically viable [22].

16.3.2 Chemical Pretreatment

Among the several pretreatment categories that were initially developed, chemical pretreatment has received the most research to date. In order to delignify cellulosic materials, it has been widely employed [63]. Cellulose and hemicellulose polymers are subjected to chemical hydrolysis, which recovers sugar monomers from the polymers. This is a vital treatment method for lignocellulosic biomass. The most common used chemical pretreatments include acid and alkali-based hydrolysis techniques and peroxide [2]. Although it has been found that chemical pretreatments have greater efficiency compared to non-chemical pretreatments, they do have drawbacks, including the inability to synthesize new compounds, the need to neutralize by-products of downstream processes, and environmental concerns. Acid pretreatment is the most widely used chemical procedure for handling lignocellulosic materials, and sulfuric acid is one of the best solvents for hemicellulose dissolution [79]. During the acid pretreatment, hemicelluloses are broken down into soluble sugars, which improves cellulose accessibility [22].

16.3.3 Physico-chemical Pretreatments

Physical–chemical preparation techniques are noticeably more efficient than physical ones. Ozone, acids, alkalis, peroxide, and organic solvents are a few chemical agents used in these processes [42]. Pretreatment methods used to prepare biomass for saccharification include autohydrolysis (steam explosion), ammonia fiber explosion (AFEX), liquid hot water (LHW), steam explosion, steam explosion of SO_2, acid treatment, and alkali treatment [35, 55].

Regularly utilized to treat lignocellulosic biomass, steam explosion pretreatment has a reasonable energy need and cost. 177 °C, 5 min, and the best steam explosion pretreatment conditions for pseudo stem were discovered to be 2.2% H_2SO_4 (v/v), which produced a high total glucose yield of 91.0%. The optimal conditions for rachis were 198 °C, 5 min, 1.5% H_2SO_4 (v/v), and a total glucose yield of 87.1% [20]. The steam explosion uses high-temperature and high-pressure steam (between 160 and 260 °C) to heat the biomass for a brief period of time (a few seconds to a few minutes) [22].

By making cellulose more accessible to enzymes than other pretreatment techniques, LHW pretreatment limits the generation of breakdown products that restrict the growth of microorganisms that cause aeration [41]. It also lowers downstream pressure. After being processed by liquid hot water and enzyme hydrolysis, the majority of people think that lignocellulose should be thrown away, burned [80].

16.3.4 Biological Pretreatment

Biological pretreatment requires less energy and chemical input, produces fewer inhibitors, and uses microorganisms and/or enzymes in a milder environment (around ambient temperature and pressure) [53]. To be a useful alternative to thermochemical pretreatment, biological pretreatment still needs to be improved in a number of areas, including incubation time and overall performance (i.e., sugar production) [74]. By utilizing ligninolytic enzymes or microorganisms that release these enzymes, biological pretreatment can break down some lignin and alter its structure. In order to release sugars from holocellulose, hydrolytic enzymes are frequently utilized in the biorefinery process [31].

16.4 Cereal By-Products Processing for Bioethanol Production

The following are examples of cereal grains: wheat (*Triticum* spp.), rye (*Secale* spp.), barley (*Hordeum* spp.), oats, millet (*Pennisetum* spp.), corn (*Zea* spp.), sorghum (*Avena* spp.), and rice, all members of the Gramineae plant family (*Sorghum* spp.). In 2018, the three most popular cereals produced globally were maize (38.7%), rice and potatoes (26.4%), and wheat (24.8%), together making up 90% of all grain output. The myriad nutrient-rich chemicals or fractions found in cereal by-products, which can be exploited as novel materials to produce both food and non-food products, are still an untapped resource. Beyond the food business, the bioethanol industry provides a significant alternative to the use of grain by-products [62]. To extract starch from the substrate, the grains are either placed through dry grinding or wet grinding as the initial stage in the manufacturing of ethanol. The process of "gelatinization" involves heating the starch to a high temperature [30].

The main source of cereal by-products is the milling sector. The two most common techniques for making cereal are dry and wet grinding [18]. The waste products from the starchy endosperm known as bran, germ, and polish are also eliminated during dry milling. Wet milling produces bran, steep solids, germ, and gluten (used for oil extraction), as well as the most intact starch granules conceivable [8]. The procedures carried out prior to milling produce a by-product known as "grain screens," which mostly include all the cereal seeds that don't meet grading specifications. Along with the refined cereal flours, additional, cereal germ and bran are produced during the milling process, with the latter being used to create cereal germ oil. Most by-products of grain milling are thought to be cereal bran. After the endosperm has been sieved out of the crushed cereal grains, it comprises the coat seed and aleurone layers that are still there [47]. Before breaking down into smaller bits and sifting to form distinct fractions based on size-exclusion separation in sieves, cereals must first be graded and washed. The grain sizes of the cereals vary depending on the mill used [62]. During the gelatinization process, the starch is heated to a high temperature after

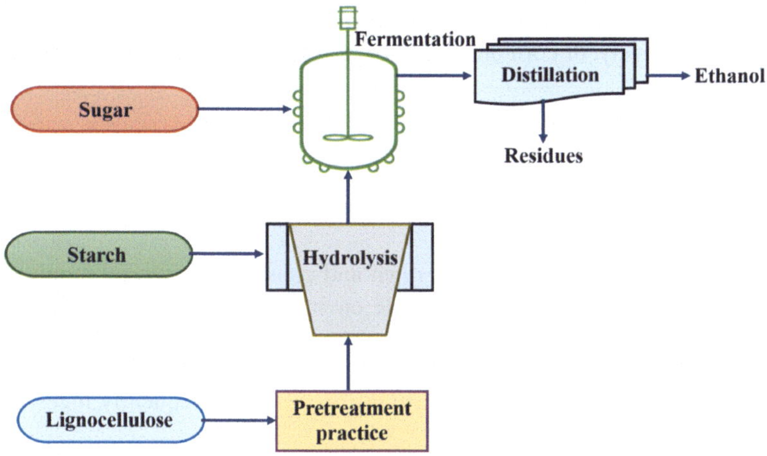

Fig. 16.1 Schematic representation of manufacturing ethanol from lignocellulosic feedstocks

milling. After gelatinization, amylolytic enzymes liquefy and saccharify the thick slurry. Then, they release simple sugars that yeast or other microorganisms use to perform anaerobic fermentation to create ethanol.

The processes of distillation, rectification, and dehydration separate and concentrate the CO_2 and ethanol produced. The amount of starch in the substrate, the operating conditions, and the ethanol production method all affect how much ethanol is created. An integrated process is being used to boost ethanol output while reducing production costs and process time, alongside simultaneous saccharification and co-fermentation, separate hydrolysis and fermentation, and simultaneous saccharification and co-fermentation [30]. Overall process depicted in Fig. 16.1.

16.4.1 Corn By-Products

There are two milling processes used for corn: dry and wet grinding. The process of wet milling softens the grain's kernels and makes it easier to separate their constituent parts by soaking them in water and SO_2. Wet milling produces a mixture of gluten, bran, step particles, and germ as a by-product after removing as many intact starch granules as possible (for oil extraction) [77]. The most by-products of corn grain are corn soak alcohol, corn germ supper, maize bran, and corn gluten supper. The most conclusion items of corn grain are starch (endosperm divisions) and oil (germ) [8]. Corn steep liquor is a suitable option for fermentation since it includes vitamins, minerals, and nitrogen (especially amino nitrogen) [67].

Usually, the pericarp is where maize fiber and bran are found (bran), one way the two differ from one another is that corn bran's cell walls originate from the endosperm component as opposed to maize fiber [65]. About 60–70 g/kg and 80–110 g/kg, respectively, of corn bran and fiber are present in each kilo of maize grain [8].

16.4.2 Rice By-Products

It is one of the cereal crops that are grown and consumed most commonly around the globe [70]. Because of its remarkable environmental adaptability and human success in changing the surrounding agroecosystem, rice can currently be produced in a broad range of environments and climates. Because of its variety of nutritional benefits and ability to enhance calorie intake, rice is most frequently ingested in its cooked form [68].

The majority of the paddy rice's rice bran, which makes up about 10% of the total, is used as feedstock, with the remainder going toward the extraction of bran oil. When brown rice is turned into white rice during the milling process, endosperm and germ are created in quantities that are mixed with the pericarp and aleurone of the grain's outer layers [17]. Rice bran, which is created, contains various levels of endosperm and germ in the outer layers of brown rice milled into white rice. In terms of composition, rice bran includes sizeable amounts of oil, dietary fibers, proteins, and extremely valuable bioactive phytochemicals [16].

16.4.3 Wheat By-Products

Because it performs better theologically than other crops, wheat is the most crucial crop for manufacturing bread [50]. The most popular industrial method of processing wheat is known as multiple grinding processes, which are used in roller milling and sifting procedures to gradually fracture the wheat kernel and extract the starchiest endosperm possible so as to produce the most flour with a high degree of purity. The by-products of this process include bran, germ, and aleurone, the outside kernel layers. Since the commonplace flour extraction rate is lies between the range of 73 and 77%, the amount of milling by-products (23–27%) represents a crucial economic factor in the production of wheat flour [51].

Wheat bran, which accounts for approximately 25% of the total weight of all wheat kernels, is a crucial by-product of milling [24]. When making breakfast cereals, nibble suppers, child nourishment, chilled nourishment, dairy nourishment, sauces, dressings, and condiments, wheat bran can be utilized as a fixing or as creature nourish. Wheat bran is widely utilized in food as a result of its technological characteristics, which make it straightforward to experience innovative forms that will make strides its functionalization and utilize in expansion to its dietary benefits. Bran is frequently added to cereal-based dishes, which diminishes the final product's quality and sensory appeal [23]. Wheat bran can benefit from enzymatic treatment, extrusion, extraction, fermentation, milling, and thermal treatments to improve its nutritive, sensory, and physical qualities for culinary uses. Wheat bran, however, is a one of the most significant feedstock for fermentation techniques that result in pure chemicals because of its composition and accessibility [78] (Table 16.2).

16.5 Applications of Cereal By-Products

The grains' cereal bran is a reservoir of nutrients. Cereal bran's chemical makeup is extremely complex, and by including it in your diet on a regular basis, you may take advantage of all of its health benefits [56]. As found in people consuming diets based on cereal grains, it contains a variety of bioactive components, such as dietary fiber, phytosterols, polyphenols, and phenolic acids, which may provide a range of biological activity and other health advantages in addition to the typical nutrients like protein, vitamins, minerals, and fats [33].

Because of its plentiful reserves and low cost of collection and transportation, maize fiber, a by-product of the corn processing industry, is a feasible raw material for the production of cellulosic ethanol and value-added commodities [22]. Worldwide, large amounts of maize are used to make ethanol, but particularly in the USA, where over 95% of the country's ethanol is generated along with being used as animal feed and food for humans. Both wet milling and dry grinding methods can be used to turn corn into ethanol [39]. It was discovered that rice bran oil, an important by-product of the extraction of rice bran, is an excellent source of a number of bioactive substances, including sterols, tocopherols, and tocotrienols. Numerous bioactive compounds with anti-inflammatory and anti-cancer characteristics may be found in rice bran oil, which is a great source of this oil [8].

16.6 Future Prospects and Challenges

The parts that came before them have amply shown that bioethanol made from biomass wastes, such as rice, corn, and sugarcane; sweet sorghum; cassava; sugar beets and wheat; witch grass; short-rotation woody crops; pulp and paper mill waste; forestry wastes; cornstalks; waste paper; and so forth, has served as a reliable supplier for numerous essential applications. However, there is a need for further research on the synthesis routes of bioethanol from different potential biomasses that require low-cost production and maximum results.

For the synthesis of bioethanol, there are several pretreatments available, among them are biological, chemical, physicochemical, and physical techniques. However, selecting the best one from the ones mentioned is difficult based on different criteria, such as economic and environmental viewpoints, and selecting the maximum conversion method is a problem. Therefore, future research is mandatory to fulfill the above

Table 16.2 Processing sugar crops and grain by-products' useful components

Cereal	By-product	Functional compounds	Health benefits	Examples	References
Rice	Bran and husk	Vitamins, proteins, dietary fiber and oil	Proteins exhibit hypoallergenic qualities, vitamins have antioxidant characteristics, and fiber lowers the risk of cardiovascular disease	179–389 mg/kg of tocopherols and tocotrienols (compounds made up of vitamin E) are found in rice bran	[18]
Sorghum and millet	Bran	Phenolic compounds, phytosterols and policosanons	Policosanoids and phytosterols, which are present in phenolic compounds, have the ability to lower cholesterol and act as antioxidants	Policosanol content, hexane extracts significantly reduced cholesterol absorption by up to 17%	[7]
Corn		Dietary fiber, Corn fiber gum, Corn fiber oil and Xylitol	Dietary fiber helps to improve health by preventing and controlling disease. As a result, gum can be replaced with corn fiber gum in products including adhesives, thickeners, stabilizers, and emulsifiers. Phytosterols included in corn fiber oil reduce serum cholesterol levels. Xylitol is a pricey natural sweetener with sweetness comparable to sucrose, yet it is safe for diabetics and does not promote tooth damage	Dietary fiber produced at its highest rate at 86.84%, increased the corn fiber gum yield by 12% and results showed that after treatment with sulfuric acid at 1210 C in comparison to untreated maize fiber, the oil concentration in the residual solids increased from 1.4 to 12.2% and the total phytosterol level increased from 176.9 to 1433.1 mg/g and the production of xylitol was minimal (0.43 g/g xylose)	[22]

(continued)

Table 16.2 (continued)

Cereal	By-product	Functional compounds	Health benefits	Examples	References
Wheat	Bran and germ	Fiber, arabinoxylans	Bran fiber aids in the speeding of intestinal transit and the expansion of feces. Arabinoxylans aid in the blood's glucose level being reduced	The prebiotic effects of arabinoxylans for obesity and other metabolic disorders, as well as their capacity to reduce blood cholesterol and the post-prandial glycemic response, are credited with providing health benefits	[44, 64]
Oat	Bran and oat mill scraps	Food fiber that is soluble, β-glucan	Dietary fiber increases fecal volume, and studies have shown that -glucan lowers blood cholesterol	Oat bran has a total dietary fiber content of 16.0% dry matter and at least 5.5% of -glucan per dry matter	[18]
Barley	Spent grains	β-glucan and dietary fiber	Dietary fiber lowers blood sugar spikes after meals, which is a result of -glucan	Phytosterols in barley oils are able to at safe dosages, dramatically reduced levels of low-density lipoprotein cholesterol	[38]
Rye	Bran	Flavonoids, coumarins, lignans, stilbenes, and condensed tannins	The addition of grain flour and bran to food products enhances the health-promoting properties of those goods as they are sources of numerous beneficial bioactive components	In rye grain has been found to be between 65 and 300 mg/100 g of grains	[12]

criteria. Additionally, a thorough analysis of the two or more pretreatment techniques combined is advised for maximal biomass breakdown. Additionally, focused efforts are required to address the need for low-cost, economically viable biomass pretreatment techniques. It is recommended to regulate energy-intensive preparation procedures.

16.7 Conclusions

It is clear from the supplied data that bioethanol can be a different approach to the current fuel problem. In recent decades, significant advancements have been made in the pretreatment of renewable biomass, the production of cellulase, the process of separating and purifying bioethanol, and the co-fermentation of sugars (pentose and hexose). However, depending on the cost of manufacture alone, bioethanol is still more expensive than fossil fuels, excluding sugar cane-based bioethanol synthesis. The major issue is still how to make bioethanol cheaper to produce. The biorefinery concept is necessary to reduce the cost of producing bioethanol by more completely utilizing renewable feedstocks and producing extra value-added by-products (such as bio-based components from lignin). As a result, bioethanol will be more economically competitive.

References

1. Amornraksa, S., et al. (2020). Systematic design of separation process for bioethanol production from corn Stover. *BMC Chemical Engineering, 2*(1), 1–16.
2. Anwar, Z., Gulfraz, M., & Irshad, M. (2014). Agro-industrial lignocellulosic biomass a key to unlock the future bio-energy: A brief review. *Journal of Radiation Research and Applied Sciences, 7*(2), 163–173.
3. Asghari, M., & Al-e, S. M. J. M. (2021). Green vehicle routing problem: A state-of-the-art review. *International Journal of Production Economics, 231*, 107899.
4. Balat, M., & Balat, H. (2009). Recent trends in global production and utilization of bio-ethanol fuel. *Applied Energy, 86*(11), 2273–2282.
5. Bhattacharjee, C., Saxena, V. K., & Dutta, S. (2017). Watermelon juice concentration using ultrafiltration: Analysis of sugar and ascorbic acid. *Food Science and Technology International, 23*(7), 637–645.
6. Bušić, A., et al. (2018). Bioethanol production from renewable raw materials and its separation and purification: A review. *Food Technology and Biotechnology, 56*(3), 289–311.
7. Carr, S. N., et al. (2005). Effects of different cereal grains and ractopamine hydrochloride on performance, carcass characteristics, and fat quality in late-finishing pigs. *Journal of Animal Science, 83*(1), 223–230.
8. Dapčević-Hadnađev, T., Hadnađev, M., & Pojić, M. (2018). The healthy components of cereal by-products and their functional properties. In *Sustainable recovery and reutilization of cereal processing by-products* (pp. 27–61). Elsevier.
9. Dhaliwal, S. S., et al. (2011). Enhanced ethanol production from sugarcane juice by galactose adaptation of a newly isolated thermotolerant strain of *Pichia kudriavzevii*. *Bioresource Technology, 102*(10), 5968–5975.

10. Dotaniya, M. L., et al. (2016). Use of sugarcane industrial by-products for improving sugarcane productivity and soil health. *International Journal of Recycling of Organic Waste in Agriculture, 5*, 185–194.
11. Duraisam, R., Salelgn, K., & Berekete, A. K. (2017). Production of beet sugar and bio-ethanol from sugar beet and it bagasse: A review. *International Journal of Engineering Trends and Technology (IJETT), 43*(4), 222–233.
12. Dziki, D. (2022). Rye flour and rye bran: New perspectives for use.
13. Edeh, I. (2021). Bioethanol production: An overview. In *Bioethanol technologies* (p. 1).
14. Edgerton, M. D. (2009). Increasing crop productivity to meet global needs for feed, food, and fuel. *Plant Physiology, 149*(1), 7–13.
15. Ensinas, A. V., et al. (2009). Reduction of irreversibility generation in sugar and ethanol production from sugarcane. *Energy, 34*(5), 680–688.
16. Fărcaş, A., et al. (2021). Cereal processing by-products as rich sources of phenolic compounds and their potential bioactivities. *Nutrients, 13*(11), 3934.
17. Ferreira, S. C., et al. (2018). New functional ingredients from agroindustrial by-products for the development of healthy foods. In *Encyclopedia of food security and sustainability*. Elsevier.
18. Galanakis, C. M. (2022). Sustainable applications for the valorization of cereal processing by-products. *Foods, 11*(2), 1–15. https://doi.org/10.3390/foods11020241
19. Gómez Millán, G., et al. (2019). Recent advances in the catalytic production of platform chemicals from holocellulosic biomass. *ChemCatChem, 11*(8), 2022–2042.
20. Guerrero, A. B., Ballesteros, I., & Ballesteros, M. (2017). Optimal conditions of acid-catalysed steam explosion pretreatment of banana lignocellulosic biomass for fermentable sugar production. *Journal of Chemical Technology & Biotechnology, 92*(9), 2351–2359.
21. Gundupalli, M. P., et al. (2022). Hydrothermal liquefaction of lignocellulosic biomass for production of biooil and by-products: current state of the art and challenges. In *Biofuels and bioenergy* (pp. 61–84).
22. Guo, Y., et al. (2022). Production of cellulosic ethanol and value-added products from corn fiber. *Bioresources and Bioprocessing, 9*(1), 1–18.
23. Heiniö, R. L., et al. (2016). Sensory characteristics of wholegrain and bran-rich cereal foods—A review. *Trends in Food Science and Technology, 47*, 25–38. https://doi.org/10.1016/j.tifs.2015.11.002
24. Higuchi, M. (2014). Antioxidant properties of wheat bran against oxidative stress. In *Wheat and rice in disease prevention and health* (pp. 181–199). Elsevier.
25. Hunsigi, G., Yekkeli, N. R., & Kongawad, B. Y. (2010). Sweet stalk sorghum: An alternative sugar crop for ethanol production. *Sugar Tech, 12*(1), 79–80.
26. Jayakumar, M., Kuppusamy Vaithilingam, S., Karmegam, N., & Jabesa, A. (2022). Bioethanol production from green biomass resources: Emerging technologies. In *Encyclopedia of green materials* (pp. 1–12).
27. Jayakumar, M., Kuppusamy Vaithilingam, S., Karmegam, N., Gebeyehu, K. B., et al. (2022). Fermentation technology for ethanol production: current trends and challenges. In B. Gurunathan & R. B. T.-B. Sahadevan (Eds.), *Biofuels and bioenergy* (pp. 105–131). Elsevier. https://doi.org/10.1016/b978-0-323-90040-9.00015-1
28. Jayakumar, M., Gebeyehu, K. B., et al. (2022). Waste ox bone based heterogeneous catalyst synthesis, characterization, utilization and reaction kinetics of biodiesel generation from *Jatropha curcas* oil. *Chemosphere, 288*, 132534.
29. Jayakumar, M., et al. (2023). Bioethanol production from agricultural residues as lignocellulosic biomass feedstock's waste valorization approach: A comprehensive review. In *Science of the total environment* (p. 163158).
30. Jeevan Kumar, S. P., Sampath Kumar, N. S., & Chintagunta, A. D. (2020). Bioethanol production from cereal crops and lignocelluloses rich agro-residues: Prospects and challenges. *SN Applied Sciences, 2*(10), 1673.
31. Keshav, P. K., et al. (2021). Lignocellulosic ethanol production from cotton stalk: an overview on pretreatment, saccharification and fermentation methods for improved bioconversion process. In *Biomass conversion and biorefinery* (pp. 1–17).

32. Limtong, S., Sringiew, C., & Yongmanitchai, W. (2007). Production of fuel ethanol at high temperature from sugar cane juice by a newly isolated *Kluyveromyces marxianus*. *Bioresource Technology, 98*(17), 3367–3374.
33. Luithui, Y., Baghya Nisha, R., & Meera, M. S. (2019). Cereal by-products as an important functional ingredient: Effect of processing. *Journal of Food Science and Technology, 56*, 1–11.
34. Machado, C. F. R., et al. (2018). Carbon dioxide and ethanol from sugarcane biorefinery as renewable feedstocks to environment-oriented integrated chemical plants. *Journal of Cleaner Production, 172*, 1232–1242.
35. Mankar, A. R., et al. (2021). Pretreatment of lignocellulosic biomass: A review on recent advances. *Bioresource Technology, 334*, 125235.
36. Marzo, C., et al. (2019). Status and perspectives in bioethanol production from sugar beet. In *Bioethanol production from food crops* (pp. 61–79). Elsevier.
37. de Matos, M., Santos, F., & Eichler, P. (2020). Sugarcane world scenario. In *Sugarcane biorefinery, technology and perspectives* (pp. 1–19). Elsevier.
38. Moreau, R. A., Flores, R. A., & Hicks, K. B. (2007). Composition of functional lipids in hulled and hulless barley in fractions obtained by scarification and in barley oil. *Cereal Chemistry, 84*(1), 1–5. https://doi.org/10.1094/CCHEM-84-1-0001
39. Mosier, N. S., & Ileleji, K. E. (2020). How fuel ethanol is made from corn. In *Bioenergy* (pp. 539–544). Elsevier.
40. Mujtaba, M., et al. (2023). Lignocellulosic biomass from agricultural waste to the circular economy: A review with focus on biofuels, biocomposites and bioplastics. *Journal of Cleaner Production*, 136815.
41. Nahak, B. K., et al. (2022). Advancements in net-zero pertinency of lignocellulosic biomass for climate neutral energy production. *Renewable and Sustainable Energy Reviews, 161*, 112393.
42. Naji, S. Z., & Tye, C. T. (2022). A review of the synthesis of activated carbon for biodiesel production: Precursor, preparation, and modification. *Energy Conversion and Management: X, 13*, 100152.
43. Negrão, D. R., et al. (2021). Inorganics in sugarcane bagasse and straw and their impacts for bioenergy and biorefining: A review. *Renewable and Sustainable Energy Reviews, 148*, 111268.
44. Onipe, O. O., Jideani, A. I. O., & Beswa, D. (2015). Composition and functionality of wheat bran and its application in some cereal food products. *International Journal of Food Science and Technology, 50*(12), 2509–2518. https://doi.org/10.1111/ijfs.12935
45. Owusu, P. A., & Asumadu-Sarkodie, S. (2016). A review of renewable energy sources, sustainability issues and climate change mitigation. *Cogent Engineering, 3*(1), 1167990.
46. Oyedeji, O., et al. (2020). Understanding the impact of lignocellulosic biomass variability on the size reduction process: A review. *ACS Sustainable Chemistry & Engineering, 8*(6), 2327–2343.
47. Papageorgiou, M., & Skendi, A. (2018). Introduction to cereal processing and by-products. In *Sustainable recovery and reutilization of cereal processing by-products* (pp. 1–25). https://doi.org/10.1016/B978-0-08-102162-0.00001-0
48. Park, S. C., & Baratti, J. (1991). Comparison of ethanol production by *Zymomonas mobilis* from sugar beet substrates. *Applied Microbiology and Biotechnology, 35*, 283–291.
49. Pensri, B., et al. (2016). Potential of fermentable sugar production from Napier cv. Pakchong 1 grass residue as a substrate to produce bioethanol. *Energy Procedia, 89*, 428–436.
50. Podolska, G., Aleksandrowicz, E., & Szafrańska, A. (2020). Bread making potential of *Triticum aestivum* and *Triticum spelta* species. *Open Life Sciences, 15*(1), 30–40.
51. Prueckler, M., et al. (2014). Wheat bran-based biorefinery 1: Composition of wheat bran and strategies of functionalization. *LWT-Food Science and Technology, 56*(2), 211–221.
52. Raspor, P., & Goranovič, D. (2008). Biotechnological applications of acetic acid bacteria. *Critical Reviews in Biotechnology, 28*(2), 101–124.
53. Rastogi, M., & Shrivastava, S. (2017). Recent advances in second generation bioethanol production: An insight to pretreatment, saccharification and fermentation processes. *Renewable and Sustainable Energy Reviews, 80*, 330–340.

54. Rony, Z. I., et al. (2023). Alternative fuels to reduce greenhouse gas emissions from marine transport and promote UN sustainable development goals. *Fuel, 338*, 127220.
55. Saini, J. K., Saini, R., & Tewari, L. (2015). Lignocellulosic agriculture wastes as biomass feedstocks for second-generation bioethanol production: Concepts and recent developments. *3 Biotech, 5*, 337–353.
56. Seal, C. J., et al. (2021). Health benefits of whole grain: Effects on dietary carbohydrate quality, the gut microbiome, and consequences of processing. *Comprehensive Reviews in Food Science and Food Safety, 20*(3), 2742–2768.
57. Sebayang, A. H., et al. (2016). A perspective on bioethanol production from biomass as alternative fuel for spark ignition engine. *RSC Advances, 6*(18), 14964–14992.
58. Selvakumar, P., et al. (2022). Optimization of binary acids pretreatment of corncob biomass for enhanced recovery of cellulose to produce bioethanol. *Fuel, 321*, 124060. https://doi.org/10.1016/j.fuel.2022.124060
59. Shrestha, S., et al. (2021). Different facets of lignocellulosic biomass including pectin and its perspectives. *Waste and Biomass Valorization, 12*, 4805–4823.
60. Siddiqui, A. J., et al. (2022). Enzymes in food fermentations. In *African fermented food products-new trends* (pp. 101–133). Springer.
61. Sitotaw, Y. W., et al. (2021). Ball milling as an important pretreatment technique in lignocellulose biorefineries: a review (pp. 1–24). In *Biomass conversion and biorefinery*.
62. Skendi, A., et al. (2020). Advances on the valorisation and functionalization of by-products and wastes from cereal-based processing industry. *Foods, 9*(9), 1243.
63. Stanley, J. T., et al. (2022). Potential pre-treatment of lignocellulosic biomass for the enhancement of biomethane production through anaerobic digestion—A review. *Fuel, 318*, 123593.
64. Stevenson, L., et al. (2012). Wheat bran: Its composition and benefits to health, a European perspective. *International Journal of Food Sciences and Nutrition, 63*(8), 1001–1013. https://doi.org/10.3109/09637486.2012.687366
65. Sui, W., et al. (2018). Effect of wheat bran modification by steam explosion on structural characteristics and rheological properties of wheat flour dough. *Food Hydrocolloids, 84*, 571–580.
66. Tabatabaei, M., et al. (2015). Renewable energy and alternative fuel technologies. In *BioMed research international*.
67. Tan, J. P., et al. (2016). Use of corn steep liquor as an economical nitrogen source for biosuccinic acid production by *Actinobacillus succinogenes*. In *IOP conference series: earth and environmental science* (p. 12058). IOP Publishing.
68. Tegegne, B., Belay, A., & Gashaw, T. (2020). Nutritional potential and mineral profiling of selected rice variety available in Ethiopia. *Chemistry International, 6*(1), 21–29.
69. Tse, T. J., Wiens, D. J., & Reaney, M. J. T. (2021). Production of bioethanol—A review of factors affecting ethanol yield. *Fermentation, 7*(4), 268. https://doi.org/10.3390/fermentation7040268
70. Wang, J., et al. (2018). Effect of climate change on the yield of cereal crops: A review. *Climate, 6*(2), 41.
71. Watt, S., et al. (2010). Analysis of a model for ethanol production through continuous fermentation: Ethanol productivity. *International Journal of Chemical Reactor Engineering, 8*(1).
72. Zabed, H., et al. (2014). Bioethanol production from fermentable sugar juice. *The Scientific World Journal*.
73. Zabed, H., et al. (2017). Bioethanol production from renewable sources: Current perspectives and technological progress. *Renewable and Sustainable Energy Reviews, 71*, 475–501.
74. Zabed, H. M., et al. (2019). Recent advances in biological pretreatment of microalgae and lignocellulosic biomass for biofuel production. *Renewable and Sustainable Energy Reviews, 105*, 105–128.
75. Zabochnicka-Świątek, M., & Sławik, L. (2010). Bioethanol-production and utilization. *Archivum Combustionis, 30*(3), 237–246.

76. Zhang, H., Han, L., & Dong, H. (2021). An insight to pretreatment, enzyme adsorption and enzymatic hydrolysis of lignocellulosic biomass: Experimental and modeling studies. *Renewable and Sustainable Energy Reviews, 140*, 110758. https://doi.org/10.1016/j.rser.2021.110758
77. Zhang, R., et al. (2021). Comprehensive utilization of corn starch processing by-products: A review. *Grain & Oil Science and Technology, 4*(3), 89–107.
78. Zhao, H.-M., Guo, X.-N., & Zhu, K.-X. (2017). Impact of solid state fermentation on nutritional, physical and flavor properties of wheat bran. *Food Chemistry, 217*, 28–36.
79. Zhou, Z., et al. (2018). Lignocellulosic biomass to biofuels and biochemicals: A comprehensive review with a focus on ethanol organosolv pretreatment technology. *Biotechnology and Bioengineering, 115*(11), 2683–2702.
80. Zhuang, X., et al. (2016). Liquid hot water pretreatment of lignocellulosic biomass for bioethanol production accompanying with high valuable products. *Bioresource Technology, 199*, 68–75.

Chapter 17
Waste Biomass Supply Chain for Sustainable Bioenergy Production

C. Nirmala, M. Sridevi, P. Loganathan, Mani Jayakumar, and Gurunathan Baskar

Abstract The advancement of ecologically benign, renewable, and sustainable alternative energy sources is vital for the current global economic theories. The largest source of renewable energy, bioenergy podia advance the economy and the circular economy by serving as a strategic component. To fulfill the expanding demands of humanity, biomass materials such as agricultural products, alcohol fuels industrial waste, landfill gas, solid waste, and wood have attracted interest for use in the synthesis of biofuels. The energy generated from biomass is less harmful to the environment, inexpensive, readily accessible locally in great quantities, and offers job opportunities for employees in diverse global boundaries. Sustainability, the decline

---

C. Nirmala
Department of Biotechnology, Paavai Engineering College, Paavai Institutions, Namakkal, Tamil Nadu, India

M. Sridevi
Department of Biotechnology, Vinayaka Missions Kirupananda Variyar Engineering College, Vinayaka Missions Research Foundation (Deemed to be University), Salem, Tamil Nadu, India

P. Loganathan
Department of Electrical and Electronics Engineering, Vinayaka Missions Kirupananda Variyar Engineering College, Vinayaka Missions Research Foundation (Deemed to be University), Salem, Tamil Nadu, India

M. Jayakumar (✉)
Department of Chemical Engineering, Haramaya Institute of Technology, Haramaya University, Dire Dawa, Ethiopia
e-mail: drjayakumarmani@haramaya.edu.et

Department of Biotechnology, Faculty of Engineering, Karpagam Academy of Higher Education, Coimbatore 641 021, Tamilnadu, India

Centre for Natural Products and Functional Foods, Karpagam Academy of Higher Education, Coimbatore 641 021, Tamilnadu, India

G. Baskar
Department of Biotechnology, St. Joseph's College of Engineering, Chennai 600119, India

School of Engineering, Lebanese American University, Byblos 1102 2801, Lebanon

G. Baskar et al. (eds.), *Circular Bioeconomy Perspectives in Sustainable Bioenergy Production*, Energy, Environment, and Sustainability,
https://doi.org/10.1007/978-981-97-2523-6_17

in emissions of greenhouse gases and an upsurge in manufacturing jobs are the benefits of using renewable sources to manufacture energy, chemicals, and fuels. These elements promote the development of local economic and social systems. The current chapter focuses on the availability of numerous waste biomass materials and their characteristics as viable renewable energy sources. The analysis also includes the current state of the world's energy supply, the need for sustainable alternative energy sources, established and developing technologies for biomass conversion, and the circular economy of waste biomass management. From the review, it is possible to infer that utilizing cost-effective sources for implementation in energy platforms and biofuels will increase the competitiveness of bio-based companies and boost the economics of existing value chains. Understanding the characteristics of biomass materials enables us to determine the best approach to use them for energy in the future and cut down on our reliance on fossil fuels.

Keywords Biomass · Waste · Bioenergy · Pretreatment · Conversion · Technologies

17.1 Introduction

Around the world, there is a continually growing demand for bioenergy (biofuels, heat, and power), which makes up about one-tenth of the total primary energy supply [1]. At the global scale, there is a steady increase in the population, and by 2050, it is foreseen that it reaches a maximum of 9.7 billion. The sinewy rise in population leads to rise in the global energy demand with a yearly surge of 1.5%. Currently, the fuels from fossil origin provide exceptionally 84% of the world's energy needs, which has an adverse milieu effect [2]. The majority of nations promote the use of renewable energy such as hydro, wind, solar, and bioenergy to meet energy requirement with lower greenhouse gas discharge. Such issues consequently prompted the prudent use of accessible earthy and waste materials for bioenergy production [3]. The current global energy system depends extensively on energy derived from biomass, such as living beings, wood, food crops like maize, energy crops, and agricultural residues, tangential residuals such as manure, wastes from food process, forests, yards, and farms [4, 5].

Utilizing diverse forms of biomass, biorefinery operate in an environmentally responsible manner and create a variety of biofuels, chemicals, and bioproducts [6]. The general cycle of biomass energy is shown in Fig. 17.1. Based on the feedstocks utilized in their manufacture, the biofuels are categorized as first, second, and third generation. The biofuels at initial phases (first generation) are produced using the plants edible sections as a raw material. The manufacturing of second phase biofuels uses lignocellulosic resources. Microalgae are utilized as a feedstock to manufacture the third phase biofuels. First-generation biofuel manufacture is well-developed and known technology. However, because of the economic and environmental constraints, their manufacture has come under scrutiny. Plantations with a high rate of biomass

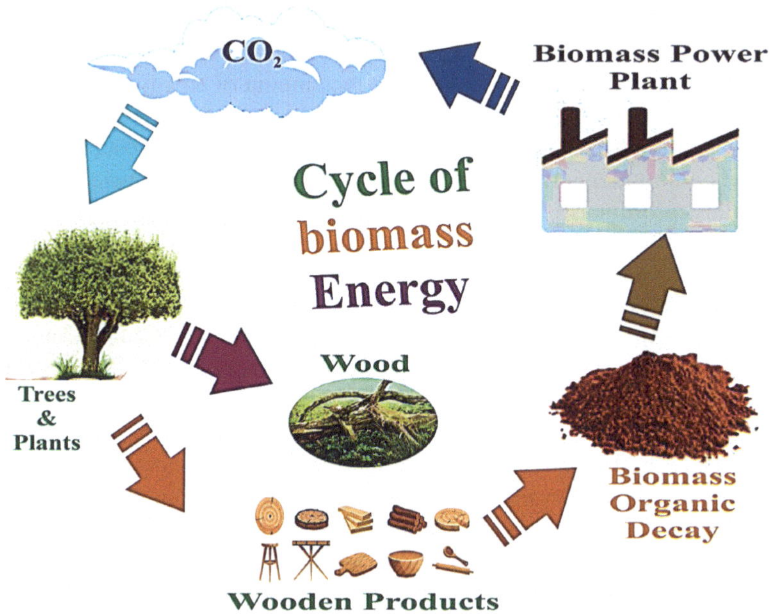

Fig. 17.1 Cycle of biomass energy

growth on non-arable land use agro-industrial waste products such as group 1 terrestrial biomass feedstock's (sugarcane, corn grain, oil seed, soy bean, etc.,) and group 2 feedstocks that are neither starchy, edible, or food-related like cellulosic biomass have less impact on the availability of food and can significantly reduce carbon emissions [7].

In order to attain environmental sustainability for the production of bioenergy, the biomass supply chain would need to be thoughtfully and sensibly designed in order to accommodate the still-evolving but promising idea of lignocellulosic and algal biorefineries. Although there is enough biomass potential in the world to meet the expanding demand, the distribution of different biomass resources is uneven. The potential annual supply of biomass on a global scale is anticipated between 97 and 147 oxyjoules (EJ) by 2030 that accounts for about 40% agricultural trash and 37–66 EJ of debris. The potential supply still available for various vegetation products varies, including 24–43 EJ of forest residues, and 33–39 EJ of energy crops [8].

17.1.1 Present Global Bioenergy Framework

The biomass supply chain encompasses multiple stages in the bioenergy production process with numerous alternate approaches [9]. The logistics of biomass, the generation of biomass as a feedstock for bioenergy, the transition of biomass to energy, and the distribution of bioenergy or its carriers for use are all different parts of the biomass supply chain. Rational design and comprehensive assimilation of various biomass supply chain activities are likely to increase the amount of energy returned, ameliorate the greenhouse gas equilibrium, and decrease the water footprint of the bioenergy producing facility [10]. Current biomass policy, which is for the most part persuaded by divided commitments to overarching segment goals, with climate alter and the vitality industry ordinarily being the driving ones, falls flat to completely capture the benefits of biomass esteem chains inside their territorial or national setting [11, 12]. Suboptimal interventions as a result of this process have prevented the market from adopting solutions that are well-unified, efficiently used resources, and domestically sourced with regional differences. Modern, sustainable biomass technologies are expected to play a significant part in efforts to bipartite the amount of renewable energies, according to REmap 2030, the worldwide strategy created by the International Renewable Energy Agency (IRENA). Supply and demand forecasts for bioenergy globally through 2030 looks at the potentials for many technologies and fields, as well as the renewable energy resource to be scaled up. By 2030, biomass use may double from its current level and account for 108 EJ, or 60% of the total net consumption of renewable energy and 20% of all original energy supply, if all technology options envisioned in the REmap analysis are implemented [13].

Currently, two-thirds (35 EJ) of biomass consumption are accounted by conventional space heating and cooking techniques like burning logs [14]. This would be replaced by contemporary biomass which would include significantly higher shares for power and transportation applications. Up to 41 EJ of heat might be used by constructions and commercial enterprise, with only 6 EJ coming from conventional, less eco-friendly uses, accounting for 36 EJ, or thirty percent of all biomass used in 2030, 36 EJ for power and district heating, and 31 EJ (nearly 29%) for transportation. Policies pertaining to biofuels are crucial for the growth of the energy sector, particularly in emerging nations. A number of countries have also conducted important policy-related actions in the last few decades to support the adoption and development of bioenergy. Policies affecting a variety of industries, including agriculture, research, industry, and commerce, have a substantial impact on how profitable is the bioenergy and biofuel production [15]. Given the variety of tools available for implementing policy such as subsidies, taxes, etc. and how it is employed, it is challenging to recognize germane approaches and measure their exact impacts [16].

The biomass policy framework combines modeling, analytical, and participatory methods and requires substantial involvement from all parties engaged at every level of the value chain [17]. In order to build and maintain market acceptance, it is necessary to evaluate difficulties and gaps in the availability of feedstock, resource competency, and legitimacy across the biomass supply chain. Their participation diligently pertains in the managerial process and makes it easier for particular actions and planned interventions to be communicated [18].

Creating a reliable and effective supply network governance system for the production of resilient bioenergy and biofuels is one of the primary problems. Although there are several research projects on the production of bioenergy and biofuels, there are few commercial cases for their adoption in developing nations like

India. Within the current investigation, the preferences for pretreatment of lignocellulosic biomass (LB) for the creation of bioenergy are assessed. The technological viability and challenges carried out for improving the biorefinery processing from biomass and biowaste supply chain are discussed.

17.2 Sources and Classification of Waste Biomass

Waste biomasses, which incorporate rural waste, wood squander, food squander, metropolitan strong waste, sewage slop squander, lignocellulosic particulates includes rural and backwoods buildups, seaweeds, crops residues, fertilizer, food misspend, and natural part of metropolitan strong waste had an enormous flair to give energy and worth add upped items by means of various transformation innovations (Fig. 17.2). These squanders are inescapable on the grounds that they are produced straightforwardly or by implication by giving the human culture food, energy, and water. One of the foremost imperative viewpoints of accomplishing an economical future is viable administration of these unavoidable biomass squanders. Numerous non-edible oil seed crops are utilized for bioenergy production [19, 20]. Carbonization, catalytic forms, gasification, liquefaction, pyrolysis, transesterification, and torrefaction are cases of thermochemical forms. On the other hand, biomethanation, fermentation, and enzymatic forms are common natural advances for changing over biomass [20]. Chemical and enzymatic methods are used for hydrolysis of biomass residues to release fermentable sugars, which are then converted by microorganisms into of high-value goods such as biofuels and biochemicals. To make biofuels and biochemical structural blocks, the feedstocks go through a heated breakdown of the natural components, paying little attention to microbial fermentable materials (i.e., saccharides) and non-fermentable portions (i.e., lignin) [21]. In comparison to natural change processes, thermochemical strategies react quicker due to the joining of tall temperatures, weights, and impetuses [22].

Fig. 17.2 Categorization of biogenic materials and waste biomass

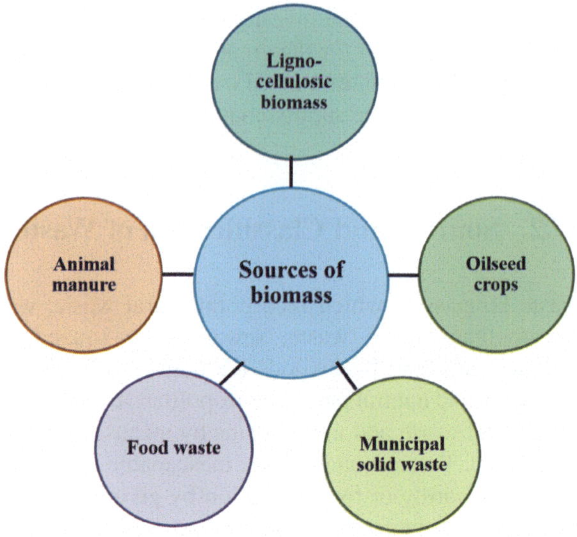

The synthesis and features of biomass find out the interaction boundaries, response rate, yields, and character of the transformation items in general. The primary features of biomass are not completely resolved by general along with extreme arrangement [23].

17.2.1 Lignocellulosic Biomass

A feedstock that exists as carbon–neutral, viable, and economical is LB, which typically contains lignin (10.24 wt%), hemicellulose (20.40 wt%), and cellulose (40.60 wt%) [24]. Crop residues from agriculture, biomass from forestry, and energy crops are all examples of LB sources. Rural squanders are the byproducts of the collecting and preparing of rural vegetative crops. Crop wastes are typically fall into two distinct categories: field residues and residues from agro-food processing. These agro-deposits do not contain food; thus, they do not compete with the food supply or ripe agricultural land. Despite its favorable benefits on society, the environment, and profitability in a range of valuable products for the domestic, industrial, building, and power industries, the bulk of forestry biomass is still poorly used. The availability and expense of some of the barriers that restrict the use of agricultural and forestry biomass for the production of bioenergy are correlated with their seasonal, regional, and climatic fluctuations. Corn stover, sugarcane bagasse, rice straw, and wheat straw are the most common crop wastes, accounting for 731.3 million tons, 354.3 million tons, 203.6 million tons, and 180.7 million tons, respectively [25]. In the agricultural waste (AW) category, these agricultural residues are the most common LBs. As a rule, scientists would utilize the two terms "horticultural buildup" and "farming waste" reciprocally [26–28].

Energy crops have prompted a rise in research interest globally due to their diversity, quick development, high rate of production, ability to fix CO_2 during photosynthesis, cost-effectiveness, and capability to thrive on unfavorable soils [29]. As a result, their cultivation may have a significant impact on the biofuel industry's ability to meet the demand for clean energy. Most common energy crops incorporate miscanthus, elephant grass, crossover poplar, switch grass, and timothy grass.

17.2.2 Oilseed Crops

Plant derived oils have historically been a major agricultural product. Oil can be found in the form of triglycerides, triacylglycerol, lipids, and fatty acids in a number of plant species [30]. As stores of energy and carbon for the improvement of seedlings, these elements are stored in cells and plant seeds. Triacylglycerol's structural resemblance to long-chain hydrocarbons serves as the basis for a practical substitute for goods that use hydrocarbons as their primary ingredient. Additionally, due to the physicochemical characteristics of fatty acids, non-edible plant derived oils are utilized to create biodiesel and are essential constituents of paints, varnishes, lubricants, and inks [31]. Plant oils are in high demand across a variety of sectors, including agriculture, nutraceuticals, cosmetics, pharmaceuticals, food, and biorefining. Mustard oil, olive oil, almond oil, corn oil, canola oil, coconut oil, grapeseed oil, sunflower oil, soybean oil, and vegetable oil are just a few of the plant derived oils used in cooking and food preparation [32]. The following plants are used to make non-edible plant derived oils: microalgae, silk cotton tree, rubber seed, *Azadirachta indica* (Neem), *Ailanthus altissima* (heaven tree), *Madhuca longifolia* (Mahua), *Jatropha*, *Pongamia pinnata* (Karanja), *Linum usitatissimum* (flax), *Ricinus communis* (Castor), *Sapindus mukorossi* (Soapnut), Toona sinensis (Juss), *Vernicia fordii* (Tung), etc. [33]. Different oilseed crops exhibit noticeable variations in the oil percentage. The oilseed plants might potentially be genetically modified to produce oils with a higher concentration, it can be deduced [34]. Oil seeds may be used to produce biofuels and biodiesel, which is a great alternative to petroleum-based energy [35].

17.2.3 Municipal Solid Waste

Municipal solid waste (MSW) refers to waste products collected from homes and municipalities (such as rubbish, recyclable and non-recyclable leftovers). MSW might also contain industrial, commercial, and institutional (ICI) garbage, which includes leftovers from enterprises, big industries, hospitals, and educational institutions. It ought to be referenced that the organization of MSW and ICI shifts relying upon the beginning, creation designs, family pay and geological area [36]. MSW includes items such as cooking and home equipment, rubber and plastic components, metallic substances, cardboard and paper scraps, inert materials, digital trash, and so on [37]. There are also biodegradable and non-biodegradable fractions in MSW. It is important to note that approximately 15% of municipal waste is recycled, while the remaining waste is dumped in open areas or landfills [38].

The key drivers of the global increase in MSW creation are population growth, gentrification, and industry. By 2025, it is anticipated that the amount of MSW produced worldwide will exceed 2 billion tons annually [39]. It is projected that in a few emerging nations and different areas of the planet, the MSW age could reach or surpass that of created countries without legitimate guidelines and arrangements for landfilling and squander reusing offices [40]. Food preferences, consumption patterns, behavior among consumers, and living conditions in both rural and urban locations are all changing as the rate of MSW generation rises.

MSW age makes serious ecological contamination when unmanaged. Besides, its change into esteem added items could give an answer for the difficulties of energy lack and maintainable waste administration. MSW can be discarded, repurposed, or hailed into fossil fuels in a variety of ways, including garbage dumps, soil fertilization, anaerobic assimilation, pyrolysis, and gasification [37]. The cremation of 1 ton of MSW could emanate 1.3 lots of CO2 comparable outflows, which is like how much CO_2 discharges from oil based power plants [41]. In addition, burning MSW releases a significant amount of pollutants into the atmosphere, including fly ash and particulate matter, refuse management strategy becoming unsustainable. However, it has been demonstrated that MSW fly and bottom ash contains heavy metals that pose a threat to ecosystems [42]

For the burial of non-recyclable waste, many municipalities worldwide prefer to dispose of MSW in landfills. Despite its potential, MSW disposal in landfills is fraught with difficulties, including groundwater contamination caused by methane gas emissions and landfill leachate. MSW landfill leachate is both acute and chronically toxic, and it frequently enters groundwater biomagnification. Besides, leachate could likewise sully the progression of water streams [43]. The generation of energy from MSW contributes to the reduction of pollution and has the potential to support a nation's economic development by enhancing energy security and the management of waste.

17.2.4 Food Waste

The term "food waste" refers to the organic and biodegradable garbage that is created from a range of sources, such as food processing facilities, eateries, kitchens, and residences [44]. This material is a part of the "organic fraction" of MSW. Due to improper handling and usage of food, a lot of food waste is sent every year. Food waste can also be caused by overproduction, contamination of produce by bacteria, pests, and insects, excessive shopping, and postponed eating [45]. Every year, the food retail network loses over 1.3 billion tons of food, including handled meat, dairy products, organic foods, and vegetables [46].

Management of waste by numerous strategies for food waste includes landfilling and burning. Food waste disposal can result in methane emissions, a more potent GHG than CO_2 [47]. The dry biomass is best disposed of by burning or cremation, on the other side. Therefore, high-dampness food waste may result in the incinerator using more energy, which would result in high operating costs. An effective substitute for methanogenic bacteria producing biogas (or biomethane) through biomethanation is anaerobic digestion of food waste [48].

Food waste comprises of cellulose, carbohydrates, proteins, lipids, fatty acids, and organic acids. Also, the carbs available in food waste could go through hydrolysis to deliver monosaccharides and oligosaccharides reasonable for organic transformation [46]. Food waste can also be a valuable resource for the fermentation of bioethanol [49], biobutanol [50], and biohydrogen [51] due to its organic composition. As biohydrogen is acquiring prominence, food waste can end up being an eco-accommodating and financially savvy feedstock for its creation through photograph/dim maturation and gasification. By pyrolysis [52] and hydrothermal gasification [53], discarded waste food can also be transformed thermochemical into bio-oil, biochar, and hydrogen-rich syngas. Activated carbon and biochar production from waste food has also demonstrated promising results [54]. The boundaries, for example, the arrangement of food squander, pretreatment techniques, and handling boundaries impact the development of biofuels.

17.2.5 Cattle Manure

Cattle excrement refers to the metabolic and waste byproducts of poultry and animal farming. Fertilizer is an important ingredient that contains natural matter and nutrients that are essential for crop development [55]. Furthermore, animal faces may include a number of diseases that pose environmental concerns. Waste feed, waste feed water, and metabolic waste or faces are all common components of animal manure. In spite of being a significant wellspring of horticultural supplements, domesticated animals compost can likewise add to the outflow of GHGs, for example, CH4 by microbial decay [56]. It is important to note that animal manure emissions account for 10% of all emissions from agricultural production [57]. The United Nation defined AW as waste from various agricultural operations, such as manure and other wastes from farms, poultry houses, and slaughterhouses; collect waste; compost overflow from fields; pesticides that enter the water, air, or soils; and fields were drained of salt and silt. The aforementioned definition includes waste streams produced by all modern agricultural processes, including animal and animal processing waste. To put it plainly, AW contains four important parts: animal waste, crop waste, hazardous waste (like pesticides), and processing waste [58].

Composting and vermicomposting are two other methods of traditional manure treatment [57]. These medicines are extremely normal in agricultural nations on account of their straightforwardness and cost-adequacy. Composting also ensures that plants will have access to nutrients. Clay soils also get more aeration and water infiltration from composting. Creature fertilizer could be valorized by a few techniques, for example, anaerobic processing [59], dull maturation [60], maturation [61], pyrolysis [62], aqueous liquefaction [63], aqueous gasification and torrefaction [64] to deliver bio-oil, bioethanol, biomethane, bio-raw petroleum, biohydrogen, syngas and roasted biomass, separately. After the anaerobic digestion of compost, the digestate left over might be utilized as a raw material for the production of biochar, bio-oil, and syngas by pyrolysis, liquefaction, and gasification, individually.

Cattle manure has proven tremendous promise for the generation of biochemicals and biofuels through biological conversion and thermochemical practices. Horse dung has been shown to be an effective fuel for hydrothermal gasification-based hydrogen generation by [45]. In a different research, ethanol plant wastewater and chicken manure were digested together anaerobically to produce feedstock [65]. Catalytic hydrothermal gasification was used to produce hydrogen using a mixture of poultry manure and eucalyptus wood [66]. The chemical composition of different waste biomass was illustrated in Table 17.1.

Table 17.1 Chemical composition of waste biomass

S. No.	Biomass sources	(%) Cellulose	(%) Hemicellulose	(%) Lignin	References
1	Live-oak-leaves	19.2	10.9	23.7	[67]
2	Wire-grass	34.8	22.8	18.7	[67]
3	Corn-stover	37.5	22.4	17.6	[68]
4	Paddy straw	32.0	24.0	13.0	[68]
5	Food waste (household)	18.30	7.55	2.16	[69]
6	Dairy cow manure	17.1	18.1	5.17	[70]
7	Hybrid poplar	43.8	14.7	25.7	[71]
8	Whole tree	51.2	23.4	25.4	[72]
9	Pine	42	21	30	[73]
10	Swine manure	12.2	34.0	5.38	[70]
11	Beech wood	44.2	33.5	21.8	[74]
12	Water hyacinth	24.5	34.1	8.6	[75]
13	Napier grass	47	31	22	[76]

According to [77], cellulose, a polymer of glucose ($C_6H_{10}O_5$) linked by 1,4-glycosidic bonds, is one of the most abundant organic compound present on earth. The structure of cellulose molecule is crystalline and linear, with a couple of nebulous zones and completely breaks down at 300–400 °C [78]. Hemicellulose, in contrast to cellulose, has a structure that is composed of several monomers like mannose, xylose, galactose, glucose, and others. ranging from 500 to 3000° of polymerization and highly branched [79]. Coniferyl alcohol (G), sinapyl alcohol (S), and P-coumaryl alcohol (H) are the three aromatic subunits (monolignols) that make up lignin, which may be a complex cross-linked polymer of arbitrary structure in differentiate to hemicellulose and cellulose. The fundamental distinction among the three is the quantity of methoxy gatherings.

17.3 Waste Biomass Pretreatment

Agriculture wastes and other biomass can be turned into valuable goods that are economical, sustainable, and good for the environment [80]. One of the crucial components of integrated conversion processes of biorefineries is the pretreatment, a standalone procedure that comes before a specific process. Figure 17.3 illustrates the overall process of waste biomass pretreatment for bioenergy conversion.

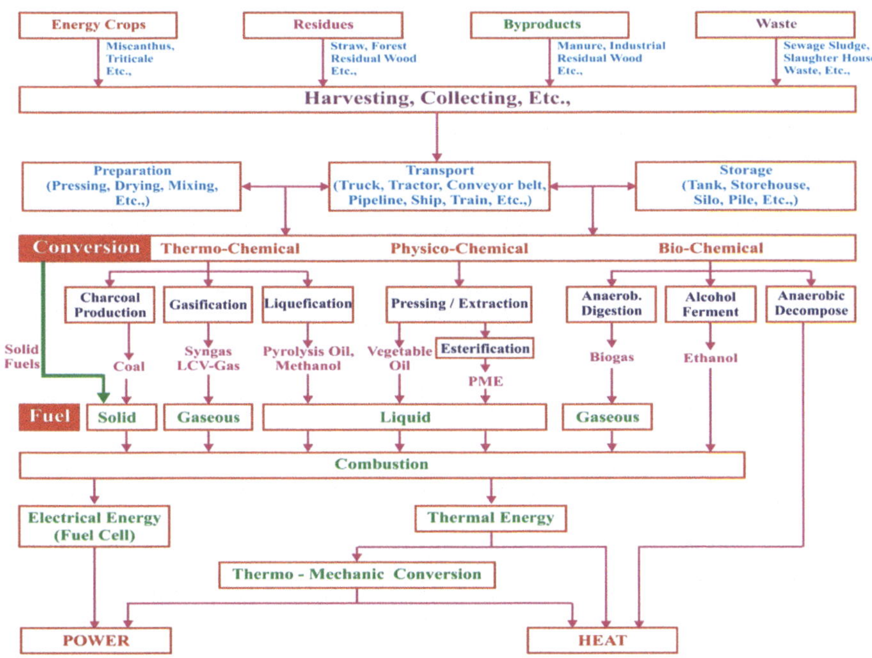

Fig. 17.3 Overall process of waste biomass pretreatment for bioenergy conversion

Effectiveness, technical, profitable, and environmental viability of pretreatments are the critical factors that make the primordial biomass structure more simple and accessible to the enzymes and reagents needed to create high added-value products thus increase the biodegradability and digestibility of agricultural LB. Significant lignocellulosic material size reduction under mesophilic conditions also boosts the production of volatile fatty acids during anaerobic digestion. In addition, structural changes such as a decrease in cellulose's crystallinity and rise in surface area result from the shrinking of lignocellulosic material. The amount of energy needed to shrink the size of lignocellulosic contents varies on the biomass type and humidity content [81, 82]. The study emphasizes the widely used pretreatment techniques, including physical, chemical, and biological techniques.

17.3.1 Physical Pretreatment

Physical pretreatment processes enlarge the surface area and decrease the particle size of waste biomass that results in improved production yield, reduced cellulose structure, improved transition effectiveness, and higher microbial and enzyme accessibility during processing. The common physical pretreatment methods are mechanical, ultrasonic, microwave, and thermal pretreatment technique.

Cutting, shearing, milling, agitation, and chipping are some of the other mechanical pretreatment techniques that lowers the amount of lignin, the level of crosslinking and the dimension of the particulates thus enables the easy degradation of cellulose and hemicellulose [83]. Comparatively mechanical pretreatment is simple and has no negative impact on the environment. In general, mechanical pretreatment also includes the use of vibrating, hammering, two-rolling, milling, chipping, and grinding procedures.

The movement of a screw inside a small container, pretreatment of lignocellulose materials is accomplished through a combination of heat and mechanical activities known as Extrusion pretreatment. Heat can be applied to lignocellulosic substances utilizing irradiation pretreatment techniques such an electron beam, microwave, ultrasonic, and gamma rays, which dissolve hemicellulose, reduce cellulose crystallinity, and depolymerize lignin [84, 85].

17.3.2 Chemical Pretreatment

To debilitate and break up the strong lignocellulose gem structure, increment the surface region, and upgrade the biodegradability of farm land residues, special chemicals are utilized in chemical pretreatments. To reduce the handling, corrosion, and risk of toxicity, diluted version of acids such as HCl, H_2SO_4, H_3PO_4, HNO_3, Na_2CO_3, CH_3COOH, CH_2O_2, $C_4H_4O_4$, caustic agents such as NaOH, NH_4OH, KOH, $NH_3{\cdot}H_2O$, $Ca(OH)_2$ are used to break the glycosidic bonds, disrupt the cellulose and hemicellulose in biomass, and consequently promote the release of fermentable sugar [86]. Oxidative pretreatment is made easier by the hemicellulose and lignin in the biomass being broken down by ozone gas and hydrogen peroxide (H_2O_2). The chemical breakdown of H_2O_2 into –OH and O_2 helps to destroy the lignin's structure while preventing the release of any inhibiting byproducts [87].

To ensure efficient transition of cellulose, lignocellulosic material is first pretreated with organic solvents such as acetone, ethanol, methanol, ethylene glycol, and tetrahydrofuranol to disrupt the internal linkages of hemicelluloses and lignin. Occasionally, catalysts are utilized to speed up the chemical process, such as HCl, H_2SO_4, or bases like NaOH, NH_3, and $CaCO_3$ [88].

The exothermic wet oxidation pretreatment is carried out at lower temperatures. This process involves mixing water with lignocellulosic material before adding an oxidizing substance, such as air, oxygen, or hydrogen peroxide [89]. Following, the feedstock is heated to a temperature of 125–300 °C and a pressure of 0.5–20 MPa. The three key variables influencing the pretreatment methods are pressure, temperature, and residence time. The inclusion of oxidizers in pretreatment results in the synthesis of free radicals and affects the reaction rate. The rate of reaction is proportional to the oxygen levels, and the cost of pretreatment might increase depending on the oxygen purity used. Alkyl aryl ether bond cleavage, electrophilic replacements, and oxidative cleavage of aromatic nuclei are the main processes [90].

The lignocellulosic material is oxidized using a sophisticated wet explosion pretreatment, which is followed by physical disruption and depends on temperature, pressure, and residence duration [91]. The pretreatment takes place at 140 and 220 °C temperature and a 0 and 3.5 MPa pressure. In ozonolysis, lignocellulosic material has been pretreated with ozone, a potent oxidizing agent that influences the lignin fraction during the process. The cellulose remained unaltered after the pretreatment; although it was observed that the hemicellulose was partly changed. The ozone concentrations, biomass particle size, and water content of the material also affect the pretreatment process [92].

17.3.3 Biological Pretreatment

Utilizing enzymes, aerobic bacteria, and fungi to degrade lignin and facilitate the conversion of lignocellulose into usable products is known as biological pretreatment [93]. This technique is environmentally friendly, creates little to no inhibitor, uses less energy, and discharges no harmful substances into the environment [94]. Fungi belonging to white-rot fungus, soft-rot, and brown-rot classes are known as fungal-based biological pretreatment approach to break down lignin. For the biodegradation of lignin, white-rot fungi from the class *Basidiomycetes Phanerochaete chrysosporium* are frequently utilized. Termites are another organism utilized to break down lignocellulose-containing materials. According to reports, *Termes hospes, Nasutitermes ephratae, and Microcerotermes parvus* release lignocellulose-degrading enzymes that help wheat straw break down before being digested anaerobically.

Organisms from soil, broiler, wastewater, etc., are utilized as microbial consortium pretreatment specialists for farming wastes [95].

Agricultural wastes and other LB can be pretreated using enzymes such as cellulases and hemicelluloses. To improve their conversion efficiency for the manufacture of biobutanol, farming wastes like sugarcane bagasse, Napier grass, corn stover, cassava pulp, rice straw, etc. have undergone pretreatment with enzymes. By using bacteria that produce enzymes, bacterial pretreatment slows down the polymerization of lignocellulose [96].

17.3.4 Hybrid Pretreatment

The bioenergy researchers are currently working to create new pretreatment techniques that will effectively separate LB into streams of cellulose, hemicellulose, and lignin [97]. Numerous researchers have employed hybrid pretreatment techniques in this area, and the outcomes are encouraging. Various pretreatment techniques are combined in order to reduce pretreatment costs and increase its efficacy. Liquid hot water is processed with chemical-free in thermochemical pretreatments at higher pressures and temperatures. When pressurized water enters the biomass, it causes cellulose to hydrate, hemicellulose to be hydrolyzed, and the lignin component of the biomass to be removed. Since the pretreatment increases the lignocellulosic materials' methane output, it is frequently employed in anaerobic digestion processes [24]. Pretreatment with alkaline solution at pH 11 and 13 increased biodegradability and biogas production. Another combination that increases methane yield while reducing pretreatment energy use is physical treatment mixed with biological process. The formation of inhibitors is decreased when wet oxidation and alkali pretreatment are combined. Besides, steam explosion combined with wet oxidation overcomes the huge particle size of biomass. Steam explosion and wet oxidation are used to handle hard biomass as well as biomass with greater lignin levels. The enzymatic hydrolysis of cellulose is improved by pretreatment with the ionic liquids combined with ultrasonic and microwave pretreatment [91].

Numerous studies with combination pretreatment strategies are rather low than mono pretreatment strategies. In order to minimize pretreatment costs and improve biogas yields, it is important to further investigate the applicability of combined pretreatment practices for numerous class of biomass [98]. The overall waste biomass pretreatment practices are shown in Fig. 17.4. Merits and demerits of various pretreatment practices are listed in Table 17.2.

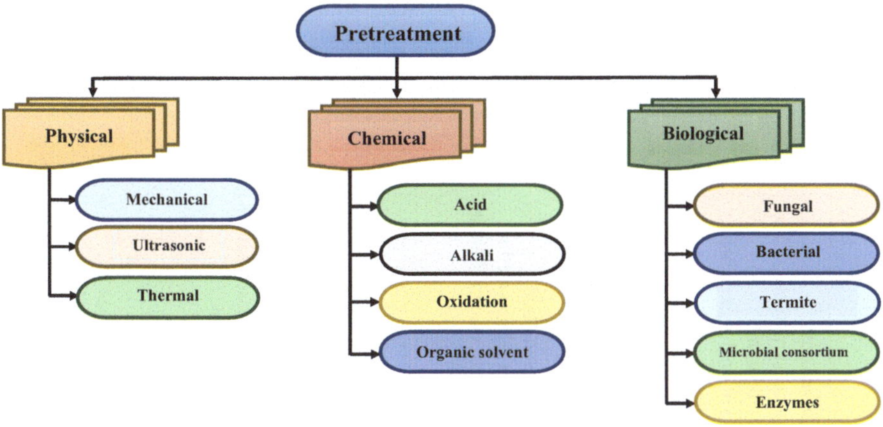

Fig. 17.4 Waste biomass pretreatment practices functions, merits, and demerits

17.4 Traditional Approaches of Waste Biomass Conversion to Bioenergy

The biological origin is one of several variables that affect the composition of biomass. The abundance of biomass as trash in many forms, including agricultural waste, forest residue, industrial waste, MSW and wood waste, has made its acquisition simpler, more affordable, and environmentally benign [99]. If biomass is not processed, it may cause a number of social and environmental issues. The socioeconomic and environmental effects of first-generation biofuels derived from edible crops on food security, land use, and biodiversity have also been challenged [100].

These concerns are allayed by the use of waste biomass for energy harvesting [39]. However, due to its intrinsic chemical and physical characteristics, without pretreatment, biomass is insufficiently valuable and less effective for direct use. Biomass has some characteristics that make it unsuitable for direct use as fuel, including high oxygen concentration, hydrophilic nature, uneven composition, moisture content, low lower grindability, calorific value, and fibrous structure [101]. Because of this, converting biomass into higher-quality biofuel and other products with added value is essential. To this end, extensive research has been done on a variety of biomass conversion processes, which are broadly split into two classes: biochemical conversion and thermochemical conversion [102]. Inherently slower and less effective than thermochemical processes, biochemical conversion techniques use bacteria and enzymes to break down biomass into smaller molecules [103]. Thermochemical conversion techniques, on the other hand, use heat to break down biomass into low-molecular compounds through a sequence of regulated physicochemical reactions to produce the required results. Although the thermochemical reactions require outside heat at least initially, they are quite simple to regulate [104]. Gasification, combustion, pyrolysis, hydrothermal liquefaction (HTL), hydrothermal carbonization (HTC), and

Table 17.2 Merits and demerits of various pretreatment practices

S. No.	Pretreatment practices	Merits	Demerits
1	Mechanical	• No inhibitors • Significant reduction in the size • Easy operation and simple • More CH_4 production • Minimized crystallinity • Enzyme digestibility induced • Higher porosity • Biomass size reduction • Low investment	• High operational cost • High energy consumption • Cannot degrade lignin • Required additional pretreatments • Low sugar yield
2	Ultrasonic	• Low energy utilization • Diminish the degree of polymerization • No formation of inhibitors • Increased hydrogen fermentation • Minimal waste production • Multiple product formation • Intensified processing	• Low selectivity • High energy consumption • Loss of energy in diluted media • Large-scale reactor design challenges
3	Thermal	• Non-toxic • Environment friendly • Highly reliable method	• Generation of toxins • Generation of inhibitors • High temperature
4	Acid	• Short residence time • High enzymatic deconstruction reaction rates • Efficient lignin removal • Decomposition of hemicellulose • Low cost	• High toxicity and corrosiveness • Formation of poisonous substances and inhibitors • Corrosion in equipment • Difficult in recycling

(continued)

Table 17.2 (continued)

S. No.	Pretreatment practices	Merits	Demerits
5	Alkali	• Low inhibitors • Increased surface area • Decrease in the crystallinity of biomass • Delignification rate is high • Utilized milder conditions • Operating temperature is less • Diminished polymerization • Low degree of cellulose crystallinity	• Environmental pollution • Residence time is long • Consumption-condensation if high energy • Redistribution of lignin • Economically non-viable
6	Oxidation	• Ecologically friendly practice • Highly selective • Can operate and room temperature • Hemicellulose and lignin elimination • The structure of cellulose is not changed	• Formation of toxic and inhibitor • High oxidant required • Expensive process • Special instrument required
7	Organic solvent	• Reaction time is short	• High corrosive • High toxicity • Environment pollution
8	Fungal	• Mild conditions • Low energy consumption • Energy efficient • No pollution	• Long reaction period • Low cellulose recovery • Low efficiency • Long residence time
9	Bacterial	• Low energy consumption • Less production of hazardous chemicals • Eco friendly	• Increased incubation period • Lengthy preparation period
10	Termite	• Mild action circumstances • Low energy consumption • Environment friendly	• Long pretreatment time

(continued)

Table 17.2 (continued)

S. No.	Pretreatment practices	Merits	Demerits
11	Microbial consortium	• Environment friendly • Dependable operability	• Needs further research and investigation
12	Enzymes	• Higher cellulose digestibility • Low cost • Rapid response time • Extremely effective • No corrosive problems • A minimal energy footprint	• Long incubation time • Require more investigations

torrefaction are the main thermochemical conversion processes. These processes produce a variety of energy outputs, including bio-oil, biochar, combustible gases, and syngas. Additionally, hydrothermal processes are best suited for biomass with high moisture content because they use subcritical or supercritical conditions of water at high pressures, whereas other thermal processes are better suited for biomass with relatively lower moisture content [105]. Advantages, limitations and challenges of traditional approaches used in waste biomass conversion to bioenergy are detailed in Table 17.3.

17.4.1 Innovative Techniques for Waste Biomass Conversion

Rapid industrialization and population growth and increased awareness in global warming prioritize emerging technologies and procedures that quicken the switch from fossil fuels to renewable energy sources. It is essential to meet future global energy demand and to safeguard and/or restoring the environment. The innovative techniques for waste biomass conversion are listed in Table. 17.4.

17.5 Bioenergy Applications

From the stoves used in homes for cooking to huge power plants are used to generate electricity [127]. New woodstove designs increase the effectiveness of the heating or cooling system and reduce the fuel usage. Biomass is used for businesses and in industry for a range of functions, including energy production, space heating, and hot water heating [128]. Many industrial environments, including lumber factory, naturally produce organic waste [129].

17.5.1 Biodiesel

Biodiesel is an effective replacement for diesel; it is a sustainable, fueled source that can improve the environment, replace petroleum thus develop job opportunities. Biodiesel is produced from a various feedstocks mixture that includes reprocessed soybean oil, cooking oil, and animal fats [1, 81]. Biodiesel initially concerned with specialized quality and engine operation conditions and may be used in diesel engines without alteration in a mixes of 5% or 20% biodiesel [130].

Table 17.3 Traditional approaches of waste biomass conversion to bioenergy, advantages, limitations, and challenges

S. No.	Name of approach	Advantages	Limitations	Challenges	References
1	Combustion	A typical combustion process has four steps with increasing temperature, i.e., drying, devolatilization, char oxidation, and gaseous product combustion which can be expressed in terms of biomass weight loss as a function of temperature	Combustion is not suitable for wet biomass wastes such as animal manure, sewage, raw digestates, etc	The combustion cannot initiate because the temperature cannot increase high enough to enter the subsequent combustion stages	[106]
2	Gasification	Compared to combustion, for electricity generation can achieve higher efficiency using biomass wastes The primary step in the production of a variety of bio-based chemicals, such as green synthesis, which can begin with syngas, is also provided by biomass gasification	This method has the intrinsic limitation in obtaining reasonable information about dispersed particles, such as particle size distribution and particle shrinkage. In contrast, the Eulerian–Lagrangian method tracks the trajectory of each particle and features the inherent information of dispersed particles, thus has emerged as a promising way to simulate dense gas–solid reacting flow	Biomass gasification is comparatively a more complex process since it involves multi-scale and multi-physics processes, and it is influenced by various process parameters. The majority of these determining factors are interdependent, and as a result, their effects may vary depending on the stage of a reaction. Consequently, it is challenging to have a pure discussion of the effects of process parameters separately	[107, 108]

(continued)

Table 17.3 (continued)

S. No.	Name of approach	Advantages	Limitations	Challenges	References
3	Pyrolysis	The merit of the pyrolysis process is that it is a zero waste process The final purpose is to achieve energy-efficient biomass wastes conversion to produce tailored products effectively (i.e., to enhance product selectivity). Due to their inherent renewable qualities, unconventional heating sources like solar-thermal-driven biomass pyrolysis are also gaining popularity	The demerit of the process includes high energy consumption due to its endothermic nature, and further research is needed before its industrial-scale implementation	The removal of char is challenging. Due to scale-up constraints, large-scale systems must be thoroughly researched	[109, 110]
4	Torrefaction	Torrefied biomass, which is composed of a modified polymeric structure compared to the raw biomass, has better fuel properties such as low moisture content, higher heating value, lower volatile content, and better grindability, making it suitable for direct usage as a fuel. Torrefaction has the advantage of producing no greenhouse gases. Instead, it fixed the carbon as densified and dehydrated biomass for further thermochemical processing	The main demerit of torrefaction is Volume of torrefied biomass is reduced only slightly	The system's drawback is the substantial quantity of heat that is applied to the product or material throughout the entire process, which causes flavor, color, and clouding degradation upon storage and shortens the product's overall shelf life	[111, 112]

(continued)

Table 17.3 (continued)

S. No.	Name of approach	Advantages	Limitations	Challenges	References
5	Hydrothermal processing	Water participates in the process as a reactant as well as a solvent, eliminating the need for the time-consuming pre-drying. Since the water is still liquid, the latent heat is preserved, increasing the thermal efficiency of the process. Furthermore, compared to most thermal processes, the hydrothermal conversion uses low temperatures. Additionally, because gaseous SO_x and NO_x easily dissolve in water, there is a relative decrease in the production of hazardous gases	It has a number of drawbacks, including expensive initial equipment costs because of the specialized machinery needed and severe corrosion that requires maintenance because of the extreme reaction conditions	The main difficulties with hydrothermal processing include a lack of fundamental thermodynamic data under extreme settings, a lack of understanding of reaction kinetics, and the impact of mass transfer on the process	[113–115]

Table 17.4 Innovative techniques for waste biomass conversion

S. No.	Innovative technology	References
1	Self-energized pine dust pyrolysis system for reduction in external electrical energy supply and to improve the yield of dry bio-oil	[116]
2	Used a rectification column as the pyrolysis reactor and a series of differently arranged condensers to model a pyrolysis system for waste plastics. The condensation and fractionation units were then simulated using ChemCad software using this model oil. The separation of near-boiling components is improved for only up to 10 theoretical stages	[117]
3	Advanced analytical methods like Fourier transform ion cyclotron resonance mass spectrometry (FT-ICR MS) and 1H and 13C nuclear magnetic resonance (NMR) spectroscopy were used to characterize tire pyrolysis oil produced in a lab-scale twin-auger reactor. This research will aid in developing the upgrading tactics	[118]
4	To recover benzene, toluene, ethylbenzene, xylene chemicals, and limonene from tire pyrolysis oil produced by an industrial-scale pyrolysis plant, a pilot-scale distillation plant was created, built, and put into operation	[119]
5	Plastic waste can be catalytically pyrolyzed to produce liquid oil and other products with added value using natural zeolite that has undergone modification. The catalytic properties of new thermal (TA) and acidic (AA) activation are improved	[120]
6	The amount of biomass production was first estimated for 20 years using the LandGEM and EViews software in order to determine the amount of electrical energy needed (in the specific study area), and the economic, environmental, and energy analysis was completed using the Homer software. The estimated payback period was 10 years, and the annual electricity production was estimated to be 11,658.265 MWh	[121]
7	Temperature, residence time, catalyst loading, and feed ratio are examples of operational parameters and are the only factors that affect the quality and kind of product forms throughout the pyrolysis process. The thermal behavior caused by the structural makeup of hemicellulose showed that it has a less impact on the generation of bio-oils during the pyrolysis process	[122]
8	Hydrothermal liquefaction (HTL) is a method of producing bio-oil from an aqueous slurry of biomass or organic materials at temperatures between 280 and 370 °C (lower than the pyrolysis process), high pressures of 10–25 MPa, and water contentment	[123]
9	Bioelectricity is mostly generated by burning lignocellulose feedstock derived from biomass sources such as farm waste, plantation forests, sawdust from sawmills, and natural forests. The use of lignocellulose from forestry and agricultural biomass to generate electricity by direct burning was reported in an Australian study on the viability of producing bioelectricity from biomass to reduce greenhouse gas emissions	[124]

(continued)

Table 17.4 (continued)

S. No.	Innovative technology	References
10	Fourth-Generation of biofuels is a general term for the application of genetic engineering, molecular biology, and transdisciplinary physicochemical techniques to optimize and boost the yield in the process of biofuel production. In order to increase biofuel output, the fourth generation of biofuels uses genetically engineered algae that accumulate large levels of lipid and carbohydrates	[125]
11	The biomass of *Graesiella emersonii* microalga serves as feedstock for biodiesel production as well as producing valuable ω-6 and carotenoids	[126]

17.5.2 Biofuels

Using conventional technologies, biofuels are produced from vegetable oils and sugars [131]. Advanced biofuels, in contrast, are produced from agricultural waste, cellulosic biomass, and woody crops, that makes it challenging to excerpt the necessary fuel [1]. Due to these significant limitations in first-generation biofuel production and threatening in food supply and biodiversity, advanced biofuel technologies are developed. Many first-generation biofuels depend on subsidies, have high production and transportation emissions, are expensive and only partially reduce greenhouse gas emissions. Advanced biofuels can resolve these issues and provide a larger share of the world's fuel supply in a cheaper, sustainable, and environmentally friendly way [132].

17.5.3 Energy for Transportation

Animal waste and plant biomass are recycled to produce fuel for transportation and energy, heat, power, fuel and steam. Biomass is also used in the production of chemicals such as detergents, bio-fertilizers, and erosion control products, as well as the food processing, animal feed, and wood products industries, which include construction and fiber products like paper and derivatives, all use [133].

17.5.4 Biogas

Biogas is a mixture of gases, also known as landfill gas (LFG) produced by the breakdown of organic matter by anaerobic digestion using anaerobic bacteria in a closed system. Biogas includes methane (CH_4), trace amounts of hydrogen sulfide (H_2S), carbon dioxide (CO_2), siloxanes and moisture [134]. It is an inexhaustible source and exerts a very small carbon footprint [135].

17.6 Circular Economy of Waste Biomass Management

A global movement that offers hope for the sustainable utilization of natural resources is the circular economy. The biomass ramifications in utility must be understood by contributors from the initial design of products up to waste governance that includes all the stages in value chain, in order to develop a circular bioeconomy [136]. A circular economy depends on resources being used to their fullest potential forever and nearly eliminating unrecoverable trash [137]. An efficient response to the combined problem of rising waste production and rising energy demand is waste-to-energy (Fig. 17.5).

However, the variety of waste poses a significant management difficulty. Because of its affordability and accessibility of raw materials, biomass waste valorization is a desirable energy alternative [138]. The following issues must be addressed: (1) Using renewable fertilizers to decouple the petrochemical industry from biomass production; (2) increasing stakeholder cooperation across value chains; (3) minimizing waste from food and agriculture; and (4) providing an abundant supply of biomass for bio-based products. Based on the final analysis utilizing linear regression and artificial neural network, new models are created to predict biomass wastes higher heating value [138]. Biomass is essential for the creation of energy and material items in a circular economy.

The bioeconomy includes primary production, industrial sectors specifically biological-oriented forestry, agricultural, aquaculture and fisheries, and economic

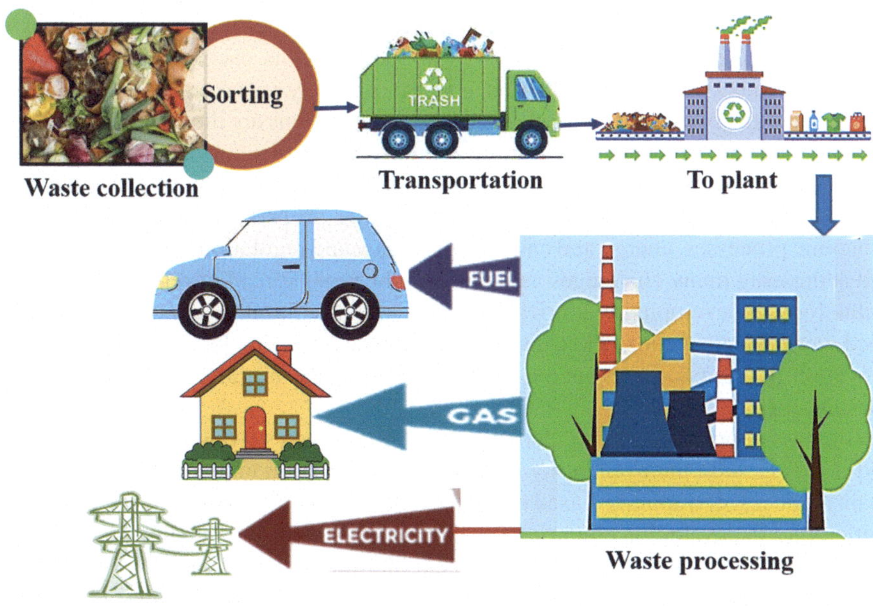

Fig. 17.5 Waste-to-energy conversion process

sectors, to create feed, food, energy, bio-based goods, and services [11]. Primary production includes carbon farming and separation and is the cost-efficient and sustainable use of natural resources management activity. It includes biodiversity, water, land, and other environmental resources organization. Industrial processing and distribution encompass a product's design and manufacturing procedure in terms of both the product's primary purpose and the possibility for waste reduction and recycling. Additionally, cascading has the prospective to upgrade the coherence of processing and utilizing waste and residue from forestry and agriculture. Food waste can be reduced by increasing circularity in packaging and product delivery, as well as by assuring recyclable materials and reducing environmental effect. Consumption includes the distribution of food and bio-based products as well as their reuse, recycling, and disposal. End-of-life refers to the point at which the waste produced from the biomass production, processing, consumption, and bioenergy production that are. For trash generated from biomass and bio-based goods to be recycled and reused, waste sorting must be improved that leads to better recycling technologies and procedures as well as the extraction of valuable compounds as processing byproducts. Furthermore, essential inputs for the generation of bioenergy include biomass and organic waste. Energy recovery, however, should only be used when lower-level options in the waste hierarchy are not viable [139].

17.7 Challenges, Futuristic Concerns, and Motivation

Energy efficiency and renewable energy are referred to as the "twin pillars" of energy policy. To manage and cut back on carbon dioxide emissions, each resource should be enhanced. The exploration, production, and use of energy are governed by numerous distinct energy policies on a global scale. These regulations are the result of a variety of sources, including trade associations, commodity companies, automobiles, and producers of wind and solar energy. The study of energy economic science has recently focused on a number of topics, including property, energy markets and economic processes, energy and environmental law and regulations, global warming, and addressing many challenges associated to the rapid spread of renewable energy technology in developing countries. The bulk of agricultural facilities in the developed world are mechanized because rural regions have been electrified. Rural electrification has considerably increased production, but it has also increased energy consumption.

Biomass supply chains are challenged in six key ways. Among the difficulties and issues include the availability of supplies, digitization and automation, supply chain alliances, community-based energy generation, social opportunities, and laws. Agriculture in developing countries generates a lot of biomass, which has an impact on the economic stability. The geographical and unpredictable availability, where some biomass can only be harvested during specific seasons in certain places, is a key factor influencing changes in the supply of biomass. The limited availability of biomass will have an adverse effect on the project's financial performance. On the other hand, developing countries continue to use biomass for low-value purposes, which is frequently a cheap replacement for investing a sizable sum of money in establishing a local biomass sector. Unfortunately, if substantial volumes of biomass were locked in these low-value activities, its availability would become extremely rarer.

Effects of climate change on how some crops are harvested. Changes in rainfall, average temperature, and humidity, for example, might affect crop dispersion and output as a result of climate change. As a result, the supply chain for biomass will become more vulnerable and its availability of biomass will be impacted. In fact, supply uncertainty is one of the main issues hindering the growth of the bioeconomy [140]. In order to lessen the vulnerability of the biomass supply chain, future crops are expected to breed climate-resilient crops that are resilient to heat stress, biotic stress, and salinity stress. In the near future, meticulous planning will be necessary to make the transition from conventional crops to productively modified ones. By increasing the efficiency of conversion through chemical addition, the product yield of biological and thermochemical conversion processes to turn biomass into value-added products can be increased, particularly the addition of nanoparticles or the introduction of various novel catalysts [for instance, single-atom catalysts (SACs), transition metal dichalcogenides (TMDC) catalysts, and metal–organic framework (MOF)] catalysts [140].

Second, industry actors can lower the supply risk by emphasizing suitable management and scheduling strategies. In general, additional biomass feedstocks and products can be stored for future use, especially when there is a lack of biomass available that enables the ongoing production of biomass [141]. Energy storage mechanisms, such as thermal or battery storage, can be used, for instance, in biomass-to-power pathways. Finally, diversifying the biomass utilized in each biomass conversion process is yet another possible tactic to lower supply risk. For downstream operations in biomass conversion plants, process automation utilizing big data, cloud computing, and IoT is essential. In addition to monitoring and controlling the process, the acquired data can be used for scheduling and storage management [141].

A virtual platform or data storage area that links all supply chain participants and offers a transparent information source about the supply chain might be imagined as block chain technology. It is possible for it to function without outside assistance. Block chain technology can provide suppliers for biomass supply chains with real-time information on the quality and origin of the biomass. Centralized power generation systems could be helpful for efficient management. However, they typically cost too much to scale up. Labor shortages are the key challenges in biomass supply networks. The day-to-day operations of biomass supply chains have been considerably impacted by limitations on movement and other steps to curb the virus's spread. This is a result of the labor-intensive tasks involved in the biomass supply chain, particularly those involving biomass gathering and harvesting [141].

17.8 Conclusions

Utilizing waste is undoubtedly the most economical method for producing renewable energy, which also benefits the environmental cleaning. The present review has uncovered the conventional and innovative techniques for waste-to-bioenergy conversion from waste substrates/feedstock. Nevertheless, when compared to conventional food crop-based biofuels, the production of biofuels from biomass waste is still regarded as having more robust transportation and conversion technology, etc. Converting wastes might assist strengthen a circular economy model by lowering waste burden and the negative environmental effect of waste disposal while creating many bioproducts in a clean and sustainable way. The use of bioenergy is an intriguing application, but it should not be employed at the expense of waste avoidance and/or reduction. Although bioenergy cannot be viewed as a saving solution, it is an alternative, a supplement to a set of waste management initiatives that governments must do. Research being done now aims to address the shortcomings of currently used technologies and enhance their effectiveness and economics.

Still, more techniques are needed

- To improve the digestibility and biodegradability of LB into biofuel and other valuable products and to decrease wastewater, animal waste, etc., innovative pretreatment strategies should be developed.
- It is advised that advanced technologies like artificial intelligence, robots, smart meters, and machine learning be widely used to reduce energy usage, reaction times, and monitoring of the numerous pretreatment operations' activities and outputs.
- For biomass utilization, further research on the techno-economic, environmental, and life cycle assessments of various pretreatment techniques is required.
- To exhibit the potential of the biorefinery concepts, focus is required.

Nevertheless, constant research is being conducted in order to address the flaws in the most recent technological advancements and increase their usefulness and economic viability.

References

1. Jayakumar, M., Gebeyehu, K. B., Abo, L. D., et al. (2023). A comprehensive outlook on topical processing methods for biofuel production and its thermal applications: Current advances, sustainability and challenges. *Fuel, 349*, 128690.
2. Neupane, D., Adhikari, P., Bhattarai, D., et al. (2022). Does climate change affect the yield of the top three cereals and food security in the world? *Earth, 3*, 45–71.
3. Jayakumar, M., Gindaba, G. T., Gebeyehu, K. B., et al. (2023). Bioethanol production from agricultural residues as lignocellulosic biomass feedstock's waste valorization approach: A comprehensive review. *Science of The Total Environment*, 163158.
4. Tesfaye Gari, M., Tessema Asfaw, B., Arumugasamy, S. K., et al. (2023). Natural resources-based activated carbon synthesis. In *Encyclopedia of green materials* (pp. 1–11). Springer.

5. Jayakumar, M., Vaithilingam, S. K., Karmegam, N., et al. (2022). Bioethanol production from green biomass resources: Emerging technologies.
6. Perlack. R. D. (2005). *Biomass as feedstock for a bioenergy and bioproducts industry: The technical feasibility of a billion-ton annual supply*. Oak Ridge National Laboratory.
7. Mahapatra, S., Kumar, D., Singh, B., & Sachan, P. K. (2021). Biofuels and their sources of production: A review on cleaner sustainable alternative against conventional fuel, in the framework of the food and energy nexus. *Energy Nexus, 4*, 100036.
8. Bioenergy IEA. (2009). *Bioenergy—A sustainable and reliable energy source*. International Energy Agency Bioenergy.
9. Hiloidhari, M., Sharno, M. A., Baruah, D. C., & Bezbaruah, A. N. (2023). Green and sustainable biomass supply chain for environmental, social and economic benefits. *Biomass and Bioenergy, 175*, 106893.
10. Panoutsou, C., & Singh, A. (2020). A value chain approach to improve biomass policy formation. *GCB Bioenergy, 12*, 464–475.
11. Ludlow, P. (2018). The European Commission. In *The New European Community* (pp. 85–132). Routledge.
12. Havelaar, A. H., Ivarsson, S., Löfdahl, M., & Nauta, M. J. (2013). Estimating the true incidence of campylobacteriosis and salmonellosis in the European Union, 2009. *Epidemiology & Infection, 141*, 293–302.
13. Corbelle-Rico, E., & Crecente-Maseda, R. (2014). Evaluating IRENA indicator "risk of farmland abandonment" on a low spatial scale level: The case of Galicia (Spain). *Land Use Policy, 38*, 9–15.
14. Rosillo-Calle, F., & Woods, J. (2012). *The biomass assessment handbook*. Routledge.
15. Ghosh, S. K. (2016). Biomass & bio-waste supply chain sustainability for bio-energy and bio-fuel production. *Procedia Environmental Sciences, 31*, 31–39.
16. Edler, J. (2013). Review of policy measures to stimulate private demand for innovation. Concepts and effects. *Compendium of Evidence on the Effectiveness of Innovation Policy Intervention, 13*, 44.
17. Karlsson, I., Rootzén, J., & Johnsson, F. (2020). Reaching net-zero carbon emissions in construction supply chains—Analysis of a Swedish road construction project. *Renewable and Sustainable Energy Reviews, 120*, 109651.
18. Bellù, L. G. (2013). *Value chain analysis for policy making. Methodological Guidelines and country cases for a quantitative approach Roma*. Food and Agriculture Organization.
19. Wang, K., & Tester, J. W. (2023). Sustainable management of unavoidable biomass wastes. *Green Energy and Resources*, 100005.
20. Lee, S. Y., Sankaran, R., Chew, K. W., et al. (2019). Waste to bioenergy: A review on the recent conversion technologies. *BMC Energy, 1*, 1–22.
21. Kushwaha, D., Srivastava, N., Mishra, I., et al. (2019). Recent trends in biobutanol production. *Reviews in Chemical Engineering, 35*, 475–504.
22. Council, N. R. (2012). *Renewable fuel standard: Potential economic and environmental effects of US biofuel policy*. National Academies Press.
23. Xing, J., Luo, K., Wang, H., et al. (2019). A comprehensive study on estimating higher heating value of biomass from proximate and ultimate analysis with machine learning approaches. *Energy, 188*, 116077.
24. Gundupalli, M. P., Gundupalli, S. P., & Thomas, A. S. S., et al. (2022). Hydrothermal liquefaction of lignocellulosic biomass for production of biooil and by-products: current state of the art and challenges. *Biofuels and Bioenergy*, 61–84.
25. Sarkar, N., Ghosh, S. K., Bannerjee, S., & Aikat, K. (2012). Bioethanol production from agricultural wastes: An overview. *Renewable Energy, 37*, 19–27.
26. Jayakumar, M., Kuppusamy Vaithilingam, S., Karmegam, N., et al. (2022). Fermentation technology for ethanol production: current trends and challenges. In B. Gurunathan, & R. B. T.-B. Sahadevan (Eds.), *Biofuels and bioenergy* (pp. 105–131). Elsevier.
27. Duque-Acevedo, M., Belmonte-Ureña, L. J., Cortés-García, F. J., & Camacho-Ferre, F. (2020). Agricultural waste: Review of the evolution, approaches and perspectives on alternative uses. *Global Ecology and Conservation, 22*, e00902.

28. Obi, F. O., Ugwuishiwu, B. O., & Nwakaire, J. N. (2016). Agricultural waste concept, generation, utilization and management. *Nigerian Journal of Technology (NIJOTECH), 35*, 957–964. https://doi.org/10.4314/njt.v35i4.34
29. Singh, A., Nanda, S., Guayaquil-Sosa, J. F., & Berruti, F. (2021). Pyrolysis of Miscanthus and characterization of value-added bio-oil and biochar products. *The Canadian Journal of Chemical Engineering, 99*, S55–S68.
30. Singh, B., Guldhe, A., Rawat, I., & Bux, F. (2014). Towards a sustainable approach for development of biodiesel from plant and microalgae. *Renewable and Sustainable Energy Reviews, 29*, 216–245.
31. Kumar, A., Sharma, A., & Upadhyaya, K. C. (2016). Vegetable oil: nutritional and industrial perspective. *Current Genomics, 17*, 230–240.
32. Lavenburg, V. M., Rosentrater, K. A., & Jung, S. (2021). Extraction methods of oils and phytochemicals from seeds and their environmental and economic impacts. *Processes, 9*, 1839.
33. Khan, I. U., Chen, H., Yan, Z., & Chen, J. (2021). Extraction and quality evaluation of biodiesel from six familiar non-edible plants seeds. *Processes, 9*, 840.
34. Blackshaw, R., Johnson, E., Gan, Y., et al. (2011). Alternative oilseed crops for biodiesel feedstock on the Canadian prairies. *Canadian Journal of Plant Science, 91*, 889–896.
35. Ajala, O. E., Aberuagba, F., Odetoye, T. E., & Ajala, A. M. (2015). Biodiesel: Sustainable energy replacement to petroleum-based diesel fuel—A review. *ChemBioEng Reviews, 2*, 145–156.
36. Kumar, A., & Samadder, S. R. (2017). An empirical model for prediction of household solid waste generation rate—A case study of Dhanbad, India. *Waste Management, 68*, 3–15.
37. Nanda, S., & Berruti, F. (2021). Municipal solid waste management and landfilling technologies: A review. *Environmental Chemistry Letters, 19*, 1433–1456. https://doi.org/10.1007/s10311-020-01100-y
38. Rajendran, N., Gurunathan, B., Han, J., et al. (2021). Recent advances in valorization of organic municipal waste into energy using biorefinery approach, environment and economic analysis. *Bioresource Technology, 337*, 125498.
39. Gunarathne, V., Ashiq, A., Ramanayaka, S., et al. (2019). Biochar from municipal solid waste for resource recovery and pollution remediation. *Environmental Chemistry Letters, 17*, 1225–1235.
40. Fazeli, A., Bakhtvar, F., Jahanshaloo, L., et al. (2016). Malaysia's stand on municipal solid waste conversion to energy: A review. *Renewable and Sustainable Energy Reviews, 58*, 1007–1016.
41. Ouda, O. K. M., Cekirge, H. M., & Raza, S. A. R. (2013). An assessment of the potential contribution from waste-to-energy facilities to electricity demand in Saudi Arabia. *Energy Conversion and Management, 75*, 402–406.
42. Clavier, K. A., Paris, J. M., Ferraro, C. C., & Townsend, T. G. (2020). Opportunities and challenges associated with using municipal waste incineration ash as a raw ingredient in cement production—A review. *Resources, Conservation and Recycling, 160*, 104888.
43. Mishra, S., Tiwary, D., Ohri, A., & Agnihotri, A. K. (2019). Impact of Municipal Solid Waste Landfill leachate on groundwater quality in Varanasi, India. *Groundwater for Sustainable Development, 9*, 100230.
44. Tong, H., Shen, Y., Zhang, J., et al. (2018). A comparative life cycle assessment on four waste-to-energy scenarios for food waste generated in eateries. *Applied Energy, 225*, 1143–1157.
45. Nanda, S., Dalai, A. K., Gökalp, I., & Kozinski, J. A. (2016). Valorization of horse manure through catalytic supercritical water gasification. *Waste Management, 52*, 147–158.
46. Kiran, E. U., Trzcinski, A. P., Ng, W. J., & Liu, Y. (2014). Bioconversion of food waste to energy: A review. *Fuel, 134*, 389–399.
47. Thapa, P., Hasnine, M. D. T., Zoungrana, A., et al. (2022). Food waste treatments and the impact of composting on carbon footprint in Canada. *Fermentation, 8*, 566.
48. Ren, Y., Wang, C., He, Z., et al. (2022). Enhanced biomethanation of lipids by high-solid co-digestion with food waste: Biogas production and lipids degradation demonstrated by long-term continuous operation. *Bioresource Technology, 348*, 126750.

49. Kiran, E. U., & Liu, Y. (2015). Bioethanol production from mixed food waste by an effective enzymatic pretreatment. *Fuel, 159*, 463–469.
50. Qin, Z., Duns, G. J., Pan, T., & Xin, F. (2018). Consolidated processing of biobutanol production from food wastes by solventogenic Clostridium sp. strain HN4. *Bioresource Technology, 264*, 148–153.
51. Yasin, N. H. M., Mumtaz, T., & Hassan, M. A. (2013). Food waste and food processing waste for biohydrogen production: A review. *Journal of Environmental Management, 130*, 375–385.
52. Patra, B. R., Nanda, S., Dalai, A. K., & Meda, V. (2021). Slow pyrolysis of agro-food wastes and physicochemical characterization of biofuel products. *Chemosphere, 285*, 131431.
53. Nanda, S., Isen, J., Dalai, A. K., & Kozinski, J. A. (2016). Gasification of fruit wastes and agro-food residues in supercritical water. *Energy Conversion and Management, 110*, 296–306.
54. Patra, B. R., Nanda, S., Dalai, A. K., & Meda, V. (2021). Taguchi-based process optimization for activation of agro-food waste biochar and performance test for dye adsorption. *Chemosphere, 285*, 131531.
55. Chen, J.-H. (2006). The combined use of chemical and organic fertilizers and/or biofertilizer for crop growth and soil fertility. In *International workshop on sustained management of the soil-rhizosphere system for efficient crop production and fertilizer use* (pp. 1–11). Citeseer.
56. Nguyen, B. T., Trinh, N. N., & Bach, Q.-V. (2020). Methane emissions and associated microbial activities from paddy salt-affected soil as influenced by biochar and cow manure addition. *Applied Soil Ecology, 152*, 103531.
57. Khoshnevisan, B., Duan, N., Tsapekos, P., et al. (2021). A critical review on livestock manure biorefinery technologies: Sustainability, challenges, and future perspectives. *Renewable and Sustainable Energy Reviews, 135*, 110033.
58. Nagendran, R. (2011). Agricultural waste and pollution. In *Waste* (pp. 341–355). Elsevier.
59. Yao, Y., Huang, G., An, C., et al. (2020). Anaerobic digestion of livestock manure in cold regions: Technological advancements and global impacts. *Renewable and Sustainable Energy Reviews, 119*, 109494.
60. Dareioti, M. A., Vavouraki, A. I., Tsigkou, K., et al. (2021). Dark fermentation of sweet sorghum stalks, cheese whey and cow manure mixture: Effect of pH, pretreatment and organic load. *Processes, 9*, 1017.
61. Yan, Q., Liu, X., Wang, Y., et al. (2018). Cow manure as a lignocellulosic substrate for fungal cellulase expression and bioethanol production. *AMB Express, 8*, 1–12.
62. Su, G., Ong, H. C., Zulkifli, N. W. M., et al. (2022). Valorization of animal manure via pyrolysis for bioenergy: A review. *Journal of Cleaner Production, 343*, 130965.
63. Liu, Q., Xu, R., Yan, C., et al. (2021). Fast hydrothermal co-liquefaction of corn stover and cow manure for biocrude and hydrochar production. *Bioresource Technology, 340*, 125630.
64. Itoh, T., Iwabuchi, K., Maemoku, N., et al. (2019). A new torrefaction system employing spontaneous self-heating of livestock manure under elevated pressure. *Waste Management, 85*, 66–72.
65. Cheong, D.-Y., Harvey, J. T., Kim, J., & Lee, C. (2019). Improving biomethanation of chicken manure by co-digestion with ethanol plant effluent. *International Journal of Environmental Research and Public Health, 16*, 5023.
66. Yong, T.L.-K., & Matsumura, Y. (2012). Catalytic gasification of poultry manure and eucalyptus wood mixture in supercritical water. *Industrial & Engineering Chemistry Research, 51*, 5685–5690.
67. Matt, F. J., Dietenberger, M. A., & Weise, D. R. (2020). Summative and ultimate analysis of live leaves from southern US forest plants for use in fire modeling. *Energy & Fuels, 34*, 4703–4720.
68. Sawatdeenarunat, C., Surendra, K. C., Takara, D., et al. (2015). Anaerobic digestion of lignocellulosic biomass: Challenges and opportunities. *Bioresource Technology, 178*, 178–186.
69. Matsakas, L., Kekos, D., Loizidou, M., & Christakopoulos, P. (2014). Utilization of household food waste for the production of ethanol at high dry material content. *Biotechnology for Biofuels, 7*, 1–9.

70. Lu, J., Li, H., Zhang, Y., & Liu, Z. (2018). Nitrogen migration and transformation during hydrothermal liquefaction of livestock manures. *ACS Sustainable Chemistry & Engineering, 6*, 13570–13578.
71. Wang, K., Zhang, J., Shanks, B. H., & Brown, R. C. (2015). The deleterious effect of inorganic salts on hydrocarbon yields from catalytic pyrolysis of lignocellulosic biomass and its mitigation. *Applied Energy, 148*, 115–120.
72. Vassilev, S. V., Baxter, D., Andersen, L. K., & Vassileva, C. G. (2010). An overview of the chemical composition of biomass. *Fuel, 89*, 913–933.
73. Sannigrahi, P., Miller, S. J., & Ragauskas, A. J. (2010). Effects of organosolv pretreatment and enzymatic hydrolysis on cellulose structure and crystallinity in Loblolly pine. *Carbohydrate Research, 345*, 965–970.
74. Demirbaş, A. (2005). Thermochemical conversion of biomass to liquid products in the aqueous medium. *Energy Sources, 27*, 1235–1243.
75. Ruan, T., Zeng, R., Yin, X.-Y., et al. (2016). Water hyacinth (*Eichhornia crassipes*) biomass as a biofuel feedstock by enzymatic hydrolysis. *BioResources, 11*, 2372–2380.
76. Reddy, K. O., Maheswari, C. U., Dhlamini, M. S., et al. (2018). Extraction and characterization of cellulose single fibers from native african napier grass. *Carbohydrate Polymers, 188*, 85–91.
77. Heinze, T. (2016). Cellulose: structure and properties. In *Cellulose chemistry and properties: Fibers, nanocelluloses and advanced materials*, 1–52.
78. Acharya, B., Sule, I., & Dutta, A. (2012). A review on advances of torrefaction technologies for biomass processing. *Biomass Conversion and Biorefinery, 2*, 349–369.
79. Phan, P. T., Nguyen, B.-S., Nguyen, T.-A., et al. (2023). Lignocellulose-derived monosugars: A review of biomass pre-treating techniques and post-methods to produce sustainable biohydrogen. *Biomass Conversion and Biorefinery, 13*, 8425–8439.
80. Koul, B., Yakoob, M., & Shah, M. P. (2022). Agricultural waste management strategies for environmental sustainability. *Environmental Research, 206*, 112285.
81. Jayakumar, M., Gebeyehu, K. B., Selvakumar, K. V., et al. (2022). Waste Ox bone based heterogeneous catalyst synthesis, characterization, utilization and reaction kinetics of biodiesel generation from Jatropha curcas oil. *Chemosphere, 288*, 22–24. https://doi.org/10.1016/j.chemosphere.2021.132534
82. Awogbemi, O., & Von Kallon, D. V. (2022). Pretreatment techniques for agricultural waste. *Case Studies in Chemical and Environmental Engineering*, 100229.
83. Zhang, H., Han, L., & Dong, H. (2021). An insight to pretreatment, enzyme adsorption and enzymatic hydrolysis of lignocellulosic biomass: Experimental and modeling studies. *Renewable and Sustainable Energy Reviews, 140*, 110758. https://doi.org/10.1016/j.rser.2021.110758
84. Tsapekos, P., Kougias, P. G., Frison, A., et al. (2016). Improving methane production from digested manure biofibers by mechanical and thermal alkaline pretreatment. *Bioresource Technology, 216*, 545–552.
85. Niu, P., Zhang, L., Liu, G., & Cheng, H. (2012). Graphene-like carbon nitride nanosheets for improved photocatalytic activities. *Advanced Functional Materials, 22*, 4763–4770.
86. Rezania, S., Oryani, B., Cho, J., et al. (2020). Different pretreatment technologies of lignocellulosic biomass for bioethanol production: An overview. https://doi.org/10.1016/j.energy.2020.117457
87. Tan, H., Li, J., He, M., et al. (2021). Global evolution of research on green energy and environmental technologies: A bibliometric study. *Journal of Environmental Management, 297*, 113382.
88. Kumari, D., & Singh, R. (2018). Pretreatment of lignocellulosic wastes for biofuel production: A critical review. *Renewable and Sustainable Energy Reviews, 90*, 877–891. https://doi.org/10.1016/J.RSER.2018.03.111
89. Chaturvedi, V., & Verma, P. (2013). An overview of key pretreatment processes employed for bioconversion of lignocellulosic biomass into biofuels and value added products. *3 Biotech, 3*, 415–431.

90. Bhaskar, T., Bhavya, B., Singh, R., et al. (2011). Thermochemical conversion of biomass to biofuels. In *Biofuels* (pp. 51–77). Elsevier.
91. Khan, M. U., Usman, M., Ashraf, M. A., et al. (2022). A review of recent advancements in pretreatment techniques of lignocellulosic materials for biogas production: Opportunities and Limitations. *Chemical Engineering Journal Advances, 10*, 100263. https://doi.org/10.1016/j.ceja.2022.100263
92. Brodeur, G., Yau, E., Badal, K., et al. (2011). Chemical and physicochemical pretreatment of lignocellulosic biomass: A review. *Enzyme Research, 2011.*
93. Iram, A., Berenjian, A., & Demirci, A. (2021). A review on the utilization of lignin as a fermentation substrate to produce lignin-modifying enzymes and other value-added products. *Molecules, 26*, 2960.
94. Hasson, D., Shemer, H., & Sher, A. (2011). State of the art of friendly "green" scale control inhibitors: A review article. *Industrial & Engineering Chemistry Research, 50*, 7601–7607.
95. Chaurasia, B. (2019). Biological pretreatment of lignocellulosic biomass (water hyacinth) with different fungus for enzymatic hydrolysis and bio-ethanol production resource: Advantages, future work and prospects. *Acta Science Agriculture, 5.*
96. He, C.-R., Kuo, Y.-Y., & Li, S.-Y. (2017). Lignocellulosic butanol production from Napier grass using semi-simultaneous saccharification fermentation. *Bioresource Technology, 231*, 101–108.
97. Jayakumar, M., Kuppusamy Vaithilingam, S., Karmegam, N., et al. (2022). Fermentation technology for ethanol production: Current trends and challenges. *Biofuels and Bioenergy, 105–131.* https://doi.org/10.1016/b978-0-323-90040-9.00015-1
98. Zheng, Y., Zhao, J., Xu, F., & Li, Y. (2014). Pretreatment of lignocellulosic biomass for enhanced biogas production. *Progress in Energy and Combustion Science, 42*, 35–53.
99. Yang, H., Zhou, Y., & Liu, J. (2009). Land and water requirements of biofuel and implications for food supply and the environment in China. *Energy Policy, 37*, 1876–1885.
100. Kurowska, K., Marks-Bielska, R., Bielski, S., et al. (2020). Food security in the context of liquid biofuels production. *Energies, 13*, 6247.
101. van der Stelt, M. J. C. (2011). Chemistry and reaction kinetics of biowaste torrefaction.
102. Porpatham, E., Ramesh, A., & Nagalingam, B. (2012). Effect of compression ratio on the performance and combustion of a biogas fuelled spark ignition engine. *Fuel, 95*, 247–256.
103. Fatehi, H., Weng, W., Li, Z., et al. (2021). Recent development in numerical simulations and experimental studies of biomass thermochemical conversion. *Energy & Fuels, 35*, 6940–6963.
104. Anca-Couce, A. (2016). Reaction mechanisms and multi-scale modelling of lignocellulosic biomass pyrolysis. *Progress in Energy and Combustion Science, 53*, 41–79.
105. Wijekoon, P., Wickramasinghe, C., Athapattu, B. C. L., et al. (2021). Biomass valorization and phytoremediation as integrated technology for municipal solid waste management for developing economic context. *Biomass Conversion and Biorefinery, 11*, 363–382.
106. Koppejan, J., & Van Loo, S. (2012). *The handbook of biomass combustion and co-firing.* Routledge.
107. Wang, S., Hu, C., Luo, K., et al. (2023). Multi-scale numerical simulation of fluidized beds: Model applicability assessment. *Particuology, 80*, 11–41.
108. Wang, S., & Shen, Y. (2020). CFD-DEM study of biomass gasification in a fluidized bed reactor: Effects of key operating parameters. *Renewable Energy, 159*, 1146–1164.
109. Scott, D. S., Majerski, P., Piskorz, J., & Radlein, D. (1999). A second look at fast pyrolysis of biomass—The RTI process. *Journal of Analytical and Applied Pyrolysis, 51*, 23–37.
110. Jahirul, M. I., Rasul, M. G., Chowdhury, A. A., & Ashwath, N. (2012). Biofuels production through biomass pyrolysis—A technological review. *Energies, 5*, 4952–5001.
111. Sarker, T. R., Nanda, S., Dalai, A. K., & Meda, V. (2021). A review of torrefaction technology for upgrading lignocellulosic biomass to solid biofuels. *BioEnergy Research, 14*, 645–669.
112. Chiou, B.-S., Cao, T., Valenzuela-Medina, D., et al. (2018). Torrefaction kinetics of almond and walnut shells: Effects of inorganic species. *Journal of Thermal Analysis and Calorimetry, 131*, 3065–3075.

113. Rabenau, A. (1985). The role of hydrothermal synthesis in preparative chemistry. *Angewandte Chemie International Edition in English, 24*, 1026–1040.
114. Gao, Y., Yu, B., Wu, K., et al. (2016). Physicochemical, pyrolytic, and combustion characteristics of hydrochar obtained by hydrothermal carbonization of biomass. *BioResources, 11*, 4113–4133.
115. Munir, M. T., Mansouri, S. S., Udugama, I. A., et al. (2018). Resource recovery from organic solid waste using hydrothermal processing: Opportunities and challenges. *Renewable and Sustainable Energy Reviews, 96*, 64–75.
116. Charis, G., Danha, G., Muzenda, E., & Nhubu, T. (2021). Modeling a sustainable, self-energized pine dust pyrolysis system with staged condensation for optimal recovery of bio-oil. *Frontiers in Energy Research, 8*, 594073.
117. Krzywda, R., & Wrzesińska, B. (2021). Simulation of the condensation and fractionation unit in waste plastics pyrolysis plant. *Waste and Biomass Valorization, 12*, 91–104.
118. Campuzano, F., Abdul Jameel, A. G., Zhang, W., et al. (2020). Fuel and chemical properties of waste tire pyrolysis oil derived from a continuous twin-auger reactor. *Energy & Fuels, 34*, 12688–12702.
119. Martínez, J. D., Veses, A., Callén, M. S., et al. (2023). On fractioning the tire pyrolysis oil in a pilot-scale distillation plant under industrially relevant conditions. *Energy & Fuels, 37*, 2886–2896.
120. Miandad, R., Rehan, M., Barakat, M. A., et al. (2019). Catalytic pyrolysis of plastic waste: Moving toward pyrolysis based biorefineries. *Frontiers in Energy Research, 7*, 27.
121. Alayi, R., Sadeghzadeh, M., Shahbazi, M., et al. (2022). Analysis of 3E (energy-economical-environmental) for biogas production from landfill: A case study. *International Journal of Chemical Engineering, 2022.*
122. Varma, A. K., Shankar, R., & Mondal, P. (2018). A review on pyrolysis of biomass and the impacts of operating conditions on product yield, quality, and upgradation. In *Recent advancements in biofuels and bioenergy utilization* (227–259).
123. Biller, P., & Ross, A. B. (2016). Production of biofuels via hydrothermal conversion. In *Handbook of biofuels production* (pp. 509–547). Elsevier.
124. Farine, D. R., O'Connell, D. A., John Raison, R., et al. (2012). An assessment of biomass for bioelectricity and biofuel, and for greenhouse gas emission reduction in Australia. *Gcb Bioenergy, 4*, 148–175.
125. Mat Aron, N. S., Khoo, K. S., Chew, K. W., et al. (2020). Sustainability of the four generations of biofuels—A review. *International Journal of Energy Research, 44*, 9266–9282.
126. Kang, N. S., Cho, K., An, S. M., et al. (2022). Taxonomic and biochemical characterization of Microalga *Graesiella emersonii* GEGS21 for its potential to become feedstock for biofuels and bioproducts. *Energies, 15*, 8725.
127. Hossain, M. F., Hossain, S., & Uddin, M. J. (2017). Renewable energy: Prospects and trends in Bangladesh. *Renewable and Sustainable Energy Reviews, 70*, 44–49.
128. Zhang, X., Ruoshui, W., Molin, H., & Martinot, E. (2010). A study of the role played by renewable energies in China's sustainable energy supply. *Energy, 35*, 4392–4399.
129. Väisänen, T., Haapala, A., Lappalainen, R., & Tomppo, L. (2016). Utilization of agricultural and forest industry waste and residues in natural fiber-polymer composites: A review. *Waste Management, 54*, 62–73.
130. Degfie, T. A., Mamo, T. T., & Mekonnen, Y. S. (2019). Optimized biodiesel production from waste cooking oil (WCO) using calcium oxide (CaO) nano-catalyst. *Scientific Reports, 9*, 1–8. https://doi.org/10.1038/s41598-019-55403-4
131. Demirbas, A. (2007). Progress and recent trends in biofuels. *Progress in Energy and Combustion Science, 33*, 1–18.
132. Zhang, W., Elaine, A. Y., Rozelle, S., et al. (2013). The impact of biofuel growth on agriculture: Why is the range of estimates so wide? *Food Policy, 38*, 227–239.
133. (US) EIA. (2012). *Annual energy review 2011*. Government Printing Office.
134. Chozhavendhan, S., Karthigadevi, G., Bharathiraja, B., et al. (2023). Current and prognostic overview on the strategic exploitation of anaerobic digestion and digestate: A review. *Environmental Research, 216*, 114526.

135. Kafarov, V., & Rosso-Cerón, A. M. (2021). Biomass as a source for heat, power and chemicals. *Advances in Carbon Management Technologies: Biomass Utilization, Manufacturing, and Electricity Management, 2*, 1.
136. Romero-Perdomo, F., & González-Curbelo, M. Á. (2023). Integrating multi-criteria techniques in life-cycle tools for the circular bioeconomy transition of agri-food waste biomass: A systematic review. *Sustainability, 15*, 5026.
137. Vela, I. C., Vilches, T. B., Berndes, G., et al. (2022). Co-recycling of natural and synthetic carbon materials for a sustainable circular economy. *Journal of Cleaner Production, 365*, 132674.
138. Awogbemi, O., & Von Kallon, D. V. (2022). *Valorization of agricultural wastes for biofuel applications.*
139. Hailemariam, A., & Erdiaw-Kwasie, M. O. (2023). Towards a circular economy: Implications for emission reduction and environmental sustainability. *Business Strategy and the Environment, 32*, 1951–1965.
140. Andiappan, V., How, B. S., & Ngan, S. L. (2021). A perspective on post-pandemic biomass supply chains: Opportunities and challenges for the new norm. *Process Integration and Optimization for Sustainability, 5*, 1003–1010.
141. Lim, C. H., Lim, S., How, B. S., et al. (2021). A review of industry 4.0 revolution potential in a sustainable and renewable palm oil industry: HAZOP approach. *Renewable and Sustainable Energy Reviews, 135*, 110223.

Chapter 18
Manifesting Sustainability Toward Food Waste into Bioenergy: Biorefinery in a Circular Economic Approach

Devi Sri Rajendran, Swethaa Venkatraman, R. Rahul, M. Afrrin, P. Karthik, and Vinoth Kumar Vaidyanathan

Abstract The global issue of excessive food waste has raised significant concerns, underscoring the urgent need for sustainable interventions incorporating eco-friendly technologies. Around 1.3 billion tons of food are discarded every year, representing a one-third of the world total food production. This food wastage comes at a considerable cost to the world economy, estimated at around $750 billion. The escalating generation of food waste highlights the pressing need for effective measures to address this issue with a dual purpose. Utilization of food waste includes distillery waste, feed and poultry waste, oil processing waste, cooking oil waste, seafood waste, dairy waste, vegetable and fruit waste for production of various biobased products reducing economic losses and promoting sustainability by harnessing. This encompasses the utilization of diverse bioprocesses, including acidogenesis, fermentation, photosynthesis, solventogenesis, methanogenesis, bio-electrogenesis, and oleaginous processes, each capable of yielding a wide spectrum of valuable products. Within this book chapter, the primary emphasis on products, specifically bioelectricity, biodiesel, and biofuels, as crucial elements for achieving sustainability and advancing the objectives outlined in sustainable development goals (SDG)-12. SDG-12 leads to responsible consumption and production, particularly SDG 12.3, which aims to halve global per capita food waste. To ensure the economic viability of these bioprocesses and align with sustainable goals, it is imperative to adopt a biorefinery strategy that optimizes the utilization of residual organic waste for the recovery of a diverse range of green products within a circular economy framework.

D. S. Rajendran · S. Venkatraman · V. K. Vaidyanathan (✉)
Integrated Bioprocessing Laboratory, Department of Biotechnology, Faculty of Engineering and Technology, School of Bioengineering, SRM Institute of Science and Technology (SRMIST), Tamil Nadu, Kattankulathur, Chengalpattu District 603203, India
e-mail: vinothkv@srmist.edu.in

R. Rahul · M. Afrrin · P. Karthik
Department of Food Technology, Faculty of Engineering, Karpagam Academy of Higher Education, Coimbatore, India

G. Baskar et al. (eds.), *Circular Bioeconomy Perspectives in Sustainable Bioenergy Production*, Energy, Environment, and Sustainability,
https://doi.org/10.1007/978-981-97-2523-6_18

Keywords Food waste · Bioprocess · Circular economy · Sustainable development goals · Biofuels

18.1 Introduction

Fossil fuels are being the primary energy source, leading to the emission of substantial quantities of greenhouse gases (GHG) and contributing to approximately 65% of the elevated mortality rate attributed to toxic air pollutants [1]. The carbon dioxide liberated during the burning of fossil fuels accounts for approximately two-thirds of GHG emissions. The anthropogenic activity since the beginning of industrialization has elevated atmospheric carbon dioxide levels from 280 ppm to about 417 ppm in the last 151 years. In order to conform to the Paris Agreement by 2030, countries are required to cut back 30 gigatons of GHG emissions per year (UNEP 2021 data). The world faces a climate emergency with impending disasters such as high-intensity hurricanes, heat waves, and drought. In order to substantially derail global temperature trends and reduce the burden on the environment, a shift toward sustainable biofuels such as bio-oil and bioethanol as an energy source is imperative [2].

Bio-oils are renewable substitutes for fuel oil and diesel, which can be utilized as combustion material for power and heat generation as a feedstock for various chemicals production such as polyols, carboxylic acids, phenolics, and levulinates [3]. Due to increasingly strict regulatory standards and rising global awareness of environmental concerns, the market of bio-lubricants is projected to grow from $ 2.0 billion in 2020 to $ 2.4 billion by 2025, reflecting a 4.1% compound annual growth rate during this period [4]. Recent developments with bio-oil have opened up its application in polyurethane foam formation, asphalt binder production, plastics, carbonaceous materials, pesticides, and fertilizers [5].

On the other hand, the bioethanol market is growing exponentially and is extrapolated to reach $ 64.8 billion by 2025 with 14% of compound annual growth rate. Many countries have shifted to bioethanol blended fuels for power generation and transport fuel to reduce their GHG emissions. The last three decades have recorded extensive research on the utilization of lignocellulosic biomass for ethanol production [6]. The United Nations (UN) established a comprehensive framework called the Sustainable Development Goals (SDGs) 2030, consisting of 17 overarching goals and 169 specific targets in September 2015 [7]. These SDGs collectively represent a globally recognized reflection of the needs and aspirations of stakeholders, encompassing social, economic, and environmental dimensions [8].

SDG target 12.3 aims to lower the food waste by 50% at consumption and retail stages. This SDG 12.3 also the minimize loss of food during production, such as post-harvest waste, by 2030 through the promotion of 3R (reduce, reuse, recycle) practices [9]. It is important to note that the global's population is projected to reach approximately nine billion people by 2050, which will inevitably lead to an increased demand for food. This rise in demand will necessitate increased food production, potentially resulting in even more food waste [10]. To provide context, it is worth noting that around 1.3 billion metric tons of food are discarded annually on a global scale, equivalent to one-third of the production of total food (FAO 2019).

A little more than a quarter, specifically 27% of food waste mainly occurs during the consumption phase [11]. The worldwide energy consumption associated with the food system is estimated to exceed 70 exajoules (1 exajoule = 10^{18} J) [12]. Notably, food production contributes towards the emission of greenhouse gas upto 20 to 30%, a significant driver of global warming and a key factor in climate change [13].

The Paris Agreement from COP 2015 aimed to increase earth temperature reduced below 2 °C, preferably 1.5°, to reduce environmental risks. Food waste contributes towards the production of 4.4 gigatons of CO_2 and 8% of global greenhouse gas emissions. Shifting to plant-based foods from animal-based ones, like red meat, can cut the food sector's carbon footprint. Reducing food waste conserves not only resources but also energy and water. Sustainability throughout the supply chain is crucial in minimizing the environmental impact of food waste [14–16].

The fourth industrial revolution and technological progress have improved global living standards but also pose potential harm [17]. Thanks to its diverse climate and soil resources, India ranks second in global fresh fruit and vegetable production after China. This study focuses on the entire fresh produce supply chain, from farming to consumption. The waste generated in this chain, including land, water, and energy misuse, has adverse environmental impacts [18].

Every stage of the food supply chain contributes to greenhouse gas emissions, causing environmental pollution that persists even after food waste disposal [19, 20]. Addressing food waste, especially in the realm of fresh produce, presents a distinct challenge due to factors such as its perishable nature, technological constraints, communication gaps among supply chain stakeholders, and mismanagement [21]. The current disorderly patterns of consumption and production contribute to a worldwide crisis. However, reshaping these patterns has the potential to boost economic growth, improve human well-being, and mitigate detrimental environmental effects [22]. From 2017 to 2020, 83 countries, the European Union and territories, documented their efforts to transition toward sustainable consumption and production (SCP) practices. Additionally, 136 policies and 27 activities were registered in 2020 aimed at implementing SCP practices [9].

Despite progress, there is a lack of specific area-wise data, highlighting the need for effective policies to enhance resource efficiency [23]. On the basis of available information as of 2016, an average of food (14%) went to waste before reaching the retail market. It is important to note that achieving SDG targets often present synergies, reducing food loss can contribute to addressing multiple goals such as poverty eradication (target 1.5), zero hunger (target 2.4), promoting healthy lifestyles (target 3.9), and improving freshwater resources (target 6.4) within the UN Sustainable Development Goals (UN-SDGs 2019) [24].

This study primarily centers around the 2030 SDG 12, which promotes sustainable production and consumption practices. It is worth noting that Goal 12 has been identified as having the highest number of trade-offs [8]. The study chiefly delves into the difficulties linked to the monitoring of S12.3 SDG, which involves reducing food waste by half, and it explores potential strategies to attain this target by 2030.

18.2 Food Waste

Food waste is a pressing global problem affecting nearly every country, squandering approximately more than a billion tons of food annually, which includes household waste of 61%, food service sector of 26%, and retail of 13% [25]. At present, malnutrition affects two billion individuals across the globe, and it is estimated that this number will continue to rise by 2050. To fulfill the requirements of the expanding global population, a 60% increase in food production will be necessary. With food insecurity affecting millions worldwide, addressing food waste becomes a crucial endeavor in sustainable, healthy, and resilient is a crucial goal [26]. One of the key benefits of reducing food waste is its positive impact on individuals, the environment, and economic well-being. It enhances food security and tackles the global challenges, including biodiversity loss, climate change, and pollution [27]. Additionally, it helps alleviate the burden on waste management systems.

The UN-SDGs have set a target to reduce food waste by half by 2030. In March 2021, the Food Waste Index 2021 report was published by the United Nations Environment Program (UNEP), presenting a standardized approach to measure and report food waste, in accordance with SDG target 12.3 [9]. This groundbreaking index reveals that global food waste is more than double according to the previous estimates. As per the report, an astounding one billion tons of food are wasted annually at the retail and consumer levels. This excessive food waste generates 8–10% of global GHG emissions, causing adverse impacts on biodiversity, water and land resources and imposing high financial costs on governments, businesses, and households, totaling nearly a trillion US dollars [28].

18.2.1 The Food Waste Index and Sustainable Development Goal 12.3

Accurate and standardized method indicating fundamental starting point for effectively reducing food waste. However, there are research gaps, a notable one being the lack of information regarding the proportion of inedible parts in food waste. The Food Waste Index, as it is currently measured, encompasses the food meant for human consumption along with edible components. The food waste is divided into specific sectors between what's edible and what's not is essential. This knowledge is crucial for stakeholders as it aids in comprehending the issue and crafting solutions [29]. Assessing food wastage within the retail and food service sectors will enhance estimates in many countries and contribute to the development of national strategies for preventing food waste. This report plays a vital role in tracking national progress toward the 2030 goal and the achievement of SDG 12.3 [30].

SDG 12.3 focuses on reducing global food waste. Its primary objective is to minimize global per capita food waste by half at both the retail and consumer levels, and to reduce food waste in production and supply chains by 2030. SDG 12.3 aims to promote efficient food management, improve resource sustainability, and address hunger and food security issues by minimizing waste along with food supply chain. It contributes to the broader sustainable development agenda by advancing responsible consumption and production practices, with the aim of creating a more sustainable and equitable future.

Improving the collection, analysis, and modeling of food waste data provides a good understanding of the global food waste problem and its significant potential for prevention in lower, middle, and high-income countries [31]. They are promoting a standardized methods for nations to gauge food waste across households, food service establishments, and retail outlets. Nations that embrace this method will produce comprehensive data to inform their domestic food waste reduction plans and enable meaningful international comparisons [32]. Requests for data related to the food waste index will be made biennially, aligning with data requisitions from the United Nations statistics division [33].

18.2.2 Types of Food Waste

18.2.2.1 Fruit and Vegetable Waste

The production and processing of fruits and vegetables result in biomass losses that are both unavoidable and potential environmental hazard [34]. These losses pose environmental risks and lead to the loss of valuable nutrients. Extracting juices and its derivatives from fruits and vegetables results in a significant volume of waste, which includes peels from fruits such as bananas, apples, carrots and oranges [35]. Numerous research studies have explored biogas production, primarily focusing on food waste alone or its co-digestion with other substrates [36]. For instance, Mu et al. [37] investigated the anaerobic co-digestion of cabbage waste and potato waste, and the methane production potential of pumpkin peels was investigated, along with an analysis of the energy and economic aspects of anaerobic co-digestion involving fluted pumpkin and poultry manure. It is worth noting that there is a scarcity of research dedicated to the sole digestion of individual fruits or vegetables [37]. Yan et al. [38] conducted a study in which they compared methane generation from the waste of 20 leafy plant species, particularly emphasizing herbaceous plants commonly used in China.

18.2.2.2 Seafood Waste

The fisheries industry, once viewed as a source of waste, now holds significant economic potential. The production of fish has surged from 37 million tons in 1961 to 200 million tons in 2016, with per capita fish consumption of fish rising from 9 to 20 kg [39]. However, a concerning practice in this industry is fish discarding, where unwanted catches are returned to the sea, dead or alive, with severe consequences for fish species and marine ecosystems [40]. Efficient utilization of fish offers various benefits. Fish protein can be used for collagen, gelatin, protein hydrolysates, peptides, and bioactive peptides [41]. Lipids can be processed into fish oil, and fish waste can yield enzymes, pigments, calcium, and flavor compounds [42]. This promotes a circular economy and reduces environmental burdens from fish waste. However, proper regulation is vital due to food safety-associated risks in valorizing fish by-products.

18.2.2.3 Dairy Waste

A fully operational dairy production facilityhas potential of processing around 500,000 liters of milk daily, may yield an estimated 200 to 350 Kg of waste [43]. Beyond this, dairy waste offers potential for the production of biosurfactants [44]. The dairy sector generates various by-products, such as buttermilk, whey, and their derivatives. Whey is a major environmental contaminant because of its high organic matter content [45]. It poses a significant threat to the environment because of its elevated levels of both chemical and biological oxygen demand. Whey typically contains components such as 0.6–0.7% protein, 4.5–5.0% lactose, and 0.4–0.5% lipids, along with minerals and vitamins [46]. Utilizing whey sustainably for the production of natural pigments has the potential to enhance the bioeconomy across various industries and, consequently, boost the national bioeconomy [47].

18.2.2.4 Cooking Oil Waste

The total waste frying oils is produced globally, particularly in densely populated nations like China produces approximately 5.6 million tons of waste frying oils. In comparison, the USA contributes 1.2 million tons, India generates 1.1 million tons, and the European Union (EU) generates around 0.9 million tons. The prevalent approach for managing waste vegetable oil is its conversion into biodiesel, a process that enhances the chemical and physical properties of raw waste vegetable oil for practical use [48]. However, it is essential to note that this conversion process involves substantial energy consumption and carries environmental impacts [49].

18.2.2.5 Oil Processing Waste

Globally, an estimated oil waste disposed each year was 200 million gallons. Intensive oil processing at a significant scale led to the production of substantial waste streams. These waste streams include tallow, soap stock, marine oils, lard and fatty acids obtained during seed oil extraction [50]. The economics of biodiesel production can be enhanced by utilizing residual seedcakes as a primary co-product. Although some oilcakes contain toxic compounds that render them unfit for consumption, they hold potential for energy production [51]. However, it is important to note that the high protein content in oilseed cakes can present challenges during the pyrolysis process, potentially slowing down the degradation rate [52].

18.2.2.6 Feed and Poultry Waste

Organic solid waste is a problem with negative externalities, and it is imperative for the agroindustry, especially in nations heavily engaged in chicken meat production, consumption, and exports, to assume responsibility for its proper management [53]. Within the poultry production chain (PPC), four key stages contribute to generating organic waste: reproductive poultry, hatchery operations, and broiler growth [54]. This waste comprises a variety of elements, including reproductive hatchery waste, poultry litter, feed wastage, solid residues from flotation sludge, carcasses, truck cleaning, charcoal, ash, and sausage casings [55]. These residues exhibit diverse characteristics, including total organic carbon ranging from 4.4 to 56.22% and nitrogen content ranging from 0.2 to 9.33% [56]. It is evident that in the PPC, there is a significant mismanagement of raw materials, energy, and nutrient resources. This is a concern, especially in developing countries, as decreasing these valuable resources is unsustainable [56].

18.2.2.7 Distillery Waste

Distillery waste comprises the assortment of by-products and residues that stem from the distillation process employed in producing alcoholic beverages like whiskey, vodka, rum, or gin [57]. These waste materials manifest in various forms, each possessing its distinct composition and characteristics. Firstly, there exist wash waters, which contain dissolved sugars, alcohols, and assorted organic compounds. These typically originate during the initial fermentation of grains or fruits. Secondly, there is the presence of stillage, alternatively referred to as pot ale or slop. This constitutes the liquid residue left in the still following the distillation process, characterized by an abundance of non-fermentable solids, organic matter, and, at times, residual alcohol [58].

Additionally, the distillation procedure frequently yields dried distillers' grains, the residual grains left over from fermentation [59]. These grains are often dehydrated and subsequently repurposed as animal feed, owing to their enriched protein and nutrient content. Spent lees represent sediment or residue that settles at the bottom of fermentation tanks or barrels during aging [60]. Efficiently managing distillery waste assumes paramount importance in mitigating environmental pollution and optimizing the utilization of valuable resources. Various techniques, such as anaerobic digestion, composting, and biogas or biofuels production, can be employed to treat and repurpose distillery waste ecologically and sustainably [61].

18.3 Necessary Action Taken by Various Countries and Organization

Several countries and organizations have taken proactive steps to address the issue of food waste. Norway, the United States and Australia have committed to reducing food waste generation by 50% by 2030 in alignment with SDG 12.3 [62]. Then, France and Italy also enacted regulations prohibiting supermarkets from discarding unsold food and mandating that they donate excess food to charitable organizations and food banks [63]. Collaborative efforts are essential in combating unnecessary food waste and safeguarding the environment. To encourage global cooperation, UNEP has initiated regional task forces dedicated to addressing the issue of food waste in regions such as the Pacific and Asia, Western Asia, Latin America, Africa, and the Caribbean [64]. These groups aim to support countries with limited data on food waste by enhancing their capacity to measure and reduce it. This initiative fosters peer-to-peer collaboration and facilitates the establishment of reports on SDG 12.3 progress in 2022, ultimately helping nations develop national strategies to prevent food waste [65].

18.4 Sustainable Food Future

Food waste substantially squanders precious resources, time, and finances. Effecting substantial transformation necessitates collaborative actions aimed at establishing a more comprehensive, sustainable, and robust food system [66]. Reducing food waste across all tiers of society presents noteworthy advantages in terms of environmental preservation, social well-being, and economic gains [67]. Accurate measurement of food waste, including non-edible portions, and monitoring per capita food waste generation nationally are crucial for tracking progress toward SDG 12.3 [68]. The food waste index provides a methodology for countries to measure food waste, spanning from households to retail sectors and food service. The FAO recommends various practices to combat food waste at the individual or household level, such as adopting healthier diets, buying what's necessary, and using effective food storage methods [69]. Figure 18.1 explains different methods used for the biological valorization of food waste into different bioenergy products.

18.4.1 Electro-fermentation

18.4.1.1 Bioelectricity

Microbial fuel cells (MFCs) are an innovative and sustainable technology that directly converts chemical energy from waste into bioelectricity [70, 71]. Microorganisms

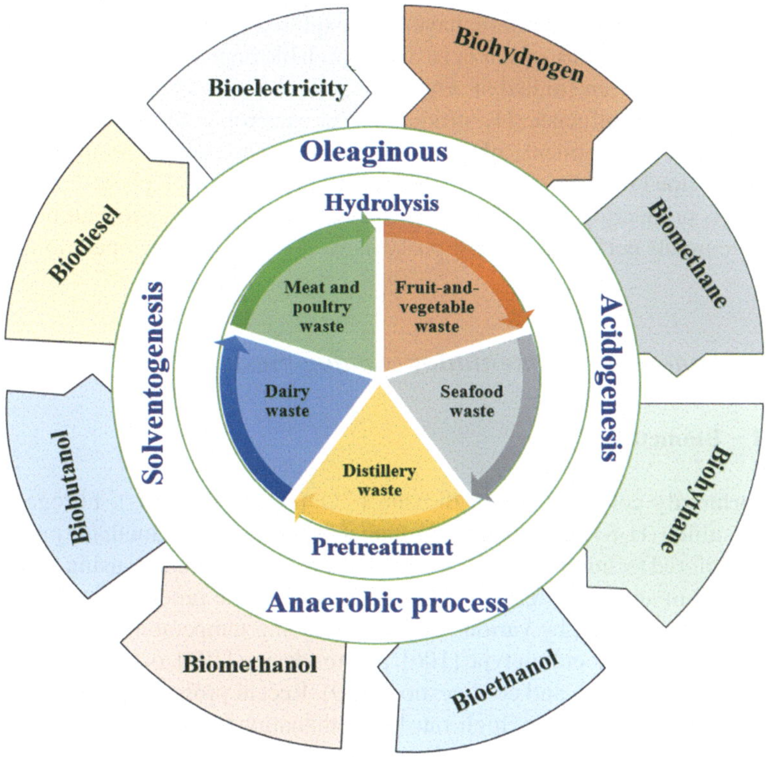

Fig. 18.1 Various methods of biological valorization of food waste into different bioenergy products

present an anode that acts as a biocatalyst and generates electrons and protons to facilitate the organic matter electrochemical oxidation. The cathode contains a terminal electron acceptor, driving electron transfer through an external circuit for bioelectricity generation, while protons pass through a separating membrane [72]. In the cathode chamber, electrons and protons undergo reduction processes, producing valuable products and treating wastewater [73]. Similar technologies, MFCs have the potential to become hybrid systems, offering not only bioelectricity but also efficient value-added products and waste treatment [6].

18.4.1.2 Biohydrogen

Biohydrogen (H_2) exhibits sustainable and green fuel, drawing global research attention for its production from waste sources. H_2 can be biologically generated through methods like photo-fermentation, anaerobic fermentation, and bio-photolysis, either individually or in combination. Acetogenic bacteria are essential for H_2 production in these processes. Various waste materials, including vegetable market waste, food waste, wastewater, and more, have been explored for biological H_2 production (Table 18.1) [6]. Food waste, rich in easily degradable organic matter, is particularly promising and has been studied at semi-pilot and pilot scales [74].

Several factors influence H_2 production in anaerobic fermentation, such as inoculum type, pretreatment, pH, co-substrates, reactor setup, temperature, and feed composition [104]. Integrating anaerobic fermentation with photo-fermentation enhances H_2 production, with the metabolites from anaerobic fermentation serving as valuable inputs performed in two-phase as well as single-stage operations [105].

18.4.2 Anaerobic Fermentation/Acidogenesis

18.4.2.1 Biomethane

Biogas primarily contains carbon dioxide (CO_2), methane (CH_4), nitrogen (N_2), hydrogen sulfide (H_2S) and moisture. Among these components, methane production is a well-explored technology for generating bioenergy, particularly using food waste as a substrate in anaerobic fermentation (Table 18.2). The production of CH_4 from food waste is influenced by various factors, including temperature, pH, carbon-to-nitrogen ratio [101], inoculum type [106], reactor design [107], organic loading rate, volatile fatty acids [108], and co-digestion [109]. Recent progress in biogas production has spurred the creation of high-rate biomethanation technologies, subsequently enhancing process efficiency [72]. Methane production, which forms biogas, from food waste has progressed to a commercial level in India and other countries, with various industries adopting and implementing full-scale technologies.

18.4.2.2 Biohythane

Methane (CH_4) and hydrogen (H_2), each have their own limitations. H_2 is reactive and has flammability characteristics, posing storage challenges, while CH_4 has low flammability. Combining CH_4 and H_2 in a 1:4 ratio results in biohythane, an emerging fuel with clean attributes and high calorific efficiency [128, 129]. This blend enhances combustion rates, significantly improving the efficacy of CH_4-fueled vehicles. The specific hydrogen (H_2) production rate of 66.7 L/kg of total volatile solid (TVS) feed was reported, along with biogas production rate of 0.72 m^3/kg of TVS fed during the second phase of standard biohytane production [130]. In a reactor scale study involving distillery spent wash, the highest recorded biohythane production was found to be 147.5 ± 2.4 L [74]. The highest H_2 yield was reported in the continuous stirred tank reactor, and the highest CH_4 yield was observed in the anaerobic fluidized bed reactor [131].

Table 18.1 Various food wastes used for the production of different biodiesel and biohydrogen products

System	Waste	Yield	Reference
Biodiesel			
Solvent extraction	Citrus peel	96.30%	[75]
Chlorella vulgaris	Fruit and vegetable waste		[76]
Auxenochlorella protothecoides	Food waste digestate	49.89 ± 0.31%	[77]
Solvent extraction	Waste cooking oil	90%	[78]
Cryptococcus curvatus *Rhodotorula glutinis*	Food waste and municipal wastewater	28.6 ± 2.2% 19.6 ± 0.2%	[79]
Candida lipolytica *Yarrowia lipolytica*	Corn cobs hydrolysate	19.4 g/L 11.3 g/L	[80]
Cryptococcus curvatus	Sweet sorghum bagasse	63.98–73.26%	[81]
Chlorella vulgaris	Agro-industrial coproducts, ethanol thin stillage and soy whey	43%	[79]
Chlorella sorokiniana	Food waste and municipal wastewater	20–22.9%	[79]
Schizochytrium mangrovei and *Chlorella pyrenoidosa*	Rice, noodles, meat, and vegetables	300 mg/g	[82]
Torulaspora maleeae *Torulaspora globosa*	Sugarcane molasses	0.31 g/L 0.2 g/L	[83]
Acid and base catalyzed transesterification and oil extraction	Spent coffee grounds	82%	[84]
Cryptococcus curvatus	Cheese whey	30–35%	[85]

(continued)

Table 18.1 (continued)

System	Waste	Yield	Reference
Aspergillus niger	Waste cooking olive oil	0.49 g/g	[86]
Yarrowia lipolytica	Waste oil from chicken products, frying fish, frying vegetables	34.02–57.89%	[87]
Hydrolysis and esterification	Waste cooking oil	91.8% fatty acid	[88]
Ultrasonication oil extraction, hydrolysis and fermentation	Spent coffee grounds	97% 0.5 g/g	[89]
Biohydrogen			
Clostridium	Vegetable waste	7.09 L/L-d	[90]
Hydrothermal treatment	Vegetable waste	38.1 mL/gVS	[91]
Steam gasification	Household food waste	73.6%	[92]
Microwave co-pyrolysis	Kitchen food waste	43.02%	[93]
Purple non-sulfur bacteria	Sugar cane bagasse	1.96 mol of H_2/mol sugar	[94]
Sewage sludge	Fallen leaves	37.8 mL/g of VS	[95]
Anaerobic activated sludge	Vegetable waste	85.65 mL-H_2 g^{-1} VS	[96]
Sludge	Kitchen waste and white rice	1.27 mmol/g of COD	[97]
Pre-treated anaerobic activated sludge	Glucose	8.65 mol/kg CODR	[98]
Anaerobic activated sludge	Municipal food waste, kitchen	245 mL/g-COD	[99]
Anaerobic activated sludge	Food waste	9.67 L/h	[100]
Scenedesmus sp. *R-16*	Starch-rich food waste	1643.5 mL/L	[100]
Anaerobic activated sludge	Vegetable waste	23.96 mmol/day	[101]

(continued)

Table 18.1 (continued)

System	Waste	Yield	Reference
Anaerobic activated sludge	Municipal food waste	370 mL/g of VS	[99]
Clostridium, *Prevotella*, *Paludibacter*, *Ensifer*, and *Petrimonas*	Rice straw	1.19 mol H_2/mol glucose	[99]
Digested sludge	POME	2.45 mol/mol-sugar	[102]
Anaerobic mixed culture	Waste wheat	654.7 L/kg	[103]

Table 18.2 Various food wastes used for the production of biohythane and biomethane products

System	Waste	Yield	Reference
Biohythane			
Dark fermentation and second-stage anaerobic digestion	Food waste	CH_4 (70–90%, v/v) + H_2 (10–30%, v/v)	[110]
Alkali extraction	Vegetable waste	Hydrogen and methane at 31.08 and 66.17%	[111]
Anaerobic digestion	Food waste	90% methane	[112]
Anaerobic digestion	Pineapple waste	Hydrogen: 1240 mL/L-d Methane: 812 mL/L-d	[113]
Phragmites australis	Municipal food waste	Hydrogen: 184.7 ± 1.83 mL/g-VS added Methane: 31.2 kJ/g-VS added	[114]
Hydrogenogenic method	Organic waste	Hydrogen: 215 mL/g VS added Methane: 676 mL/g VS added	[115]
Anaerobic sludge	Food waste		[6]
Digested sludge	Food waste	115.2 L of H_2/kg VS and 334.7 L of CH_4/kg COD	[116]
Anaerobic consortium	Distillery spent wash	147.5 ± 2.4 L	[74]
Biomethane			
Anaerobic digestion	Spent coffee grounds	36.0 mL-CH_4/g-VS added	[117]
Anaerobic digestion	Vegetable and meat waste	439.6 mL/gVS	[91]
Anaerobic digestion	Food and plastic waste	520.8 mL CH_4/gVS	[118]
Spontaneous acidification	Canteen food waste	80%	[119]
Anaerobic digestion	Municipal solid waste	429.16 mL/g VS	[120]
Sargassum spp.	Food waste	615.5 L of CH_4/kg	[121]
Anaerobic digester	Sugarcane waste	36.24	[122]
Anaerobic digester	Vegetable waste	247.3 mL/g of VS	[123]
Anaerobic activated sludge	Food waste	18.4 g/L (as acetic acid)	[124]

(continued)

Table 18.2 (continued)

System	Waste	Yield	Reference
Anaerobic pretreated sludge	Food waste	11.1 g/L	[101]
Anaerobic activated sludge	Food waste	29.1 g-COD/L	[125]
Anaerobic activated sludge	Food waste	0.918 g/g	[126]
Anaerobic activated sludge	Food waste	6.3 g/L	[127]

18.4.3 Solventogenesis

18.4.3.1 Bioethanol

Global demand for ethanol production reached 100 billion L in 2015 and continues to rise, driven by increasing needs. To address this increasing demand, a critical transition occurs from fossil-derived ethanol production to renewable resource-based methods [132]. Presently, pilot-scale bioethanol production is predominantly centered around sugarcane and corn. However, a significant challenge arises because sugarcane and corn are used for bioethanol and serve as food and animal feed, creating competition between food and fuel resources. Research has focused on utilizing agricultural waste such as rice straw, sugar cane bagasse, wheat straw, and other biomass (Table 18.3) [133]. One of the major hurdles in this transition is the interference of lignin converting cellulose and hemicelluloses into monosaccharides, posing a significant limitation to ethanol production. The implementation of simultaneous saccharification and fermentation in a single stage resulted in diminished ethanol yields, specifically amounting to 0.31 g of ethanol per gram of total solids [134]. The high organic content found in food waste makes it a promising option for bioethanol production, and it offers a potential solution to the food-versus-fuel challenge.

18.4.3.2 Biobutanol

Biobutanol (butyl alcohol), serves various purposes as a extractant, solvent, supplement, eluent, and precursor for other compounds. The biobutanol blended with gasoline used as a transportation fuel. Biobutanol possesses advantageous properties such as low vapor pressure, insensitivity to water, lower toxicity, reduced volatility, and flammability. Notably, biobutanol can replace gasoline, making it a drop-in fuel of significant interest. Biobutanol production primarily relies on starch-rich crops like wheat, rice, maize, energy crops, cassava, trees, grasses, and algae (Table 18.3). Therefore, bio-butanol is categorized as first, second, or third-generation depending on the substrate utilized [157]. The main production process involves extracting

Table 18.3 Various food wastes used for the production of different biobutanol and bioethanol products

System	Waste	Yield	Reference
Biobutanol			
Clostridium beijerinckii	Carrot waste	6.4 g/L	[135]
Clostridium saccharoperbutylacetonicum N1-4	Starchy food waste	12.1 g/L	[136]
Clostridium sp. strain BOH_3	Orange peel	17.7 g/L	[137]
Clostridium beijerinckii P260	Vegetable waste	59%	[138]
Clostridium sp. strain BOH_3	Starchy food waste	14.1 g/L	[139]
Clostridium acetobutylicum	Vegetable waste	1.5 g/L	[140]
C. acetobutylicum CICC 8012	Meat waste	12.8 g/L	[123]
Clostridium acetobutylicum YM1	Cooked rice	6.01 g/L	[141]
C. acetobutylicum DSM 792	Apple peel waste	14 g/L	[142]
C. acetobutylicum NCIM 2877	Orange peel waste	19.5 g/L	[143]
C. acetobutylicum ATCC 824	Sweet potato vines	6.4 g/L	[144]
Bioethanol			
Saccharomyces cerevisiae	Municipal food waste	77%	[145]
Acid extraction	Municipal food waste	1.29 kg CO_2 equiv. per 1 gal	[146]
Enzymatic hydrolysis	Vegetable food waste	73.2 g/L	[147]
Saccharomyces cerevisiae	Potato and wheat bran	7.52% (v/v)	[148]
Enzymatic saccharification	Carbohydrate food waste	75.70%	[149]
Batch fermentation	Carrot discard juices	11.98 g/L	[150]
Ensiling + centrifugation	Orange peel waste	120 g/kg TS	[151]
Saccharomyces cerevisiae	Spent coffee grounds	8.3%v/v	[152]
Saccharomyces cerevisiae 424A	Corn stover	68.30%	[153]
Saccharomyces cerevisiae Pe-2	Sugarcane bagasse	85.41–88.12	[154]
Hydrolysis and anaerobic fermentation	Waste hamburger	0.271 g ethanol/g WH	[155]
Fast pyrolysis	Spent coffee grounds	29%	[156]

sugars from the substrate through various pretreatment methods and then utilizing these extracts with strains from the *Clostridiaceae* family. Interestingly, domestic food waste, which is rich in carbon, can also be utilized for producing biobutanol [158].

18.4.4 Oleaginous Metabolism

Oleaginous metabolism holds substantial importance within the realm of biorefinery operations. Biorefineries are establishments dedicated to the transformation of diverse biomass sources, including agricultural and forestry residues, as well as algae and other organic materials like food waste. The objective is to convert these resources into a broad spectrum of valuable end products, encompassing biofuels, biochemicals, and various other bioproducts. Oleaginous microorganisms, which include specific yeasts, algae, and bacteria, have garnered notable attention in the biorefinery context due to their remarkable capacity to accumulate significant quantities of lipids, which are essentially fats and oils. This lipid accumulation serves as an effective energy storage mechanism. This process offers the potential for lipid production, biofuel generation (such as biodiesel), waste reduction, resource recovery, and environmental benefits, aligning with sustainability goals by efficiently reusing and converting organic waste into useful products, thus promoting a more circular and eco-friendly approach to resource utilization.

18.4.4.1 Biodiesel

Biodiesel gives a sustainable energy source derived from natural renewable materials. Chemically, it comprises long-chain fatty acids of methyl esters. In acid-catalyzed transesterification, acid initiates the reaction by donating a proton to the oxygen atom of the ester functional group, which then undergoes nucleophilic attack by an alcohol molecule. This leads to the formation of the desired ester products (biodiesel) and glycerol as a by-product. Base-catalyzed transesterification, the base removes a proton from the alcohol, creating an alkoxide ion that is more nucleophilic. This ion then attacks the ester, leading to the formation of biodiesel and glycerol. Traditionally, biodiesel is manufactured using various feedstocks such as waste cooking oils, vegetable oils, and animal fats. However, the limited availability of these feedstocks hampers the further growth of the biodiesel industry. A range of cost-effective carbon sources, including molasses [144], flour extracts, grape must, radish brine, hydrolyzed agricultural residues [52], municipal wastewater [79], distillery wastewater, cheese whey [159], food and feed waste [160], and municipal wastewater have been successfully employed as primary feedstock [79].

18.5 Circular Bioeconomy and Biorefinery

The increasing global demand for energy and resources is driving a shift from a fossil-based linear economy to a sustainable circular bioeconomy. The concept of a circular economy aims to modernize the traditional recycling of materials within natural ecosystems. The bioeconomy relies on renewable feedstocks that can produce a wide array of biobased products, effectively functioning as a self-contained refinery. It introduces the notion of cascading, involving the diversified utilization of materials through a series of production processes. Cascading is achieved through interconnected networks of factories, where organic by-products, instead of being disposed of in landfills or returned directly to the soil, are repurposed as inputs for producing value-added products. This process transforms food waste into a valuable resource for biorefineries, where biological materials are progressively converted into bio-based chemicals, plastics, pharmaceuticals, and fuel. Also, this transition encompasses a broad spectrum of scientific, managerial, and engineering disciplines [161]. One intriguing approach to maximize the value derived from food waste is to reintegrate it into the food supply chain. This concept modernizes the traditional practice of soil restoration by repurposing food residues as inputs for the agri-food industry.

The biogenic waste emerges as a promising feedstock for fostering a bioeconomy [44]. The approach involves the integration of biogenic waste into an integrated biorefinery framework to produce bio-based products, ultimately leading to the establishment of a sustainable waste-based bioeconomy (Fig. 18.2). This integrated approach combines various enabling bioprocesses within a cascading loop, including acidogenesis, fermentation, methanogenesis, oleaginous processes, solventogenesis, photosynthesis, bio-electrogenesis, and more. Such integration facilitates the efficient conversion of waste into economically valuable products.

Food waste is generated throughout the entire food supply chain (FSC), with a significant portion originating from the consumption stage (56%) and the food

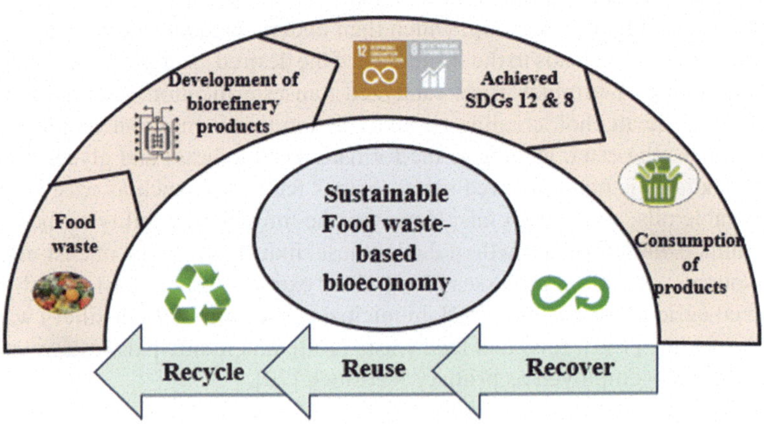

Fig. 18.2 Sustainability toward food waste into bioenergy in a circular economic approach

manufacturing stage (around 38%). While food waste from the consumption stage varies widely in composition, food waste from food processing is notably consistent in composition, making it a valuable resource for extracting high-value products. Regrettably, this type of food waste is often directed to landfills, incurring economic and environmental costs, or at best, repurposed for animal feed or energy recovery processes like anaerobic digestion. To prevent this loss or underutilization of valuable resources, the food waste can serve as a feedstock in biorefineries specializing in value-added products and chemicals. The biorefinery concept can yield a diverse array of products, including biofuels, platform chemicals, biomass, bioelectricity biomaterials, animal feed, biofertilizers and more through food waste. Importantly, it enhances the efficiency of waste treatment processes. The food waste biorefinery concept can serve as a valuable complement to fossil-based refineries, addressing critical drivers for the bioeconomy, including climate concerns, ecosystem services, and resource security [161]. In the near future, the food waste biorefinery approach can potentially establish a sustainable, environmentally friendly pathway with minimal environmental impact.

18.6 Techno-economic Analysis

A techno-economic analysis of food waste entails an evaluation that combines technical and economic factors to assess the feasibility and sustainability of various processes and technologies related to the management and utilization of food waste. Research assesses the environmental and economic consequences of extracting energy and material resources from food waste. The findings reveal that, per ton of waste treated, anaerobic digestion exhibits the lowest environmental impacts in 13 of the 19 categories analyzed, including a net-negative global warming potential. In-vessel composting is found to be the least environmentally sustainable option, despite being favored over incineration in alignment with circular economy principles. Incineration emerges as the most cost-effective option (£71/t), while landfilling proves to be the most expensive (£123/t). Managing the annual collection of food waste around 4.9 million tons from UK households through these methods results in the generation of 340,000 tons of CO_2 equivalent and costs £452 million, accompanied by various other environmental problems. This conserves energy around 1.9 PJ, primarily through electricity generation via incineration. They were incinerated, savings of £103 million and 360,000 tons of CO_2 equivalent per year could be achieved compared to current waste management, rendering incineration a carbon-negative technology. The environmental benefits of incineration would be exceeded if all food waste was processed through anaerobic digestion. This method could save 490,000 tons of CO_2 and generate 50% more electricity per ton of waste compared to incineration [162].

Another study conducted a techno-economic analysis of an integrated system for the biofuels and polyhydroxyalkanoates (PHAs) production, including bioethanol, biohydrogen,, and 2,3-butanediol, using FW as the feedstock. The integrated process was developed, resulting in a daily production plan of 2.01 million tons (MT) of PHAs, biohydrogen (0.29 MT), bioethanol (4.79 MT), and 2,3-butanediol (6.79 MT), all from food waste of 50 MT. The economic analysis revealed a positive net present value (NPV) of $4.47 million, a return on investment (ROI) of 13.68%, a payback period of 7.31 years, and an internal rate of return (IRR) of 11.95%. Sensitivity analysis was employed to assess economic viability. The minimum selling price (MSP) of PHAs was found to be 4.83 dollars per kilogram, with the lowest achievable MSP (2.41 dollars per kilogram) achieved with a 30% solid loading [75].

18.7 Conclusions

The global issue of excessive food waste is a pressing concern, not only in terms of economic losses but also for its environmental impact. The annual discard of 1.3 billion tons of food worldwide emphasizes the urgency of sustainable interventions. Eco-friendly technologies and bioprocesses, such as acidogenesis, fermentation, and bio-electrogenesis, offer promising solutions for transforming food waste into valuable bio-based products like bioelectricity, biodiesel, and biofuels. These products align with SDG, particularly SDG 12.3 and SDG 8. Adopting a biorefinery strategy that embraces a circular economy framework is imperative to maximize economic viability and sustainability. By converting the food waste as a potential resource, one can reduce waste and contribute to a more sustainable and prosperous future for all.

The following recommendations have been put forth in relation to the implementation of SDG 12.3. Firstly, the integration of advanced technologies like data analytics and blockchain in the food supply chain should be encouraged to enhance traceability and reduce losses. Second, comprehensive consumer education and awareness campaigns are vital to promote responsible food consumption and minimize waste at the household level. Third, governments and international organizations should establish and enforce regulations targeting food waste reduction, including the setting of targets and policies. Embracing the principles of a circular economy, optimizing food systems, and repurposing food waste should be a priority. Collaborative partnerships among governments, businesses, and civil society, knowledge sharing, and investments in innovative solutions for waste repurposing are necessary. Sustainable packaging materials and cross-border cooperation should also be promoted. Furthermore, standardizing metrics and data collection for food waste is essential, and support for local and community initiatives addressing food waste is crucial. By implementing these recommendations, SDG 12.3 can progress toward a more sustainable and responsible approach to food management, significantly reducing food waste and its global consequences.

The integration of biorefinery and bioeconomy principles can be effectively realized, thereby fostering a closed-loop, zero-waste manufacturing system centered around the high-value products generation from food waste. To achieve this, further research is imperative to fine-tune process efficiency and explore avenues for scalability and commercialization, ensuring that forthcoming ventures are economically viable, socially acceptable, and environmentally sustainable. Additionally, it would be advantageous to investigate the interconnectedness between food waste-to-bioenergy conversion, sustainable food waste management, and the integrated biorefinery and bioeconomy. This holistic approach offers a sustainable means of valorizing food waste with minimal long-term environmental repercussions. Furthermore, addressing the issue necessitates a concerted effort to enhance public awareness through training, socialization, and education. This is pivotal in shifting societal perspectives and attitudes toward food waste and its potential for valorization.

Declaration of Competing Interest The authors declare that they have no known competing financial interests or personal relationships that could have appeared to influence the work reported in this paper.

Funding This research did not receive any specific grant from funding agencies in the public, commercial, or not-for-profit sectors.

References

1. Lelieveld, J., Evans, J. S., Fnais, M., Giannadaki, D., & Pozzer, A. (2015). The contribution of outdoor air pollution sources to premature mortality on a global scale. *Nature, 525*, 367–371.
2. Osman, A. I., Qasim, U., Jamil, F., Ala'a, H., Jrai, A. A., Al-Riyami, M., Al-Maawali, S., Al-Haj, L., Al-Hinai, A., & Al-Abri, M. (2021). Bioethanol and biodiesel: Bibliometric mapping, policies and future needs. *Renewable and Sustainable Energy Reviews 152*, 111677.
3. Jiang, W., Kumar, A., & Adamopoulos, S. (2018). Liquefaction of lignocellulosic materials and its applications in wood adhesives—A review. *Industrial Crops and Products, 124*, 325–342.
4. Spekreijse, J., Lammens, T., Parisi, C., Ronzon, T., & Vis, M. (2019). Insights into the European market for bio-based chemicals. Analysis Based on Ten Key Product Categories.
5. Hu, K., Peris, A., Torán, J., Eljarrat, E., Sarrà, M., Blánquez, P., & Caminal, G. (2020). Exploring the degradation capability of *Trametes versicolor* on selected hydrophobic pesticides through setting sights simultaneously on culture broth and biological matrix. *Chemosphere, 250*, 126293.
6. Sarkar, O., & Venkata Mohan, S. (2016). Deciphering acidogenic process towards biohydrogen, biohythane, and short chain fatty acids production: Multi-output optimization strategy. *Biofuel Research Journal, 3*, 458–469.
7. Anastas, P., Nolasco, M., Kerton, F., Kirchhoff, M., Licence, P., Pradeep, T., Subramaniam, B., & Moores, A. (2021). The power of the United Nations sustainable development goals in sustainable chemistry and engineering research. *ACS Sustainable Chemistry & Engineering, 9*, 8015–8017.
8. Fonseca, L. M., Domingues, J. P., & Dima, A. M. (2020). Mapping the sustainable development goals relationships. *Sustainability, 12*, 3359.
9. Ardra, S., & Barua, M. K. (2022). Halving food waste generation by 2030: The challenges and strategies of monitoring UN sustainable development goal target 12.3. *Journal of Cleaner Production, 380*, 135042.
10. Rohr, J. R., Barrett, C. B., Civitello, D. J., Craft, M. E., Delius, B., DeLeo, G. A., Hudson, P. J., Jouanard, N., Nguyen, K. H., & Ostfeld, R. S. (2019). Emerging human infectious diseases and the links to global food production. *Nature Sustainability, 2*, 445–456.

11. Coudard, A., Corbin, E., de Koning, J., Tukker, A., & Mogollón, J. M. (2021). Global water and energy losses from consumer avoidable food waste. *Journal of Cleaner Production, 326*, 129342.
12. Usubiaga-Liaño, A., Behrens, P., & Daioglou, V. (2020). Energy use in the global food system. *Journal of Industrial Ecology, 24*, 830–840.
13. Binns, C. W., Lee, M. K., Maycock, B., Torheim, L. E., Nanishi, K., & Duong, D. T. T. (2021). Climate change, food supply, and dietary guidelines. *Annual Review of Public Health, 42*, 233–255.
14. Bechthold, A., Boeing, H., Tetens, I., Schwingshackl, L., & Nöthlings, U. (2018). Perspective: Food-based dietary guidelines in Europe—Scientific concepts, current status, and perspectives. *Advances in Nutrition, 9*, 544–560.
15. Al-Ansari, T., Korre, A., Nie, Z., & Shah, N. (2015). Development of a life cycle assessment tool for the assessment of food production systems within the energy, water and food nexus. *Sustainable Production and Consumption, 2*, 52–66.
16. Joshi, P., & Visvanathan, C. (2019). Sustainable management practices of food waste in Asia: Technological and policy drivers. *Journal of Environmental Management, 247*, 538–550.
17. Stanujkic, D., Popovic, G., Zavadskas, E. K., Karabasevic, D., & Binkyte-Veliene, A. (2020). Assessment of progress towards achieving Sustainable Development Goals of the "Agenda 2030" by using the CoCoSo and the Shannon Entropy methods: The case of the EU Countries. *Sustainability, 12*, 5717.
18. Gokarn, S., & Kuthambalayan, T. S. (2017). Analysis of challenges inhibiting the reduction of waste in food supply chain. *Journal of Cleaner Production, 168*, 595–604.
19. Gokarn, S., & Choudhary, A. (2021). Modeling the key factors influencing the reduction of food loss and waste in fresh produce supply chains. *Journal of Environmental Management, 294*, 113063.
20. Porter, L. (2018). From an urban country to urban country: Confronting the cult of denial in Australian cities. *Australian Geographer., 49*, 239–246.
21. Gardas, B. B., Raut, R. D., & Narkhede, B. (2018). Evaluating critical causal factors for post-harvest losses (PHL) in the fruit and vegetables supply chain in India using the DEMATEL approach. *Journal of Cleaner Production, 199*, 47–61.
22. Schulz, C., & Bailey, I. (2014). The green economy and post-growth regimes: Opportunities and challenges for economic geography. *Geografiska Annaler: Series A, Physical Geography, 96*, 277–291.
23. Pal, M. S., & Bhatia, M. (2022). Current status, topographical constraints, and implementation strategy of municipal solid waste in India: A review. *Arabian Journal of Geosciences, 15*, 1176.
24. Alawneh, R., Ghazali, F., Ali, H., & Sadullah, A. F. (2019). A novel framework for integrating United Nations Sustainable Development Goals into sustainable non-residential building assessment and management in Jordan. *Sustainable Cities and Society, 49*, 101612.
25. Capanoglu, E., Nemli, E., & Tomas-Barberan, F. (2022). Novel approaches in the valorization of agricultural wastes and their applications. *Journal of Agriculture and Food Chemistry, 70*, 6787–6804.
26. Movilla-Pateiro, L., Mahou-Lago, X. M., Doval, M. I., & Simal-Gandara, J. (2021). Toward a sustainable metric and indicators for the goal of sustainability in agricultural and food production. *Critical Reviews in Food Science and Nutrition, 61*, 1108–1129.
27. Johnston, J. L., Fanzo, J. C., & Cogill, B. (2014). Understanding sustainable diets: A descriptive analysis of the determinants and processes that influence diets and their impact on health, food security, and environmental sustainability. *Advances in Nutrition, 5*, 418–429.
28. Yang, N., Zhang, H., Chen, M., Shao, L.-M., & He, P.-J. (2012). Greenhouse gas emissions from MSW incineration in China: Impacts of waste characteristics and energy recovery. *Waste Management, 32*, 2552–2560.
29. Thamagasorn, M., & Pharino, C. (2019). An analysis of food waste from a flight catering business for sustainable food waste management: A case study of halal food production process. *Journal of Cleaner Production, 228*, 845–855.

30. Fattibene, D., Recanati, F., Dembska, K., & Antonelli, M. (2020). Urban food waste: A framework to analyse policies and initiatives. *Resources, 9*, 99.
31. Bond, M., Meacham, T., Bhunnoo, R., & Benton, T. (2013). Food waste within global food systems. Global Food Security Swindon, UK.
32. Annosi, M. C., Brunetta, F., Bimbo, F., & Kostoula, M. (2021). Digitalization within food supply chains to prevent food waste. Drivers, barriers and collaboration practices. *Industrial Marketing Management, 93*, 208–220.
33. van den Bos Verma, M., de Vreede, L., Achterbosch, T., & Rutten, M. M. (2020). Consumers discard a lot more food than widely believed: Estimates of global food waste using an energy gap approach and affluence elasticity of food waste. *PLoS One, 15*, e0228369.
34. Babu, S., Rathore, S. S., Singh, R., Kumar, S., Singh, V. K., Yadav, S. K., Yadav, V., Raj, R., Yadav, D., & Shekhawat, K. (2022). Exploring agricultural waste biomass for energy, food and feed production and pollution mitigation: A review. *Bioresource Technology*, 127566.
35. Sánchez, M., Laca, A., Laca, A., & Díaz, M. (2021). Value-added products from fruit and vegetable wastes: A review. *CLEAN–Soil, Air, Water, 49*, 2000376.
36. Kunatsa, T., & Xia, X. (2022). A review on anaerobic digestion with focus on the role of biomass co-digestion, modelling and optimisation on biogas production and enhancement. *Bioresource Technology, 344*, 126311.
37. Mu, H., Zhao, C., Zhao, Y., Hua, D., Li, Y., Zhang, X., & Cao, Y. (2019). Optimizing methane production from anaerobic batch co-digestion of apple pomace and vegetable waste. *Journal of Biobased Materials and Bioenergy, 13*, 508–516.
38. Yan, F., Hu, H., Lu, L., & Zheng, X. (2016). Rhamnolipids induce oxidative stress responses in cherry tomato fruit to Alternaria alternate. *Pest Management Science, 72*, 1500–1507. https://doi.org/10.1002/ps.4177
39. Tlusty, M. F., Tyedmers, P., Bailey, M., Ziegler, F., Henriksson, P. J. G., Béné, C., Bush, S., Newton, R., Asche, F., & Little, D. C. (2019). Reframing the sustainable seafood narrative. *Global Environmental Change, 59*, 101991.
40. Calderwood, J., Pedreschi, D., & Reid, D. G. (2021). Technical and tactical measures to reduce unwanted catches in mixed fisheries: Do the opinions of Irish fishers align with management advice? *Marine Policy, 123*, 104290.
41. Marti-Quijal, F. J., Remize, F., Meca, G., Ferrer, E., Ruiz, M.-J., & Barba, F. J. (2020). Fermentation in fish and by-products processing: An overview of current research and future prospects. *Current Opinion in Food Science, 31*, 9–16.
42. Venugopal, V. (2021). Valorization of seafood processing discards: Bioconversion and biorefinery approaches. *Frontiers in Sustainable Food Systems, 5*, 611835.
43. Pérez-Marroquín, X. A., Estrada-Fernández, A. G., García-Ceja, A., Aguirre-Álvarez, G., & León-López, A. (2023). Agro-food waste as an ingredient in functional beverage processing: Sources. *Functionality, Market and Regulation, Foods, 12*, 1583.
44. Kumar, P. S., Mohanakrishna, G., Hemavathy, R. V., Rangasamy, G., & Aminabhavi, T. M. (2022). Sustainable production of biosurfactants via valorisation of industrial wastes as alternate feedstocks. *Chemosphere*, 137326.
45. Hameed, A., Anwar, M. J., Perveen, S., Amir, M., Naeem, I., Imran, M., Hussain, M., Ahmad, I., Afzal, M. I., & Inayat, S. (2023). Functional, industrial and therapeutic applications of dairy waste materials. *International Journal of Food Properties, 26*, 1470–1496.
46. Gutierrez-Hernandez, C. A., Hernandez-Almanza, A., Hernandez-Beltran, J. U., Balagurusamy, N., & Hernandez-Teran, F. (2022). Cheese whey valorization to obtain single-cell oils of industrial interest: An overview. *Food Bioscience*, 102086.
47. Poonia, A., & Pandey, S. (2023). Production of microbial pigments from whey and their applications: A review. *Nutrition & Food Science, 53*, 265–284.
48. Hosseinzadeh-Bandbafha, H., Li, C., Chen, X., Peng, W., Aghbashlo, M., Lam, S. S., & Tabatabaei, M. (2022). Managing the hazardous waste cooking oil by conversion into bioenergy through the application of waste-derived green catalysts: A review. *Journal of Hazardous Materials, 424*, 127636.

49. Di Fraia, S., Massarotti, N., Prati, M. V., & Vanoli, L. (2020). A new example of circular economy: Waste vegetable oil for cogeneration in wastewater treatment plants. *Energy Conversion and Management, 211*, 112763. https://doi.org/10.1016/j.enconman.2020.112763
50. Sen Siow, H., Sudesh, K., Murugan, P., & Ganesan, S. (2021). Mealworm (*Tenebrio molitor*) oil characterization and optimization of the free fatty acid pretreatment via acid-catalyzed esterification. *Fuel, 299*, 120905.
51. Cao, Y., Doustgani, A., Salehi, A., Nemati, M., Ghasemi, A., & Koohshekan, O. (2020). The economic evaluation of establishing a plant for producing biodiesel from edible oil wastes in oil-rich countries: Case study Iran. *Energy, 213*, 118760.
52. Yu, X., Zheng, Y., Dorgan, K. M., & Chen, S. (2011). Oil production by oleaginous yeasts using the hydrolysate from pretreatment of wheat straw with dilute sulfuric acid. *Bioresource Technology, 102*, 6134–6140.
53. Kee, S. H., Chiongson, J. B. V., Saludes, J. P., Vigneswari, S., Ramakrishna, S., & Bhubalan, K. (2021). Bioconversion of agro-industry sourced biowaste into biomaterials via microbial factories—A viable domain of circular economy. *Environmental Pollution, 271*, 116311.
54. Chiarelotto, M., Restrepo, J. C. P. S., Lorin, H. E. F., & Damaceno, F. M. (2021). Composting organic waste from the broiler production chain: A perspective for the circular economy. *Journal of Cleaner Production, 329*, 129717.
55. Tabler, T. (2019). Environment-friendly management of livestock and poultry mortalities. In *Animal environment and welfare.*
56. Chiarelotto, M., Restrepo, J. C. P. S., Lorin, H. E. F., & Damaceno, F. M. (2021). Composting organic waste from the broiler production chain: A perspective for the circular economy. *Journal of Cleaner Production, 329*, 129717. https://doi.org/10.1016/j.jclepro.2021.129717
57. Piyani, H. O., & Chandrarin, G. (2023). Analysis of the influence of financial literacy on business sustainability through the utilization of e-commerce: A study of MSMEs in the food and beverage industry sector in Balikpapan City. *European Journal of Business and Management Research, 8*, 306–314.
58. Salanţă, L. C., Coldea, T. E., Ignat, M. V., Pop, C. R., Tofană, M., Mudura, E., Borşa, A., Pasqualone, A., & Zhao, H. (2020). Non-alcoholic and craft beer production and challenges. *Processes, 8*, 1382.
59. Iram, A., Cekmecelioglu, D., & Demirci, A. (2020). Distillers' dried grains with solubles (DDGS) and its potential as fermentation feedstock. *Applied Microbiology and Biotechnology, 104*, 6115–6128.
60. Portarapillo, M., Danzi, E., Sanchirico, R., Marmo, L., & Di Benedetto, A. (2021). Energy recovery from vinery waste: Dust explosion issues. *Applied Sciences, 11*, 11188.
61. Bhatnagar, N., Ryan, D., Murphy, R., & Enright, A. M. (2022). A comprehensive review of green policy, anaerobic digestion of animal manure and chicken litter feedstock potential—Global and Irish perspective. *Renewable and Sustainable Energy Reviews, 154*, 111884.
62. Orr, L., Goossens, Y., Heinrich, M., & Brüggemann, N. (2023). Jointly reducing food waste—The experiences of the German discussion forum for wholesale and retail. *Sustainability, 15*, 12289.
63. Wikurendra, E. A., Abdeljawad, N. S., & Nagy, I. (2023). A review of municipal waste management with zero waste concept: Strategies, potential and challenge in Indonesia. *International Journal of Environmental Science and Development, 14*.
64. Ögel, İY., Ecer, F., & Özgöz, A. A. (2023). Identifying the leading retailer-based food waste causes in different perishable fast-moving consumer goods' categories: Application of the F-LBWA methodology. *Environmental Science and Pollution Research, 30*, 32656–32672.
65. Bos-Brouwers, H., Achterbosch, T. J., Castelein, B., & Cloutier, J. M. I. (2023). Theory of change: Acceleration agenda for reducing food waste 2022–2025: Recommendations to the Ministry of Agriculture, Nature and Food Quality (LNV). Wageningen Food & Biobased Research.
66. Redman, E., & Redman, A. (2014). Transforming sustainable food and waste behaviors by realigning domains of knowledge in our education system. *Journal of Cleaner Production, 64*, 147–157.

67. Onoh, C. E., & Ogbuagu, T. C. (2023). The impact of social work intervention on sustainable consumption through food waste reduction in Gävle Sweden: A qualitative study on the environmental and socio-economic benefits.
68. Munir, K. (2022). Sustainable food waste management strategies by applying practice theory in hospitality and food services—A systematic literature review. *Journal of Cleaner Production, 331*, 129991.
69. Graham-Rowe, E., Jessop, D. C., & Sparks, P. (2014). Identifying motivations and barriers to minimising household food waste. *Resources, Conservation and Recycling, 84*, 15–23.
70. Goud, R. K., Arunasri, K., Yeruva, D. K., Krishna, K. V., Dahiya, S., & Mohan, S. V. (2017). Impact of selectively enriched microbial communities on long-term fermentative biohydrogen production. *Bioresource Technology, 242*, 253–264.
71. Ishii, S., Suzuki, S., Norden-Krichmar, T. M., Wu, A., Yamanaka, Y., Nealson, K. H., & Bretschger, O. (2013). Identifying the microbial communities and operational conditions for optimized wastewater treatment in microbial fuel cells. *Water Research, 47*, 7120–7130.
72. Dahiya, S., Kumar, A. N., Sravan, J. S., Chatterjee, S., Sarkar, O., & Mohan, S. V. (2018). Food waste biorefinery: Sustainable strategy for circular bioeconomy. *Bioresource Technology, 248*, 2–12.
73. Xia, X., Tokash, J. C., Zhang, F., Liang, P., Huang, X., & Logan, B. E. (2013). Oxygen-reducing biocathodes operating with passive oxygen transfer in microbial fuel cells. *Environmental Science and Technology, 47*, 2085–2091.
74. Pasupuleti, S. B., & Mohan, S. V. (2015). Single-stage fermentation process for high-value biohythane production with the treatment of distillery spent-wash. *Bioresource Technology, 189*, 177–185.
75. Rajendran, N., & Han, J. (2022). Techno-economic analysis of food waste valorization for integrated production of polyhydroxyalkanoates and biofuels. *Bioresource Technology, 348*, 126796.
76. Mahmoud, A. H., Hussein, M. Y., Ibrahim, H. M., Hanafy, M. H., Salah, S. M., El-Bassiony, G. M., & Abdelfattah, E. A. (2022). Mixed microalgae-food waste cake for feeding of Hermetia illucens larvae in characterizing the produced biodiesel. *Biomass and Bioenergy, 165*, 106586. https://doi.org/10.1016/j.biombioe.2022.106586
77. Talapatra, N., & Ghosh, U. K. (2022). New concept of biodiesel production using food waste digestate powder: Co-culturing algae-activated sludge symbiotic system in low N and P paper mill wastewater. *Science of The Total Environment, 844*, 157207
78. Suzihaque, M. U. H., Syazwina, N., Alwi, H., Ibrahim, U. K., Abdullah, S., & Haron, N. (2022). A sustainability study of the processing of kitchen waste as a potential source of biofuel: Biodiesel production from waste cooking oil (WCO). *Materials Today: Proceedings, 63*, S484–S489.
79. Chi, Z., Zheng, Y., Jiang, A., & Chen, S. (2011b). Lipid production by culturing oleaginous yeast and algae with food waste and municipal wastewater in an integrated process. *Applied Biochemistry and Biotechnology, 165*, 442–453.
80. Kahr, H., Pointner, M., Krennhuber, K., Wallner, B., & Jäger, A. (2015). Lipid production from diverse oleaginous yeasts from steam exploded corn cobs. *Agronomy Research, 13*(2).
81. Liang, Y., Tang, T., Siddaramu, T., Choudhary, R., & Umagiliyage, A. L. (2012). Lipid production from sweet sorghum bagasse through yeast fermentation. *Renewable Energy, 40*(1), 130–136.
82. Pleissner, D., Lam, W. C., Sun, Z., & Lin, C. S. K. (2013). Food waste as nutrient source in heterotrophic microalgae cultivation. *Bioresource Technology, 137*, 139–146.
83. Papone, T., Kookkhunthod, S., Paungbut, M., & Leesing, R. (2016). Producing of microbial oil by mixed culture of microalgae and oleaginous yeast using sugarcane molasses as carbon substrate. *J. Clean Energy Technol, 4*(4), 253–256.
84. Haile, M. (2014). Integrated volarization of spent coffee grounds to biofuels. *Biofuel Research Journal, 1*, 65–69.
85. Takakuwa, N., & Saito, K. (2010). Conversion of beet molasses and cheese whey into fatty acid methyl esters by the yeast Cryptococcus curvatus. *Journal of Oleo Science, 59*(5), 255–260.

86. Papanikolaou, S., Dimou, A., Fakas, S., Diamantopoulou, P., Philippoussis, A., Galiotou-Panayotou, M., & Aggelis, G. (2011). Biotechnological conversion of waste cooking olive oil into lipid-rich biomass using Aspergillus and Penicillium strains. *Journal of Applied Microbiology, 110*(5), 1138–1150.
87. Bialy, H. El, Gomaa, O. M., & Azab, K. S. (2011). Conversion of oil waste to valuable fatty acids using oleaginous yeast. *World Journal of Microbiology and Biotechnology, 27*(12), 2791–2798.
88. Vescovi, V., Rojas, M. J., Baraldo, A., Botta, D. C., Santana, F. A. M., Costa, J. P., Machado, M. S., Honda, V. K., de Lima Camargo Giordano, R., & Tardioli, P. W. (2016). Lipase-catalyzed production of biodiesel by hydrolysis of waste cooking oil followed by esterification of free fatty acids. *Journal of the American Oil Chemists' Society, 93*, 1615–1624.
89. Rocha, M. V. P., de Matos, L. J. B. L., de Lima, L. P., da Silva Figueiredo, P. M., Lucena, I. L., Fernandes, F. A. N., & Gonçalves, L. R. B. (2014). Ultrasound-assisted production of biodiesel and ethanol from spent coffee grounds. *Bioresource Technology, 167*, 343–348
90. Jung, J.-H., Sim, Y.-B., Ko, J., Park, S. Y., Kim, G.-B., & Kim, S.-H. (2022). Biohydrogen and biomethane production from food waste using a two-stage dynamic membrane bioreactor (DMBR) system. *Bioresource Technology, 352*, 127094.
91. Ding, L., Cheng, J., Qiao, D., Li, H., & Zhang, Z. (2022a). Continuous co-generation of biohydrogen and biomethane through two-stage anaerobic digestion of hydrothermally pretreated food waste. *Energy Conversion and Management, 268,* 116000.
92. Moogi, S., Lam, S. S., Chen, W.-H., Ko, C. H., Jung, S.-C., & Park, Y.-K. (2022). Household food waste conversion to biohydrogen via steam gasification over copper and nickel-loaded SBA-15 catalysts. *Bioresource Technology, 366*, 128209
93. Nyambura, S. M., Jufei, W., Hua, L., Xuebin, F., Xingjia, P., Bohong, L., Ahmad, R., Jialiang, X., Bertrand, G. V, Ndiithi, J., & Xuhui, L. (2022). Microwave co-pyrolysis of kitchen food waste and rice straw for waste reduction and sustainable biohydrogen production: Thermo-kinetic analysis and evolved gas analysis. *Sustainable Energy Technologies and Assessments, 52,* 102072
94. Mirza, S. S., Qazi, J. I., Liang, Y., & Chen, S. (2019). Growth characteristics and photofermentative biohydrogen production potential of purple non sulfur bacteria from sugar cane bagasse. *Fuel, 255,* 115805.
95. Yang, G., Hu, Y., & Wang, J. (2019). Biohydrogen production from co-fermentation of fallen leaves and sewage sludge. *Bioresource Technology, 285*, 121342.
96. Marone, A., Massini, G., Patriarca, C., Signorini, A., Varrone, C., & Izzo, G. (2012). Hydrogen production from vegetable waste by bioaugmentation of indigenous fermentative communities. *International Journal of Hydrogen Energy, 37*(7), 5612–5622.
97. Wang, Y.-H., Li, S.-L., Chen, I.-C., & Cheng, S.-S. (2009). Starch hydrolysis characteristics of hydrogen producing sludge in thermophilic hydrogen fermentor fed with kitchen waste. *International Journal of Hydrogen Energy, 34*(17), 7435–7440.
98. Goud, R. K., & Mohan, S. V. (2011). Pre-fermentation of waste as a strategy to enhance the performance of single chambered microbial fuel cell (MFC). *International Journal of Hydrogen Energy, 36*(21), 13753–13762.
99. El-Bery, H., Tawfik, A., Kumari, S., & Bux, F. (2013). Effect of thermal pre-treatment on inoculum sludge to enhance bio-hydrogen production from alkali hydrolysed rice straw in a mesophilic anaerobic baffled reactor. *Environmental Technology, 34*(13–14), 1965–1972.
100. Ren, H.-Y., Kong, F., Cui, Z., Zhao, L., Ma, J., Ren, N.-Q., & Liu, B.-F. (2019). Cogeneration of hydrogen and lipid from stimulated food waste in an integrated dark fermentative and microalgal bioreactor. *Bioresource Technology, 287*, 121468.
101. Karthikeyan, O. P., & Visvanathan, C. (2013). Bio-energy recovery from high-solid organic substrates by dry anaerobic bio-conversion processes: A review. *Reviews in Environmental Science & Biotechnology, 12*, 257–284.
102. Mahmod, S. S., Azahar, A. M., Tan, J. P., Jahim, J. M., Abdul, P. M., Mastar, M. S., Anuar, N., Yunus, M. F. M., Asis, A. J., & Wu, S.-Y. (2019). Operation performance of up-flow anaerobic sludge blanket (UASB) bioreactor for biohydrogen production by self-granulated

sludge using pre-treated palm oil mill effluent (POME) as carbon source. *Renewable Energy, 134*, 1262–1272.

103. Gorgeç, F. K., & Karapinar, I. (2019). Production of biohydrogen from waste wheat in continuously operated UPBR: the effect of influent substrate concentration. *International Journal of Hydrogen Energy, 44*(32), 17323–17333.
104. Mohan, S. V., Mohanakrishna, G., Goud, R. K., & Sarma, P. N. (2009). Acidogenic fermentation of vegetable based market waste to harness biohydrogen with simultaneous stabilization. *Bioresource Technology, 100*, 3061–3068.
105. Mohan, S. V., Chiranjeevi, P., Chandrasekhar, K., Babu, P. S., & Sarkar, O. (2019). Acidogenic biohydrogen production from wastewater. In *Biohydrogen* (pp. 279–320). Elsevier.
106. Deepanraj, B., Sivasubramanian, V., & Jayaraj, S. (2017). Multi-response optimization of process parameters in biogas production from food waste using Taguchi-Grey relational analysis. *Energy Conversion and Management, 141*, 429–438.
107. Kondusamy, D., & Kalamdhad, A. S. (2014). Pre-treatment and anaerobic digestion of food waste for high rate methane production—A review. *Journal of Environmental Chemical Engineering, 2*, 1821–1830.
108. Xu, Z., Zhao, M., Miao, H., Huang, Z., Gao, S., & Ruan, W. (2014). In situ volatile fatty acids influence biogas generation from kitchen wastes by anaerobic digestion. *Bioresource Technology, 163*, 186–192.
109. Agyeman, F. O., & Tao, W. (2014). Anaerobic co-digestion of food waste and dairy manure: Effects of food waste particle size and organic loading rate. *Journal of Environmental Management, 133*, 268–274.
110. Bolzonella, D., Battista, F., Cavinato, C., Gottardo, M., Micolucci, F., Lyberatos, G., & Pavan, P. (2018). Recent developments in biohythane production from household food wastes: a review. *Bioresource Technology, 257*, 311–319.
111. Deheri, C., & Acharya, S. K. (2022). Purified biohythane (biohydrogen+biomethane) production from food waste using CaO2+CaCO3 and NaOH as additives. *International Journal of Hydrogen Energy, 47*(5), 2862–2873.
112. Chen, G., Li, H., Song, Y., Mu, L., Pei, L., Zhou, T., Qing, Z., & Zeng, Y. (2023). Optimization of food-to-microorganism ratio and addition of Tween 80 to enhance biohythane production via two-stage anaerobic digestion of food waste. *Journal of Chemical Technology & Biotechnology, 98*(10), 2477–2488.
113. Nguyen, T.-T., Ta, D.-T., Lin, C.-Y., Chu, C.-Y., & Ta, T.-M.-N. (2022). Biohythane production from swine manure and pineapple waste in a single-stage two-chamber digester using gel-entrapped anaerobic microorganisms. *International Journal of Hydrogen Energy, 47*(60), 25245–25255.
114. Alavi-Borazjani, S. A., Tarelho, L. A. da C., & Capela, M. I. (2022). Addition of biomass ash as a promising strategy for high-value biohythane production from the organic fraction of municipal solid waste. *Biomass and Bioenergy, 159*, 106392.
115. Alavi-Borazjani, S. A., da Cruz Tarelho, L. A., Marques Ruivo, L. C., Paula da Silva Seabra, M., & Capela, M. I. (2023). Effects of adding micro- and nano-sized biomass fly ash on two-stage biohythane production from the urban organic solid waste. *International Journal of Hydrogen Energy*.
116. Yeshanew, M. M., Frunzo, L., Pirozzi, F., Lens, P. N. L., & Esposito, G. (2016b). Production of biohythane from food waste via an integrated system of continuously stirred tank and anaerobic fixed bed reactors. *Bioresource Technology, 220*, 312–322.
117. Lee, M., Yang, M., Choi, S., Shin, J., Park, C., Cho, S.-K., & Kim, Y. M. (2019). Sequential production of lignin, fatty acid methyl esters and biogas from spent coffee grounds via an integrated physicochemical and biological process. *Energies, 12*(12), 2360.
118. Zhang, L., Yao, D., Tsui, T.-H., Loh, K.-C., Wang, C.-H., Dai, Y., & Tong, Y. W. (2022). Plastic-containing food waste conversion to biomethane, syngas, and biochar via anaerobic digestion and gasification: Focusing on reactor performance, microbial community analysis, and energy balance assessment. *Journal of Environmental Management, 306*

119. Wehner, M., Kleidorfer, I., Whittle, I., Bischof, D., Bockreis, A., Insam, H., Mueller, W., & Hupfauf, S. (2023). Decentralised system for demand-oriented collection of food waste—Assessment of biomethane potential, pathogen development and microbial community structure. *Bioresource Technology, 376*, 128894.
120. Agrawal, A. V., Chaudhari, P. K., & Ghosh, P. (2023). Effect of mixing ratio on biomethane potential of anaerobic co-digestion of fruit and vegetable waste and food waste. *Biomass Conversion and Biorefinery.*
121. Castro, Y. A., Rodríguez, A., & Rivera, E. (2022). Biomethane production kinetics during the anaerobic co-digestion of Sargassum spp. and food waste using batch and fed-batch systems in Punta Cana, Dominican Republic. *Materials for Renewable and Sustainable Energy, 11*(3), 287–297.
122. Alruqi, M., & Sharma, P. (2023). Biomethane Production from the Mixture of Sugarcane Vinasse, Solid Waste and Spent Tea Waste: A Bayesian Approach for Hyperparameter Optimization for Gaussian Process Regression. *Fermentation, 9*(2), 120.
123. Liu, J., Fan, S., Bai, K., & Xiao, Z. (2021). Combining acetone-butanol-ethanol production and methyl orange decolorization in wastewater by fermentation with solid food waste as substrate. *Renewable Energy, 179,* 2246–2255
124. Li, Y., Su, D., Feng, H., Yan, F., Liu, H., Feng, L., & Liu, G. (2017). Anaerobic acidogenic fermentation of food waste for mixed-acid production. *Energy Sources, Part A: Recovery, Utilization, and Environmental Effects, 39*(7), 631–635.
125. Hong, C., & Haiyun, W. (2010). Optimization of volatile fatty acid production with co-substrate of food wastes and dewatered excess sludge using response surface methodology. *Bioresource Technology, 101*(14), 5487–5493
126. Wang, K., Yin, J., Shen, D., & Li, N. (2014). Anaerobic digestion of food waste for volatile fatty acids (VFAs) production with different types of inoculum: effect of pH. *Bioresource Technology, 161*, 395–401.
127. Dahiya, S., & Joseph, J. (2015). High rate biomethanation technology for solid waste management and rapid biogas production: an emphasis on reactor design parameters. *Bioresource Technology, 188*, 73–78.
128. Pasupuleti, S. B., Sarkar, O., & Mohan, S. V. (2014). Upscaling of biohydrogen production process in semi-pilot scale biofilm reactor: Evaluation with food waste at variable organic loads. *International Journal of Hydrogen Energy, 39*, 7587–7596.
129. Sen, B., Aravind, J., Kanmani, P., & Lay, C.-H. (2016). State of the art and future concept of food waste fermentation to bioenergy. *Renewable and Sustainable Energy Reviews, 53*, 547–557.
130. Cavinato, C., Giuliano, A., Bolzonella, D., Pavan, P., & Cecchi, F. (2012). Bio-hythane production from food waste by dark fermentation coupled with anaerobic digestion process: A long-term pilot scale experience. *International Journal of Hydrogen Energy, 37*, 11549–11555.
131. Yeshanew, M. M., Frunzo, L., Pirozzi, F., Lens, P. N. L., & Esposito, G. (2016). Production of biohythane from food waste via an integrated system of continuously stirred tank and anaerobic fixed bed reactors. *Bioresource Technology, 220*, 312–322.
132. Sarkar, N., Ghosh, S. K., Bannerjee, S., & Aikat, K. (2012). Bioethanol production from agricultural wastes: An overview. *Renewable Energy, 37*, 19–27.
133. Humbird, D., Davis, R., Tao, L., Kinchin, C., Hsu, D., Aden, A., Schoen, P., Lukas, J., Olthof, B., & Worley, M. (2011). Process design and economics for biochemical conversion of lignocellulosic biomass to ethanol: Dilute-acid pretreatment and enzymatic hydrolysis of corn stover. National Renewable Energy Laboratory (NREL), Golden, CO (United States).
134. Walker, K., Vadlani, P., Madl, R., Ugorowski, P., & Hohn, K. L. (2013). Ethanol fermentation from food processing waste. *Environmental Progress & Sustainable Energy, 32*, 1280–1283.
135. López-Linares, J. C., Coca, M., Plaza, P. E., Lucas, S., & García-Cubero, M. T. (2023). Waste-to-fuel technologies for the bioconversion of carrot discards into biobutanol. *Renewable Energy, 202*, 362–369.
136. Wang, X., Sun, H., Wang, Y., Wang, F., Zhu, W., Wu, C., Wang, Q., & Gao, M. (2023). Feasibility of Efficient, Direct, Butanol Production from Food Waste without Nutrient Supplement by Clostridium saccharoperbutylacetonicum N1-4. *Sustainability, 15*(7), 6061

137. Su, G., Chan, C., & He, J. (2022). Enhanced biobutanol production from starch waste via orange peel doping. *Renewable Energy, 193,* 576–583.
138. Teresa Sponza, D., & Öztekin, R. (2022). 1-Butanol Production from the Food Wastes in Dokuz Eylül University Campus Buca-Izmir, Turkey. *Open J Case Reports.*
139. Zhang, C., Li, T., Su, G., & He, J. (2020). Enhanced direct fermentation from food waste to butanol and hydrogen by an amylolytic Clostridium. *Renewable Energy, 153,* 522–529
140. Zhang, C., Ling, Z., & Huo, S. (2021). Anaerobic fermentation of pretreated food waste for butanol production by co-cultures assisted with in-situ extraction. *Bioresource Technology Reports, 16,* 100852. https://doi.org/10.1016/j.biteb.2021.100852
141. Ozturk, A. B., Arasoglu, T., Gulen, J., Cheng, S., Al-Shorgani, N. K. N., Habaki, H., Egashira, R., Kalil, M. S., Yusoff, W. M. W., & Cross, J. S. (2021). Techno-economic analysis of a two-step fermentation process for bio-butanol production from cooked rice. *Sustainable Energy & Fuels, 5*(14), 3705–3718.
142. Raganati, F., Procentese, A., Olivieri, G., Russo, M. E., & Marzocchella, A. (2016). Butanol production by fermentation of fruit residues. *Chemical Engineering Transactions, 49,* 229–234.
143. Joshi, S. M., Waghmare, J. S., Sonawane, K. D., & Waghmare, S. R. (2015). Bio-ethanol and bio-butanol production from orange peel waste. *Biofuels, 6*(1–2), 55–61.
144. Alvarez, R. M., Rodriguez, B., Romano, J. M., Diaz, A. O., Gomez, E., Miro, D., Navarro, L., Saura, G., & Garcia, J. L. (1992). Lipid accumulation in *Rhodotorula glutinis* on sugar cane molasses in single-stage continuous culture. *World Journal of Microbiology & Biotechnology, 8,* 214–215.
145. Passadis, K., Christianides, D., Malamis, D., Barampouti, E. M., & Mai, S. (2022). Valorisation of source-separated food waste to bioethanol: pilot-scale demonstration. *Biomass Conversion and Biorefinery, 12*(10), 4599–4609. https://doi.org/10.1007/s13399-022-02732-6
146. Byun, J., Kwon, O., Kim, J., & Han, J. (2022). Carbon-Negative Food Waste-Derived Bioethanol: A Hybrid Model of Life Cycle Assessment and Optimization. *ACS Sustainable Chemistry & Engineering, 10*(14), 4512–4521
147. Zhou, H., Jiang, J., Zhao, Q., Wang, Z., Li, L., Gao, Q., & Wang, K. (2023). Performance of high solids enzymatic hydrolysis and bioethanol fermentation of food waste under the regulation of saponin. *Bioresource Technology, 387,* 129486.
148. Gherbi, Y., Boudjema, K., Djeziri, M., & Fazouane–Naimi, F. (2023). Isolation and identification of thermotolerant yeast strains producing bioethanol from agro-food wastes. *Biomass Conversion and Biorefinery.*
149. Lee, J., Kim, S., Lee, K. H., Lee, S. K., Chun, Y., Kim, S. W., Park, C., & Yoo, H. Y. (2022). Improvement of bioethanol production from waste chestnut shells via evaluation of mass balance-based pretreatment and glucose recovery process. *Environmental Technology & Innovation, 28,* 102955
150. Clementz, A. L., Manuale, D., Sanchez, E., Vera, C., & Yori, J. C. (2019). Use of discards of bovine bone, yeast and carrots for producing second generation bio-ethanol. *Biocatalysis and Agricultural Biotechnology, 22,* 101392.
151. Fazzino, F., Mauriello, F., Paone, E., Sidari, R., & Calabrò, P. S. (2021). Integral valorization of orange peel waste through optimized ensiling: Lactic acid and bioethanol production. *Chemosphere, 271,* 129602.
152. Karmee, S. K., Niemeijer, B., Casiraghi, L., Mlambo, B., Lapkin, A., & Greiner, L. (2014). Facile biocatalytic synthesis of a macrocyclic lactone in sub-and supercritical solvents. *Biocatalysis and Biotransformation, 32*(2), 125–131.
153. Jin, M., Gunawan, C., Uppugundla, N., Balan, V., & Dale, B. E. (2012). A novel integrated biological process for cellulosic ethanol production featuring high ethanol productivity, enzyme recycling and yeast cells reuse. *Energy & Environmental Science, 5*(5), 7168–7175.
154. Silva, C. O. G., Vaz, R. P., & Filho, E. X. F. (2018). Bringing plant cell wall-degrading enzymes into the lignocellulosic biorefinery concept. *Biofuels, Bioproducts and Biorefining, 12*(2), 277–289.

155. Han, W., Liu, Y., Xu, X., Huang, J., He, H., Chen, L., Qiu, S., Tang, J., & Hou, P. (2020). Bioethanol production from waste hamburger by enzymatic hydrolysis and fermentation. *Journal of Cleaner Production, 264*, 121658.
156. Ktori, R., Kamaterou, P., & Zabaniotou, A. (2018). Spent coffee grounds valorization through pyrolysis for energy and materials production in the concept of circular economy. *Materials Today: Proceedings, 5*(14), 27582–27588.
157. Stoeberl, M., Werkmeister, R., Faulstich, M., & Russ, W. (2011). Biobutanol from food wastes—Fermentative production, use as biofuel an the influence on the emissions. *Procedia Food Science, 1*, 1867–1874.
158. Huang, H., Singh, V., & Qureshi, N. (2015). Butanol production from food waste: A novel process for producing sustainable energy and reducing environmental pollution. *Biotechnology for Biofuels, 8*, 1–12.
159. Castanha, R. F., Mariano, A. P., de Morais, L. A. S., Scramin, S., & Monteiro, R. T. R. (2014). Optimization of lipids production by *Cryptococcus laurentii* 11 using cheese whey with molasses. *Brazilian Journal of Microbiology, 45*, 379–387.
160. Schneider, T., Graeff-Hönninger, S., French, W. T., Hernandez, R., Claupein, W., Holmes, W. E., & Merkt, N. (2012). Screening of industrial wastewaters as feedstock for the microbial production of oils for biodiesel production and high-quality pigments. *Journal of Combustion.*
161. Amulya, K., Jukuri, S., & Mohan, S. V. (2015). Sustainable multistage process for enhanced productivity of bioplastics from waste remediation through aerobic dynamic feeding strategy: Process integration for up-scaling. *Bioresource Technology, 188*, 231–239.
162. Slorach, P. C., Jeswani, H. K., Cuéllar-Franca, R., & Azapagic, A. (2019). Environmental and economic implications of recovering resources from food waste in a circular economy. *Science of the Total Environment, 693*, 133516. https://doi.org/10.1016/j.scitotenv.2019.07.322
163. Rajendran, N., Kang, D., Han, J., & Gurunathan, B. (2022). Process optimization, economic and environmental analysis of biodiesel production from food waste using a citrus fruit peel biochar catalyst. *Journal of Cleaner Production, 365*, 132712.
164. Mitra, D., van Leeuwen, J. H., & Lamsal, B. (2012). Heterotrophic/mixotrophic cultivation of oleaginous Chlorella vulgaris on industrial co-products. *Algal Research, 1*(1), 40–48.

SPRINGER NATURE

GPSR Compliance

The European Union's (EU) General Product Safety Regulation (GPSR) is a set of rules that requires consumer products to be safe and our obligations to ensure this.

If you have any concerns about our products, you can contact us on ProductSafety@springernature.com

In case Publisher is established outside the EU, the EU authorized representative is:

Springer Nature Customer Service Center GmbH
Europaplatz 3
69115 Heidelberg, Germany

The manufacturer's authorised representative in the EU is Springer Nature Customer Service Centre GmbH, Europaplatz 3, 69115 Heidelberg, Germany. If you have any concerns regarding our products, please contact ProductSafety@springernature.com

Printed and bound by CPI Group (UK) Ltd, Croydon, CR0 4YY
01/10/2026
02225863-0015